T0341235

FOUNDATIONS
of
CRYSTALLOGRAPHY
with
Computer Applications

SECOND EDITION

FOUNDATIONS
of
CRYSTALLOGRAPHY
with
Computer Applications
SECOND EDITION

Maureen M. Julian

Department of Materials Science and Engineering
Virginia Tech, USA

CRC Press is an imprint of the
Taylor & Francis Group, an **informa** business

MATLAB® is a trademark of The MathWorks, Inc. and is used with permission. The MathWorks does not warrant the accuracy of the text or exercises in this book. This book's use or discussion of MATLAB® software or related products does not constitute endorsement or sponsorship by The MathWorks of a particular pedagogical approach or particular use of the MATLAB® software.

CRC Press
Taylor & Francis Group
6000 Broken Sound Parkway NW, Suite 300
Boca Raton, FL 33487-2742

© 2015 by Taylor & Francis Group, LLC
CRC Press is an imprint of Taylor & Francis Group, an Informa business

No claim to original U.S. Government works

Printed on acid-free paper
Version Date: 20140227

International Standard Book Number-13: 978-1-4665-5291-3 (Hardback)

This book contains information obtained from authentic and highly regarded sources. Reasonable efforts have been made to publish reliable data and information, but the author and publisher cannot assume responsibility for the validity of all materials or the consequences of their use. The authors and publishers have attempted to trace the copyright holders of all material reproduced in this publication and apologize to copyright holders if permission to publish in this form has not been obtained. If any copyright material has not been acknowledged please write and let us know so we may rectify in any future reprint.

Except as permitted under U.S. Copyright Law, no part of this book may be reprinted, reproduced, transmitted, or utilized in any form by any electronic, mechanical, or other means, now known or hereafter invented, including photocopying, microfilming, and recording, or in any information storage or retrieval system, without written permission from the publishers.

For permission to photocopy or use material electronically from this work, please access www.copyright.com (http://www.copyright.com/) or contact the Copyright Clearance Center, Inc. (CCC), 222 Rosewood Drive, Danvers, MA 01923, 978-750-8400. CCC is a not-for-profit organization that provides licenses and registration for a variety of users. For organizations that have been granted a photocopy license by the CCC, a separate system of payment has been arranged.

Trademark Notice: Product or corporate names may be trademarks or registered trademarks, and are used only for identification and explanation without intent to infringe.

Library of Congress Cataloging-in-Publication Data

Julian, Maureen M.
    Foundations of crystallography with computer applications / Maureen M. Julian. -- Second edition.
        pages cm
    "A CRC title."
    Includes bibliographical references and indexes.
    ISBN 978-1-4665-5291-3 (hardcover : alk. paper)
    1. Crystallography--Data processing. 2. Crystallography, Mathematical. I. Title.

QD906.7.E4J85 2015
548--dc23   2014006992

Visit the Taylor & Francis Web site at
http://www.taylorandfrancis.com

and the CRC Press Web site at
http://www.crcpress.com

This book is dedicated to the future of all young people, particularly my grandchildren

Carla Itzel Julian

Isaac Batian Mikanatha

Jonathan Rey Julian

and all the students who have taken, or will take, the crystallography journey with me.

Preface

This book presents the fundamentals of crystallography to university and college students in biology, chemistry, engineering, geological sciences, materials science, and physics. It is also appropriate for scientists who want to teach themselves.

The chapters are

- Chapter 1: Lattices

- Chapter 2: Unit Cell Calculations

- Chapter 3: Point Groups

- Chapter 4: Space Groups

- Chapter 5: Reciprocal Lattice

- Chapter 6: Properties of X-rays

- Chapter 7: Electron Density Maps

- Chapter 8: Introduction to the Seven Crystals Exemplifying the Seven Crystal Systems

- Chapter 9: Triclinic System: DL Leucine

- Chapter 10: Monoclinic System: Sucrose

- Chapter 11: Orthorhombic System: Polyethylene

- Chapter 12: Tetragonal System: α-Cristobalite

- Chapter 13: Trigonal System: $H_{12}B_{12}^{-2}, 3K^+, Br^-$

- Chapter 14: Hexagonal System: Magnesium

- Chapter 15: Cubic System: Acetylene

This second edition includes new material on the color coding of the symbol and general position diagrams. In the first seven chapters many more molecular examples have been added, including a detailed description of a third model crystal—caffeine monohydrate. There are eight new chapters. In seven of these chapters, one for each crystal system, the reader is introduced to a detailed crystallographic analysis of a crystal illustrating its crystal system. Many new topics have been added, such as neutron diffraction. A new appendix on point groups gives detailed information including a representative matrix for each symmetry operation and the color-coded stereographic projections. The *Brief Teaching Edition of Volume A of the International Tables for Crystallography* is the companion book from which almost all the examples and exercises are based as is appropriate. The Starter Programs have been updated.

GOALS OF THE BOOK

The goal of this book is to describe the symmetry tools used to interpret the crystal structure, that is, the spatial arrangement of atoms in a crystal.

The material is given in a logical order with an emphasis on how crystal systems are related to each other. The ideas proceed from simple to complex. The theoretical material is developed extensively in two dimensions and then extended to three dimensions.

An emphasis is on crystallographic interrelationships. For example, many point group trees have been drawn, not only with their point group names, but also with the general position and symbol stereographic projection icons.

This book aims to communicate an understanding of how to read reports on the arrangement of atoms in any crystal that may be of interest. Once a crystallographer has analyzed the arrangement of atoms, the results are reported in a code that is published in the *International Tables for Crystallography*. The code is terse, but it captures the essence of a body of theory that deals with all the ways atoms can be arranged in a crystal.

Finally, topics are included to show how crystallography can be applied. For example, thermal expansion, compressibility, and piezoelectricity are discussed.

MATHEMATICS PREREQUISITES AND COMPUTERS

The mathematics prerequisite for this book is an introductory knowledge of linear algebra including matrices and determinants, although these matters are explained and illustrated in the course of this book.

Linear algebra allows exploitation of the power of the metric matrix. The metric matrix incorporates the information for the lattice—the lattice constants a, b, c, α, β, and γ—into a single matrix. All the crystal systems from triclinic to cubic are accommodated in one technique. This allows easy computation of important crystallographic quantities, such as bond lengths, bond angles, unit cell volumes, reciprocal lattice constants, interplanar angles, and d-spacings in all crystal systems.

Computers are an essential part of crystallography, and appropriate computer-based exercises are integrated into this book. The computing requirement is a familiarity with an advanced computer language like MATLAB. However, Starter Programs are supplied to help considerably with the details of the MATLAB language. Videos are on the book website https://sites.google.com/a/vt.edu/foundations_of_crystallography/.

TEACHING WITH COLOR CODING OF SYMMETRY DIAGRAMS

This second edition includes new material on the color coding of the symbol and general position diagrams. In the first edition of this book many diagrams were colored to distinguish crystallographically distinct graphical symbols, such as the four crystallographically distinct rotations in *p*2. Colors were also used to distinguish guidelines from the more significant crystallographic symbols. Now a big step has been taken to insert more information into the diagrams and to increase student participation. These color diagrams are all related to the change of handedness of the operations.

The color coding began as a way to illuminate the information in the point group and space group diagrams. A system was needed to test easily and quickly the understanding of the diagrams. Black and white diagrams were handed out to the class to color code as an exercise. As a result, the class became actively involved and generally comprehended these diagrams more fully.

MODEL CRYSTALS EMPHASIZED IN CHAPTERS 1 THROUGH 7

The first model crystal, hexamethylbenzene, $C_6(CH_3)_6$ (HMB), is triclinic. The procedures and programs developed for a triclinic crystal can handle crystals in all crystal systems. In HMB, the 12 carbon atoms lie in a plane and can be treated as a two-dimensional entity. This small organic structure can be considered to be a model polymer with both covalent and Van der Waals bonds. The structure is of historical importance as it proved for the first time that the six carbon atoms of the benzene ring lie in a plane.

The second model crystal, anhydrous alum, $KAl(SO_4)_2$ (AA), is a small framework structure typical of ceramics and metals. The AA crystal is noncentrosymmetric in contrast to the HMB crystal, which is centrosymmetric. The AA crystal is used to explain how the symmetry operations work in a crystal. For example, the structure shows that the environment of the oxygen atoms near the Al^{3+} ions is different from the environment near the K^+ ions, and this explains the different lengths of the S–O bonds in the SO_4 tetrahedrons. AA is part of an isotypic series.

This second edition introduces a third model crystal, caffeine monohydrate, to provide a better, more complex prototype for the calculations of the crystals in Chapters 9 through 15 as well as for the student projects suggested in this book. For caffeine monohydrate, a detailed description is given of the application of the Starter Program to produce the populated unit cell. The Cambridge Structure Database can be used as an alternative to the Starter Program, as explained in Chapter 8.

In my course, I have each student select a crystal structure of personal interest to pursue a semester-long preparation of a presentation of that crystal to the class. Students have chosen a wide range of crystals, including superconductors, polymers, ceramics, gems, metals, and biologically important organic materials. Graduate students pick crystals related to their theses. The book is designed so instructors may, for example, (a) allow each student complete freedom in selecting a crystal, (b) assign a crystal from a selection of worked-out examples made available on the web page, https://sites.google.com/a/vt.edu/foundations_of_crystallography/, (c) divide the class into groups and assign a crystal structure for each group, or (d) omit this exercise.

OVERALL ORGANIZATION: CHAPTERS 1 THROUGH 7: FUNDAMENTALS OF CRYSTALLOGRAPHY

The book is organized into 15 chapters, of which the first 7 are enhanced from the first edition and the rest are new.

Chapter 1 is on lattices, including basis vectors, unit cells, handedness, and transformations between basis vectors. The Starter Program transforms nonorthogonal coordinates to Cartesian coordinates, which are needed for a graphical interface. A three-dimensional triclinic unit cell is constructed. The model unit cell is rotated and projections are examined. Already the idea of going from the general to the specific is in place, since once the triclinic cell is drawn, the student can then go to the literature and construct any unit cell. I bring in practical examples as early as possible, so I include temperature and pressure variations in the lattice parameters.

Chapter 2 forms the mathematical basis for the rest of the book. The metric matrix is introduced in the derivation of the length of a vector in the unit cell. From now on the metric matrix is used in the calculation of interatomic distances, interatomic angles, and volumes of unit cells. Unit cell transformations are derived. The computer project for this chapter is the two-dimensional construction of a unit cell of HMB populated with carbon atoms. Note that the concepts of symmetry have not been introduced. This powerful exercise is a foretaste of what is to come in Chapter 4, when an arbitrary cell is populated. HMB was chosen because all 12 of its carbon atoms lie in the xy plane.

Chapter 3, on point groups, begins the study of symmetry. The point groups are presented first in two dimensions, where they are easy to visualize, and then in three dimensions. The task here is not only to understand the individual symmetry operations that make up the point groups but also to use a holistic approach to see how the point groups are constructed and related. This relationship is seen in the point group tree and in the increasing size of the multiplication tables as the order of the point groups increases. The symmetry operations are illustrated both with common objects, such as horses, and with molecules. The stereographic projections are color coded. The main computer exercise for this chapter is the construction of multiplication tables. All the notations in this chapter are consistent with the *International Tables for Crystallography*. An extensive appendix is included in the back of the book to facilitate the construction of matrices for the multiplication table for each of the point groups.

Chapter 4, on space groups, is where Bravais lattices are combined with point groups to produce space groups in both two and three dimensions. Symmorphic and nonsymmorphic space groups are compared. Color coding and molecular examples continue in the style of Chapter 3. The main computer exercise is a three-dimensional model of a populated unit cell. The asymmetric unit and the three special projections are analyzed.

Chapter 5 focuses on the reciprocal lattice, including calculations of the reciprocal cell parameters, the volume of the reciprocal cell, and the $\mathbf{G}^*$ matrix. The computer exercise is to superimpose the reciprocal unit cell on the crystal's unit cell. The interfacial angles are derived using $\mathbf{G}^*$. The relationship of the reciprocal lattice to the diffraction pattern is explored.

Chapter 6 introduces the experimental side of crystallography, including the discovery of x-rays, properties of waves, the first x-ray diffraction picture, and the identification of materials by powder diffraction. The chapter ends with a description of the production of x-rays from a synchrotron with a very large protein analyzed by synchrotron radiation as an example.

Chapter 7 discusses how the contents of the unit cell influence the intensities of the diffraction maxima. Scattering by a single electron, by a single atom, and finally by the whole crystal are discussed. The structure factors are calculated from the positions of the atoms, the reciprocal lattice, and the atomic scattering curves. The structure factors are proportional to the amplitudes of the Fourier series that are used in the chapter to calculate the electron density map of HMB.

OVERALL ORGANIZATION: CHAPTER 8: THE BRIDGE

Chapter 8 forms a bridge between the first seven chapters and the remainder of the book. It is an introduction to the analysis of the seven crystals used to illustrate the seven crystal systems in Chapters 9 through 15. Section 8.2 gives the crystallographic data of the seven crystals. Section 8.3 shows how each of the chapters is constructed containing *parallel topics* and *special topics*. The *parallel topics* are the fundamental crystallographic topics, for example, the unit cell and the asymmetric unit. Additionally each crystal has its own unique challenges that give an opportunity to introduce relevant *special topics*. Section 8.4 presents the color coding of the general position diagrams and the symbol diagrams. Section 8.5 gives the criteria used for selecting the crystals. For example, the space groups are limited to those found in the *Brief Teaching Edition of the International Tables for Crystallography* (Hahn, 2005). Section 8.6 gives suggestions for student projects and ideas for instructors who do not wish to use the Starter Programs. Finally, the chapter ends with a discussion of the distribution of crystal structures by space group and crystal system.

OVERALL ORGANIZATION: CHAPTERS 9 THROUGH 15: THE CRYSTAL SYSTEM EXAMPLES

Chapters 9 through 15, each illustrating a crystal system, are listed here with the chosen crystal:

- Chapter 9: Triclinic System: DL-Leucine

- Chapter 10: Monoclinic System: Sucrose

- Chapter 11: Orthogonal System: Polyethylene

- Chapter 12: Tetragonal System: α-Cristobalite

- Chapter 13: Trigonal System: $H_{12}B_{12}{}^{2-},3K^+,Br^-$

- Chapter 14: Hexagonal System: Magnesium

- Chapter 15: Cubic System: Acetylene

Chapters 9 through 15 are run in parallel and are designed to be read independently and in any order. The *parallel topics* are an armature that is repeated in each chapter, and the *special topics* are related to the crystal of choice. The *parallel topics* are given in the following list:

- Point group properties
  - Multiplication table
  - Stereographic projections
- Space group properties
  - Space group diagrams
  - Maximal subgroups and minimal supergroups
  - Asymmetric unit
- Direct and reciprocal lattices
- Fractional coordinates and other data for the crystal structure
- Crystal structure
  - Symmetry of the three special projections
- Reciprocal lattice and *d*-spacings
  - Powder diffraction pattern
- Atomic scattering curves
- Structure factor
  - Calculation of the structure factor at 000
  - Calculation of the contribution of one atom to the structure factor

Also included in each of Chapters 9 through 15 are *special topics*, as follows:

- Chapter 9: *CIF*, chiral molecules, enantiomers, racemic mixtures
- Chapter 10: Proper point groups, proper space groups, scanning electron microscope
- Chapter 11: Low-temperature crystallography, short and full Hermann–Mauguin symbols, relating the Hermann–Mauguin symbol to the cell choice, relating the Hermann–Mauguin symbol to the asymmetric unit, special projections and the Hermann–Mauguin symbols
- Chapter 12: Organizing crystal structures with polyhedrons, enantiomorphic space group pairs, experimental detection of enantiomers
- Chapter 13: Rhombohedral and hexagonal axes, transforming crystallographic directions, space group diagrams for combined hexagonal and rhombohedral cells, boron icosahedron, space-filling model, $H_{12}B_{12}{}^{2-}$,$3K^+$,Br^- isotypic series
- Chapter 14: Close-packed structures, model of close-packed spheres, hexagonal close-packed structure (*hcp*), cubic close-packed structure (*ccp*), comparing *hcp*

and *ccp* structures, interstitial spaces in a close-packed structure, sixfold axes, shifting the origin of the crystal, coordination number 12, site symmetry

- Chapter 15: Comparison of x-ray, neutron, and electron diffraction, neutron diffraction, neutron structure factors, and symbols for threefold symmetry axes parallel to body diagonals

The colors on the cover, orange and maroon, are the school colors of Virginia Tech.

The author would like to hear from readers of this book. Please contact her at Maureen.Julian@taylorandfrancis.com with comments, suggestions, or corrections. For additional student and instructor materials associated with this book, please see https://sites.google.com/a/vt.edu/foundations_of_crystallography/. My colleague, Sean Corcoran, sgc@vt.edu, will provide, on request, Starter Programs for this book written in *Mathematica*. These programs are also on the above website.

MATLAB® and Simulink® are registered trademarks of The MathWorks, Inc. For product information, please contact:

The MathWorks, Inc.
3 Apple Hill Drive
Natick, MA 01760-2098 USA
Tel: 508 647 7000
Fax: 508-647-7001
E-mail: info@mathworks.com
Web: www.mathworks.com

Acknowledgments

My interest in crystallography was sparked by Dr. Enid W. Silverton, who convinced me that crystallography had the right mix of mathematics, molecules, and computer science; by Dr. J. L. Hoard, my thesis advisor at Cornell, who insisted on a broad background in both chemistry and crystallography; and by Dr. Kathleen Lonsdale of University College, London, from whom I received a deep love for symmetry.

Two special correspondents have been Dr. Theo Hahn of the Institut fur Kristallographie, Germany, and former editor of the *International Tables for Crystallography* and Dr. Mois I. Aroyo, Universidad del Pais Vasco, Bilbao, Spain, and the Bilbao Crystallographic Server. Both have made insightful comments that have influenced the manuscript.

Dr. Norman Dowling and Dr. David Clark of Virginia Tech have enthusiastically supported my crystallography course in the Department of Materials Science and Engineering. Dr. Sean Corcoran, also of this department, particularly encouraged the work and helped create parallel Starter Programs in *Mathematica*.

Several persons have facilitated the extensive use of crystallographic databases. Dr. Peter A. Wood of the Cambridge Structural Database (CSD) has encouraged this project in many ways, especially by allowing inclusion of molecular and other illustrations in the book and by adding more crystal examples of my choice to the free Teaching Subset of the CSD so there could be an overlap between that database and the *Brief Teaching Edition of the International Tables for Crystallography*. Dr. Robert T. Downs of the American Mineralogist Crystal Structure Database and the University of Arizona generously gave permission to reproduce the picture of α-cristobalite used in the study referred to in Chapter 12. Dr. Klaus Brandenburg of the Pearson's Crystal Structure Database for Inorganic Compounds compiled the distribution of inorganic crystals among the 230 space groups. Dr. David S. Goodsell of the Protein Data Bank allowed reproduction of the molecular diagrams of the protein crystals.

Many contributed material to the book. Dr. Doletha Marian Szebenyi of Cornell University took photographs of the synchrotron and suggested an appropriate protein crystal. Dr. William Reynolds and Dr. Mitsuhiro Murayama of Virginia Tech took the SEM photographs of sucrose. Dr. Colin Kennard, retired from the University of Queensland, Australia, contributed the picture of the statue of β-boron in Brisbane, Australia. Dr. Carla Slebodnick of the Chemistry Department at Virginia Tech read and commented on parts of the manuscript and also allowed me to photograph her huge single crystal of NaCl. Dr. James Silverton suggested the propeller as an example of point group 32 symmetry.

Dr. Slavik Vlado Jablan of the Mathematical Institute in Yugoslavia generously encouraged me to use his Neolithic ornamental art patterns to illustrate many of the space groups in this book. Dr. William J. Floyd of the Mathematics Department, Virginia Tech, named the polyhedron describing the twelve nearest neighbors in HCP. Anna McNider painted the cats and mice picture in Figure 1.1 as part of her senior chemistry project of art and crystallography at Hollins College. Dr. Byron Rubin, a crystallographer and sculptor, invited me to include his metal sculpture of the double helix, which I photographed at the Florence IUCr meeting. Elena Leshyn took the quartz picture. Dr. Kevin Roberts of the University of Leeds provided me with his urea data for the section on thermal expansion. Dr. M. Oehzelt of the Institute of Solid State Physics at the Graz University of Technology in Austria provided me with his anthracene data for the section on compressibility. Dr. Ross Angel of the University of Padova, Italy, took the quartz diffraction patterns. Dr. Don Bloss of Virginia Tech let me use the illustration of a room with the inversion point. Dr. Kathleen Julian, MD, of the Pennsylvania State University, drew the picture of Ruffo, the Spanish-speaking dog, with the mirror plane. She and Dr. Nkuchia Mikanatha, of the University of Pennsylvania, reviewed and helped organize the preface. Francis Julian is my computing consultant.

I have had useful conversations with Dr. Carolyn P. Brock of the University of Kentucky, the editor-in-chief of the *International Tables for Crystallography*, and with Peter Strickland, the managing editor of the International Union of Crystallography. Dr. Michael Glazer of Oxford University has been a very helpful longtime correspondent. Dr. K. K. Kantardjieff of the California State University, San Marcos, and an editor of the *Journal of Applied Crystallography* has made useful comments. Dr. Elinor Spencer of the Geosciences Department, Virginia Tech, has read and made helpful comments on the manuscript.

My wonderful editor, Lance Wobus, acquiring editor, at Taylor & Francis supported publication of the first edition from the moment of the original submission. He encouraged and fostered this second edition. After his promotion he was succeeded by Barbara Glunn, who is presently guiding the book as editor. I also thank Ed Curtis, project editor at Taylor & Francis, and Syed Mohamad Shajahan, deputy manager at Techset Composition.

I thank Dr. Carl L. Julian, an enthusiastic theoretical physicist, for his work as principal editor and being my all-around best friend. Our son Francis, his wife Eli, their children Carla and Jonathan, our daughter Kathleen, her husband Nkuchia, and their son Isaac have joined the cheering squad.

Maureen Julian
Blacksburg, Virginia
Montauk, New York
2014 International Year of Crystallography

Author

Maureen M. Julian earned an AB from Hunter College, New York City, with a double major in physics and mathematics, and a PhD from Cornell University in physical chemistry with a thesis in crystallography. She was a research fellow at University College, London, with Professor Dame Kathleen Lonsdale, a founder of the *International Tables for Crystallography*. The author has given several series of crystallography workshops for undergraduate and graduate students in various departments, including chemistry, geology, and materials science. Her interests include *ab initio* calculations, molecular bonding, and group theory.

Since 2000, she has been teaching the crystallography course in the Materials Science and Engineering Department at Virginia Tech, Blacksburg, Virginia, where this book was developed. Every year she presides over a Festival of Crystals, which has become a rite of passage where students present a crystallographic analysis of a crystal of their choice. She resides in Blacksburg, Virginia, and Montauk, New York.

Contents

CHAPTER 1 ▪ Lattices   1

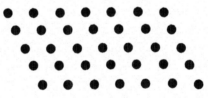

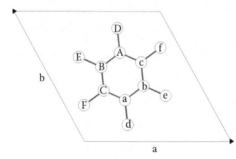

CHAPTER 3 ■ Point Groups 69

CHAPTER 5 ◼ The Reciprocal Lattice

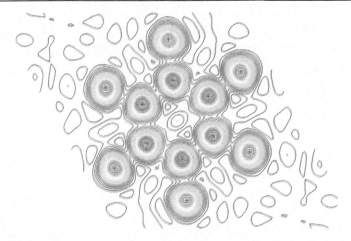

CHAPTER 8 ◾ Introduction to the Seven Crystals Exemplifying the Seven Crystal Systems 359

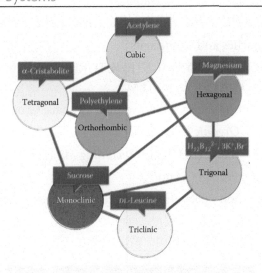

CHAPTER 9 ■ Triclinic Crystal System: DL-Leucine

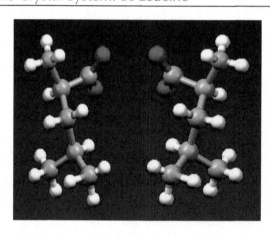

CHAPTER 10 ■ Monoclinic System: Sucrose

CHAPTER 11 ■ Orthorhombic Crystal System: Polyethylene 425

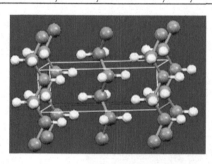

CHAPTER 13 ■ Trigonal Crystal System: $H_{12}B_{12}^{-2}, 3K^+, Br^-$　471

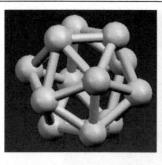

CHAPTER 14 ■ Hexagonal System: Magnesium

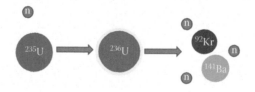

List of Symbols

CIF Crystallographic Information File
f atomic scattering factor
$F(hkl)$ structure factor
G metric matrix in direct space
G* metric matrix in reciprocal space
H$(h,k,l) = h\mathbf{a}^* + k\mathbf{b}^* + l\mathbf{c}^*$ reciprocal lattice vector
hcp hexagonal close-packed
hkl Laue indices or Miller indices
I identity matrix
P, Q transformation matrices
$\mathbf{r}_{12}$ interatomic bond between atoms 1 and 2.
$\mathbf{r} = (\mathbf{a}\ \mathbf{b}\ \mathbf{c})\ X = x\mathbf{a} + y\mathbf{b} + z\mathbf{c}$ coordínate vector
$\mathbf{t}(u,v,w) = u\mathbf{a} + v\mathbf{b} + w\mathbf{c}$ direct lattice vector
$[u\ v\ w]$ crystallographic direction
V volume of unit cell in direct space
V^* volume unit cell in reciprocal space

$$X = \begin{pmatrix} x \\ y \\ z \end{pmatrix} = x/y/z \text{ coordinate matrix}$$

$$X_{12} = \begin{pmatrix} x_2 - x_1 \\ y_2 - y_1 \\ z_2 - z_1 \end{pmatrix} \text{ where } x_1, y_1, z_1 \text{ are fractional coordinates of atom 1 and}$$

x_2, y_2, z_2 are fractional coordinates of atom 2
Z' number of molecules of the compound in the asymmetric unit
Z number of molecules of the compound in the unit cell

List of Tables

List of Figures

List of Starter Programs

Introduction

A stylistic representation of a snow-flake on a flag.

Crystallography is the science of finding the locations of atoms in crystals. Perhaps the most familiar crystals are diamonds, which are valued for their sparkle, which is normally enhanced by cutting them to display regular planar surfaces (technically called "faces"). Other spectacular crystals are perfect snowflakes, which have a delightful symmetry known in crystallography as "6mm." Many common materials are crystalline—for example, table salt and copper wires. Huge crystals of silicon are made to serve as the basis for the majority of personal computers. Historically, minerals were the first crystals to fascinate people, having faces that were consistent from one sample to another. Quartz crystals are excellent examples.

Crystallography goes far beyond crystals of these kinds. An excellent way to study the arrangement of the atoms in a *molecule*—for example, in a molecule of caffeine—is to persuade a billion or so of these molecules to form a crystal. One would certainly prefer to look at a single molecule by itself. However, if any sort of light shines on a single molecule, the reflected light is too meager to detect. When a billion or so molecules are inserted in a beam of x-rays, the reflections are bright enough to detect, though challenging to understand.

The goal of this book is to describe the tools used to interpret the x-ray reflections from a crystal. Once a crystallographer has analyzed the arrangement of the atoms, the results are recorded in a code, published in the *International Tables for Crystallography*. The code is terse, but it captures the essence of a body of theory that deals with all the possible ways that atoms can be arranged in a crystal. This book aims to communicate an understanding of how to read the reports on the arrangement of the atoms in any crystals that may be of interest.

The manner in which the atoms are arranged in a crystal affects macroscopic properties, such as piezoelectricity and the electrical conductivity of a semiconductor. Thus, crystallography provides information that is vital to science and engineering.

This book presumes that the reader has had an introduction to linear algebra and to the algebra of vectors in space, especially the dot product and cross product. This book also presumes that the reader has had an introduction to a mathematical computer programming language, such as MATLAB. A course based on this book then gives the reader many opportunities to apply the algebraic tools in the easily visualized context of crystals and to use those tools in combination with a personal computer to work out specific examples that otherwise, without a personal computer, would be too laborious to be worthwhile to attempt.

While this book aims to communicate useful information, the author hopes that the reader will share her delight in seeing profound mathematical principles used in very practical ways.

Crystallography is a grand revelation of the mysteries of nature in the same sense that Newton's interpretation of the motion of the planets and of the tides is such a revelation.

Carl L. Julian
Blacksburg

## CHAPTER 1

Lattices

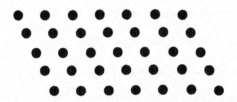

The first step in the analysis of a crystal is the identification of the lattice. A lattice is a regular array of points in space. Above is a two-dimensional set of lattice points. Ideally these points go on forever, filling the entire plane.

CONTENTS

CHAPTER OBJECTIVES

- Recognize a crystal as a repeating pattern.

- Associate a unique lattice with every repeating pattern.

- Describe a lattice with its basis vectors.

- Distinguish the handedness of basis vectors.

- Mathematically describe a lattice.

- Realize that the basis vectors need not be orthogonal.

- Describe a lattice with more than one set of basis vectors.

- Associate a unit cell with basis vectors.

- Transform between sets of basis vectors.

- Expand concepts of lattices from two to three dimensions.

- Construct a transformation matrix between crystallographic and Cartesian coordinates for use in computer programs.

- Construct, using a computer program, a unit cell of any crystal given the lattice parameters from the literature.

- Observe variations in lattice parameters with temperature and pressure.

1.1 INTRODUCTION

The ideal crystal has perfect periodicity. The implications of periodicity permeate this book. This chapter describes periodicity in terms of lattices. A lattice is described mathematically by basis vectors, which in turn produce a unit cell. In a crystal there can be one atom associated with each lattice point, as in copper, or thousands of atoms associated with each lattice point, as in some proteins. The lattice is chosen to be infinite, as an idealization, to postpone concern about what happens at the surfaces of a crystal. For other references in crystallography see Lonsdale (1948), Boisen and Gibbs (1985), Clegg et al. (2001), Cullity and Stock (2001), Golubitsky and Stewart (2002), Ladd and Palmer (2003), Liboff (2004),

McKie and McKie (1986), Prince (2004), Senechal (1990), Stróz (2003), Tiekink and Vittal (2006), Tilley (2006), and Zachariasen (1967).

In this book, the crystallographic ideas are first developed in two dimensions for clarity, simplicity, and overall understanding. Later, the transition to three dimensions is made. Throughout the first seven chapters of this book computer-based calculations are done on three study crystals: hexamethylbenzene, $C_6(CH_3)_6$, (HMB); anhydrous alum, $KAl(SO_4)_2$, (AA); and caffeine monohydrate. HMB is a relatively small molecule. In this book, it is considered to be a prototype of a polymer because it has two types of bonding, covalent bonds within the molecule and weaker van der Waals bonds between molecules. AA is an ionic framework structure typically found in ceramics. Caffeine monohydrate with more atoms per unit cell is more complex than either HMB or AA. See Section 4.14. These three crystals have complementary crystallographic features.

1.2 TWO-DIMENSIONAL LATTICES

A *crystal* is a solid composed of atoms arranged in a periodic array. Before analyzing crystals, consider abstract wallpaper patterns. Look at the pattern of cats and mice in Figure 1.1. Assume that all the cats are identical and that all the mice are identical, too. Instead of cats and mice, there could be cations and anions. Imagine that the pattern extends forever in two dimensions, filling the entire plane. In crystals, the electron density fills all of space and is greatest at the centers of the atoms, as in the molecule in Figure 7.21. The ball-and-stick models are useful, but do not show how space is filled. Look at this pattern carefully and observe the repetitions. The unit that is repeated consists of one mouse and one cat. This unit is repeated again and again to fill the plane.

FIGURE 1.1 Two-dimensional repeating pattern of cats and mice. (Painted by Anna McNider.)

(a)   (b)

FIGURE 1.2   (a) Selection and labeling of one point in the pattern. (b) Labeling of all points with identical neighborhoods.

Put a piece of tracing paper over the pattern. Select any point, such as the left paw of the cat, and label it with a dot on the tracing paper. See Figure 1.2a. Now find all the points with identical neighborhoods, that is, the left paws; and label those points on the tracing paper with dots. See Figure 1.2b.

Take away the colored pattern, and just the points are left on the tracing paper. See Figure 1.3a. These points, extended forever in two dimensions, form a lattice. Thus, the *lattice of a pattern* is the array of points in space with identical neighborhoods.

The lattice is unique to the pattern.
There is one repeating unit for each lattice point.

The lattice does not depend on the selection of the starting point. Take the tracing paper and move it back onto the pattern. Only this time, instead of lining the paper up with the original selected point, select the nose of the mouse. A new set of identical neighborhoods is found. See Figure 1.3b.

The lattice is independent of the origin.

The idea of a vector is suggested. A vector can be translated or moved parallel to itself and still remain unchanged.

Consider the two repeating patterns in Figure 1.4. On a piece of tracing paper draw the lattice points of the figure of the fish and boats. Perhaps select the eye of the fish. Now move

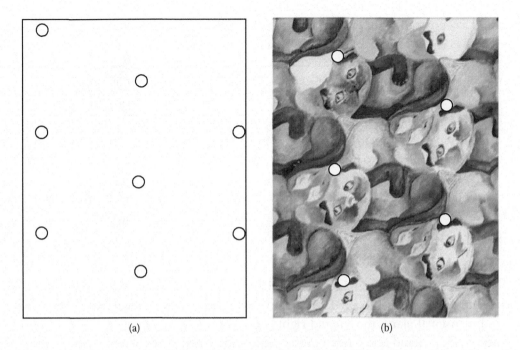

(a) (b)

FIGURE 1.3   (a) Lattice of points. (b) Lattice with a new starting origin, the nose of the mouse.

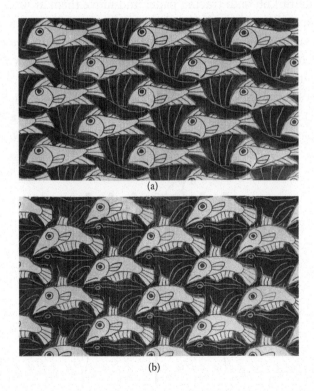

(a)

(b)

FIGURE 1.4   Different patterns with the same lattice. (a) Fish and boats, (b) fish and frogs. (From MacGillavry, C.H. 1976. *Fantasy and Symmetry: Periodic Drawings of M. C. Escher*, Harry N. Abrams, Inc., New York.)

that lattice, on the tracing paper, to the second picture of the fish and frogs. The second pattern has the same lattice associated with it as the first pattern.

While the lattice is unique to the pattern, the pattern is not unique to the lattice.

1.3 TWO-DIMENSIONAL BASIS VECTORS AND UNIT CELLS

A lattice is described mathematically by basis vectors. Consider an abstract lattice in Figure 1.5. Choose any starting point on the lattice as the origin 0 (yellow circle). Choose a second lattice point; let vector **a** connect these two lattice points. Be sure that there are no lattice points on **a** except the two end points. Select another vector, **b**, linearly independent of **a**. Also be sure that there are no lattice points on **b** except the two end points. The *basis vectors* **a** and **b** define the lattice.

1.3.1 Handedness of Basis Vectors

The basis vectors **a** and **b** can be either left-handed or right-handed. Figure 1.6 shows a left hand and a right hand with a red vertical mirror between them. If the hands are kept in the two-dimensional plane, the left and right hands cannot fit exactly on top of one another. However, if one hand is taken out of the plane and turned over, then the hands can fit on top of one another. The same can be said of the two sets of basis vectors **a** and **b**. Copy the basis vectors in Figure 1.6b onto tracing paper and move them around to show that they cannot be superimposed on Figure 1.6a without going out of the plane. The pair, **a** and **b** in that order, on the right is defined to be right-handed and the pair, **a′** and **b′** in that order, on the left is defined to be left-handed. Note that the lattices formed by the respective basis vectors are different and are mirror images of each other.

Another way of examining the problem is to consider the labeling of a single lattice. Figure 1.7a shows a left-handed set of basis vectors. The cross product **a** × **b** points into the plane of the paper away from the reader. By contrast, Figure 1.7b shows a right-handed set

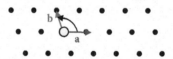

FIGURE 1.5   Lattice with basis vectors **a** and **b**.

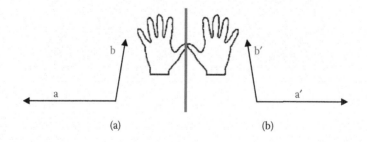

FIGURE 1.6   (a) Left-handed systems, (b) right-handed systems, separated by a vertical mirror in red.

(a)   (b)

FIGURE 1.7 Lattice with basis vectors, (a) left-handed, (b) right-handed.

of basis vectors. The cross product $\mathbf{a}' \times \mathbf{b}'$ points out of the plane of the paper toward the reader. Usually right-handed sets of basis vectors are used.

1.3.2 Describing the Lattice Mathematically

Every point in the lattice can be specified mathematically. Figure 1.8 shows each point labeled with an ordered pair of integers, u and v. The integer u labels the columns and v labels the rows. In general, $-\infty < u < \infty$ and $-\infty < v < \infty$. In Figure 1.8, u goes from $\bar{2}$ to 4 and v goes from $\bar{1}$ to 1. For a book on applied linear algebra see Olver and Shakiban (2006).

In crystallography negative one is written as $\bar{1}$ and is read "one bar." This labeling applies to all negative numbers.

Vector $\mathbf{t}$ describes a plane *lattice* with basis vectors $\mathbf{a}$ and $\mathbf{b}$:

$$\mathbf{t}(u,v) = u\mathbf{a} + v\mathbf{b} \tag{1.1}$$

where u and v are integers such that $-\infty < u < \infty$ and $-\infty < v < \infty$

Basis vectors $\mathbf{a}$ and $\mathbf{b}$ partition space into identical unit cells. Twelve unit cells are shown. Every point in the lattice is reached by an integral multiple of basis vectors $\mathbf{a}$ and $\mathbf{b}$. These basis vectors form a primitive unit cell. Each unit cell is associated with one and only one lattice point. For example, every cell could be uniquely identified by labeling its lattice point in the lower left-hand corner. A *primitive unit cell* is a unit cell with only one lattice point.

Every lattice has a primitive unit cell.

Shift the basis vectors parallel to themselves, so that the lattice points, instead of being at the vertices of the unit cells, are interior to the cells, as in Figure 1.9. Now it is clear that one lattice point is associated with each cell.

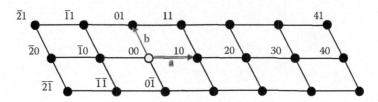

FIGURE 1.8 Lattice with origin 0 0, yellow circle, basis vectors $\mathbf{a}$ and $\mathbf{b}$ and ordered number pairs u v.

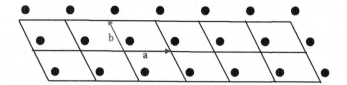

FIGURE 1.9   Shifted lattice.

1.3.3 The Unit Cell

The planar *unit cell* is a parallelogram whose sides are the basis vectors **a** and **b**, as shown in Figure 1.10. In general, the magnitudes of vectors **a** and **b** are different. The vectors are not necessarily orthogonal. The angle between **a** and **b** is γ, which is the Greek letter gamma. The scalar values a, b, and γ are called the *lattice parameters* or *lattice constants*. Figure 1.11a shows the fish and boat pattern with a unit cell drawn in red. Figure 1.11b isolates the unit cell from the rest of the pattern. Look carefully at the isolated unit cell. All the information needed to draw both the fish and the boat is in the unit cell, even though the parts of the fish or boat are not connected as expected. For example, parts of the face of the fish are located in every one of the corners of the isolated unit cell. Every part of both figures is contained in the unit cell. See Exercise 1.1.

The only information needed to construct the pattern is contained in the unit cell, which is repeated multiple times.

The basis vectors are not unique. Figure 1.12 shows different sets of basis vectors with different origins. The lattice is labeled with respect to a chosen origin and chosen basis vectors. See Exercise 1.2. Each set of basis vectors generates a primitive unit cell.

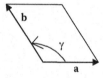

FIGURE 1.10   Unit cell defined by basis vectors **a** and **b**. The angle between **a** and **b** is γ.

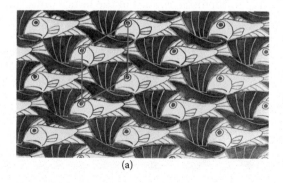

(a)   (b)

FIGURE 1.11   (a) Unit cell drawn on the pattern, (b) isolated unit cell.

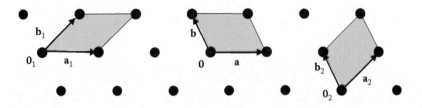

FIGURE 1.12 Lattice with three different combinations of origins and basis vectors forming primitive unit cells.

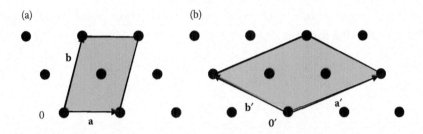

FIGURE 1.13 Multiple unit cells containing (a) two lattice points and (b) three lattice points.

For each lattice there is an infinite number of choices of primitive unit cells. Sometimes a nonprimitive cell is appropriate.

Figure 1.13 shows basis vectors **a** and **b** and origin O. The unit cell for these basis vectors contains one interior lattice point as well as one lattice point associated with the four vertices of the unit cell. This particular cell contains two lattice points. Another multiple unit cell with basis vectors **a′** and **b′** is also shown. It has three lattice points. A *multiple unit cell* contains more than one lattice point. An infinite number of multiple cells may be constructed. There is also no limit to the number of lattice points that can be contained in a unit cell. In practice, however, the number of lattice points in the unit cell rarely exceeds four. For multiple cells, Equation 1.1 is modified so that some of the values of u and v are fractions. See Exercise 1.3.

1.4 TWO-DIMENSIONAL TRANSFORMATIONS BETWEEN SETS OF BASIS VECTORS

Since the basis vectors are not unique, consider a transformation between sets of basis vectors. Sometimes for computational purposes it is more convenient to use a new set of basis vectors. A linear transformation relates any two sets of basis vectors. Figure 1.14 shows a lattice with two sets of basis vectors, $\mathbf{a}_1$, $\mathbf{b}_1$ and also $\mathbf{a}_2$, $\mathbf{b}_2$. The unit cell for each set of basis vectors is drawn. The relationship between the two sets of basis vectors is

$$\mathbf{a}_2 = \mathbf{a}_1 - \mathbf{b}_1 \tag{1.2a}$$

$$\mathbf{b}_2 = 2\mathbf{a}_1 + \mathbf{b}_1 \tag{1.2b}$$

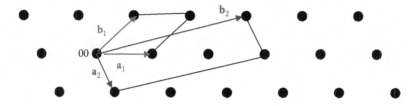

FIGURE 1.14   Lattice with two sets of basis vectors $\mathbf{a}_1$, $\mathbf{b}_1$ and $\mathbf{a}_2$, $\mathbf{b}_2$.

Vectors $\mathbf{a}_1$, $\mathbf{b}_1$ form a primitive unit cell and vectors $\mathbf{a}_2$, $\mathbf{b}_2$ form a multiple cell containing three lattice points. The transformation is from $\mathbf{a}_1$, $\mathbf{b}_1$ to $\mathbf{a}_2$, $\mathbf{b}_2$.

In accordance with the *International Tables for Crystallography*, such relationships are written in the following form:

$$(\mathbf{a}_2 \quad \mathbf{b}_2) = (\mathbf{a}_1 \quad \mathbf{b}_1)P = (\mathbf{a}_1 \quad \mathbf{b}_1)\begin{pmatrix} 1 & 2 \\ \bar{1} & 1 \end{pmatrix}$$

where

$$P = \begin{pmatrix} 1 & 2 \\ \bar{1} & 1 \end{pmatrix} \tag{1.3}$$

See Exercise 1.4. In this case the determinant of P is positive and $\det(P) = 3$. If the determinant of P is positive, then both sets of basis vectors have the same handedness. In this case they are both right-handed. By contrast, if the determinant of P is negative, then the handedness changes between the two sets of basis vectors. Thus, if the original set of basis vectors is right-handed and $\det(P)$ is negative, then P takes the original right-handed set into a left-handed one. If $\det(P) = 0$, then the new vectors, $\mathbf{a}_2$ and $\mathbf{b}_2$, are not linearly independent, and hence are not a set of basis vectors. Under this condition $\mathbf{a}_2$ and $\mathbf{b}_2$ are parallel.

The unit cell formed by $\mathbf{a}_2$, $\mathbf{b}_2$ is a multiple cell with three lattice points and the unit cell formed by $\mathbf{a}_1$, $\mathbf{b}_1$ is primitive. The ratio of the number of lattice points between the $\mathbf{a}_2$, $\mathbf{b}_2$ set of basis vectors and the $\mathbf{a}_1$, $\mathbf{b}_1$ set of basis vectors is 3. The magnitude of the determinant of P, $|\det(P)|$, gives the ratio of the number of lattice points in the new set of basis vectors to the number in the old set. See Table 1.1 and Exercise 1.5.

TABLE 1.1   Properties of the Transformation Matrix P

	Information Relating the Old Set of Basis Vectors to the New Set
$\det(P) > 0$	Keeps handedness
$\det(P) < 0$	Changes handedness
$\det(P) = 0$	The new vectors are not linearly independent; they are not a valid set of basis vectors
$\|\det(P)\|$ (if $\neq 0$)	$\dfrac{\text{Number of lattice points in new unit cell}}{\text{Number of lattice points in old unit cell}}$

EXAMPLE 1.1

Write the transformation matrix P, if

$$\mathbf{a}_2 = 7\mathbf{a}_1 - 3\mathbf{b}_1$$

$$\mathbf{b}_2 = 2\mathbf{a}_1 - \mathbf{b}_1$$

Also determine the handedness of $\mathbf{a}_2$, $\mathbf{b}_2$ if $\mathbf{a}_1$, $\mathbf{b}_1$ is right-handed. Demonstrate that the new basis vectors are linearly independent. Give the ratio of the number of lattice points in the new unit cell to the number in the old unit cell.

Solution

The transformation matrix P is made from the column vectors $\begin{pmatrix} 7 \\ 3 \end{pmatrix}$ and $\begin{pmatrix} 2 \\ 1 \end{pmatrix}$. Thus,

$$P = \begin{pmatrix} 7 & 2 \\ 3 & 1 \end{pmatrix}$$

The $\det(P) = -1$. Therefore, the set of basis vectors $\mathbf{a}_2$, $\mathbf{b}_2$ is left-handed. Since the $\det(P) \neq 0$, then the new vectors $\mathbf{a}_2$, $\mathbf{b}_2$ are linearly independent. Both sets of basis vectors $\mathbf{a}_2$, $\mathbf{b}_2$ and $\mathbf{a}_1$, $\mathbf{b}_1$ have the same number of lattice points in the unit cells because the magnitude of $\det(P)$ is 1.

In general,

$$(\mathbf{a}_2 \quad \mathbf{b}_2) = (\mathbf{a}_1 \quad \mathbf{b}_1)P = (\mathbf{a}_1 \quad \mathbf{b}_1)\begin{pmatrix} P_{11} & P_{12} \\ P_{21} & P_{22} \end{pmatrix} = (P_{11}\mathbf{a}_1 + P_{21}\mathbf{b}_1 \quad P_{12}\mathbf{a}_1 + P_{22}\mathbf{b}_1)$$

where P is the transformation matrix. Thus,

$$\mathbf{a}_2 = P_{11}\,\mathbf{a}_1 + P_{21}\,\mathbf{b}_1$$

$$\mathbf{b}_2 = P_{12}\,\mathbf{a}_1 + P_{22}\,\mathbf{b}_1$$

1.5 THREE-DIMENSIONAL BASIS VECTORS, UNIT CELLS, AND LATTICE TRANSFORMATIONS

The ideas developed so far are easily extended to three dimensions.

1.5.1 Basis Vectors

In three dimensions, as in two, the lattice of a crystal is an array of points with identical neighborhoods. The lattice is unique to the pattern. There is one repeat unit for each lattice

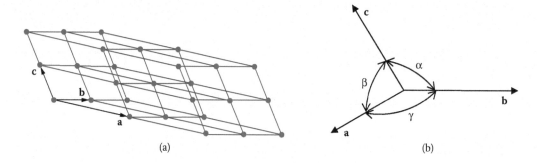

FIGURE 1.15 (a) Three-dimensional lattice, (b) lattice parameters.

point. Linearly independent vectors **a**, **b**, and **c** are the basis vectors that describe the lattice. Mathematically, vector $\mathbf{t}(u,v,w)$ produces a lattice, Figure 1.15a, with basis vectors **a**, **b**, and **c**

$$\mathbf{t}(u,v,w) = u\mathbf{a} + v\mathbf{b} + w\mathbf{c}$$

where u, v, and w are integers such that $-\infty < u < \infty$, $-\infty < v < \infty$, and $-\infty < w < \infty$.

1.5.2 Unit Cell

The three-dimensional primitive *unit cell* is a parallelepiped formed by the basis vectors **a**, **b**, and **c**. The angle between **a** and **b** is called γ, the angle between **a** and **c** is called β, and the angle between **b** and **c** is called α. See Figure 1.15b. A *primitive unit cell* has one lattice point. A *multiple unit cell* has more than one lattice point. The unit cell contains all the information needed to describe the three-dimensional pattern.

1.5.3 Three-Dimensional Transformations between Sets of Basis Vectors

A linear transformation relates any two sets of basis vectors in three dimensions.

For example, consider the transformation

$$\mathbf{a}_2 = \mathbf{a}_1 + \mathbf{b}_1 - \mathbf{c}_1$$
$$\mathbf{b}_2 = -2\mathbf{a}_1 - \mathbf{b}_1 + \mathbf{c}_1$$
$$\mathbf{c}_2 = -\mathbf{a}_1 + \mathbf{b}_1 + \mathbf{c}_1$$

In accordance with the notation of the *International Tables for Crystallography*,

$$(\mathbf{a}_2 \quad \mathbf{b}_2 \quad \mathbf{c}_2) = (\mathbf{a}_1 \quad \mathbf{b}_1 \quad \mathbf{c}_1) \begin{pmatrix} 1 & \bar{2} & \bar{1} \\ 1 & \bar{1} & 1 \\ \bar{1} & 1 & 1 \end{pmatrix}$$

Thus,

$$P = \begin{pmatrix} 1 & \bar{2} & \bar{1} \\ 1 & \bar{1} & 1 \\ \bar{1} & 1 & 1 \end{pmatrix}$$

and det $(P) = 2$. Since the determinant of P is positive, then if $\mathbf{a}_1$, $\mathbf{b}_1$, $\mathbf{c}_1$ is a right-handed set of basis vectors, then $\mathbf{a}_2$, $\mathbf{b}_2$, $\mathbf{c}_2$ is also right-handed. Since det$(P) \neq 0$, then the new vectors $\mathbf{a}_2$, $\mathbf{b}_2$, $\mathbf{c}_2$ are linearly independent. And finally, the number of lattice points in the unit cell formed by the $\mathbf{a}_2$, $\mathbf{b}_2$, $\mathbf{c}_2$ is twice the number of lattice points in the unit cell formed by the $\mathbf{a}_1$, $\mathbf{b}_1$, $\mathbf{c}_1$. These ideas will be expanded in Chapter 2. See Exercise 1.6.

In general,

$$(\mathbf{a}_2 \quad \mathbf{b}_2 \quad \mathbf{c}_2) = (\mathbf{a}_1 \quad \mathbf{b}_1 \quad \mathbf{c}_1)P = (\mathbf{a}_1 \quad \mathbf{b}_1 \quad \mathbf{c}_1) \begin{pmatrix} P_{11} & P_{12} & P_{13} \\ P_{21} & P_{22} & P_{23} \\ P_{31} & P_{32} & P_{33} \end{pmatrix}$$

where P is the transformation matrix.

In addition,

$$\mathbf{a}_2 = P_{11} \mathbf{a}_1 + P_{21} \mathbf{b}_1 + P_{31} \mathbf{c}_1$$

$$\mathbf{b}_2 = P_{12} \mathbf{a}_1 + P_{22} \mathbf{b}_1 + P_{32} \mathbf{c}_1$$

$$\mathbf{c}_2 = P_{13} \mathbf{a}_1 + P_{23} \mathbf{b}_1 + P_{33} \mathbf{c}_1$$

EXAMPLE 1.2

Consider the transformation in Figure 1.16. Calculate the transformation matrix P from $\mathbf{a}_1$, $\mathbf{b}_1$, $\mathbf{c}_1$ to $\mathbf{a}_2$, $\mathbf{b}_2$, $\mathbf{c}_2$. Give the handedness of the new set of basis vectors and the ratio of the number of lattice points in the new cell to the number in the old.

Solution

The relationship between the two sets of vectors is

$$\mathbf{a}_2 = \frac{1}{2}(\mathbf{a}_1 + \mathbf{b}_1)$$

$$\mathbf{b}_2 = \frac{1}{2}(-\mathbf{a}_1 + \mathbf{b}_1)$$

$$\mathbf{c}_2 = \mathbf{c}_1$$

(continued)

EXAMPLE 1.2 (continued)

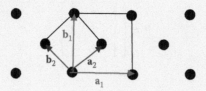

FIGURE 1.16   Transformation from $\mathbf{a}_1, \mathbf{b}_1, \mathbf{c}_1$ to $\mathbf{a}_2, \mathbf{b}_2, \mathbf{c}_2$ where $\mathbf{c}_1 = \mathbf{c}_2$ and both are perpendicular to the plane of the page and point in the direction of the reader.

The transformation matrix P is composed of the column vectors $\begin{pmatrix} \frac{1}{2} \\ \frac{1}{2} \\ 0 \end{pmatrix}$, $\begin{pmatrix} \frac{\bar{1}}{2} \\ \frac{1}{2} \\ 0 \end{pmatrix}$, and $\begin{pmatrix} 0 \\ 0 \\ 1 \end{pmatrix}$; thus,

$$P = \begin{pmatrix} \frac{1}{2} & \frac{\bar{1}}{2} & 0 \\ \frac{1}{2} & \frac{1}{2} & 0 \\ 0 & 0 & 1 \end{pmatrix}.$$

The $\det(P) = 1/2$. As can be seen from Figure 1.16, both sets of basis vectors are right-handed. The handedness is not changed, as is confirmed by the positive determinant of P. The number of lattice points in the unit cell formed by the $\mathbf{a}_2, \mathbf{b}_2, \mathbf{c}_2$ is 1/2 the number of lattice points in the unit cell formed by the $\mathbf{a}_1, \mathbf{b}_1, \mathbf{c}_1$.

1.6 CONVERSION INTO CARTESIAN COORDINATES

Computer programs are usually written in Cartesian coordinates. This fact is especially appropriate when a drawing is made because the computer language knows how to represent Cartesian coordinates in drawings.

1.6.1 Two-Dimensional Conversion into Cartesian Coordinates

Crystallographic vectors $\mathbf{a}$ and $\mathbf{b}$ will be referred to orthogonal unit vectors $\mathbf{e}_1$ and $\mathbf{e}_2$. As in Figure 1.17a choose $\mathbf{e}_1$ along the crystallographic $\mathbf{a}$ basis vector. Thus,

$$\mathbf{a} = a\,\mathbf{e}_1$$

Choose $\mathbf{e}_2$ perpendicular to $\mathbf{e}_1$ and such that $\mathbf{e}_2$ is obtained from $\mathbf{e}_1$ by a counterclockwise rotation of 90°. The basis vector $\mathbf{b}$ is now decomposed into projections along $\mathbf{e}_1$ and $\mathbf{e}_2$ as in Figure 1.17b.

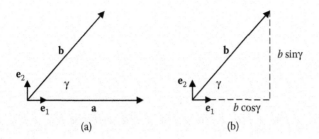

FIGURE 1.17   (a) Basis vectors **a**, **b**. (b) Decomposition of basis vector **b**.

Thus,

$$\mathbf{b} = b_1\,\mathbf{e}_1 + b_2\,\mathbf{e}_2 = b\cos\gamma\,\mathbf{e}_1 + b\sin\gamma\,\mathbf{e}_2.$$

Thus,

$$(\mathbf{a}\quad \mathbf{b}) = (\mathbf{e}_1\quad \mathbf{e}_2)\begin{pmatrix} a & b\cos\gamma \\ 0 & b\sin\gamma \end{pmatrix} = (\mathbf{e}_1\,\mathbf{e}_2)\,\mathbf{C} \tag{1.4}$$

Thus, the transformation matrix **C** is

$$\mathbf{C} = \begin{pmatrix} a & b\cos\gamma \\ 0 & b\sin\gamma \end{pmatrix} \tag{1.4a}$$

Consider vector **t**, which is written as $\mathbf{t} = u\,\mathbf{a} + v\,\mathbf{b}$ in crystallographic coordinates with the basis vectors **a**, **b**, and lattice points u,v and the same vector $\mathbf{t} = x\,\mathbf{e}_1 + y\,\mathbf{e}_2$ with Cartesian coordinates x, y. Then,

$$\mathbf{t} = x\,\mathbf{e}_1 + y\,\mathbf{e}_2 = u\,\mathbf{a} + v\,\mathbf{b}$$

or

$$(\mathbf{e}_1\quad \mathbf{e}_2)\begin{pmatrix} x \\ y \end{pmatrix} = (\mathbf{a}\quad \mathbf{b})\begin{pmatrix} u \\ v \end{pmatrix} \tag{1.5}$$

Substitute Equation 1.4 into Equation 1.5

$$(\mathbf{e}_1\quad \mathbf{e}_2)\begin{pmatrix} x \\ y \end{pmatrix} = (\mathbf{e}_1\quad \mathbf{e}_2)\,\mathbf{C}\begin{pmatrix} u \\ v \end{pmatrix}$$

or

$$\begin{pmatrix} x \\ y \end{pmatrix} = \mathbf{C}\begin{pmatrix} u \\ v \end{pmatrix} \tag{1.5a}$$

EXAMPLE 1.3

Find the Cartesian coordinates of lattice point 4,1 (see Figure 1.18).

Solution

$$\mathbf{C}\begin{pmatrix} 4 \\ 1 \end{pmatrix} = \begin{pmatrix} a & b\cos\gamma \\ 0 & b\sin\gamma \end{pmatrix}\begin{pmatrix} 4 \\ 1 \end{pmatrix} = \begin{pmatrix} 4a + b\cos\gamma \\ b\sin\gamma \end{pmatrix} = \begin{pmatrix} x \\ y \end{pmatrix}$$

The computer plots $(4a + b\cos\gamma, b\sin\gamma)$ for the coordinates of lattice point 4, 1. See Exercise 1.7.

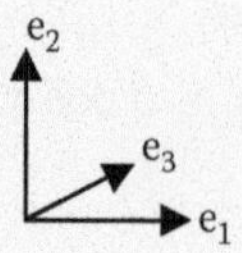

FIGURE 1.18 Relationship among three-dimensional unit vectors with $\mathbf{e}_3$ perpendicular to the plane of the paper and pointing toward the reader.

1.6.2 Three-Dimensional Conversion into Cartesian Coordinates

Crystallographic vectors **a**, **b**, and **c** will be referred to orthogonal unit vectors $\mathbf{e}_1$, $\mathbf{e}_2$, $\mathbf{e}_3$. Choose $\mathbf{e}_1$ and $\mathbf{e}_2$ in the same way that they were chosen for the two-dimensional case. Choose unit vector $\mathbf{e}_3$ to be perpendicular to both $\mathbf{e}_1$ and $\mathbf{e}_2$ with $\mathbf{e}_3 = \mathbf{e}_1 \times \mathbf{e}_2$ as shown in Figure 1.18. Basis vector **c** can be written as

$$\mathbf{c} = c_1\,\mathbf{e}_1 + c_2\,\mathbf{e}_2 + c_3\,\mathbf{e}_3$$

Take the dot products, remembering that

$$\mathbf{e}_1 \cdot \mathbf{e}_1 = \mathbf{e}_2 \cdot \mathbf{e}_2 = \mathbf{e}_3 \cdot \mathbf{e}_3 = 1 \quad \text{and} \quad \mathbf{e}_1 \cdot \mathbf{e}_2 = \mathbf{e}_2 \cdot \mathbf{e}_3 = \mathbf{e}_1 \cdot \mathbf{e}_3 = 0$$

Then

$$\mathbf{a} \cdot \mathbf{c} = a\,c\,\cos\beta = a\,\mathbf{e}_1 \cdot (c_1\,\mathbf{e}_1 + c_2\,\mathbf{e}_2 + c_3\,\mathbf{e}_3) = a\,c_1$$

$$c_1 = c\cos\beta$$

$$\mathbf{b} \cdot \mathbf{c} = b\,c\,\cos\alpha = (b\cos\gamma\,\mathbf{e}_1 + b\sin\gamma\,\mathbf{e}_2) \cdot (c\cos\beta\,\mathbf{e}_1 + c_2\,\mathbf{e}_2 + c_3\,\mathbf{e}_3)$$

$$c_2 = \frac{c(\cos\alpha - \cos\gamma\cos\beta)}{\sin\gamma}$$

Finally,

$$c_3 = +\sqrt{c^2 - c_1^2 - c_2^2} = +c\sqrt{1 - \cos^2\beta - \frac{(\cos\alpha - \cos\gamma\cos\beta)^2}{\sin^2\gamma}}$$

This choice of a plus sign for the square root, makes the **a**, **b**, **c** set of basis vectors right-handed.

Thus, the matrix **C** to convert into Cartesian coordinates is

$$\mathbf{C} = \begin{pmatrix} a & b\cos\gamma & c_1 \\ 0 & b\sin\gamma & c_2 \\ 0 & 0 & c_3 \end{pmatrix} \tag{1.6}$$

Consider vector **t**, which is written either as $\mathbf{t} = u\,\mathbf{a} + v\,\mathbf{b} + w\,\mathbf{c}$ with basis vectors **a**, **b**, **c** and lattice points u,v,w or as $\mathbf{t} = x\,\mathbf{e}_1 + y\,\mathbf{e}_2 + z\,\mathbf{e}_3$ with Cartesian coordinates x, y, z. Then

$$\mathbf{t} = x\,\mathbf{e}_1 + y\,\mathbf{e}_2 + z\,\mathbf{e}_3 = u\,\mathbf{a} + v\,\mathbf{b} + w\,\mathbf{c}$$

and

$$\begin{pmatrix} x \\ y \\ z \end{pmatrix} = \mathbf{C} \begin{pmatrix} u \\ v \\ w \end{pmatrix}$$

See Exercise 1.8.

⌨ 1.7 A CRYSTAL: HEXAMETHYLBENZENE

HMB is the first case-study crystal. Throughout this book, many calculations are carried out on this crystal. This computer icon, ⌨, means that a computer language designed for numerical analysis should be used.

In the 1920s, the structure of diamond was known. Since diamond has a puckered six-membered ring, W. H. Bragg and many of the leading scientists were convinced, by analogy, that the six-membered benzene ring would be puckered. Since benzene is not crystalline at room temperature, the six-membered benzene ring was studied via related molecules.

In 1929 Kathleen Lonsdale proved that the benzene ring was planar. She did this by solving the structure of HMB, $C_6(CH_3)_6$. This molecule crystallizes into beautiful transparent crystals with one molecule to the unit cell. Figure 1.19a shows what the molecule looks like. "Me" stands for methyl group or $-CH_3$. Six carbon atoms, "C," form the benzene ring. A methyl group, consisting of a carbon atom and three hydrogen atoms, is attached to each carbon atom in the benzene ring. The 12 carbon atoms all lie in a single plane. The HMB molecules are distinct entities with internal covalent bonds. Van der Waals bonds exist between the individual molecules.

Figure 1.19b shows a three-dimensional stereoscopic drawing of HMB including the hydrogen atoms on the methyl groups. These hydrogen atoms are not in the plane of the 12 carbon atoms.

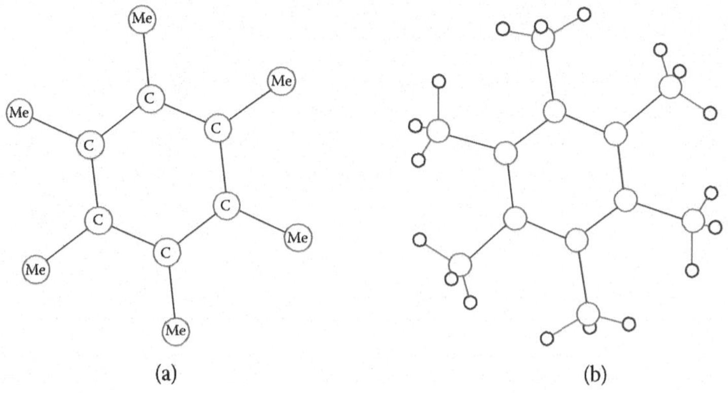

FIGURE 1.19 HMB molecule: (a) carbon atoms only, (b) carbon and hydrogen atoms. (Fractional coordinates from Stride, 2005.)

1.7.1 Definition of Angstrom

The *angstrom*, Å, is a unit of length used with x-rays and spectroscopy. An angstrom is approximately the radius of a hydrogen atom. The unit is named after Anders Jöns Ångström (1814–1874), a Swedish physicist, mathematician, and astronomer who measured lines in the solar spectrum.

---

1 Å = 10^{-10} m = 0.1 nm = 10^{-8} cm

1.7.2 Two-Dimensional Unit Cell for HMB

▣ EXAMPLE 1.4

The lattice parameters of HMB are

$$a = 9.010 \text{ Å}, b = 8.926 \text{ Å}, c = 5.344 \text{ Å},$$

$$\alpha = 44° \; 27', \beta = 116° \; 43', \gamma = 119° \; 34' \text{ (Lonsdale, 1929)}$$

Write a program to draw a two-dimensional unit cell, as in Figure 1.20, with sides a and b at angle γ to each other. Write it in general terms so that it can be used for other crystals. The Starter Program in MATLAB, given at the end of the chapter, may be used. Starter Programs reduce the burden of the coding, and at the same time allow rapid progress in understanding the crystallography. For useful books on MATLAB see Chapman (2002), Higham and Higham (2000), and Recktenwald (2000).

Solution

The problem is to convert into a Cartesian coordinate system (for the computer graphics) from a more general system where γ may lie anywhere between 0° and 180°. This

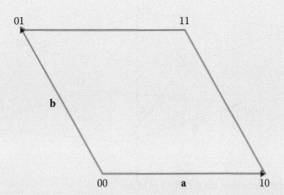

FIGURE 1.20 Two-dimensional unit cell for HMB with lattice points at vertices.

is a two-dimensional case where the crystallographic coordinates of the unit cell are used.

The Cartesian coordinates of the lattice points are generated by the **C** matrix, in Equation 1.4a, multiplied by a matrix containing the lattice points as column vectors as in Equation 1.5a. Begin with the origin and go counterclockwise until returning to the origin.

$$\text{C} \quad * \quad \begin{matrix}\text{(lattice points as} \\ \text{column vectors)}\end{matrix} = \begin{matrix}\text{lattice points as column vectors} \\ \text{in Cartesian coordinates}\end{matrix}$$

$$\begin{pmatrix} a & b\cos\gamma \\ 0 & b\sin\gamma \end{pmatrix} \begin{pmatrix} 0 & 1 & 1 & 0 & 0 \\ 0 & 0 & 1 & 1 & 0 \end{pmatrix} = \begin{pmatrix} 0 & a & a+b\cos\gamma & b\cos\gamma & 0 \\ 0 & 0 & b\sin\gamma & b\sin\gamma & 0 \end{pmatrix}$$

The Cartesian coordinates for the lattice point

$$\begin{pmatrix} 1 \\ 1 \end{pmatrix} \text{ are } \begin{pmatrix} a+b\cos\gamma \\ b\sin\gamma \end{pmatrix}$$

The computer plots point $(a+b\cos\gamma, b\sin\gamma)$ for that lattice point. See Exercise 1.9.

1.7.3 Three-Dimensional Unit Cell of HMB

The technique for drawing the unit cell in two dimensions is now extended to three. Use the **C** matrix defined in Equation 1.6. In Figure 1.21, the lattice points have three coordinates, and the matrix containing the lattice-point information is adjusted accordingly.

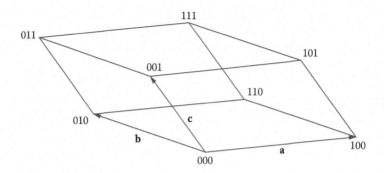

FIGURE 1.21 Three-dimensional unit cell of HMB.

Thus,

$$\text{C} * (\text{lattice points as column vectors}) = \frac{\text{lattice points as column vectors}}{\text{in Cartesian coordinates}}$$

$$\begin{pmatrix} a & b\cos\gamma & c_1 \\ 0 & b\sin\gamma & c_2 \\ 0 & 0 & c_3 \end{pmatrix} \begin{pmatrix} 0 & 1 & 1 & 0 & ... \\ 0 & 0 & 0 & 0 & ... \\ 0 & 0 & 1 & 0 & ... \end{pmatrix} = \text{lattice points to be gotten by computer}$$

Note that retracing a line does not affect the figure. See Exercise 1.10.

Once the exercise is done carefully label the file and save it for future use.

1.8 A CRYSTAL: ANHYDROUS ALUM

In addition to HMB, another case-study crystal emphasized throughout the book is the framework structure AA, whose chemical formula is $KAl(SO_4)_2$. See Lipson (1935), Manoli et al. (1970), and Cromer et al. (1967). The lattice parameters are

$$a = 4.709 \text{ Å}; \quad b = 4.709 \text{ Å}; \quad c = 7.984 \text{ Å}$$

$$\alpha = 90°; \quad \beta = 90°; \quad \gamma = 120°.$$

Now the power of an advanced computer language like MATLAB is seen. The computer program written for HMB is used for AA just by changing the lattice parameters. Immediately the unit cell for AA is drawn.

The three-dimensional computer-generated figure can be viewed from every possible angle, giving a better understanding of the shape of the unit cell. Figure 1.22 shows the completed unit cell. See Exercises 1.11 through 1.14.

In three-dimensional graphing, the views or projections along various directions can be captured as the figure is rotated. For example, Figure 1.23 shows both the planar projections

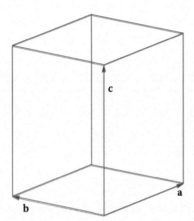

FIGURE 1.22   Three-dimensional unit cell of AA.

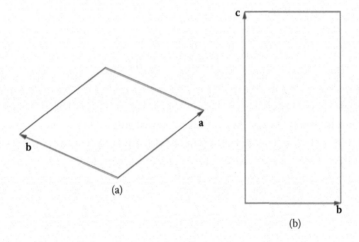

FIGURE 1.23   (a) Projection down **c**, (b) projection down **a**.

down **c** and down **a**. In the projection down **c**, the angle between **a** and **b** is 120°. When the unit cell is positioned so that the projection is down **a**, then **b** and **c** are perpendicular. A view down **b** is similar to Figure 1.23b. This information is shown in Figure 1.22, but is less obvious.

The plane of a projection is always perpendicular to the direction of the projection.

1.9 EFFECTS OF TEMPERATURE AND PRESSURE ON THE LATTICE PARAMETERS

Lattice parameters a, b, c, α, β, and γ are usually given at room temperature and pressure. Two cases are now considered. First, if the temperature of the crystal is increased, thermal energy converts, in part, into vibrational energy; and the unit cell expands. Second, if the crystal is put under pressure, the unit cell contracts. See Hazen (1982) and Chung et al. (1993).

Isotropic properties of a material are those properties that are independent of direction. On the other hand, *anisotropic properties* are those properties that depend on direction. For example, some crystals have isotropic thermal expansion properties and others are anisotropic. The unit cell parameters a, b, and c often change at different rates.

1.9.1 Temperature Variations and Thermal Expansion

In this section, the effect of an increase in temperature on urea is examined. Urea is an important commercial chemical as well as a product of nitrogen metabolism in mammals. The lattice parameters for urea are a and c, with $a = b$ and $\alpha = \beta = \gamma = 90°$. Table 1.2 shows that as the temperature rises, the lattice parameters a and c expand. On the other hand, the lattice parameters a and b remain equal to each other and the angles α, β, and γ remain unchanged; they are not included in the table.

An important material property can be derived from the data in Table 1.2. The *linear coefficient of thermal expansion*, α_l, at constant pressure for a particular lattice parameter is defined as

$$\alpha_l = \frac{1}{l}\left(\frac{\partial l}{\partial T}\right)_P = \left(\frac{\partial \ln l}{\partial T}\right)_P \tag{1.7}$$

where l is the lattice parameter, T is the temperature, and P is the pressure. The data are recorded at a single pressure, so the pressure is constant.

See Exercise 1.15. The linear coefficients of thermal expansion of urea are anisotropic because they are different for lattice parameters a and c.

TABLE 1.2   Temperature Dependence of the Lattice Parameters of Urea

T, Temperature (K)	a (Å)	c (Å)
188	5.61365	4.69524
198	5.61652	4.69680
208	5.61902	4.69698
218	5.62233	4.69706
228	5.62423	4.69821
238	5.62731	4.69882
248	5.63022	4.69916
258	5.63642	4.69939
268	5.63913	4.70015
278	5.64140	4.70053
288	5.64554	4.70081
298	5.64492	4.70204
308	5.64854	4.70249
318	5.65101	4.70253
328	5.65360	4.70312

Source: Data kindly provided by K. Roberts, also see Hammond, R. et al. 2005. *Journal of Applied Crystallography*, 38(Part 6), 1038–1039.

🖳 EXAMPLE 1.5

For urea calculate the linear coefficient of thermal expansion for the lattice parameter a.

Solution

From the data in Table 1.2, plot the natural logarithm of the lattice parameter a versus temperature. The equation of the least-squares line that fits the data in Figure 1.24 is

$$\ln a = 5.23 \times 10^{-5}\, T + 1.7154 \tag{1.8}$$

where a is the lattice parameter and T is the temperature in K.

The slope in Equation 1.8 is the linear coefficient of thermal expansion. Thus,

$$\alpha_l = 5.23 \times 10^{-5}\ \mathrm{K}^{-1}$$

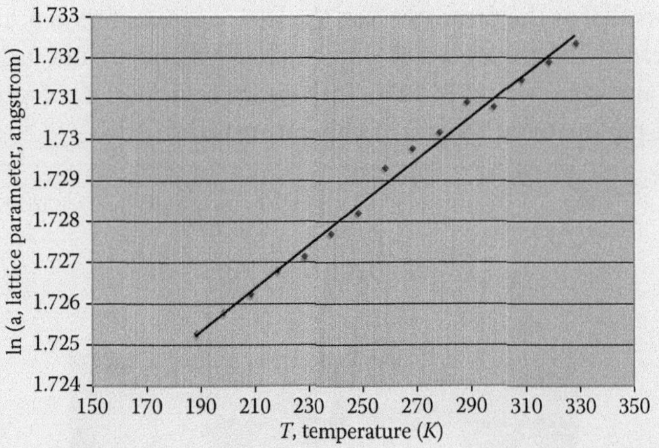

FIGURE 1.24 Temperature dependence of the natural logarithm of the lattice parameter a for urea.

1.9.2 Pressure Variations and Compressibility

Next, consider the effect of pressure on the lattice parameters of anthracene. Anthracene, a product of coal tar, has lattice parameters a, b, c, and β with $\alpha = \gamma = 90°$. The anthracene sample is put in a diamond–anvil cell, which consists of two diamonds clamped together with the sample between them as in Figure 1.25. The sample (green) is inside a gasket (blue) which contains a fluid (pink). The fluid allows the cell to be under hydrostatic pressure. The anvils are made of diamond because diamond is one of the hardest materials known and because it has special properties for the diffraction of x-rays. The diamond–anvil cell is then mounted on an x-ray device. The pressure is varied and a sequence of x-ray patterns is recorded from which the lattice parameters are calculated. See Figure 1.26.

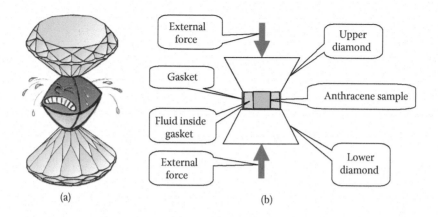

(a)   (b)

FIGURE 1.25   (a) Cartoon of diamond anvil (drawn by Ronald Miletich), (b) schematic of a diamond–anvil cell.

Table 1.3 shows that as the pressure rises, the lattice parameters *a*, *b*, and *c* decrease and the angle β increases. The angles α and γ are unchanged and are not included in the table. A pressure of 1 atm equals 101.325 kPa. The *linear coefficient of compressibility*, β_l, for a particular lattice parameter at constant temperature is defined as

$$\beta_l = -\frac{1}{l}\left(\frac{\partial l}{\partial P}\right)_T = -\left(\frac{\partial \ln l}{\partial P}\right)_T \tag{1.9}$$

FIGURE 1.26   A single-crystal diffractometer adapted for extremely precise measurements of lattice parameters of crystals held at high pressures in a diamond–anvil cell. (Photograph by Ross Angel.)

TABLE 1.3   Pressure Dependence of the Anthracene Lattice Parameters

Pressure (GPa)	$a(Å)$	$b(Å)$	$c(Å)$	$β(deg)$
0.00	8.5319	6.0044	11.1482	124.5774
1.05	8.1099	5.8435	10.8806	125.2857
2.45	7.8850	5.7461	10.7518	125.7366
3.07	7.8195	5.7170	10.7159	125.8601
4.04	7.7259	5.6736	10.6630	126.0248
5.05	7.6342	5.6296	10.6144	126.2840
6.07	7.5646	5.5947	10.5853	126.4584
6.70	7.5408	5.5839	10.5637	126.5426
7.06	7.5044	5.5641	10.5454	126.6517
8.02	7.4464	5.5348	10.5271	126.7967
9.01	7.3965	5.5092	10.4910	126.9172
9.45	7.3809	5.4971	10.4731	126.9770
10.2	7.3410	5.4779	10.4654	127.0757
12.6	7.2516	5.4270	10.4074	127.3705
14.6	7.1846	5.3852	10.3719	127.5643
16.6	7.1293	5.3520	10.3500	127.7650
18.3	7.0789	5.3249	10.3204	127.9069
20.2	7.0347	5.2990	10.2852	128.1421
22.7	6.9790	5.2619	10.2402	128.3885

Source: Data kindly provided by M. Oehzelt. See also Oehzelt, M., Resel, R. and Nakayama, A. 2002. *Physical Review*, B66, 174104.

where l is the lattice parameter, T is the temperature, and P is the pressure. The minus sign indicates that usually the lattice parameter decreases with increasing pressure.

Figure 1.27 shows that the relationship between the natural logarithm of a and the pressure is nonlinear. (Note that here the word "linear" has two different meanings. *Linear compressibility* refers to a lattice parameter that varies with pressure. The relationship between the natural logarithm of a and the pressure can be nonlinear.) Here the linear compressibility depends on the pressure. Several fits including exponential and polynomial are not satisfactory, so the *mean* linear compressibility between two pressures, $\overline{β}_{P_1,P_2}$ is useful. Thus,

$$\overline{β}_{P_1,P_2} = \frac{-2}{a_1 + a_2}\left[\frac{a_2 - a_1}{P_2 - P_1}\right] \cong β_{(P_1+P_2)/2} \tag{1.10}$$

where a_1 and a_2 are the values of the lattice parameter at the corresponding pressures, P_1 and P_2.

See Exercises 1.16 and 1.17. The linear compressibility of anthracene is an anisotropic property because it is different for lattice parameters a, b, and c.

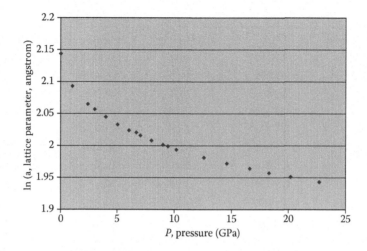

FIGURE 1.27 Pressure dependence of the natural logarithm of the lattice parameter a for anthracene.

■ EXAMPLE 1.6

Calculate the mean linear compressibility for anthracene for lattice parameter a between pressures of 6.07 and 6.70 GPa.

Solution

Use Table 1.3 and Equation 1.10 to get

$$\bar{\beta}_{6.07, 6.70 GPa} = \frac{-2}{7.5646 + 7.5408}\left[\frac{7.5408 - 7.5646}{6.70 - 6.07}\right] \cong \beta_{3.36 GPa}$$

$$= 0.005002\ \text{GPa}^{-1} = 5.0\ \text{MPa}^{-1}$$

DEFINITIONS

Angstrom, Å
Anisotropic properties
Basis vectors
Crystal
Isotropic properties
Lattice
Lattice constants
Lattice of a pattern
Lattice parameters
Linear coefficient of compressibility, β_l
Linear coefficient of thermal expansion, α_l

Mean linear compressibility
Multiple unit cell
Primitive unit cell
Unit cell

EXERCISES

💻 This computer icon means that a computer language designed for numerical analysis should be used.

1.1 Copy Figure 1.28. Draw a lattice and then draw a primitive unit cell. Is this primitive cell unique? (yes or no).

1.2 In Figure 1.29a label all the lattice points with respect to origin 0_1 and basis vectors a_1 and b_1. In Figure 1.29b, label all the lattice points with respect to origin 0_2 and basis vectors a_2 and b_2.

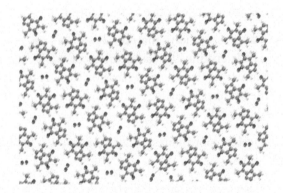

FIGURE 1.28 Projection of caffeine monohydrate. (CAFINE: CSD.)

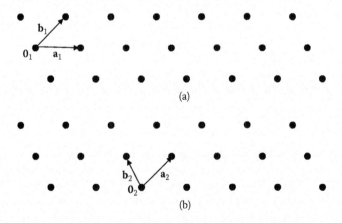

FIGURE 1.29 Two identical lattices with basis vectors (a) a_1 and b_1 and origin 0_1, (b) a_2 and b_2 and origin 0_2.

FIGURE 1.30 Lattice with basis vectors **a** and **b** and origin **0**.

1.3 Given the origin, **0**, and the basis vectors **a** and **b** of the lattice in Figure 1.30, label each point in the lattice with an ordered pair of numbers.

1.4 Verify the compatibility of Equations 1.2 and 1.3 by multiplying out the right-hand side of Equation 1.3.

1.5 Write the transformation matrix P, if $\mathbf{a}_2 = 3\mathbf{a}_1 + \mathbf{b}_1$; $\mathbf{b}_2 = -\mathbf{a}_1 + \mathbf{b}_1$. Determine the handedness of $\mathbf{a}_2$, $\mathbf{b}_2$ if $\mathbf{a}_1$, $\mathbf{b}_1$ is right-handed. Demonstrate that the new basis vectors are linearly independent. Give the ratio of the number of lattice points in the new unit cell relative to the number in the old unit cell. Draw both $\mathbf{a}_1$, $\mathbf{b}_1$ and $\mathbf{a}_2$, $\mathbf{b}_2$ on a single lattice as in Figure 1.14.

1.6 Write the transformation matrix P, if $\mathbf{a}_2 = \mathbf{a}_1 - \mathbf{b}_1 + \mathbf{c}_1$; $\mathbf{b}_2 = 2\mathbf{a}_1 + \mathbf{b}_1$; $\mathbf{c}_2 = 3\mathbf{b}_1 + 3\mathbf{c}_1$. Determine the handedness of $\mathbf{a}_2$, $\mathbf{b}_2$, $\mathbf{c}_2$, if $\mathbf{a}_1$, $\mathbf{b}_1$, $\mathbf{c}_1$, is right-handed. Demonstrate that the new basis vectors are linearly independent. Give the ratio of the number of lattice points in the new unit cell relative to the number in the old unit cell.

⌨ 1.7 Find the Cartesian coordinates of lattice point $3, \bar{1}$ for HMB. (Use data from Section 1.7.2.)

⌨ 1.8 Find the Cartesian coordinates of lattice point $3, \bar{1}, 7$ for HMB. (Use data from Section 1.7.2.)

⌨ 1.9 Complete the two-dimensional unit cell of HMB. See Figure 1.20. Include the computer code.

⌨ 1.10 Complete the three-dimensional unit cell of HMB. See Figure 1.21. Include the computer code.

⌨ 1.11 Complete the three-dimensional unit cell for AA and show projections as in Figures 1.22 and 1.23(a) projection down **c**, (b) projection down **a**. Include the computer code.

⌨ 1.12 For HMB illustrate the projections down **a**, **b**, and **c**.

⌨ 1.13 Given niobium germanide, an intermetallic type 1 semiconductor discovered in 1975 with a = 5.1692 Å = b = c and $\alpha = \beta = \gamma = 90°$. (Rasmussen and Hazell 1979). Complete the three-dimensional unit cell and show the projection down **a**. Include the computer code.

1.14 Go to the CSD and find a crystal of interest to you. Give the lattice parameters. Complete the three-dimensional unit cell and show the projections down **a**, **b**, and **c**. Include the computer code.

1.15 Calculate the linear coefficient of thermal expansion along the **c** axis for urea. Plot *ln c* versus temperature. Use data from Table 1.2. Include the computer code.

1.16 Plot the pressure dependence of the natural logarithms of the lattice parameters *b* and *c* for anthracene. Use data from Table 1.3. Include the computer code.

1.17 Calculate the mean compressibility for anthracene along the **b** and **c** axes between pressures of 8.02 and 9.01 GPa.

MATLAB CODE: STARTER PROGRAM FOR CHAPTER 1: GRAPHIC OF TRICLINIC UNIT CELL

This is the code for drawing a two-dimensional general oblique unit cell. The example is HMB. The first three lines clear the computer for MATLAB. Next is the name of the file (useful for locating file), the purpose, and the date of revisions. All variables are defined with units given. The input is the two-dimensional cell parameters, *a*, *b*, and γ. Two of the sides of the unit cell are produced by the program. The first exercise is for the student to complete the unit cell. The next exercise is to create a three-dimensional unit cell. Then the program is applied to another crystal using its cell parameters.

```
clc
close all
clear

%Script file:  HMB2Dcell.m
%
% Purpose:
% This file contains the crystallographic data for HMB
% Record of revisions:
% Date              Programmer              Description of change
% 1/27/07           M. Julian               Original Code
%
%   define variables:
%   a   = crystallographic a axis   (Angstroms)
%   b   = crystallographic b axis   (Angstroms)
%   gamma    angle between a and b  (degrees)
%   chemical_formula     chemical formula of crystal
reference ='K.Lonsdale, X-ray evidence of the Structure of the
Benzene Nucleus, Trans. Fraraday Soc. 1929, 25, 352-366';
chemical_formula = 'C_{12}H_{18}';
a=9.010;
b=8.926;
```

```
gamma=119 + 34/60;
%*********************************************************
%convert gamma to radians
gamma = gamma*pi/180;
%plot corners of the unit cell
%convert to Cartesian
conversion =[a b*cos(gamma)
             0 b*sin(gamma) ];
outline = [0 1 1
           0 0 1 ];
cc= (conversion*outline)';
%select points to plot from outline
x=cc(:,1)';
y=cc(:,2)';
%plot unit cell
plot(x,y,'LineWidth',2);
axis equal  %preserves relative axis lengths
axis off    %turns off axis
hold on
%label a axis in bold
hold on
text(4,-.5, '\bfa');
%plot arrow head on a axis select 100 from outline - column 2
hold on
plot(cc(2,1)',cc(2,2)','>','MarkerSize',8,'MarkerFaceColor','k')
%plot arrow head on b axis select 010 from outline - 4
```

## CHAPTER 2

Unit Cell Calculations

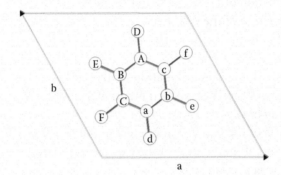

The figure above shows the 12 carbon atoms of HMB positioned within the unit cell. The carbon atoms all lie in a plane. The hydrogen atoms are omitted. Calculating detailed structural relationships, such as the distances and the angles between the atoms of a crystal, is an essential part of crystallography. The analysis of HMB by Kathleen Lonsdale was the first definitive proof of the planarity of the benzene ring.

CONTENTS

CHAPTER OBJECTIVES

- Calculate and graph fractional coordinates.

- Manipulate the atoms in the unit cell.

- Use computer graphics to display the HMB molecule in the unit cell.

- Use the metric matrix, also known as the G matrix, to calculate interatomic distances, interatomic angles, and the area or volume of the unit cell, working first in two dimensions and then expanding to three dimensions.

- Understand detailed calculations with a quartz crystal.

- Use the *P* matrix to transform fractional coordinates, the metric matrix, and area or volume.

- Understand detailed calculations with HMB.

- Define crystallographic directions and planes.

- Calculate density.

- Revisit thermal expansion and isothermal compression.

2.1 INTRODUCTION

Chapter 1 introduced the unit cell, which is the building block of the crystal. In this chapter, the mathematics is developed to place the atoms within the unit cell. As seen in Chapter 1, the coordinates are not Cartesian. Fractional coordinates are used to locate the atoms. The metric matrix is defined and used to calculate interatomic lengths and angles as well as volume and density. Crystallographic directions and planes are defined and finally thermal expansion and isothermal compressibility are revisited.

2.2 FRACTIONAL COORDINATES

All the information necessary to describe the crystal is found in the unit cell. *Fractional coordinates* are the atomic coordinates given as fractions of the basis vectors. Being fractions they are unitless. Fractional coordinates are used in the crystallographic literature.

The fractional coordinates within a single unit cell lie between 0 and 1.

The positions of the atoms in the unit cell can be described by vectors. The *coordinate vector* **r** of an atom is specified by

$$\mathbf{r} = x\,\mathbf{a} + y\,\mathbf{b} + z\,\mathbf{c}, \tag{2.1}$$

where x, y, z are the fractional coordinates of the atom (unitless) and **a**, **b**, **c** are the basis vectors, usually measured in angstroms.

EXAMPLE 2.1

Draw atom A in the **a**, **b** plane. Atom A has fractional coordinates

$$x = 0.50 \quad \text{and} \quad y = 0.67.$$

This unit cell has lattice parameters

$$a = 10.0\ \text{Å}, \quad b = 10.0\ \text{Å}, \quad \text{and} \quad \gamma = 120°.$$

(continued)

EXAMPLE 2.1 (continued)

Solution

First, draw the basis vectors to form the unit cell in Figure 2.1. Multiply the fractional coordinates, x and y, by a and b as follows:

$$x\,a = 0.50\,a = 0.50 \times 10\ \text{Å} = 5.0\ \text{Å}$$

$$y\,b = 0.67\,b = 0.67 \times 10\ \text{Å} = 6.7\ \text{Å}.$$

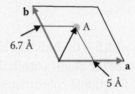

FIGURE 2.1   Atom A in the unit cell.

The atom with fractional coordinates (0.50, 0.67) is at position A. *The basis vectors are not orthogonal.* See the next section for using a computer program for drawing nonorthogonal figures.

The atom A can be referred to by its coordinates (5.0, 6.7) (Å) or its fractional coordinates (0.50, 0.67) (unitless). In crystallography, the fractional coordinates are commonly used.

If there is an atom at fractional coordinates (0.50, 0.67), then there is also an *identical* atom at the following fractional coordinates, as can be seen in Figure 2.2:

(0.50, 0.67) atom A in cell A

(1.50, 0.67) atom A in cell B

(0.50, 1.67) atom A in cell C

(2.50, 0.67) atom A in cell D

(0.50 − 1, 0.67) = (−0.50, 0.67) atom A in cell E

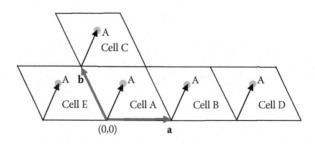

FIGURE 2.2   Five unit cells, each with a single atom, A.

Thus, integers can be added to (or subtracted from) the fractional coordinates to find an *identical* atom's location in another unit cell. To graph atom A, each of these fractional coordinates is converted into lengths along the basis vectors, as in Example 2.1, and then graphed in Figure 2.2.

2.3 PLOTTING ATOMS IN THE UNIT CELL

In this section, the general method for plotting atoms in the unit cell is worked out and then applied to HMB as an example. All that is needed is to apply to atoms the method used for lattice points.

2.3.1 Plotting a Single Atom

Computer calculations are usually done in Cartesian coordinates. The conversion of lattice points in crystallographic coordinates into Cartesian coordinates was worked out in Chapter 1. The matrix, $\mathbf{C}$, is

$$\mathbf{C} = \begin{pmatrix} a & b\cos\gamma & c_1 \\ 0 & b\sin\gamma & c_2 \\ 0 & 0 & c_3 \end{pmatrix},$$

where

$$c_1 = c\cos\beta$$

$$c_2 = \frac{c(\cos\alpha - \cos\gamma\cos\beta)}{\sin\gamma}$$

$$c_3 = c\sqrt{1 - \cos^2\beta - \frac{(\cos\alpha - \cos\gamma\cos\beta)^2}{\sin^2\gamma}}$$

If a, b, c are given in angstroms, then the units of the matrix $\mathbf{C}$ are angstroms. This matrix was derived for plotting the lattice points. This same matrix can be applied to fractional coordinates for plotting atomic positions within the unit cell. Given the fractional coordinates x, y, z then the *coordinate matrix* X is defined as the column vector

$$\begin{pmatrix} x \\ y \\ z \end{pmatrix}$$

The Cartesian coordinates, $\mathbf{CX}$, for plotting the atom are

$$\mathbf{CX} = \begin{pmatrix} a & b\cos\gamma & c_1 \\ 0 & b\sin\gamma & c_2 \\ 0 & 0 & c_3 \end{pmatrix}\begin{pmatrix} x \\ y \\ z \end{pmatrix}$$

EXAMPLE 2.2

Carbon atom A in HMB has fractional coordinates

$$X = \begin{pmatrix} 0.071 \\ 0.182 \\ 0 \end{pmatrix}$$

What are the Cartesian coordinates of atom A?

Solution

For HMB

$$C = \begin{pmatrix} a & b\cos\gamma & c_1 \\ 0 & b\sin\gamma & c_2 \\ 0 & 0 & c_3 \end{pmatrix} = \begin{pmatrix} 9.0100 & -4.4044 & -2.4025 \\ 0 & 7.7637 & 3.0230 \\ 0 & 0 & 3.6942 \end{pmatrix}$$

$$CX = \begin{pmatrix} 9.0100 & -4.4044 & -2.4025 \\ 0 & 7.7637 & 3.0230 \\ 0 & 0 & 3.6942 \end{pmatrix} \begin{pmatrix} 0.071 \\ 0.182 \\ 0 \end{pmatrix} = \begin{pmatrix} -0.1619 \\ 1.4130 \\ 0 \end{pmatrix} \mathring{A}$$

2.3.2 Plotting Atomic Coordinates Directly from the Crystallographic Literature

The coordinate matrix

$$X = \begin{pmatrix} x \\ y \\ z \end{pmatrix}$$

can be extended to include the coordinates of many atoms. For example, the coordinate matrix containing all 12 HMB carbon atoms can be written as

$$X_{HMB} = \begin{pmatrix} x_A & x_B & x_C \ldots x_f \\ y_A & y_B & y_C \ldots y_f \\ z_A & z_B & z_C \ldots z_f \end{pmatrix}$$

Then the Cartesian coordinates, CX_{HMB}, for plotting all the carbon atoms is

TABLE 2.1 Fractional Coordinates for the Carbon Atoms in HMB

Carbon Atom	x Fractional Coordinate	y Fractional Coordinate	z Fractional Coordinate
A	0.071	0.182	0
B	−0.109	0.073	0
C	−0.180	−0.109	0
D	0.145	0.371	0
E	−0.222	0.149	0
F	−0.367	−0.222	0
a	−0.071	−0.182	0
b	0.109	−0.073	0
c	0.180	0.109	0
d	−0.145	−0.371	0
e	0.222	−0.149	0
f	0.367	0.222	0

Source: Adapted from Lonsdale, K. 1929a. The structure of the benzene ring in $C_6(CH_3)_6$. *Proceedings of the Royal Society of London A* 123, 494–515.

$$CX_{HMB} = \begin{pmatrix} a & b\cos\gamma & c_1 \\ 0 & b\sin\gamma & c_2 \\ 0 & 0 & c_3 \end{pmatrix} \begin{pmatrix} x_A & x_B & x_C \dots x_f \\ y_A & y_B & y_C \dots y_f \\ z_A & z_B & z_C \dots z_f \end{pmatrix}$$

Table 2.1 gives the fractional coordinates of the carbon atoms in HMB. The hydrogen atoms are omitted. All the z fractional coordinates are zero, which means that the carbon atoms all lie in the xy-plane and the problem is two-dimensional. An examination of this table shows that the coordinates of the carbon atoms represented by the capital letters A, B, C, D, E, and F can be multiplied by −1 to produce carbon atoms represented by the lower case letters a, b, c, d, e, and f. This feature is examined in Chapter 3.

Figure 2.3 shows atomic coordinates for HMB plotted directly from the literature. Note that 8 of the 12 carbon atoms of the molecule lie outside the blue unit cell. Exercise 2.1 at the end of the chapter reproduces Figure 2.3.

2.3.3 Moving Atoms into the Unit Cell

In Figure 2.3 only 4 of the 12 carbon atoms as plotted appear in the unit cell. Nevertheless, the benzene ring is apparent as well as the six carbon atoms attached to the benzene ring. To get all the fractional coordinates of the atoms in the same unit cell, add or subtract integers until all the coordinates lie between 0 and 1; include 0, but not 1. For example, carbon atom C in Table 2.1 has

$$X_C = \begin{pmatrix} -0.180 \\ -0.109 \\ 0 \end{pmatrix}$$

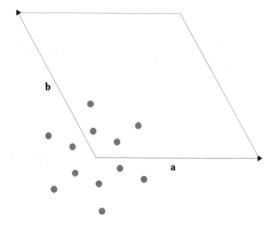

FIGURE 2.3   The HMB carbon atoms with fractional coordinates from Table 2.1.

If the vector

$$\begin{pmatrix} 1 \\ 1 \\ 0 \end{pmatrix}$$

is added then

$$X_C + \begin{pmatrix} 1 \\ 1 \\ 0 \end{pmatrix} = \begin{pmatrix} 0.820 \\ 0.891 \\ 0 \end{pmatrix}.$$

These new fractional coordinates lie between 0 and 1 and will be in the same unit cell as carbon atoms A and B from Table 2.1. In Exercise 2.2 at the end of the chapter all the carbon atoms are brought into a single unit cell (see Figure 2.4 and Exercise 2.3). Even though all the atoms are in the unit cell, the connected molecule is not seen. A similar effect is seen in Figure 1.11 in Chapter 1.

2.3.4 Translating the Molecule into the Unit Cell

A vector is unchanged if it is translated parallel to itself. Each atom is located at the end of a vector drawn from the origin to the atomic position. Add 1/2, 1/2, 0 to all the fractional coordinates in Table 2.1. All of the atoms are brought into the unit cell and their relative positions are preserved since the coordinates are all translated together (see Figure 2.5 and Exercise 2.4). All the carbon atoms are labeled and the bonds are indicated.

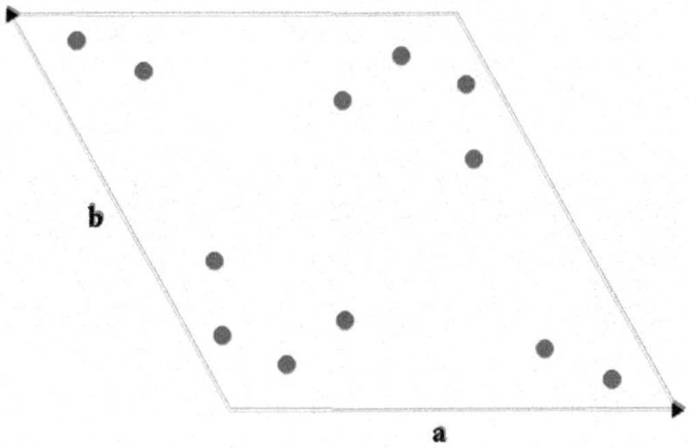

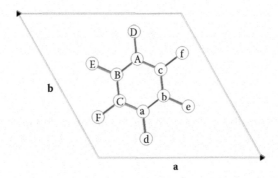

FIGURE 2.4   HMB, with fractional coordinates from Table 2.1, transposed into a single unit cell.

FIGURE 2.5   The HMB carbon atoms with 1/2, 1/2, 0 added to the fractional coordinates in Table 2.1 and all the appropriate bonds drawn.

The atoms ABCabc form a planar hexagon showing that the benzene ring is planar. The bond lengths of the hexagon are equal within experimental error. The hydrogen atoms are not shown.

2.4 CALCULATION OF INTERATOMIC BOND DISTANCES

The interatomic distance is an indication of how tightly the atoms are bound: the shorter bond length is the stronger bond. Bond strength in turn is responsible for many physical properties. In this section, the metric matrix is introduced, in part to facilitate the calculations of bond lengths. See Julian and Gibbs (1985, 1988).

2.4.1 Calculation of the Magnitude of a Vector

First, calculate the length of vector $\mathbf{r} = x\,\mathbf{a} + y\,\mathbf{b} + z\,\mathbf{c}$ with fractional coordinates x, y, z. The coordinate matrix

$$\mathbf{X} = \begin{pmatrix} x \\ y \\ z \end{pmatrix}$$

This is a vector whose origin is at the origin of the unit cell.
Take the dot product of **r** into itself

$$\mathbf{r} \cdot \mathbf{r} = (x\,\mathbf{a} + y\,\mathbf{b} + z\,\mathbf{c}) \cdot (x\,\mathbf{a} + y\,\mathbf{b} + z\,\mathbf{c})$$

Perform the multiplication:

$$\mathbf{r} \cdot \mathbf{r} = x^2\mathbf{a} \cdot \mathbf{a} + y^2\mathbf{b} \cdot \mathbf{b} + z^2\mathbf{c} \cdot \mathbf{c} + 2\,xy\,\mathbf{a} \cdot \mathbf{b} + 2\,xz\,\mathbf{a} \cdot \mathbf{c} + 2\,yz\,\mathbf{b} \cdot \mathbf{c} \qquad (2.2)$$

2.4.2 The Metric Matrix, **G**

In matrix form,

$$\mathbf{r} \cdot \mathbf{r} = (x \quad y \quad z) \begin{pmatrix} \mathbf{a} \cdot \mathbf{a} & \mathbf{a} \cdot \mathbf{b} & \mathbf{a} \cdot \mathbf{c} \\ \mathbf{a} \cdot \mathbf{b} & \mathbf{b} \cdot \mathbf{b} & \mathbf{b} \cdot \mathbf{c} \\ \mathbf{a} \cdot \mathbf{c} & \mathbf{b} \cdot \mathbf{c} & \mathbf{c} \cdot \mathbf{c} \end{pmatrix} \begin{pmatrix} x \\ y \\ z \end{pmatrix}. \qquad (2.3)$$

Carry out the matrix multiplication in Equation 2.3 to get Equation 2.2 (see Exercise 2.5).
Let

$$\mathbf{G} = \begin{pmatrix} \mathbf{a} \cdot \mathbf{a} & \mathbf{a} \cdot \mathbf{b} & \mathbf{a} \cdot \mathbf{c} \\ \mathbf{a} \cdot \mathbf{b} & \mathbf{b} \cdot \mathbf{b} & \mathbf{b} \cdot \mathbf{c} \\ \mathbf{a} \cdot \mathbf{c} & \mathbf{b} \cdot \mathbf{c} & \mathbf{c} \cdot \mathbf{c} \end{pmatrix}. \qquad (2.4)$$

The **G** matrix is the *metric matrix* and is also known as the metric tensor. It contains information about the basis vectors of the unit cell. The units of **G** are length squared.

The **G** matrix can also be written as the dyad dot product of a vector with itself.

$$\mathbf{G} = \begin{pmatrix} \mathbf{a} \\ \mathbf{b} \\ \mathbf{c} \end{pmatrix} \cdot (\mathbf{a} \quad \mathbf{b} \quad \mathbf{c}). \qquad (2.5)$$

2.4.3 Using the Metric Matrix

From now on, the metric matrix, **G**, is used extensively. The **G** matrix contains the information about the lattice parameters and allows separation of the lattice parameters from the rest of an expression. For example, from Equation 2.3 the magnitude of **r** is

$$r = \sqrt{X^t\mathbf{G}X},$$

where X^t is the transpose of X. If

$$X = \begin{pmatrix} x \\ y \\ z \end{pmatrix}, \quad \text{then } X^t = (x \quad y \quad z).$$

2.4.4 Two-Dimensional Calculation of Interatomic Bond Distances

Consider two atoms in the unit cell. These two atoms are represented by vectors $\mathbf{r}_1$ and $\mathbf{r}_2$ in Figure 2.6. The magnitude of vector $\mathbf{r}_{12}$ is the interatomic distance. This distance is to be understood as the distance from the *center* of atom 1 to the *center* of atom 2. Sometimes the situation is idealized by saying that one atom *touches* the other, as in the hard-sphere model. Vector $\mathbf{r}_{12}$ points from atom 1 to atom 2; the head is at 2 and the tail is at 1. The coordinate vectors for atoms 1 and 2 are

$$\mathbf{r}_1 = x_1 \, \mathbf{a} + y_1 \, \mathbf{b}$$

$$\mathbf{r}_2 = x_2 \, \mathbf{a} + y_2 \, \mathbf{b}$$

where $x_1 \, y_1$ are the fractional coordinates of atom 1 and $x_2 \, y_2$ are the fractional coordinates of atom 2.

The interatomic vector $\mathbf{r}_{12}$ is the vector difference between $\mathbf{r}_1$ and $\mathbf{r}_2$.

$$\mathbf{r}_{12} = \mathbf{r}_2 - \mathbf{r}_1 = (x_2 - x_1) \, \mathbf{a} + (y_2 - y_1) \, \mathbf{b}$$

To find the interatomic distance, the magnitude of $\mathbf{r}_{12}$ is calculated by taking the square root of its dot product with itself.

$$\mathbf{r}_{12} \cdot \mathbf{r}_{12} = (x_2 - x_1)^2 \, \mathbf{a} \cdot \mathbf{a} + (y_2 - y_1)^2 \, \mathbf{b} \cdot \mathbf{b} + 2 \, (x_2 - x_1) \, (y_2 - y_1) \, \mathbf{a} \cdot \mathbf{b}$$

$$\mathbf{r}_{12} \cdot \mathbf{r}_{12} = (x_2 - x_1)^2 \, a^2 + (y_2 - y_1)^2 \, b^2 + 2(x_2 - x_1) \, (y_2 - y_1) a \, b \cos \gamma$$

The magnitude of $\mathbf{r}_{12}$ is

$$r_{12} = [(x_2 - x_1)^2 a^2 + (y_2 - y_1)^2 \, b^2 + 2 \, (x_2 - x_1) \, (y_2 - y_1) \, a \, b \cos \gamma]^{1/2} \tag{2.6}$$

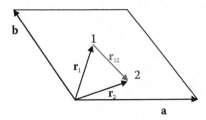

FIGURE 2.6 Atoms 1 and 2 in the unit cell with associated vectors $\mathbf{r}_1$ and $\mathbf{r}_2$ and interatomic vector $\mathbf{r}_{12}$.

After rearranging and putting into matrix form:

$$(r_{12})^2 = (x_2 - x_1 \ y_2 - y_1) \begin{pmatrix} a \cdot a & a \cdot b \\ a \cdot b & b \cdot b \end{pmatrix} \begin{pmatrix} x_2 - x_1 \\ y_2 - y_1 \end{pmatrix}. \tag{2.7}$$

Let

$$X_{12} = \begin{pmatrix} x_2 - x_1 \\ y_2 - y_1 \end{pmatrix}$$

This *coordinate matrix* contains the difference of the fractional coordinates and is unitless. Let

$$G = \begin{pmatrix} a \cdot a & a \cdot b \\ a \cdot b & b \cdot b \end{pmatrix}$$

This is the two-dimensional form of the **G** matrix. The bond length, r_{12}, between atoms 1 and 2 is

$$r_{12} = \sqrt{X_{12}^t G X_{12}},$$

where X_{12}^t is the transpose of X_{12}.
If

$$X_{12} = \begin{pmatrix} x_2 - x_1 \\ y_2 - y_1 \end{pmatrix}$$

then $X_{12}^t = (x_2 - x_1 \ y_2 - y_1)$.

2.4.5 Three-Dimensional Calculation of Interatomic Bond Distances
For three dimensions

$$\mathbf{r}_1 = x_1 \mathbf{a} + y_1 \mathbf{b} + z_1 \mathbf{c}$$

$$\mathbf{r}_2 = x_2 \mathbf{a} + y_2 \mathbf{b} + z_2 \mathbf{c}$$

where $x_1 \ y_1 \ z_1$ are the fractional coordinates of atom 1 and $x_2 \ y_2 \ z_2$ are the fractional coordinates of atom 2.
Similarly,

$$(r_{12})^2 = (x_2 - x_1 \ y_2 - y_1 \ z_2 - z_1) \begin{pmatrix} a \cdot a & a \cdot b & a \cdot c \\ a \cdot b & b \cdot b & b \cdot c \\ a \cdot c & b \cdot c & c \cdot c \end{pmatrix} \begin{pmatrix} x_2 - x_1 \\ y_2 - y_1 \\ z_2 - z_1 \end{pmatrix}$$

Let

$$X_{12} = \begin{pmatrix} x_2 - x_1 \\ y_2 - y_1 \\ z_2 - z_1 \end{pmatrix}$$

This *coordinate matrix* contains the difference of the fractional coordinates and is unitless.

Since

$$G = \begin{pmatrix} a \cdot a & a \cdot b & a \cdot c \\ a \cdot b & b \cdot b & b \cdot c \\ a \cdot c & b \cdot c & c \cdot c \end{pmatrix}$$

the formula for the interatomic bond length between atoms 1 and 2 is

$$r_{12} = \sqrt{X_{12}^t G X_{12}}, \tag{2.8}$$

where X_{12}^t is the transpose of X_{12}.

If

$$X_{12} = \begin{pmatrix} x_2 - x_1 \\ y_2 - y_1 \\ z_2 - z_1 \end{pmatrix}$$

then $X_{12}^t = (x_2 - x_1 \ y_2 - y_1 \ z_2 - z_1)$.

2.5 CALCULATION OF INTERATOMIC BOND ANGLES

Another important calculation in crystallography is the interatomic bond angle. The bond angles are needed to understand the structure of the crystal. Consider three atoms in the unit cell in Figure 2.7 forming angle 213. Atom 1 is at the apex of the angle. The angle between $\mathbf{r}_{12}$ and $\mathbf{r}_{13}$ is θ. The formula for the dot product of two vectors is used to calculate the angle between them.

$$\mathbf{r}_{12} \cdot \mathbf{r}_{13} = r_{12} \, r_{13} \cos \theta$$

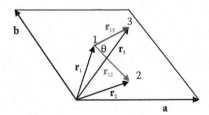

FIGURE 2.7   Three atoms in the unit cell forming angle 213.

Solve for cos θ.

$$\cos \theta = \frac{\mathbf{r}_{12} \cdot \mathbf{r}_{13}}{r_{12} \, r_{13}}, \tag{2.9}$$

where r_{12} is the magnitude of vector $\mathbf{r}_{12}$ and r_{13} is the magnitude of vector $\mathbf{r}_{13}$. Of course, the angle θ equals the inverse cosine of cos θ.

2.5.1 Two-Dimensional Atomic Bond Angles

$$\mathbf{r}_{12} = \mathbf{r}_2 - \mathbf{r}_1 = (x_2 - x_1)\,\mathbf{a} + (y_2 - y_1)\,\mathbf{b}$$

$$\mathbf{r}_{13} = \mathbf{r}_3 - \mathbf{r}_1 = (x_3 - x_1)\,\mathbf{a} + (y_3 - y_1)\,\mathbf{b}$$

The dot product is

$$\mathbf{r}_{12} \cdot \mathbf{r}_{13} = (x_2 - x_1)(x_3 - x_1)\,a^2 + (y_2 - y_1)\,(y_3 - y_1)\,b^2$$

$$+ \, [(x_2 - x_1)\,(y_3 - y_1) + (x_3 - x_1)\,(y_2 - y_1)]\,a\,b\,\cos\theta$$

Now put this into matrix form

$$\mathbf{r}_{12} \cdot \mathbf{r}_{13} = (x_2 - x_1 \;\; y_2 - y_1) \begin{pmatrix} \mathbf{a} \cdot \mathbf{a} & \mathbf{a} \cdot \mathbf{b} \\ \mathbf{a} \cdot \mathbf{b} & \mathbf{b} \cdot \mathbf{b} \end{pmatrix} \begin{pmatrix} x_3 - x_1 \\ y_3 - y_1 \end{pmatrix} = X_{12}^t G X_{13}$$

Substitute into Equation 2.9.

$$\cos\theta = \frac{X_{12}^t G X_{13}}{r_{12} \, r_{13}} = \frac{X_{12}^t G X_{13}}{\sqrt{X_{12}^t G X_{12}} \, \sqrt{X_{13}^t G X_{13}}}$$

2.5.2 Three-Dimensional Atomic Bond Angles

$$\mathbf{r}_{12} \cdot \mathbf{r}_{13} = (x_2 - x_1)(x_3 - x_1)\,a^2 + (y_2 - y_1)\,(y_3 - y_1)\,b^2 + (z_2 - z_1)(z_3 - z_1)\,c^2$$

$$+ \, [(x_2 - x_1)\,(y_3 - y_1) + (x_3 - x_1)\,(y_2 - y_1)]\,a\,b\,\cos\gamma$$

$$+ \, [(z_2 - z_1)\,(x_3 - x_1) + (z_3 - z_1)\,(x_2 - x_1)]\,c\,a\,\cos\beta$$

$$+ \, [(y_2 - y_1)\,(z_3 - z_1) + (y_3 - y_1)\,(z_2 - z_1)]\,b\,c\,\cos\alpha$$

In matrix form:

$$\mathbf{r}_{12} \cdot \mathbf{r}_{13} = (x_2 - x_1 \;\; y_2 - y_1 \;\; z_2 - z_1) \begin{pmatrix} \mathbf{a} \cdot \mathbf{a} & \mathbf{a} \cdot \mathbf{b} & \mathbf{a} \cdot \mathbf{c} \\ \mathbf{a} \cdot \mathbf{b} & \mathbf{b} \cdot \mathbf{b} & \mathbf{b} \cdot \mathbf{c} \\ \mathbf{a} \cdot \mathbf{c} & \mathbf{b} \cdot \mathbf{c} & \mathbf{c} \cdot \mathbf{c} \end{pmatrix} \begin{pmatrix} x_3 - x_1 \\ y_3 - y_1 \\ z_3 - z_1 \end{pmatrix} = X_{12}^t G X_{13}.$$

Finally,

$$\cos\theta = \frac{X_{12}^t G X_{13}}{r_{12}\, r_{13}} = \frac{X_{12}^t G X_{13}}{\sqrt{X_{12}^t G X_{12}}\ \sqrt{X_{13}^t G X_{13}}} \tag{2.10}$$

which is the same as the equation for two dimensions.

2.6 AREA AND VOLUME OF THE UNIT CELL

In two dimensions the derivation of the area of a unit cell is done by first considering the unit orthogonal vectors $\mathbf{e}_1$ and $\mathbf{e}_2$ shown in Figure 2.8.

Write the basis vectors in terms of $\mathbf{e}_1$ and $\mathbf{e}_2$

$$\mathbf{a} = a_1\, \mathbf{e}_1 + a_2\, \mathbf{e}_2$$

$$\mathbf{b} = b_1\, \mathbf{e}_1 + b_2\, \mathbf{e}_2$$

where a_1 and a_2 are the components of $\mathbf{a}$ with respect to $\mathbf{e}_1$ and $\mathbf{e}_2$ and similarly b_1 and b_2 are the components of $\mathbf{b}$ with respect to $\mathbf{e}_1$ and $\mathbf{e}_2$.

The area of the parallelogram with sides $\mathbf{a}$ and $\mathbf{b}$ is the magnitude of the cross product of $\mathbf{a}$ and $\mathbf{b}$. Thus,

$$\text{Area} = |\mathbf{a} \times \mathbf{b}| = |(a_1\, \mathbf{e}_1 + a_2\, \mathbf{e}_2) \times (b_1\, \mathbf{e}_1 + b_2\, \mathbf{e}_2)| = |(a_1\, b_2 - a_2\, b_1)\, \mathbf{e}_1 \times \mathbf{e}_2| = |(a_1\, b_2 - a_2\, b_1)|$$

since the magnitude of $\mathbf{e}_1 \times \mathbf{e}_2$ is 1. Therefore,

$$\text{Area} = \left| \det \begin{pmatrix} a_1 & a_2 \\ b_1 & b_2 \end{pmatrix} \right|.$$

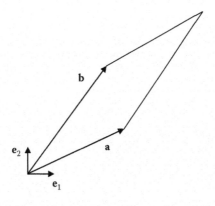

FIGURE 2.8   Orthogonal unit vectors $\mathbf{e}_1$ and $\mathbf{e}_2$, and unit cell basis vectors $\mathbf{a}$ and $\mathbf{b}$.

Squaring the area gives

$$\text{Area}^2 = \det\begin{pmatrix} a_1 & a_2 \\ b_1 & b_2 \end{pmatrix} \det\begin{pmatrix} a_1 & a_2 \\ b_1 & b_2 \end{pmatrix}.$$

The rows and column of a matrix can be interchanged without changing the value of the determinant. Thus,

$$\text{Area}^2 = \det\begin{pmatrix} a_1 & a_2 \\ b_1 & b_2 \end{pmatrix} \det\begin{pmatrix} a_1 & b_1 \\ a_2 & b_2 \end{pmatrix}.$$

In general, det(A) times det(B) equals det(AB).

EXAMPLE 2.3

Let $A = \begin{pmatrix} 3 & 2 \\ 1 & 3 \end{pmatrix}$, $B = \begin{pmatrix} 1 & 0 \\ 2 & 3 \end{pmatrix}$, then $AB = \begin{pmatrix} 7 & 6 \\ 7 & 9 \end{pmatrix}$

Demonstrate that det(A) times det(B) equals det(AB).

Solution

Since det(A) = 7, det(B) = 3, and det(AB) = 21, the equality is demonstrated.

Thus,

$$\text{Area}^2 = \det\begin{pmatrix} a_1^2 + a_2^2 & a_1 b_1 + a_2 b_2 \\ a_1 b_1 + a_2 b_2 & b_1^2 + b_2^2 \end{pmatrix} = \det\begin{pmatrix} \mathbf{a} \cdot \mathbf{a} & \mathbf{a} \cdot \mathbf{b} \\ \mathbf{a} \cdot \mathbf{b} & \mathbf{b} \cdot \mathbf{b} \end{pmatrix} = \det(\mathbf{G}).$$

The result is that the area is equal to the square root of the determinant of the metric matrix **G**.

$$\text{Area} = \sqrt{\det(\mathbf{G})}. \tag{2.11}$$

In three dimensions, the volume of the unit cell is equal to the square root of the determinant of the **G** matrix. See Exercise 2.6.

$$\text{Volume} = \sqrt{\det(\mathbf{G})}. \tag{2.12}$$

2.7 SUMMARY OF METRIC MATRIX CALCULATIONS

Table 2.2 summarizes the concepts of basis vectors, the metric matrix, volume, fractional coordinates, the coordinate vector, the coordinate matrix, bond lengths, and angles.

TABLE 2.2   Summary of the Calculations Using the Metric Matrix

Basis Vectors, Å	$\mathbf{a}, \mathbf{b}, \mathbf{c}$
Metric matrix, $\mathbf{G}$, Å^2	$\begin{pmatrix} \mathbf{a} \cdot \mathbf{a} & \mathbf{a} \cdot \mathbf{b} & \mathbf{a} \cdot \mathbf{c} \\ \mathbf{a} \cdot \mathbf{b} & \mathbf{b} \cdot \mathbf{b} & \mathbf{b} \cdot \mathbf{c} \\ \mathbf{a} \cdot \mathbf{c} & \mathbf{b} \cdot \mathbf{c} & \mathbf{c} \cdot \mathbf{c} \end{pmatrix}$
Volume, Å^3	$V = \sqrt{\det(\mathbf{G})}$
Fractional coordinates, unitless	x, y, z
Coordinate vector, Å	$\mathbf{r} = x\,\mathbf{a} + y\,\mathbf{b} + z\,\mathbf{c}$
Coordinate matrix, unitless	$X = \begin{pmatrix} x \\ y \\ z \end{pmatrix}$
Magnitude of $\mathbf{r}$, Å	$r = \sqrt{X^t G X}$
Interatomic bond length, r_{12}, Å	$r_{12} = \sqrt{X_{12}^t G X_{12}}$
Interatomic bond angle, θ, degrees, vertex at atom 1.	$\cos\theta = \dfrac{X_{12}^t G X_{13}}{r_{12}\, r_{13}} = \dfrac{X_{12}^t G X_{13}}{\sqrt{X_{12}^t G X_{12}} \sqrt{X_{13}^t G X_{13}}}$

2.8 QUARTZ EXAMPLE

The use of the metric matrix, $\mathbf{G}$, as summarized in Table 2.2, is now demonstrated for quartz.

🖳 EXAMPLE 2.4

In the mineral quartz, which has the formula unit SiO_2, each oxygen atom is connected to two silicon atoms, Si_1 and Si_2 (see Figure 2.9). Calculate (1) the $\mathbf{G}$ matrix, (2) the unit cell volume, (3) the bond length $r_{Si_1 O}$, (4) the bond length $r_{Si_2 O}$, and (5) the bond angle θ, for quartz.

Given $a = b = 4.91$ Å; $c = 5.41$ Å: $\alpha = \beta = 90°$; and $\gamma = 120°$, with fractional coordinates in Table 2.3.

Solution

1. $G = \begin{pmatrix} \mathbf{a} \cdot \mathbf{a} & \mathbf{a} \cdot \mathbf{b} & \mathbf{a} \cdot \mathbf{c} \\ \mathbf{a} \cdot \mathbf{b} & \mathbf{b} \cdot \mathbf{b} & \mathbf{b} \cdot \mathbf{c} \\ \mathbf{a} \cdot \mathbf{c} & \mathbf{b} \cdot \mathbf{c} & \mathbf{c} \cdot \mathbf{c} \end{pmatrix} = \begin{pmatrix} 24.11 & -12.05 & 0 \\ -12.05 & 24.11 & 0 \\ 0 & 0 & 29.26 \end{pmatrix}$ Å^2

2. Volume $= \sqrt{\det(\mathbf{G})} = 113.01\,$Å^3

3. $r_{Si_1 O} = (X^t_{Si_1 O} G X_{Si_1 O})^{1/2}$

$$X_{Si_1 O} = \begin{pmatrix} 0.4141 - 0.4699 \\ 0.2681 - 0 \\ 0.1188 - 0 \end{pmatrix} = \begin{pmatrix} -0.0559 \\ 0.2681 \\ 0.1188 \end{pmatrix}$$

(continued)

EXAMPLE 2.4 (continued)

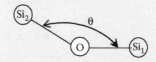

FIGURE 2.9   Two silicon atoms, Si_1 and Si_2, joined to a central oxygen atom.

TABLE 2.3   Fractional Coordinates for Quartz

Atom	x	y	z
Si_1	0.4699	0	0
Si_2	0.5301	0.5301	0.3333
Oxygen	0.4141	0.2681	0.1188

Thus,

$$r_{Si_1O} = 1.6067 \text{ Å}$$

4. Similarly, $r_{Si_2O} = (X_{Si_2O}^t G X_{Si_2O})^{1/2}$

$$X_{Si_2O} = \begin{pmatrix} 0.4141 - 0.5301 \\ 0.2681 - 0.5301 \\ 0.1188 - 0.3333 \end{pmatrix} = \begin{pmatrix} -0.1160 \\ -0.2620 \\ -0.2145 \end{pmatrix}$$

Thus,

$$r_{Si_2O} = 1.6102 \text{ Å}$$

5. $\cos \theta = \dfrac{X_{Si_1O}^t G X_{Si_2O}}{r_{Si_1O} \, r_{Si_2O}}$

$$= \frac{(-0.0559 \quad 0.2681 \quad 0.1188) \begin{pmatrix} 24.11 & -12.05 & 0 \\ -12.05 & 24.11 & 0 \\ 0 & 0 & 29.26 \end{pmatrix} \begin{pmatrix} -0.1160 \\ -0.2620 \\ -0.2145 \end{pmatrix}}{1.6067 \cdot 1.6102}$$

Thus, $\theta = 143.46°$

See Exercises 2.7 through 2.12.

2.9 TRANSFORMATION MATRICES

The transformation matrix P, defined in Chapter 1, is used to calculate the transformations of the coordinates, the G matrix and the volume. As in Chapter 1, a shift of origin is not considered. All basis sets have a common origin.

2.9.1 Transformation Matrix, P

From Chapter 1, the transformation between basis vectors $\mathbf{a}_1$, $\mathbf{b}_1$, $\mathbf{c}_1$ and another set, on the same lattice, $\mathbf{a}_2$, $\mathbf{b}_2$, $\mathbf{c}_2$ is

$$(\mathbf{a}_2 \quad \mathbf{b}_2 \quad \mathbf{c}_2) = (\mathbf{a}_1 \quad \mathbf{b}_1 \quad \mathbf{c}_1)P = (\mathbf{a}_1 \quad \mathbf{b}_1 \quad \mathbf{c}_1)\begin{pmatrix} P_{11} & P_{12} & P_{13} \\ P_{21} & P_{22} & P_{23} \\ P_{31} & P_{32} & P_{33} \end{pmatrix}. \tag{2.13}$$

where P is the transformation matrix.

Thus,

$$(\mathbf{a}_1 \quad \mathbf{b}_1 \quad \mathbf{c}_1) = (\mathbf{a}_2 \quad \mathbf{b}_2 \quad \mathbf{c}_2)\, P^{-1} \tag{2.14}$$

2.9.2 Transformation of Fractional Coordinates

Now consider an atom with coordinate vector $\mathbf{r}$ with fractional coordinates

$$X_1 = \begin{pmatrix} x_1 \\ y_1 \\ z_1 \end{pmatrix}$$

with respect to basis $\mathbf{a}_1$, $\mathbf{b}_1$, $\mathbf{c}_1$. Then

$$\mathbf{r} = x_1\mathbf{a}_1 + y_1\mathbf{b}_1 + z_1\mathbf{c}_1 = (\mathbf{a}_1 \quad \mathbf{b}_1 \quad \mathbf{c}_1)\begin{pmatrix} x_1 \\ y_1 \\ z_1 \end{pmatrix} = (\mathbf{a}_1 \quad \mathbf{b}_1 \quad \mathbf{c}_1)X_1. \tag{2.15}$$

That *same* vector $\mathbf{r}$ can be written with respect to a *second* set of basis vectors and a *second* set of fractional coordinates.

$$\mathbf{r} = x_2\mathbf{a}_2 + y_2\mathbf{b}_2 + z_2\mathbf{c}_2 = (\mathbf{a}_2 \quad \mathbf{b}_2 \quad \mathbf{c}_2)\begin{pmatrix} x_2 \\ y_2 \\ z_2 \end{pmatrix} = (\mathbf{a}_2 \quad \mathbf{b}_2 \quad \mathbf{c}_2)X_2. \tag{2.16}$$

Equate Equation 2.15 with Equation 2.16:

$$(\mathbf{a_1} \quad \mathbf{b_1} \quad \mathbf{c_1}) \, X_1 = (\mathbf{a_2} \quad \mathbf{b_2} \quad \mathbf{c_2}) \, X_2 \qquad (2.17)$$

Now substitute Equation 2.13 into Equation 2.17:

$$(\mathbf{a_1} \quad \mathbf{b_1} \quad \mathbf{c_1}) \, X_1 = (\mathbf{a_1} \quad \mathbf{b_1} \quad \mathbf{c_1}) \, \boldsymbol{P} X_2 \qquad (2.18)$$

Thus,

$$X_1 = \boldsymbol{P} X_2 \qquad (2.19)$$

or

$$\boldsymbol{P}^{-1} X_1 = \boldsymbol{P}^{-1} \boldsymbol{P} X_2 = X_2 \qquad (2.20)$$

2.9.3 Transformation of the **G** Matrix

The **G** matrix contains the information about the basis vectors **a**, **b**, and **c**.

$$G = \begin{pmatrix} \mathbf{a \cdot a} & \mathbf{a \cdot b} & \mathbf{a \cdot c} \\ \mathbf{a \cdot b} & \mathbf{b \cdot b} & \mathbf{b \cdot c} \\ \mathbf{a \cdot c} & \mathbf{b \cdot c} & \mathbf{c \cdot c} \end{pmatrix} = \begin{pmatrix} \mathbf{a} \\ \mathbf{b} \\ \mathbf{c} \end{pmatrix} (\mathbf{a} \quad \mathbf{b} \quad \mathbf{c}). \qquad (2.21)$$

For basis $\mathbf{a_1}$, $\mathbf{b_1}$, $\mathbf{c_1}$ then

$$G_1 = \begin{pmatrix} \mathbf{a_1} \\ \mathbf{b_1} \\ \mathbf{c_1} \end{pmatrix} \cdot (\mathbf{a_1} \quad \mathbf{b_1} \quad \mathbf{c_1}), \qquad (2.22)$$

and for basis $\mathbf{a_2}$, $\mathbf{b_2}$, $\mathbf{c_2}$ then

$$G_2 = \begin{pmatrix} \mathbf{a_2} \\ \mathbf{b_2} \\ \mathbf{c_2} \end{pmatrix} \cdot (\mathbf{a_2} \quad \mathbf{b_2} \quad \mathbf{c_2}). \qquad (2.23)$$

From Equation 2.8

$$(r)^2 = X_1^t G_1 X_1 = X_2^t G_2 X_2 \qquad (2.24)$$

Equation 2.19 may be written as

$$X_1^t = (\boldsymbol{P} X_2)^t = X_2^t \boldsymbol{P}^t$$

because the transpose of the product is equal to the product of the transposes in the reverse order. Equation 2.24 becomes

$$X_1^t G_1 X_1 = X_2^t G_2 X_2 = X_2^t P^t G_1 P X_2$$

Thus,

$$G_2 = P^t G_1 P \tag{2.25}$$

Since the transpose of the inverse is equal to the inverse of the transpose, write

$$G_1 = (P^{-1})^t G_2 P^{-1} \tag{2.26}$$

2.9.4 Area and Volume Transformations

In two dimensions, the area of the unit cell is the square root of the determinant of the G matrix, $A = \sqrt{\det(G)}$. In three dimensions, the volume of the unit cell is the square root of the determinant of the G matrix, $V = \sqrt{\det(G)}$.

In general, the determinant of the product of two matrices is the product of the determinants. Algebraically, $\det(AB) = \det(A) \cdot \det(B)$.

Since

$$G_2 = P^t G_1 P$$

then

$$\det(G_2) = \det(P^t G_1 P) = \det(P^t) \det(G_1) \det(P).$$

Since

$$\det(P^t) = \det(P)$$

then

$$A_2 = \sqrt{\det(G_2)} = \sqrt{\det(P^t G_1 P)} = \sqrt{\det(P^t)\det(G_1)\det(P)}$$
$$= |\det(P)|\sqrt{\det(G_1)} = |\det(P)| A_1$$

and

$$A_2/A_1 = |\det(P)| \tag{2.27}$$

EXAMPLE 2.5

Calculate the ratio of the area of basis vectors a_2, b_2 to the area of basis vectors a_1, b_1 for the transformation illustrated in Figure 2.10

$$a_2 = 3a_1 + b_1$$
$$b_2 = 2a_1 + 3b_1$$

(continued)

EXAMPLE 2.5 (continued)

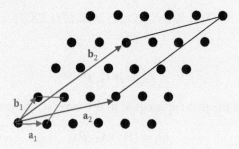

FIGURE 2.10   Basis vectors $\mathbf{a}_1$, $\mathbf{b}_1$ and basis vectors $\mathbf{a}_2$, $\mathbf{b}_2$.

Solution

The transformation matrix $\boldsymbol{P}$ is

$$P = \begin{pmatrix} 3 & 2 \\ 1 & 3 \end{pmatrix}.$$

The det ($\boldsymbol{P}$) is 7.

From Equation 2.27, the ratio of the areas is

$$A_2/A_1 = 7$$

From Chapter 1, Table 1.1, the ratio of the number of lattice points before and after the transformation is also equal to $|\det(\boldsymbol{P})|$.

Thus,

$$\left|\det(\mathbf{P})\right| = \frac{\text{number of lattice points in unit cell with basis vectors } \mathbf{a}_2, \mathbf{b}_2}{\text{number of lattice points in unit cell with basis vectors } \mathbf{a}_1, \mathbf{b}_1} = \frac{A_2}{A_1}$$

In the special case where both bases form primitive cells then

$$|\det(\mathbf{P})| = 1; \quad A_1 = A_2$$

The above arguments are extended to three dimensions in Table 2.4. See Exercise 2.13.

TABLE 2.4   Transformations Using Matrix $\boldsymbol{P}$

Basis Vectors	$(\mathbf{a}_2\,\mathbf{b}_2\,\mathbf{c}_2) = (\mathbf{a}_1\,\mathbf{b}_1\,\mathbf{c}_1)\,P$	$(\mathbf{a}_1\,\mathbf{b}_1\,\mathbf{c}_1) = (\mathbf{a}_2\,\mathbf{b}_2\,\mathbf{c}_2)\,P^{-1}$
Fractional coordinates	$X_2 = P^{-1}\,X_1$	$X_1 = P\,X_2$
G matrices	$G_2 = P^t\,G_1\,P$	$G_1 = (P^{-1})^t\,G_2 P^{-1}$
Volume	$V_2 = |\det(P)|\,V_1$	$V_1 = \dfrac{1}{|\det(P)|}V_2$

2.10 HMB EXAMPLE

⌨ Now a detailed example is provided using HMB. The exercise is to evaluate the transformation matrix P, and then transform the fractional coordinates in Table 2.1, the G matrix, the lattice parameters, and the volume all to new basis vectors.

The lattice parameters of HMB are

$$a_1 = 9.010 \text{ Å}, \quad b_1 = 8.926 \text{ Å}, \quad c_1 = 5.344 \text{ Å},$$

$$\alpha_1 = 44° 27', \quad \beta_1 = 116° 43', \quad \gamma_1 = 119° 34'.$$

The transformation is

$$\mathbf{a}_2 = -\mathbf{c}_1$$

$$\mathbf{b}_2 = \mathbf{b}_1 - \mathbf{c}_1$$

$$\mathbf{c}_2 = \mathbf{a}_1 + \mathbf{c}_1.$$

This particular transformation is used in Section 5.7.3.5.

2.10.1 Calculation of the Transformation Matrix, P

The transformation matrix P is $\begin{pmatrix} 0 & 0 & 1 \\ 0 & 1 & 0 \\ \bar{1} & \bar{1} & 1 \end{pmatrix}$. The det $(P) = 1$. Since the original basis for HMB is primitive, then the new basis is also primitive. Since the determinant of P is positive then both systems have the same handedness. Since the original basis was chosen to be right-handed, the new basis is right-handed.

2.10.2 Calculation of the Transformed Fractional Coordinates

When the basis changes then the fractional coordinates also change according to the formula $X_2 = P^{-1} X_1$.

The fractional coordinates for carbon atom A in the original basis are

$$X_{A1} = \begin{pmatrix} 0.071 \\ 0.182 \\ 0 \end{pmatrix}.$$

Calculate the transformed fractional coordinates in the new basis, $P^{-1} X_{A1}$.

First, calculate the inverse of the transformation matrix.

$$P^{-1} = \begin{pmatrix} 1 & \bar{1} & \bar{1} \\ 0 & 1 & 0 \\ 1 & 0 & 0 \end{pmatrix}.$$

Thus,

$$
X_{A2} = P^{-1}X_{A1} = \begin{pmatrix} 1 & \bar{1} & \bar{1} \\ 0 & 1 & 0 \\ 1 & 0 & 0 \end{pmatrix} \begin{pmatrix} 0.071 \\ 0.182 \\ 0 \end{pmatrix} = \begin{pmatrix} -0.111 \\ 0.182 \\ 0.071 \end{pmatrix} \text{ unitless.}
$$

As an exercise calculate the fractional coordinates of all 12 carbon atoms. See Exercise 2.14.

2.10.3 Calculation of G_2

First, G_1 is calculated from the lattice parameters. Although the details are written out here, a mathematical program should be used for such calculations.

$$
G_1 = \begin{pmatrix} a_1 \cdot a_1 & a_1 \cdot b_1 & a_1 \cdot c_1 \\ a_1 \cdot b_1 & b_1 \cdot b_1 & b_1 \cdot c_1 \\ a_1 \cdot c_1 & b_1 \cdot c_1 & c_1 \cdot c_1 \end{pmatrix}
$$

$$
G_1 = \begin{pmatrix} (9.010)^2 & 9.010 \cdot 8.926 \cdot \cos(119°34') & 9.010 \cdot 5.344 \cdot \cos(116°43') \\ 9.010 \cdot 8.926 \cdot \cos(119°34') & (8.926)^2 & 8.926 \cdot 5.344 \cdot \cos(44°24') \\ 9.010 \cdot 5.344 \cdot \cos(116°43') & 8.926 \cdot 5.344 \cdot \cos(44°24') & (5.344)^2 \end{pmatrix}
$$

$$
G_1 = \begin{pmatrix} 81.18 & -39.68 & -21.64 \\ -39.68 & 79.67 & 34.05 \\ -21.64 & 34.05 & 28.55 \end{pmatrix} \text{Å}^2.
$$

Next, calculate the related $G_2 = P^t\, G_1\, P$ where P^t is the transpose of matrix P.

$$
G_2 = \begin{pmatrix} 0 & 0 & -1 \\ 0 & 1 & -1 \\ 1 & 0 & 1 \end{pmatrix} \begin{pmatrix} 81.18 & -39.68 & -21.64 \\ -39.68 & 79.67 & 34.05 \\ -21.64 & 34.05 & 28.55 \end{pmatrix} \begin{pmatrix} 0 & 0 & 1 \\ 0 & 1 & 0 \\ -1 & -1 & 1 \end{pmatrix}
$$

$$
G_2 = \begin{pmatrix} 28.55 & -5.49 & -6.91 \\ -5.49 & 40.12 & -12.54 \\ -6.91 & -12.54 & 66.44 \end{pmatrix} \text{Å}^2
$$

Check the procedure by using $G_1 = (P^{-1})^t\, G_2\, P^{-1}$ to get G_1 back.

$$G_1 = \begin{pmatrix} 1 & 0 & 1 \\ -1 & 1 & 0 \\ -1 & 0 & 0 \end{pmatrix} \begin{pmatrix} 28.55 & -5.49 & -6.91 \\ -5.49 & 40.12 & -12.54 \\ -6.91 & -12.54 & 66.44 \end{pmatrix} \begin{pmatrix} 1 & -1 & -1 \\ 0 & 1 & 0 \\ 1 & 0 & 0 \end{pmatrix}$$

$$G_1 = \begin{pmatrix} 81.18 & -39.68 & -21.64 \\ -39.68 & 79.67 & 34.05 \\ -21.64 & 34.05 & 28.55 \end{pmatrix} Å^2.$$

This final step shows that the inverse transformation of the transformation gives the original **G** matrix. It is instructive to demonstrate that these transformations work numerically.

2.10.4 Calculation of the Lattice Parameters in the New Basis

From the G_2 matrix

$\mathbf{a_2} \cdot \mathbf{a_2} = 28.55$. Thus, $a_2 = \sqrt{28.55} = 5.34\,Å$

$\mathbf{b_2} \cdot \mathbf{b_2} = 40.12$. Thus, $b_2 = \sqrt{40.12} = 6.33\,Å$

$\mathbf{c_2} \cdot \mathbf{c_2} = 66.44$. Thus, $c_2 = \sqrt{66.44} = 8.15\,Å$

$\mathbf{a_2} \cdot \mathbf{b_2} = a_2\, b_2 \cos \gamma_2 = -5.49 = (5.34)\,(6.33)\,\cos \gamma_2$

Thus, $\cos \gamma_2 = -0.162$; $\gamma_2 = 99.33°$
Similarly, $\alpha_2 = 104.05°$; $\beta_2 = 99.33°$

2.10.5 Comparison of the Volume of the Cell in Both Bases

The volume of the unit cell is the square root of the determinant of the **G** matrix,

$$V = \sqrt{\det(\mathbf{G})}.$$

In the first basis,

$$V_1 = \sqrt{\det(\mathbf{G_1})} = \sqrt{\det \begin{pmatrix} 81.18 & -39.68 & -21.64 \\ -39.68 & 79.67 & 34.05 \\ -21.64 & 34.05 & 28.55 \end{pmatrix}} = 258.41 Å^3$$

In the second basis

$$V_2 = \sqrt{\det(\mathbf{G}_2)} = \sqrt{\det \begin{pmatrix} 28.55 & -5.49 & -6.91 \\ -5.49 & 40.12 & -12.54 \\ -6.91 & -12.54 & 66.44 \end{pmatrix}} = 258.41\,\text{Å}^3.$$

Even though the **G** matrices are different numerically, the volumes of these unit cells are identical because both the HMB unit cells are primitive having one lattice point and thus one motif to the unit cell.

2.10.6 Magnitude of a Vector **r**

The magnitude of a vector **r** is invariant under the transformation **P**. Let r_1 be the magnitude of the vector in the first basis and let r_2 be the magnitude of the same vector in the second basis. From Table 2.2

$$r_1 = \sqrt{X_1^t G_1 X_1},$$

and

$$r_2 = \sqrt{X_2^t G_2 X_2}.$$

Since $X_2 = \mathbf{P}^{-1} X_1$ and $X_2^t = (\mathbf{P}^{-1} X_1)^t = X_1^t (\mathbf{P}^{-1})^t = X_1^t (\mathbf{P}^t)^{-1}$, and $G_2 = \mathbf{P}^t\, G_1\, \mathbf{P}$ then

$$r_2 = \sqrt{X_2^t G_2 X_2} = \sqrt{X_1^t (\mathbf{P}^t)^{-1} \mathbf{P}^t\, G_1\, \mathbf{P}\,\mathbf{P}^{-1}\, X_1} = \sqrt{X_1^t G_1 X_1}.$$

This shows that the length of a vector is invariant under the transformation **P**. This can be specifically demonstrated by considering

$$X_{A1} = \begin{pmatrix} 0.071 \\ 0.182 \\ 0 \end{pmatrix},$$

then

$$r_1 = \sqrt{(0.071 \quad 0.182 \quad 0) \begin{pmatrix} 81.18 & -39.68 & -21.64 \\ -39.68 & 79.67 & 34.05 \\ -21.64 & 34.05 & 28.55 \end{pmatrix} \begin{pmatrix} 0.071 \\ 0.182 \\ 0 \end{pmatrix}}$$

$r_1 = 1.43\,\text{Å}$ answer.

Also this same coordinate matrix in the transformed coordinates is

$$X_{A2} = \begin{pmatrix} -0.111 \\ 0.182 \\ 0.071 \end{pmatrix},$$

then

$$r_2 = \sqrt{(-0.111 \quad 0.182 \quad 0.071) \times \begin{pmatrix} 28.55 & -5.49 & -6.91 \\ -5.49 & 40.12 & -12.54 \\ -6.91 & -12.54 & 66.44 \end{pmatrix} \times \begin{pmatrix} -0.111 \\ 0.182 \\ 0.071 \end{pmatrix}}$$

$r_2 = 1.43$ Å answer.

Even though the **G** matrices are different numerically, the distances are equal. This is an example of an invariant property. *Invariant properties* are those properties that are the same before and after the transformation. Invariant properties are independent of the choice of basis vectors. Another example of an invariant property is interatomic bond angle.

2.10.7 Interatomic Bond Angle

The interatomic bond angle is also invariant under the transformation **P**.

2.11 CRYSTALLOGRAPHIC DIRECTIONS

A crystallographic direction is a vector between two lattice points. Because a vector can always be translated parallel to itself, the origin is made one of the lattice points.

In Figure 2.11, two lattice points are indicated, 21 and 42. The vector from the origin at 00 to the lattice point at 21 has exactly the same direction as the vector also from the origin at 00 to lattice point 42. The crystallographic direction is indicated by integers that do not contain a common integer factor, except 1. Thus, the crystallographic direction indicated in Figure 2.11 is [21]. The square brackets indicate a crystallographic direction. The expression [21] is read as "the two one crystallographic direction." No comma separates the numbers unless the comma is needed for clarity as in the [11,1] direction. Crystallographic directions are vectors and always have an arrow head to show which way the vector points.

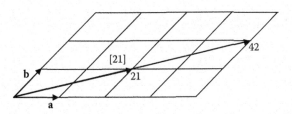

FIGURE 2.11   Crystallographic direction [21].

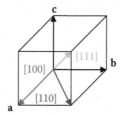

FIGURE 2.12   Crystallographic directions [100], [110], and [111].

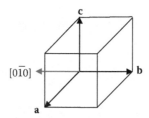

FIGURE 2.13   The [0$\bar{1}$0] crystallographic direction.

Given a lattice in three dimensions with basis vectors **a**, **b**, and **c**, as in Chapter 1, every lattice point can be reached by a vector **t** which connects it with the origin.

$$\mathbf{t}(u,v,w) = u\mathbf{a} + v\mathbf{b} + w\mathbf{c}$$

where u, v, and w are integers.

The *crystallographic direction* is a vector between two lattice points in which the direction is indicated by [$u\ v\ w$] where u, v, and w are integers that do not contain a common integer factor, except one. The integers u, v, and w are called the *indices of the crystallographic direction* and specify an infinite set of parallel vectors. See Exercises 2.15 and 2.16.

Figure 2.12 shows the three-dimensional crystallographic directions [100], [110], and [111]. In copper, Young's modulus is an anisotropic property, which is 66.7 GPa along [100], 130.3 GPa along [110], and 191.1 GPa along [111]. The direction [010] is along **b**. The opposite direction, along negative **b**, is the [0$\bar{1}$0] which is read as the "zero one bar zero direction" (see Figure 2.13).

2.12 CRYSTALLOGRAPHIC PLANES AND MILLER INDICES

Find two lattice points and connect them to form a line. Find a third lattice point, not on the line; and define a plane going through these three lattice points. This is a *lattice plane*. Given one such plane, since all lattice points have identical neighborhoods, there is a plane parallel to the given plane passing through every lattice point. A *crystallographic plane* is a set of parallel lattice planes. There are an infinite number of these parallel lattice planes.

A crystallographic plane is specified by three integers, h, k, and l called the Miller indices. Parentheses around the Miller indices indicate the crystallographic plane (hkl). Parallel planes have the same indices. The expression (123) is read as "the one two three crystallographic plane." *Miller indices* are three relatively prime integers, *hkl*, that are reciprocals of the fractional intercepts that the crystallographic plane makes with the crystallographic axes. The term "relatively prime integers" means that the integers do not contain a common factor other than 1. In the special case where the fractional intercepts are all zero, that is, the plane passes through the origin, consider a parallel plane. The reciprocals of the fractional intercepts *may* have to be multiplied or divided until they do not have a common factor.

The method for calculating the Miller indices seems somewhat clumsy at first, but it simplifies the diffraction equations which are developed in Section 5.6.

Given the Miller indices, *hkl*, there are two steps to plot the associated lattice plane:

- Take the reciprocal of the Miller indices, $1/h$, $1/k$, $1/l$.

- These are the fractional coordinates of the intercepts with the crystallographic axes of the associated plane. Now plot the plane.

This associated plane turns out to be a lattice plane.

On the other hand there are two steps to finding the Miller indices, *hkl*, of a lattice plane given the fractional coordinates of the intercepts of that plane with the crystallographic axes:

- Take the reciprocals of these fractional coordinates.

- Multiply or divide these reciprocals until a set of relatively prime integers are got. These integers, *hkl*, are the Miller indices. It turns out that the fractional coordinates are always rational.

EXAMPLE 2.6

In two dimensions, plot the (21) plane.

Solution

First, take the reciprocal of 2 which is 1/2. Then take the reciprocal of 1 which is 1. Draw a plane such that it intersects the basis vector **a** at 1/2 **a** and the basis vector **b** at 1**b**. The result is the (21) plane. See Figure 2.14.

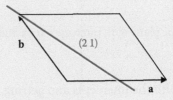

FIGURE 2.14   The (21) crystallographic plane.

The Miller indices may contain a zero.

EXAMPLE 2.7

Plot the (01) plane.

Solution

In this case, the reciprocal of 0 is taken as ∞, or in other words, this plane does not intersect the basis vector **a**. The reciprocal of 1 is 1. Now draw a plane that does not intersect basis vector **a** and intersects basis vector **b** at 1 (see Figure 2.15).

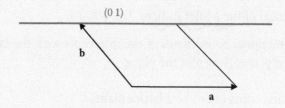

FIGURE 2.15 The (01) crystallographic plane.

EXAMPLE 2.8

Given a lattice plane that intersects basis vector **a** at 1/3 **a** and basis vector **b** at 2/5 **b** what are the Miller indices of this plane?

Solution

First draw the plane. See Figure 2.16. Now take the reciprocals of the intercepts. In this case they are 3 and 5/2, respectively. Since these are not integers, multiply by the smallest number that will make them integers, in this case 2. The Miller indices are (65). Note that the integers 6 and 5 have no common integer factor other than 1.

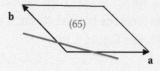

FIGURE 2.16 Crystallographic plane with intercepts 1/3 **a** and 2/5 **b**.

Now consider starting with the intercepts and getting the Miller indices.

This discussion is continued in Chapter 5. For three dimensions, Figure 2.17 shows (100), (110), and (111) planes. Parallel planes are equivalent. See Exercises 2.17 and 2.18.

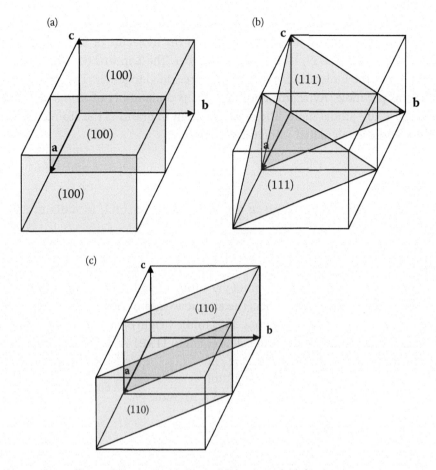

FIGURE 2.17 Crystallographic planes (a) (100), (b) (111), and (c) (110).

2.13 DENSITY

Density is a physical property that is useful in crystallography because it bridges the macroscopic world that can be seen and felt with the submicroscopic world of atomic dimensions. The density of a single crystal can be measured by the floatation method, in which a crystal is immersed in liquids of known density. When the crystal stays suspended in the liquid, neither rising nor falling, the density of the crystal is equal to the density of the liquid. The *density*, δ, is the mass per unit volume. On a macroscopic scale, for example, the mass could be in grams and the volume could be in cubic centimeters. On a crystallographic scale, the volume is the volume of the unit cell and the mass is the mass of all the atoms in the unit cell. If the material is uniform and nonporous then the two densities will be close.

$$\text{density} = \delta = \frac{\text{mass}}{\text{volume}} = \frac{Z(\Sigma A)}{V\,N_A} = \frac{Z(\Sigma A)}{\sqrt{\det(\mathbf{G})}\,N_A} \tag{2.28}$$

where Z is the number of molecules (or the number of formula units) in the unit cell, A is the atomic mass of each of the atoms or ions in the molecule or formula unit, V is the volume of the unit cell, and N_A is Avogadro's number. The Σ means that the atomic masses of all the atoms in the molecule or formula unit are added together. The atomic mass, A, is given in grams per mole. Avogadro's number, N_A, is the number of atoms per mole and is equal to 6.023×10^{23} atoms/mole. The mass per atom in grams is the atomic mass divided by N_A. The volume is calculated using the **G** matrix.

EXAMPLE 2.9

What is the density of quartz, given that there are three SiO_2 formula units to the unit cell?

Solution

In quartz there are three formula units per unit cell and each formula unit has one silicon atom and two oxygen atoms. The atomic weight of silicon is 28.09 g/mol, the atomic weight of oxygen is 16.00 g/mol. The atomic weight of SiO_2 is 60.08 g/mol. From Example 2.4, the volume of the unit cell is equal to 113.01 Å^3 = 113.01×10^{-24} cm^3.

$$\delta = \frac{3 \times 60.08 \text{ g/mole}}{113.01 \times 10^{-24} \text{cm}^3 \times 6.023 \times 10^{23} \text{mole}^{-1}} = 2.649 \text{ g/cm}^3$$

The value measured experimentally is 2.660 g/cm^3.

Crystallographers use density to calculate the number of molecules or formula units in the unit cell. The density can be experimentally measured, the volume can be calculated from the metric matrix, and the atomic mass of the molecule or formula unit can be calculated by adding up the atomic masses of all the atoms. The density equation can then be solved for Z, which is an integer. See Exercises 2.19 and 2.20.

2.14 REVISITING THERMAL EXPANSION AND ISOTHERMAL COMPRESSIBILITY

In Chapter 1, the concepts of linear thermal expansion and linear compressibility were introduced. Now that volume can be calculated, the concepts can be expanded.

2.14.1 Volumetric Thermal Expansion

The *volumetric thermal expansion coefficient* is defined as

$$\alpha_V = \frac{1}{V}\left(\frac{\partial V}{\partial T}\right)_P = \left(\frac{\partial \ln V}{\partial T}\right)_P, \tag{2.29}$$

where V is the volume of the unit cell, T is the temperature, and P is the pressure.

In Chapter 1, the lattice parameters are given for urea as a function of temperature. The **G** matrix for urea can be calculated at 188 K. The volume is then calculated with lattice parameter a equal to lattice parameter b and angles α, β, and γ remaining constant at 90°, as in Chapter 1. Plot the temperature *versus* the natural logarithm of the volume. Do a least-squares fit to the data and calculate the volumetric thermal expansion coefficient for urea. See Exercise 2.21.

2.14.2 Volumetric Compressibility

The *volumetric compressibility* is defined as

$$\beta_V = -\frac{1}{V}\left(\frac{\partial V}{\partial P}\right)_T = -\left(\frac{\partial \ln V}{\partial P}\right)_T, \tag{2.30}$$

where V is the volume of the unit cell, T is the temperature, and P is the pressure. The volume compressibility is often referred to simply as the compressibility. See Table 2.5 and Exercise 2.23.

The *mean compressibility*, $\bar{\beta}$, between any two pressures P_1 and P_2 is

$$\bar{\beta} = \frac{-2}{V_1 + V_2}\left[\frac{V_2 - V_1}{P_2 - P_1}\right]. \tag{2.31}$$

In Section 1.9.2, the crystallographic data are given for anthracene. Calculate the mean compressibility between pressures 5.05 and 6.07 GPa. See Exercises 2.22 and 2.23.

An important physical property is the *bulk modulus*, K, which is the reciprocal of the compressibility. The bulk modulus, or incompressibility, has units of pressure. The bulk modulus measures the ability of a material to resist compression.

$$K = \frac{1}{\beta_V}.$$

The bulk modulus for quartz is about 38 GPa and for steel is about 160 GPa.

TABLE 2.5   Unit Cell Volume of Quartz as a Function of Pressure

Volume, V (Å^3)	ln (Volume V, Å^3)	Pressure, P (GPa)
112.981	4.727220	0.0001
111.725	4.716040	0.429
110.711	4.706923	0.794
108.597	4.687644	1.651
108.150	4.683519	1.845
107.974	4.681890	1.933
106.467	4.667835	2.628
105.141	4.655302	3.299
104.831	4.652350	3.468

Source: Data courtesy of Ross Angel, 1977.

DEFINITIONS

Bulk modulus
Coordinate matrix
Coordinate vector
Crystallographic direction
Crystallographic plane
Density
Fractional coordinates
Interatomic bond length
Invariant properties
Lattice plane
Mean compressibility
Metric matrix, **G**
Miller indices
Volumetric compressibility
Volumetric thermal expansion coefficient

EXERCISES

2.1 Plot the fractional coordinates of HMB from Table 2.1 to produce Figure 2.3. Do the two-dimensional case, that is ignore z.

2.2 The fractional coordinates in Table 2.1 for HMB are not all in the same unit cell. Convert these fractional coordinates so that all the values are between 0 and 1. Include 0 but do not include 1.

2.3 Plot the fractional coordinates calculated in Exercise 2.2. Go back and look at the isolated unit cell in Figure 1.11. While all 12 atoms are in the unit cell, the connected molecule is not obvious. See Figure 2.4.

2.4 A vector is unchanged if the origin is translated. In Table 2.1 add the coordinate matrix (1/2 1/2) to the x and y fractional coordinates. Are all the coordinates within the range 0 to 1? (Again 0 is included and 1 is not.) Plot this set of coordinates. Connect and label the carbon atoms to look like Figure 2.5.

2.5 Perform the matrix multiplication in Equation 2.7 and compare it to the right-hand side of Equation 2.6.

2.6 Show that in three dimensions the volume of the unit cell is equal to the square root of the determinant of the **G** matrix.

2.7 Calculate the **G** matrix for $CuSO_4 \cdot 5H_2O$. The cell parameters are $a = 6.091$ Å, $b = 10.63$ Å, $c = 5.964$ Å, $\alpha = 82.410°$, $\beta = 107.50°$, $\gamma = 102.70°$.

2.8 Calculate the **G** matrix for AA. The cell constants are given in Chapter 1.

2.9 Calculate the sulfur–oxygen bond length in AA. The fractional coordinates are $X_S = (0.333\ 0.337\ 0.222)$ and $X_O = (0.667\ 0.656\ 0.317)$.

⌨ 2.10 Looking at Figure 2.5, calculate all six bond lengths of the benzene ring. Now calculate the bond distance between the methyl carbons and benzene ring carbons. From this calculate the radius of the carbon atom in the benzene ring and compare this value with a literature value for the carbon atom. Now calculate the radius of the methyl carbon.

⌨ 2.11 Look at Figure 2.5 and calculate all the bond angles around each of the carbon atoms in the benzene ring.

⌨ 2.12 Calculate the volume of the HMB unit cell.

2.13 Calculate the ratio of the area of a parallelepiped formed by basis vectors $\mathbf{a}_2$, $\mathbf{b}_2$ to the area of a parallelepiped formed by basis vectors $\mathbf{a}_1$, $\mathbf{b}_1$ for the transformation: $\mathbf{a}_2 = \mathbf{a}_1 - \mathbf{b}_1$, $\mathbf{b}_2 = \mathbf{a}_1 + 2\mathbf{b}_1$.

⌨ 2.14 Given $a_2 = 5.3434$ Å, $b_2 = 6.3347$ Å, and $c_2 = 8.1513$ Å, $\alpha_2 = 104.0589°$, $\beta_2 = 99.3386°$, $\gamma_2 = 99.3398°$ and the transformation $\mathbf{a}_1 = \mathbf{a}_2 - \mathbf{c}_2$; $\mathbf{b}_1 = \mathbf{a}_2 + \mathbf{b}_2$; and $\mathbf{c}_1 = -\mathbf{a}_2$. Calculate P, $\mathbf{G}_2$, $\mathbf{G}_1$, a_1, b_1, c_1, α_1, β_1, γ_1, V_1, and V_2.

⌨ 2.15 Compute the HMB two-dimensional unit cell and on it indicate the [01], [11], [10] directions.

⌨ 2.16 Compute the HMB three-dimensional unit cell and indicate the [010], [111], [100], [001], [101] directions on separate figures.

2.17 Given a plane that intersects the $\mathbf{a}$-axis at 1/2, the $\mathbf{b}$-axis at 2/3, and the $\mathbf{c}$-axis at 1/3, what are the Miller indices of this plane?

⌨ 2.18 Compute the HMB three-dimensional unit cell and indicate the (010), (111), and (101) planes on separate figures.

2.19 What is the density of HMB? Take the hydrogens into account.

⌨ 2.20 Calculate the number of formula units in AA. The measured density is 2.784 g/cm³. The cell parameters are given in Chapter 1.

⌨ 2.21 Using the data from Table 1.2 calculate the $\mathbf{G}$ matrix for urea at 188 K. From the $\mathbf{G}$ matrix calculate the volume assuming that the lattice parameter a equals lattice parameter b and that angles α, β, and γ remain constant at 90°. Now calculate the volume for all the temperature data points from 188 to 328 K. Plot the temperature versus the natural log of the volume. Run a least squares through the data and calculate the volumetric thermal expansion coefficient for urea.

⌨ 2.22 Using the data from Table 1.3 calculate the $\mathbf{G}$ matrix for anthracene at 5.05 and 6.07 GPa. From the $\mathbf{G}$ matrix calculate the volume assuming that the angles α and γ remain constant at 90°. Calculate the mean compressibility between pressures 5.05 and 6.07 GPa for anthracene.

⌨ 2.23 Plot the natural logarithm of the volume of the unit cell of quartz versus pressure on the same graph as the corresponding anthracene data in Table 1.3 up to 4.04 GPa. Compare the data for the two substances. See Table 2.5.

MATLAB CODE: STARTER PROGRAM FOR CHAPTER 2: GRAPHIC OF HMB PROJECTION

This program populates a two-dimensional unit cell of HMB from the fractional coordinates. The cell parameters and three atoms have been inserted. The student puts in the remaining atoms from Table 2.1 to produce Figure 2.3. Now all the atoms are moved by 1/2,1/2 and then labeled to produce Figure 2.5.

```
clc                     %clears content of command window
close all               %
clear                   %clears variables in workspace

% Script file: HMB2D.m
%
% Purpose: Draw 2D general oblique unit cell and apply it to HBM.
%
% Record of revisions:
% Date       Programmer          Description of change
% 31 Jan     M. Julian           Original Code
%
% Define Variables:
% a = crystallographic a axis (Angstroms)
% b = crystallographic b axis (Angstroms)
% c = crystallographic c axis (Angstroms)
% alpha = angle between b and c (degrees)
% beta = angle between a and c (degrees)
% gamma = angle between a and b (degrees)
% conversion = function to convert from triclinc to cartesian
% outline = function to outline cell
% cc = cartesian coordinates = conversion*outline
a = 9.010;
b = 8.926;
%c = 5.344;
%alpha = 44 + 27/60;
%beta = 116.72;
gamma = 119.57;
conversion = [a b*cosd(gamma)
              0 b*sind(gamma)];

outline = [0 1 1 0 0
           0 0 1 1 0];

cc = (conversion*outline)'
x = cc(:,1)';
```

```
y = cc(:,2)';

plot(x,y)
xlim([-5.75 10.25])
ylim([-1.25 9])
text(4.15,-0.5,'\bfa')
%plot arrow on a axis
hold on
plot(cc(2,1)',cc(2,2)','>','MarkerSize', 8,'MarkerFaceColor','k')
text(-2.8,3.8,'\bfb')
%plot arrow on b axis
hold on
plot(cc(4,1)',cc(4,2)','>','MarkerSize', 8,'MarkerFaceColor','k')
grid off
axis off
axis equal
%The 2D unit cell drawn


%%%%%%%%%%%%%%%%%%%%%%%%%%%%%%%%%%%%%%%%%%%%%%%%%%%%%%%%%%%%%%%%%%%%%%%%%%
%%%%
%put in fractional coordinates
%fractional coordinates for atom A
cA = [0.071 0.182];
%fractional coordinates for atom B
cB = [-0.109 0.072];
%fractional coordinates for atom C
cC = [-0.180 -0.109];
%put in rest of atoms to complete molecule
%collect ALL atoms in matrix, transpose to get have column
coordinates
carbon = [cA; cB; cC]'
%count the atoms,n, which is the number of columns
n = size(carbon,2)
%convert to Cartesian space using conversion, transpose to make
row vectors
%for plotting
carbonC = (conversion*carbon)'
%%% plot(carbonC(:,1),carbonC(:,2),'ro')
%%% stop
%%%%remove this stop when you understand how the red atoms are
plotted
%%%%you may put stops in to see what happens at any step,
suggested stops have
%%%%been put in
%%%%now remove these red atoms since they are not needed any more
%%%%remove them by putting a comment symbol (%)
```

```
%%%%%%%%%%%%%%%%%%%%%%%%now atoms will be moved to center of cell,
must be added to fractional
%%%%%%%%%%%%%%%%%%%%%%%%%%%%%%%%%%%%%%
%coordinates; I put in HH to mean half, half
carbonHH =.5*ones(2,n)+carbon
carbonHHC = (conversion*carbonHH)'
%%% plot(carbonHHC(:,1),carbonHHC(:,2),'ko')
%%% stop%remove this stop
%plot bonds
%Now make a list of the atoms you wish to plot
Ax = carbonHHC(1,1);Ay = carbonHHC(1,2);
Bx = carbonHHC(2,1);By = carbonHHC(2,2);
Cx = carbonHHC(3,1);Cy = carbonHHC(3,2);
%produce outline of molecule, conversion to Cartesian is already
made
%draw a continous path for the molecule, repeating if necessary
OutlineM = [Ax Bx Cx
           Ay By Cy]
xx = OutlineM(1,:);
yy = OutlineM(2,:);
plot(xx,yy,'b','LineWidth',2)
%suggested stop
%now atoms are put in place
plot(xx,yy,'ro','MarkerSize',20,'MarkerFaceColor','w')
%suggested stop
%Label individual atoms
text(Ax-.1,Ay,'A')
```

## CHAPTER 3

Point Groups

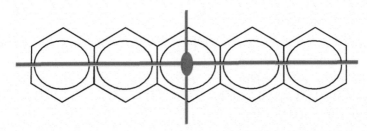

Here is a molecule of pentacene, an organic semiconductor. This conductive plastic can be used in high-performance organic thin-film transistors. Applications include smart cards, smart price, and inventory cards (Lin, 1997). This chapter studies point groups. Point groups use symmetry to classify the shapes of individual objects, such as a molecule of pentacene. The symmetries are color-coded red and purple according to change of handedness.

CONTENTS

CHAPTER OBJECTIVES

PART I: TWO DIMENSIONS

- Define a group.

- Describe symmetry operations—identity, mirror and n-fold rotations in two dimensions.

- Color-code symmetries red, blue, and purple according to change of handedness.

- Derive a matrix for each symmetry operation.

- Prove that only one-, two-, three-, four-, and sixfold rotations are possible in periodic patterns.

- Identify the point group of any two-dimensional figure.

- Draw both the point and symbol stereographic projections for the point groups.

- Classify the two-dimensional point groups according to oblique, rectangular, square, and hexagonal systems.

- Draw the point group trees.

PART II: THREE DIMENSIONS

- Introduce three-dimensional point groups as extensions of the two-dimensional case.

- Describe rotoinversion operations, found only in three dimensions.

- Classify three-dimensional point groups according to triclinic, monoclinic, ortho-rhombic, tetragonal, trigonal, hexagonal, and cubic.

- Examine point groups $2/m$ and 32 in detail.

- Classify materials as piezoelectric or nonpiezoelectric.

- Identify centrosymmetric point groups on the point group tree.

- Relate point groups and etch figures.

PART I: TWO DIMENSIONS

3.1 INTRODUCTION

The last chapter introduced calculations within the unit cell, including fractional coordinates, the metric matrix **G**, bond distances, bond angles, and volumes. In this chapter point groups classify the symmetries of individual objects, including molecules. The symmetries are color-coded red, blue, and purple according to change of handedness. The two-dimensional symmetries are the identity, mirrors, and rotations. The 10 two-dimensional point groups are explored and interrelated. Three-dimensional symmetry operations are introduced. The notation of the *International Tables for Crystallography* is followed, in order to facilitate reading the crystallographic literature.

3.2 GROUP THEORY

A *set* is a collection of elements. These elements may interact with one another by an operation. A group is defined in terms of elements and one operation. The point groups in this chapter have a finite number of elements. Terms like "multiplication table" and "inverse" have special meanings.

3.2.1 Elements in a Set

Start out with a collection of elements, in a set S. Curly braces indicate a set that can be defined by example or rule. For example, if the set S consists of three letters a, b, and d, then set S = {a, b, d}. The set could be infinite, for example, containing all the integers greater than 3. Then S = {integers greater than 3} or S = {4, 5, 6 ...}. An element could be an operation, for example, a rotation. If 3^+ is a counterclockwise rotation of 120°, then the set consisting of that one element is S = {3^+}.

EXAMPLE 3.1

Think of the clock in Figure 3.1. The clock has three numbers, 0, 1, and 2. Thus if a set G containing these three numbers is formed, then G = {0, 1, 2}. Choose the operation for this set to be addition mod 3, which works like a clock with three numbers. List all the pairs of elements joined under the operation addition mod 3 for this set. Include the result of the operation in each case.

Solution

Start at 0, add 1 to get 1, add another 1 to get 2, and finally add 1 again to get back to 0. Since this is a finite set, a list can be made of all the combinations.

$0 \oplus 0 = 0$
$0 \oplus 1 = 1$

$$0 \oplus 2 = 2$$
$$1 \oplus 0 = 1$$
$$1 \oplus 1 = 2$$
$$1 \oplus 2 = 0$$
$$2 \oplus 0 = 2$$
$$2 \oplus 1 = 0$$
$$2 \oplus 2 = 1$$

FIGURE 3.1 A clock with three numbers.

3.2.2 Operations

The operation of a group is indicated by the symbol ⊕, read "operation." To have an operation, elements are needed.

This information is conveniently listed in Table 3.1, called a multiplication table. Table 3.1 is now written in abbreviated form as Table 3.2. Careful examination of Table 3.2 reveals the following properties:

- For the operation between any two elements a and b in G, then $a \oplus b = c$ where c is also in G. This property is called *closure* meaning that the operation ⊕ does not introduce any new elements.

- There is one element E, called the *identity*, such that if a is an element in G, then $E \oplus a = a \oplus E = a$. In this example, the identity E = 0. See Exercise 3.1.

TABLE 3.1 Multiplication Table for the Operation Addition mod 3 Operating on G = {0, 1, 2}

⊕ = Addition mod 3	0	1	2
0	$0 \oplus 0 = 0$	$0 \oplus 1 = 1$	$0 \oplus 2 = 2$
1	$1 \oplus 0 = 1$	$1 \oplus 1 = 2$	$1 \oplus 2 = 0$
2	$2 \oplus 0 = 2$	$2 \oplus 1 = 0$	$2 \oplus 2 = 1$

TABLE 3.2 Multiplication Table in an Abbreviated Form

Addition mod 3	0	1	2
0	0	1	2
1	1	2	0
2	2	0	1

- For every element a in the set, there is an *inverse*, a^{-1}, such that

$$a \oplus a^{-1} = a^{-1} \oplus a = E$$

Inspection of the table shows that

0 is the inverse of 0 because $0 \oplus 0 = 0$;

1 is the inverse of 2 because $2 \oplus 1 = 1 \oplus 2 = 0$ and

2 is the inverse of 1 because $1 \oplus 2 = 2 \oplus 1 = 0$.

- There is a fourth property called the *associative law*. The associative law means that given elements a, b, and c then $(a \oplus b) \oplus c = a \oplus (b \oplus c)$. Note that the order of the elements is preserved. One example from the table is

$$(1 \oplus 2) \oplus 1 = 0 \oplus 1 = 1$$

and

$$1 \oplus (2 \oplus 1) = 1 \oplus 0 = 1.$$

See Exercise 3.2.

For a finite group, a table, called a *multiplication table*, can be constructed giving all possible combinations relating two elements by the operation of the group. The order of the elements—that is, which is first and which is second—is important for many groups.

3.2.3 Group

A *group* is a set of elements with an operation between pairs of the elements that has the following four properties:

1. Closure

2. Identity

3. Inverse element

4. Associative law

EXAMPLE 3.2

Here is another example of a group. Consider a group where the elements are operations on objects in space, specifically rotations (see Figure 3.2).

Let

E = the rotation of 0°
3^+ = the counterclockwise rotation of 120°
3^- = the counterclockwise rotation of 240°

then G = {E, 3^+, 3^-} is the group.

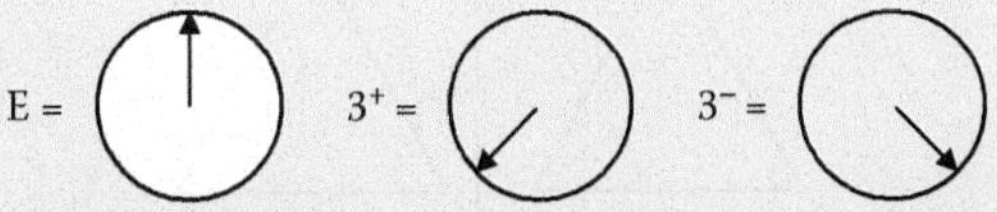

FIGURE 3.2 Symbolic illustrations of E, 3^+, and 3^-.

TABLE 3.3 Multiplication Table for G = {E, 3^+, 3^-} with the Operation of Rotation

Rotation	E	3^+	3^-
E	E	3^+	3^-
3^+	3^+	3^-	E
3^-	3^-	E	3^+

Solution

To show that G = {E, 3^+, 3^-} is a group, construct the multiplication table in Table 3.3. The operation of the group is the composition of rotations—the application of the first rotation S_1 to an object followed by the application of the second rotation S_2 to the object to obtain the result represented by $S_2 \times S_1$.

There is closure because no new elements are generated by the operation. The identity is E, or 0° rotation. Each element has an inverse and the associative law holds.

See Exercise 3.3.

The operation of a group can be defined through its multiplication table. For example, consider the group consisting of elements {E, A, B} which obey the multiplication table shown in Table 3.4. Tables 3.3 and 3.4 have identical multiplication tables when 3^+ is replaced by A and 3^- is replaced by B. This phenomenon is discussed later in the chapter. For a more advanced, but readable, treatment of group theory see *Algebra and Geometry* by Alan Beardon (2005), *Groups and Characters* by Victor Hill (2000), and *The Mathematical Theory of Symmetry in Solids* by C.J. Bradley and A. P. Cracknell (2011). For a general introduction see *Basic Concepts of Crystallography* by Emil Zolotoyabko (2011). See Exercise 3.4.

TABLE 3.4 Multiplication Table for Elements {E, A, B} Where the Operation ⊕ Is Defined by the Table

⊕	E	A	B
E	E	A	B
A	A	B	E
B	B	E	A

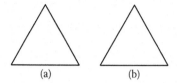

FIGURE 3.3   Equilateral triangle (a) before operation 3⁺, (b) after operation 3⁺.

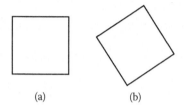

FIGURE 3.4   Square (a) before operation 3⁺, (b) after operation 3⁺.

3.3 SYMMETRY OPERATIONS

A *symmetry operation* for an object is an operation in which the object looks the same before and after the operation. For example, consider the symmetry operation 3⁺ described in Example 3.2. The operation 3⁺ rotates an object counterclockwise by 120°. Suppose the object is an equilateral triangle as in Figure 3.3a. When operation 3⁺ is applied at the center of the triangle, Figure 3.3b is produced. Since Figures 3.3a and 3.3b are indistinguishable, then 3⁺, applied at the center of the triangle, is a symmetry operation for the equilateral triangle.

On the contrary, if the operation 3⁺ is applied to the center of a square, as in Figure 3.4, there is a difference before and after a counterclockwise rotation of 120°. Thus 3⁺, located at the center of the square, is NOT a symmetry operation for a square.

3.3.1 Conventions of the *International Tables for Crystallography*

In Chapter 1 the basis vectors **a** and **b** were introduced. In this chapter the basis vectors create a framework for analytically describing the symmetry operations. The angle γ is defined as the angle between **a** and **b**. A point is located by the coordinate vector $\mathbf{r} = x$ **a** $+ y$ **b**. The variables x and y are unitless and are used to indicate the coordinates of the vector **r**. The axes are labeled x and y. In accordance with the conventions of the *International Tables for Crystallography*, the origin is placed in the upper left-hand corner of a diagram, the basis vector **b** is laid out horizontally pointing to the right, and the basis vector **a** points downward (see Figure 3.5).

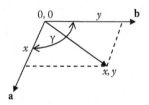

FIGURE 3.5   Basis vectors following the *International Tables for Crystallography*.

3.3.2 Mirror Symmetry Operation

A *mirror* is the symmetry operation of reflection. Figure 3.6 shows the sun and several trees with their reflected image in a still pond as the sun sets. The mirror plane is a horizontal imaginary line that is the perpendicular bisector of the line connecting the sun with its reflected image. Now look at the letter **B**. Figure 3.7a shows a mirror plane, *m*, and the top half of the letter **B**. When the mirror operates, the full letter **B** is produced. The mirror positioned in Figure 3.7 is a symmetry operation on the letter **B**. The letter **B** is said to contain a mirror plane. The *International Tables for Crystallography* uses the term "mirror line" for two dimensions and the term "mirror plane" for three dimensions. In this book the term "mirror plane" is used for both two and three dimensions.

When basis vector **b** is in the mirror plane, the mirror operation is called m_y or m 0,y. All the points on this mirror plane have an $x = 0$ and an arbitrary y. In Figure 3.8 the angle between the x and y axes is 90°, and the origin is placed on the mirror plane. Reference points 0,1 and 1,0 are labeled.

The symbol for a mirror plane is a heavy solid line.

FIGURE 3.6   Mirror in a photograph of a still pond. (Photograph by Maureen M. Julian.)

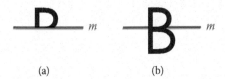

(a)   (b)

FIGURE 3.7   (a) The top half of the letter **B** with the purple mirror plane drawn, (b) when the mirror, *m*, operates on the top half of the letter **B**, the whole letter **B** is produced.

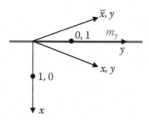

FIGURE 3.8   Mirror plane (in purple), m_y at 0, y.

Now consider what the operation of the mirror m_y does to the point at 1,0 and the point at 0,1, as drawn.

A negative 1 is read "one bar" and indicated by a bar over the 1 namely $\bar{1}$.

Thus $1, 0 \rightarrow \bar{1},0$ and in general $x,0 \rightarrow \bar{x},0$; and $0,1 \rightarrow 0,1$ and in general $0,y \rightarrow 0,y$; combining $x,y \rightarrow \bar{x},y$.

Therefore in two dimensions with orthogonal axes, the mirror m_y is a symmetry operation that takes a point with the coordinates x,y into the point with coordinates $\bar{x},y$.

A matrix is associated with every symmetry operation. For example, the mirror operation that takes a point with the coordinates x,y into the point with the coordinates $\bar{x},y$ has the matrix m_y. Thus,

$$m_y \begin{pmatrix} x \\ y \end{pmatrix} = \begin{pmatrix} \bar{1} & 0 \\ 0 & 1 \end{pmatrix} \begin{pmatrix} x \\ y \end{pmatrix} = \begin{pmatrix} \bar{x} \\ y \end{pmatrix}$$

Specifically, the matrix for m_y with orthogonal axes is

$$m_y = \begin{pmatrix} \bar{1} & 0 \\ 0 & 1 \end{pmatrix}$$

The determinant of the matrix is $\bar{1}$. The *trace* of a matrix is the sum of the diagonal elements. Here trace of $m_y = \bar{1} + 1 = 0$. In general, the matrix

$$\begin{pmatrix} a & b \\ c & d \end{pmatrix}$$

has trace $a + d$.

The mirror plane can also be vertical as in the butterfly in Figure 3.9. See Fay (1989). When the basis vector **a** is in the mirror plane, the mirror operation is called m_x or m $x,0$. All the points on this mirror plane have an arbitrary x and $y = 0$. See Exercises 3.5 and 3.6.

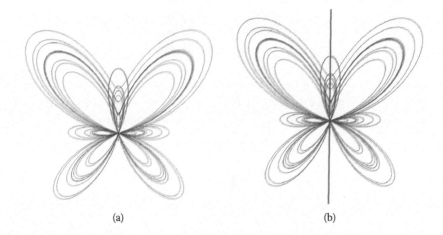

(a)   (b)

FIGURE 3.9 (a) Butterfly with vertical mirror plane, (b) the vertical mirror plane in purple.

3.3.3 Identity Symmetry Operation

The *identity operation*, 1, is a symmetry operation that takes a point with coordinates x,y into itself, or x,y. The associated matrix is the identity matrix,

$$\begin{pmatrix} 1 & 0 \\ 0 & 1 \end{pmatrix}$$

Thus,

$$1\begin{pmatrix} x \\ y \end{pmatrix} = \begin{pmatrix} 1 & 0 \\ 0 & 1 \end{pmatrix}\begin{pmatrix} x \\ y \end{pmatrix} = \begin{pmatrix} x \\ y \end{pmatrix}$$

The identity symmetry operation changes nothing. It plays a role for symmetry operations resembling the role played for numbers by zero, under the operation of addition.

3.3.4 n-Fold Rotation Symmetry Operations

In the *International Tables for Crystallography,* the motion in all the *n*-fold rotations is in the *counterclockwise* direction.

3.3.4.1 Twofold Rotation Symmetry Operation

A *twofold rotation* is a 180° rotation that is a symmetry operation. Look at the letter **S** in Figure 3.10. Figure 3.10a shows the top part of the **S**. Put the point of a pencil on the twofold rotation axis (shown here in yellow) and rotate the paper by 180°. The complete letter **S** is formed. In Figure 3.10b the **S** before the rotation is indistinguishable from the **S** after the rotation. The *International Tables for Crystallography* uses the term "twofold rotation point" for two dimensions, and the term "twofold rotation axis" for three dimensions. In this book the term "twofold rotation axis" is used for both two and three dimensions.

(a)   (b)

FIGURE 3.10 (a) The top half of the letter **S** with the twofold rotation axis in yellow. (b) When the twofold rotation, 2, operates on the top half of the letter **S**, the whole letter S is produced.

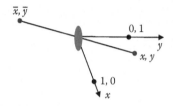

FIGURE 3.11 Twofold axis, 2, with the red symbol.

In Figure 3.11, the angle between the x and y axes is arbitrary, and the twofold rotation axis is placed at the origin. Reference points 1,0 and 0,1 are labeled. Consider what the twofold does to these points. Thus, $1,0 \rightarrow \bar{1},0$ and in general $x,0 \rightarrow \bar{x},0$; and $0,1 \rightarrow 0,\bar{1}$ and in general $0,y \rightarrow 0,\bar{y}$; combining $x,y \rightarrow \bar{x},\bar{y}$.

The symbol for the twofold axis is shown in red. The twofold operation takes the point with coordinates x,y and moves it to the point with coordinates $\bar{x},\bar{y}$.

The matrix associated with the twofold operation is

$$2\begin{pmatrix} x \\ y \end{pmatrix} = \begin{pmatrix} \bar{1} & 0 \\ 0 & \bar{1} \end{pmatrix}\begin{pmatrix} x \\ y \end{pmatrix} = \begin{pmatrix} \bar{x} \\ \bar{y} \end{pmatrix}$$

Thus, the matrix

$$2 = \begin{pmatrix} \bar{1} & 0 \\ 0 & \bar{1} \end{pmatrix}.$$

The determinant of this matrix is +1 and its trace is −2.

3.3.4.2 Threefold Rotation Symmetry Operation

A threefold rotation is a 120° rotation that is a symmetry operation. Figure 3.12 shows a figure with threefold rotation symmetry. Put the point of a pencil on the threefold rotation axis and rotate the paper by 120°. The rotated figure is indistinguishable from the original one or, in other words, *invariant* under the operation. For a *Mathematica* treatment see Shakiban web reference with no date.

In Figure 3.13 the angle between the x and y axes is 120°, and the threefold rotation axis is placed at the origin. The lengths of the basis vectors are equal to accommodate the

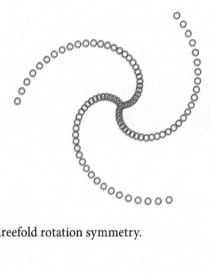

FIGURE 3.12   Figure with threefold rotation symmetry.

symmetry. Points 1,0 and 0,1 are reference points. Consider what 3^+ does to these points. Thus $1,0 \rightarrow 0,1$ and in general $x,0 \rightarrow 0,x$; and $0,1 \rightarrow \bar{1},\bar{1}$ and in general $0,y \rightarrow \bar{y},\bar{y}$; combining $x,y \rightarrow \bar{y}, x + \bar{y}$.

The symbol for the threefold axis is a triangle. The threefold moves the point with the coordinates x,y to the point with the coordinates $\bar{y}, x + \bar{y}$.

The matrix associated with 3^+ is

$$3^+ \begin{pmatrix} x \\ y \end{pmatrix} = \begin{pmatrix} 0 & \bar{1} \\ 1 & \bar{1} \end{pmatrix} \begin{pmatrix} x \\ y \end{pmatrix} = \begin{pmatrix} \bar{y} \\ x + \bar{y} \end{pmatrix}$$

The matrix

$$3^+ = \begin{pmatrix} 0 & \bar{1} \\ 1 & \bar{1} \end{pmatrix}.$$

The determinant of this matrix is +1 and the trace is −1.

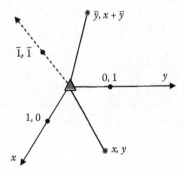

FIGURE 3.13   Threefold axis with red triangle symbol.

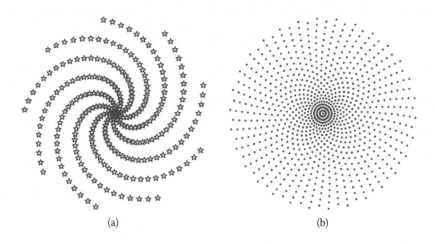

FIGURE 3.14   (a) Figure with ninefold rotation symmetry, (b) figure with 30-fold rotation symmetry.

3.3.4.3 n-Fold Rotation Symmetry Operations

In general, if *n* is an integer, an *n-fold rotation* is a 360°/*n* rotation that is a symmetry operation. There is no restriction on *n*: It could be *n* = 9 or 30, as in Figure 3.14. See Exercise 3.7.

3.4 CRYSTALLOGRAPHIC ROTATIONS

Crystallography is concerned with finding symmetry operations compatible with lattices because these are the only symmetry operations that a *crystal* can have. *The periodicity of a lattice limits the number of compatible rotation operations.* The theorem, proved in this section, is one of the most far-reaching and surprising facts in crystallography. It reduces the study of crystallography from an infinite number of symmetry operations to a finite number.

Figure 3.15a shows part of an infinite periodic pattern that has a fourfold rotation axis. See Jablan (1995, 2002). In Figure 3.15b choose a point coinciding with an axis and indicate the lattice by marking all the other points with identical neighborhoods. Now each

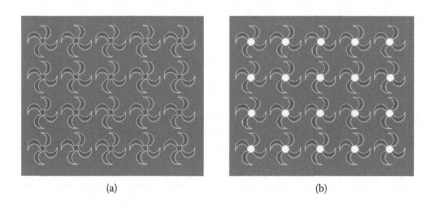

FIGURE 3.15   (a) Neolithic African ornamental art, (b) lattice points coinciding with fourfold rotation axes. (Courtesy of Slavik V. Jabla.)

lattice point contains a fourfold rotation axis. For a fourfold rotation axis, the rotation angle $\theta = 360°/4 = 90°$.

The rotation angle, θ, for an n-fold rotation is $360°/n$.

In order to show that the lattice constrains the types of rotation operations, consider the following lemma.

Lemma

Figure 3.16 shows a general lattice. Consider two vectors **a** and **b** both of which connect lattice points. Then the vector sum of **a** and **b**, that is **c**, is also a vector which connects two lattice points.

Theorem

Only one-, two-, three-, four-, and sixfold rotation operations are consistent with a lattice.

Given: A lattice that contains an n-fold rotation axis where n is an integer.

Proof: Choose the origin of the lattice such that the n-fold rotation axis goes through a lattice point. Then every lattice point contains an n-fold rotation axis.

In Figure 3.17a choose vector **a** to connect any two lattice points, with the restriction that there are no lattice points on **a** except the two end points. Except for these two lattice points in gray, all the rest of the lattice points have been omitted from the diagram.

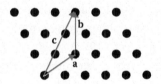

FIGURE 3.16   Sum of two vectors.

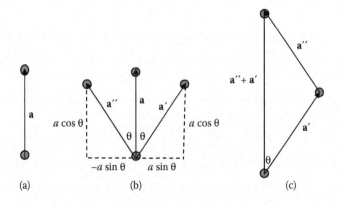

FIGURE 3.17   (a) Vector **a**, (b) vectors **a**, **a**′ and **a**″, (c) addition of **a**′ and **a**″. Except for the gray lattice points of **a** all the rest of the lattice points have been omitted from the diagram.

Since the tail of **a** is on an *n*-fold rotation axis, rotate **a** by θ degrees where $\theta = 360°/n$, as in Figure 3.17b and get vector **a′**. Necessarily **a′** ends in a lattice point. Let vector **a′** be decomposed into two perpendicular components, $a \cos \theta$ and $a \sin \theta$. Next, rotate **a** by $-\theta$ degrees and get vector **a″**, which also ends in a lattice point. Let **a″** be decomposed into two perpendicular components: $a \cos \theta$ and $-a \sin \theta$. From the construction **a**, **a′**, and **a″** all have the same length.

Figure 3.17c shows the sum of **a′** and **a″** which is **a′** + **a″**. Vector **a″** has been translated parallel to itself. From the lemma, this new vector, **a′** + **a″**, also connects two lattice points. Since the horizontal components of **a′** and **a″** are equal and in opposite directions, then the sum of **a′** and **a″** must be parallel to the original vector **a** or equal to 0. The sum of **a′** and **a″** must also be an integral multiple of the original **a**.

Thus,

$$\mathbf{a′} + \mathbf{a″} = m\,\mathbf{a} \quad \text{where } m \text{ is an integer.}$$

Along the vertical component $2a \cos \theta = m\,a$. Since $a \neq 0$, $2 \cos \theta = m \cos \theta = m/2$. Since $-1 \leq \cos \theta \leq 1$ then $-2 \leq m \leq 2$.

There are only five integers that satisfy the conditions for *m*, that is $m = 0, 1, 2, -1, -2$. Table 3.5 shows the allowed rotations—0°, 60°, 90°, 120°, and 180°. A 0° rotation is equivalent to a 360° rotation.

Conclusion: Only one-, two-, three-, four-, and sixfold rotation operations are consistent with a lattice. ■

For example, in the case where θ equals 60° for a sixfold rotation, $\cos 60° = 1/2$ and the sum of **a′** and **a″** is **a**, as in Figure 3.18a. On the other hand, for a fivefold rotation, θ equals 72°, $\cos 72° = 0.3090$. Here **a′** + **a″** = 0.6180 **a**. See Exercises 3.8 and 3.9.

Another way of looking at this is that hexagons can tile space, but pentagons cannot tile space without leaving gaps (see Figure 3.19). A single regular hexagon can be surrounded by six hexagons and the figure fills space. The interior angle of a hexagon is 120° and three times 120° is 360°. On the contrary, the interior angle of a regular pentagon is 108°. No more than three pentagons can be packed around a single point and three times 108° is 324°. Thus in packing pentagons about a single point there is a gap of 360°–324° = 36°.

Crystallographic rotations are those rotations consistent with a lattice. They are limited to five cases: 1-fold rotations, two-, three-, four-, and sixfold rotations. This restriction also holds in three dimensions.

TABLE 3.5   Crystallographic Rotations

m	$m/2 = \cos \theta$	$\theta(°)$	$n = 360°/\theta$
0	0	90	Fourfold
1	1/2	60	Sixfold
2	1	0 = 360	Identity (onefold)
−1	−1/2	120	Threefold
−2	−1	180	Twofold

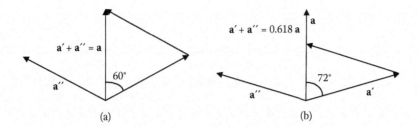

FIGURE 3.18 (a) Sixfold rotation, (b) fivefold rotation.

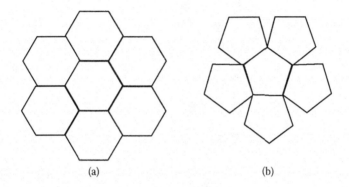

FIGURE 3.19 Packing of (a) hexagons and (b) pentagons.

3.5 SUMMARY OF THE TWO-DIMENSIONAL CRYSTALLOGRAPHIC OPERATIONS

Table 3.6 summarizes the information for the two-dimensional crystallographic operations—identity, twofold rotation, mirror, fourfold rotation, threefold rotation, and sixfold rotation. The angle between the basis vectors **a** and **b** is γ. The order of the symmetry operations in this table follows the order in the *International Tables for Crystallography*.

For both the identity and the twofold rotation axis, γ is unrestricted. The lengths of the basis vectors are also unrestricted. The identity is indicated by the integer 1, and the twofold rotation is indicated by the integer 2. See Sections 3.3.3 and 3.3.4.1.

The mirror restricts the angle γ to be 90° but leaves the lengths of basis vectors **a** and **b** unrestricted. The mirror is the symmetry operation of reflection. See Section 3.3.2.

The fourfold rotation restricts the angle γ to be 90° and the lengths of basis vectors **a** and **b** to be equal. A *fourfold rotation* is a 90° rotation that is a symmetry operation. See Table 3.6.

The threefold rotation restricts the angle γ to be 120° and the lengths of basis vectors **a** and **b** to be equal. A dashed line in the $[\bar{1}\,\bar{1}]$ direction has been inserted to clarify the coordinate transformation. A threefold rotation is a 120° rotation that is a symmetry operation. See Section 3.3.4.2.

The sixfold rotation restricts the angle γ to be 120° and the lengths of basis vectors **a** and **b** to be equal. Dashed lines in the $[\bar{1}\,\bar{1}]$, $[\bar{1}\,0]$, and $[0\,\bar{1}]$ directions are inserted to clarify the

TABLE 3.6   Summary of Some Symmetry Operations in Two Dimensions

Symmetry Operation	Assignment of Origin and Basis to Construct Coordinate System	Coordinate Transformations	Associated Symmetry Matrix
1, Identity		$1,0 \rightarrow 1,0 \quad 0,1 \rightarrow 0,1$ $x,0 \rightarrow x,0 \quad 0,y \rightarrow 0,y$ $x,y \rightarrow x,y$	$\begin{pmatrix} 1 & 0 \\ 0 & 1 \end{pmatrix}$
2, Twofold rotation with its axis at the origin		$1,0 \rightarrow \bar{1},0 \quad 0,1 \rightarrow 0,\bar{1}$ $x,0 \rightarrow \bar{x},0 \quad 0,y \rightarrow 0,\bar{y}$ $x,y \rightarrow \bar{x},\bar{y}$	$\begin{pmatrix} \bar{1} & 0 \\ 0 & \bar{1} \end{pmatrix}$
m, Mirror with **b** in its plane.		$1,0 \rightarrow \bar{1},0 \quad 0,1 \rightarrow 0,1$ $x,0 \rightarrow \bar{x},0 \quad 0,y \rightarrow 0,y$ $x,y \rightarrow \bar{x},y$	$\begin{pmatrix} \bar{1} & 0 \\ 0 & 1 \end{pmatrix}$
4^+, Fourfold rotation with its axis at the origin		$1,0 \rightarrow 0,1 \quad 0,1 \rightarrow \bar{1},0$ $x,0 \rightarrow 0,x \quad 0,y \rightarrow \bar{y},0$ $x,y \rightarrow \bar{y},x$	$\begin{pmatrix} 0 & \bar{1} \\ 1 & 0 \end{pmatrix}$
3^+, Threefold rotation with its axis at the origin		$1,0 \rightarrow 0,1 \quad 0,1 \rightarrow \bar{1},\bar{1}$ $x,0 \rightarrow 0,x \quad 0,y \rightarrow \bar{y},\bar{y}$ $x,y \rightarrow \bar{y}, x+\bar{y}$	$\begin{pmatrix} 0 & \bar{1} \\ 1 & \bar{1} \end{pmatrix}$
6^+, Sixfold rotation with its axis at the origin		$1,0 \rightarrow 1,1 \quad 0,1 \rightarrow \bar{1},0$ $x,0 \rightarrow x,x \quad 0,y \rightarrow \bar{y},0$ $x,y \rightarrow x+\bar{y}, x$	$\begin{pmatrix} 1 & \bar{1} \\ 1 & 0 \end{pmatrix}$

coordinate transformation. A *sixfold rotation* is a 60° rotation that is a symmetry operation. See Table 3.6 and Exercise 3.10.

3.6 TWO-DIMENSIONAL CRYSTALLOGRAPHIC POINT GROUPS

Many physical properties of crystals are associated with their point groups. Some of these properties are pyroelectricity, piezoelectricity, ferroelectricity, external morphology or shape of the crystal, optical activity, etch figures, and indices of refraction. Other examples of the use of point groups are the descriptions of the symmetries of a molecule and the symmetries of the orbitals forming covalent bonds.

The point groups are a subset of the general groups defined in Section 3.2. A *point group* is a group of symmetry operations that leave at least one point of an object or pattern unmoved. In two dimensions the elements of the point group include the identity and may include one or more mirrors and/or one or more *n*-fold rotations. The symmetry operations are represented by matrices, as in Table 3.6. The operation between the group elements is called *multiplication*. If S_1 is a symmetry operation and S_2 is another symmetry operation, then their product, $S_2 \times S_1$, is the symmetry operation consisting of first applying S_1 to the object and then applying S_2 to the object. A point that does not move under a symmetry operation is an *invariant* point. The identity holds all points invariant. The mirror holds all the points along the mirror plane invariant and the rotations hold only one point invariant. Translations are excluded from point groups.

It is often useful to look at the matrices associated with symmetry operations. These matrices form a group with the same multiplication table as the point group.

Point groups are divided into crystallographic point groups and noncrystallographic point groups. The *crystallographic point groups* are point groups consistent with lattices. The rotations in these point groups are limited to the crystallographic rotations. The number of crystallographic point groups is finite. In two dimensions there are 10 point groups. For example, etch figures produced on the crystal faces by chemical attack give a figure on the crystal face whose symmetry is one of the 10 two-dimensional point groups.

No rotational restrictions are placed on the *noncrystallographic point groups*, which are point groups that are not crystallographic point groups. There are an infinite number of noncrystallographic point groups. An *n*-fold rotation can have $n = 19$, 72, or any integer. Figure 3.20 shows two examples of objects, namely a pentagon and an octagon, with non-crystallographic point groups. The pentagon has a fivefold rotation and the octagon has an eightfold rotation. A circle has an ∞-fold rotation and therefore has an infinite number of symmetry operations.

The 10 two-dimensional point groups are now discussed.

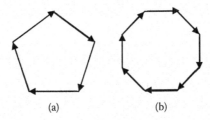

(a)   (b)

FIGURE 3.20 Examples of objects with non-crystallographic symmetry point groups. (a) Pentagon, (b) octagon.

TABLE 3.7   Multiplication
Table for Point Group 1

×	1
1	1

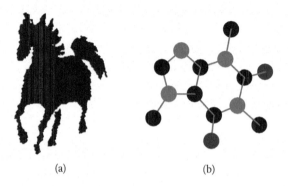

| (a) | (b) |

FIGURE 3.21   Motifs with the symmetry of point group 1. (a) Horse, (b) molecule of caffeine.

3.6.1 Point Group 1

The identity point group 1 has just one element, 1, for which the matrix

$$1 = \begin{pmatrix} 1 & 0 \\ 0 & 1 \end{pmatrix}.$$

This one element forms a group. Table 3.7 is the multiplication table for point group 1.

Since there is only one element, there is closure and the associative law is trivial. The identity element is 1 and the inverse of 1 is 1. Furthermore, the identity point group is a subgroup of every other point group. Figure 3.21a, a horse, and Figure 3.21b, a molecule of caffeine, are examples of motifs whose symmetry is point group 1.

The *order* of a group is the number of elements contained in the group. The identity group, 1, contains only one element, 1, and hence has order 1. See Exercise 3.11.

3.6.2 Point Group 2

The point group 2 starts out with one element, the twofold rotation, 2, with, as in Table 3.6, the matrix

$$2 = \begin{pmatrix} \bar{1} & 0 \\ 0 & \bar{1} \end{pmatrix}.$$

The multiplication is

$$2 \times 2 = \begin{pmatrix} \bar{1} & 0 \\ 0 & \bar{1} \end{pmatrix}\begin{pmatrix} \bar{1} & 0 \\ 0 & \bar{1} \end{pmatrix} = \begin{pmatrix} 1 & 0 \\ 0 & 1 \end{pmatrix} = 1.$$

TABLE 3.8   Multiplication Table for Point Group 2

×	1	2
1	1	2
2	2	1

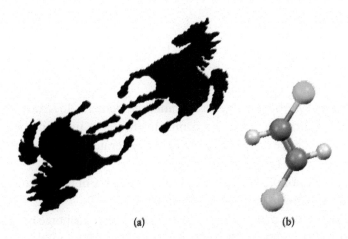

(a)   (b)

FIGURE 3.22   Motifs with the symmetry of point group 2. (a) Horses, (b) $C_2H_2Cl_2$.

The multiplication of 2 by itself generates another element, 1. Now there are two elements in the multiplication table. To complete the table using matrix multiplication, $1 \times 2 = 2 \times 1 = 2$. Since no more elements are generated, closure is satisfied. Table 3.8 is the multiplication table for point group 2.

Figure 3.22 has two examples of motifs with the symmetry of point group 2. The figure of the two horses has a twofold rotation axis located between them. The planar molecule $C_2H_2Cl_2$, contains a twofold axis between the two carbon atoms.

The determinant of the matrix 2 is +1. Note that the horse in Figure 3.22 is rotated, but the handedness of the horse is not changed: The left front hoof of each horse is touching the ground. This type of operation is called a proper operation. A *proper operation* has a matrix with a determinant of +1; the operation retains the handedness of the object. All of the rotation operations are proper.

Point group 2 has order 2 and has two symmetry operations, 1 and 2.

3.6.3 Point Group *m*

Point group *m* also starts out with one element, the mirror, *m*. A subscript is added to describe the specific mirror plane. In Table 3.6 the mirror with **b** in its plane is called m_y or *m* 0,*y*. The matrix is

$$m_y = \begin{pmatrix} \bar{1} & 0 \\ 0 & 1 \end{pmatrix}$$

TABLE 3.9   Multiplication Table for
Point Group m

×	1	m_y
1	1	m_y
m_y	m_y	1

The multiplication is

$$m_y \times m_y = \begin{pmatrix} \bar{1} & 0 \\ 0 & \bar{1} \end{pmatrix}\begin{pmatrix} \bar{1} & 0 \\ 0 & \bar{1} \end{pmatrix} = \begin{pmatrix} 1 & 0 \\ 0 & 1 \end{pmatrix} = 1.$$

The multiplication of m_y by itself generates another element, 1. Again there are two elements in the multiplication table. To complete the table using matrix multiplication, $1 \times m_y = m_y \times 1 = m_y$. Since no more elements are generated, closure is satisfied. Table 3.9 is the multiplication table for point group m.

Figure 3.23a is an example of a motif with the symmetry of point group m. The figure is two horses with a mirror plane, m_y, inserted at 0,y.

The determinant of the matrix m_y is −1. In Figure 3.23a, the horse on the top has its left front foot down and the horse on the bottom has its right front foot down. In Figure 3.23b, the planar molecule urea, $CO(NH_2)_2$, contains a mirror slicing through the oxygen (red) atom and the carbon (gray) atom. This type of operation is called an improper operation. An *improper operation* has a matrix with a determinant of −1 and the operation causes the handedness to change. The mirror operation is an improper operation.

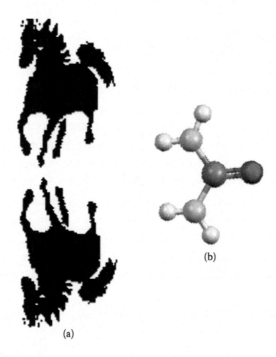

(b)

(a)

FIGURE 3.23   Motifs with the symmetry of point group m. (a) Horses, (b) urea. (CSD-AMILEN.)

Point group m has order 2 and has two symmetry operations, 1 and m_y.

Compare point group 2 and point group m. First look at their multiplication tables. The tables are identical if m_y and 2 are interchanged. These are examples of isomorphic groups. *Isomorphic groups* have the same multiplication table. It follows that isomorphic groups have the same order. In this case point group m has order 2. However, two groups can have the same order and not be isomorphic, as is shown later in the chapter.

3.6.4 Point Group 3

Point group 3 starts out with one element, 3^+, where, in Table 3.6, the matrix

$$3^+ = \begin{pmatrix} 0 & \bar{1} \\ 1 & \bar{1} \end{pmatrix}$$

The multiplication is

$$3^+ \times 3^+ = \begin{pmatrix} 0 & \bar{1} \\ 1 & \bar{1} \end{pmatrix}\begin{pmatrix} 0 & \bar{1} \\ 1 & \bar{1} \end{pmatrix} = \begin{pmatrix} \bar{1} & 1 \\ \bar{1} & 0 \end{pmatrix}.$$

The multiplication of 3^+ by itself generates a new element. This new element is called 3^-, as in *Example 3.2*. Since the identity has not been generated, multiply 3^+ by 3^-. Thus,

$$3^+ \times 3^- = \begin{pmatrix} 0 & \bar{1} \\ 1 & \bar{1} \end{pmatrix}\begin{pmatrix} \bar{1} & 1 \\ \bar{1} & 0 \end{pmatrix} = 1$$

Now there are three elements in the multiplication table. To complete the table using matrix multiplication, $1 \times 3^- = 3^- \times 1 = 3^-$ and $1 \times 3^+ = 3^+ \times 1 = 3^+$. Also $3^+ \times 3^- = 3^- \times 3^+ = 1$. Since no more elements are generated, closure is satisfied. Table 3.10 is the multiplication table for point group 3.

The multiplication table is symmetric. A general element e_{ij} is in the ith row and the jth column. For example e_{12} is 3^+ in Table 3.10. If $e_{ij} = e_{ji}$, the group is commutative. Point group 3 is a commutative group. A *commutative group* has a symmetric multiplication table. The order in which the operation is performed does not matter. Here, for example, $3^+ \times 3^- = 3^- \times 3^+ = 1$.

Figure 3.24 shows motifs with the symmetry of point group 3, an abstract figure and a planar molecule, $B(OH)_3$. Point group 3 has order 3 and has three symmetry

TABLE 3.10   Multiplication Table for Point Group 3

×	1	3^+	3^-
1	1	3^+	3^-
3^+	3^+	3^-	1
3^-	3^-	1	3^+

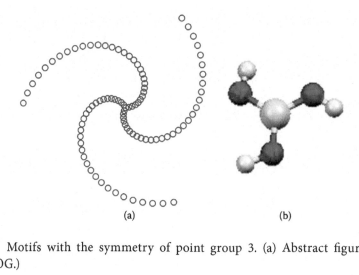

FIGURE 3.24 Motifs with the symmetry of point group 3. (a) Abstract figure, (b) $B(OH)_3$. (CSD-HABFOG.)

operations, 1, 3^+, and 3^-. Point group 3 is not isomorphic with point groups 2 or m. See Exercise 3.12.

⌨ A Starter Program is given that multiplies the matrices to form the multiplication table.

3.6.5 Point Group 4

Point group 4 starts out with one element, 4^+, where, in Table 3.6, the matrix

$$4^+ = \begin{pmatrix} 0 & \bar{1} \\ 1 & 0 \end{pmatrix}$$

The multiplication is

$$4^+ \times 4^+ = \begin{pmatrix} 0 & \bar{1} \\ 1 & 0 \end{pmatrix}\begin{pmatrix} 0 & \bar{1} \\ 1 & 0 \end{pmatrix} = \begin{pmatrix} \bar{1} & 0 \\ 0 & \bar{1} \end{pmatrix} = 2$$

The multiplication of 4^+ by itself generates an element from point group 2. The information learned about point group 2 can be applied to point group 4. Since the identity has not been generated, multiply 4^+ by 2. Thus,

$$4^+ \times 2 = \begin{pmatrix} 0 & \bar{1} \\ 1 & 0 \end{pmatrix}\begin{pmatrix} \bar{1} & 0 \\ 0 & \bar{1} \end{pmatrix} = \begin{pmatrix} 0 & 1 \\ \bar{1} & 0 \end{pmatrix}$$

and the new element is called 4^-. Since the identity has not been generated, multiply $4^+ \times 4^-$ which equals

TABLE 3.11 Multiplication Table for Point Group 4

×	1	2	4^+	4^-
1	1	2	4^+	4^-
2	2	1	4^-	4^+
4^+	4^+	4^-	2	1
4^-	4^-	4^+	1	2

$$4^+ \times 4^- = \begin{pmatrix} 0 & \bar{1} \\ 1 & 0 \end{pmatrix} \begin{pmatrix} 0 & 1 \\ \bar{1} & 0 \end{pmatrix} = 1$$

Now there are four elements in the multiplication table. To complete the table, use matrix multiplication. Since no more elements are generated, closure is satisfied. Table 3.11 is the multiplication table for point group 4. The table is symmetric; so the point group is commutative.

Point group 2 (in red) is completely contained in point group 4 and is thus a subgroup of point group 4. A *subgroup* is a group wholly contained in a larger group. The point group 1 is also a subgroup of point group 4. Point group 1 is a subgroup of every point group. A proper subgroup of a group has an order less than the order of the group. Point group 2 is a proper subgroup of point group 4.

The associative law follows but is not discussed here.

Generators are a set of elements of a group such that every group element is obtained as a product of these generators. Generators capture the essence of a group. Here 4^+ is the generator of point group 4 because all the elements in the group can be generated from 4^+. Thus $2 = 4^+ \times 4^+$; $4^- = 2 \times 4^+ = (4^+)^3$; $1 = (4^+)^4$. The *International Tables for Crystallography* lists 1, 2, and 4^+ as the generators for group 4. These generators were chosen to emphasize the subgroups.

Point group 4 has order four and has four symmetry operations, 1, 2, 4^+, and 4^-. Figure 3.25 shows an example of an abstract motif with the symmetry of point group 4 and a projection of a clathrate $(C_6D_6)_{2n}$, $n(C_4H_6N_6Ni_2)$, catena-(bis(Hexadeuterobenzene) tetrakis(m2-cyano)-tetrakis(ammonio)-di-nickel clathrate), also with the same symmetry. A *clathrate* is a structure in which one molecule traps and contains a second molecule. See Exercise 3.13.

3.6.6 Point Group 6

Point group 6 starts out with one element, 6^+ where

$$6^+ = \begin{pmatrix} 1 & \bar{1} \\ 1 & 0 \end{pmatrix}$$

Point group 6 can be generated from 6^+. However, the *International Tables for Crystallography* chooses to list the generators as 3^+, 2, and 1. The order of point group 6 is six; the six symmetry operations are 1, 3^+, 3^-, 2, 6^-, and 6^+. Figure 3.26 shows motifs with the

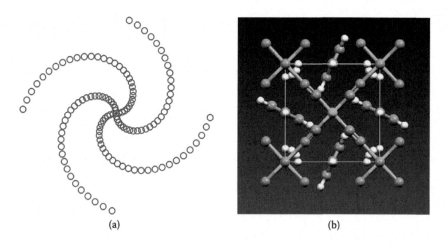

FIGURE 3.25   Motifs with the symmetry of point group 4. (a) Abstract, (b) projection of clathrate, $(C_6D_6)_{2n}$, $n(C_4H_6N_6N_{12})$. (CSD-HEMJIS.)

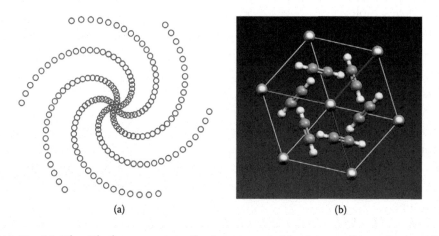

FIGURE 3.26   Motifs with the symmetry of point group 6. (a) Abstract, (b) projection of several acetylene molecules, C_2H_2, along [111]. (CSD-ACETYL03.)

symmetry of point group 6, an abstract figure and a projection of several acetylene molecules, C_2H_2, along [111], from Chapter 15. See Exercises 3.14 and 3.15.

3.6.7 Point Group 2mm

The previous six point groups, 1, 2, 3, 4, 6, and *m*, are cyclic groups. A *cyclic group* needs only one element to generate the entire group. In the next point groups, the mirror is combined with rotations. In each case two *independent* group elements are needed to generate the point group. The *International Tables for Crystallography* lists 1, 2, and *m* as the generators for *2mm*. This is not the minimal set, but gives an indication of the subgroups of the group. Point group *2mm* combines the symmetry operations of point group 2 with a mirror symmetry operation. Choose the mirror from Table 3.6.

$$\text{matrix } m_y = \begin{pmatrix} \bar{1} & 0 \\ 0 & 1 \end{pmatrix} \quad \text{and} \quad \text{matrix } 2 = \begin{pmatrix} \bar{1} & 0 \\ 0 & \bar{1} \end{pmatrix}.$$

The multiplication is

$$m_y \times 2 = \begin{pmatrix} \bar{1} & 0 \\ 0 & 1 \end{pmatrix}\begin{pmatrix} \bar{1} & 0 \\ 0 & \bar{1} \end{pmatrix} = \begin{pmatrix} 1 & 0 \\ 0 & \bar{1} \end{pmatrix}.$$

A new element is generated. This new symmetry operation takes x,y into $x,\bar{y}$ which is a mirror along $x,0$. The determinant of the matrix

$$\det\begin{pmatrix} 1 & 0 \\ 0 & \bar{1} \end{pmatrix} = -1,$$

indicates an improper operation; that is, the handedness has changed. The new symmetry operation is m_x. Table 3.12 is the multiplication table for point group $2mm$.

The molecule of pentacene (see Collins (2004)), considered here as a two-dimensional object, in Figure 3.27, has the symmetry of point group $2mm$. The two mirrors (in purple) intersect at the invariant point of the twofold rotation (in red).

The symbols for the point groups are chosen to be short descriptions of the point groups.

Point group $2mm$ has order four and has four symmetry operations: 1, 2, m_x, and m_y. In the *International Tables for Crystallography*, the three generators for $2mm$ are 1, then 2 at 0,0, and finally m at $0,y$. See Exercises 3.16 and 3.17.

TABLE 3.12 Multiplication Table for Point Group $2mm$

×	1	2	m_x	m_y
1	1	2	m_x	m_y
2	2	1	m_y	m_x
m_x	m_x	m_y	1	2
m_y	m_y	m_x	2	1

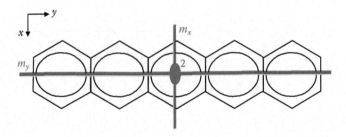

FIGURE 3.27 Molecule of pentacene with the symmetry of the 2-dimensional point group $2mm$.

3.6.8 Point Group 3m

Point group $3m$ combines the symmetry operations of point group 3 with the mirror symmetry operation, m. Given one mirror plane, the threefold rotation generates two more. Figure 3.28a shows a motif with the symmetry of point group $3m$. Figure 3.28b shows the motif with the symbol of the triangle for the threefold rotation and the symbols for the three mirror planes superimposed on the figure. Figure 3.28c shows another motif with this symmetry, the planar molecule, $C_6H_6N_6O_6$, 1,3,5-triamino-2,4,6-trinitrobenzene. See Exercise 3.18.

The *International Tables for Crystallography* give two ways to describe $3m$. In the first case the basis vectors **a** and **b** are chosen so that the mirror plane is along the crystallographic direction [11] (see Figure 3.29a). The matrix for the mirror in orthogonal coordinates, as in Table 3.6, differs from the matrix in nonorthogonal coordinates. To derive the algebraic expression, use the method of Table 3.6. The mirror operation with the mirror plane along [11] is called m_{xx}. For that mirror

$$1,0 \rightarrow 0,1 \quad \text{and} \quad x,0 \rightarrow 0,x$$

and

$$0,1 \rightarrow 1,0 \quad \text{and} \quad 0,y \rightarrow y,0$$

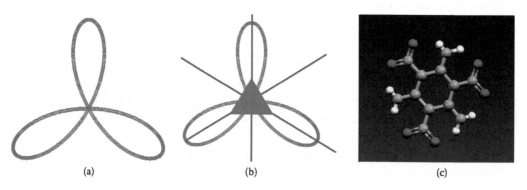

FIGURE 3.28 Motifs with point group $3m$. (a) Figure with equation $r = \cos 3\theta$ in polar coordinates, (b) purple mirrors and red threefold rotation axis, (c) $C_6H_6N_6O_6$, 1,3,5-triamino-2,4,6-trinitrobenzene. (CSD-TATNBZ.)

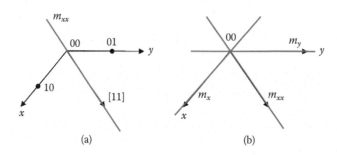

FIGURE 3.29 (a) Mirror m_{xx}, (b) three mirrors, m_{xx}, m_x, and m_y generated by threefold rotations.

and in combination

$$x,y \rightarrow y, x$$

In matrix form

$$m_{xx} = \begin{pmatrix} 0 & 1 \\ 1 & 0 \end{pmatrix}$$

The threefold rotation generates two more mirrors, m_x and m_y, as in Figure 3.29b. The mirrors are at 120° to one another. The matrices for those two mirrors can be calculated from first principles as above or generated from m_{xx}.

Using the latter method,

$$m_{xx} \times 3^+ = \begin{pmatrix} 0 & 1 \\ 1 & 0 \end{pmatrix} \begin{pmatrix} 0 & \bar{1} \\ 1 & \bar{1} \end{pmatrix} = \begin{pmatrix} 1 & \bar{1} \\ 0 & \bar{1} \end{pmatrix} = m_x.$$

Note that the m_x matrix has a determinant of −1, and is thus an improper symmetry operation. Mirror m_x is obtained as a counterclockwise rotation of m_{xx} by 120°. Here m_x takes $x, y \rightarrow x + \bar{y}, \bar{y}$ in agreement with the *International Tables for Crystallography*. See Exercise 3.19.

Complete the multiplication table using matrix multiplication. The element of the multiplication table in row i and column j is $e_{ij} = S_i \times S_j$ where S_i is the ith operation in the left-hand column and S_j is the jth operation in the top row. Use information from point group 3. Table 3.13 is the multiplication table for point group $3m$. See Exercise 3.20.

The multiplication table is not symmetric. That is, element $e_{ij} \neq e_{ji}$. The point group $3m$ is a noncommutative group. A *noncommutative group* does not have a symmetric multiplication table. The order in which the operations are performed matters. Here $m_{xx} \times 3^+ = m_x$, which is different from $3^+ \times m_{xx} = m_y$. In Figure 3.30a consider the operation $m_{xx} \times 3^+(B)$ where B is the blue ball. This expression means first apply a threefold rotation of 120° counter-clockwise to get to the yellow ball. Now apply the mirror m_{xx} to get to the green ball. Alternatively the green ball can be reached directly from the blue ball by m_x. In Figure 3.30b consider the operation of $3^+ \times m_{xx}(B)$ where B is the blue ball. This expression means first apply the m_{xx} to get to the yellow ball. Now apply a threefold rotation of 120° counter-clockwise to get to the green ball. Alternatively, the green ball can be reached directly from the blue ball by applying m_y. The position of the green ball is different depending on the order of the operations.

TABLE 3.13 Multiplication Table for Noncommutative Point Group 3m

×	1	3$^+$	3$^-$	m_{xx}	m_x	m_y
1	1	3$^+$	3$^-$	m_{xx}	m_x	m_y
3$^+$	3$^+$	3$^-$	1	m_y	m_{xx}	m_x
3$^-$	3$^-$	1	3$^+$	m_x	m_y	m_{xx}
m_{xx}	m_{xx}	m_x	m_y	1	3$^+$	3$^-$
m_x	m_x	m_y	m_{xx}	3$^-$	1	3$^+$
m_y	m_y	m_{xx}	m_x	3$^+$	3$^-$	1

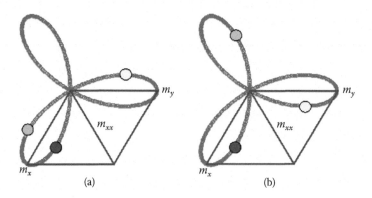

(a)   (b)

FIGURE 3.30   Illustrates (a) $m_{xx} \times 3^+ = m_x$, (b) $3^+ \times m_{xx} = m_y$.

No two mirrors are perpendicular to each other. Point group $3m$ has order six and has six symmetry operations, 1, 3^+, 3^-, m_{xx}, m_x, and m_y.

There are two versions of $3m$ because there are two ways to insert the mirror planes into the unit cell, which is used as a reference for the algebraic expressions. The version above with symmetry operations 1, 3^+, 3^-, m_{xx}, m_x, and m_y is called $31m$ and has a mirror along the short diagonal [11]. The second version is called $3m1$ and has a mirror along $[1\,\overline{1}]$. The 3^+ symmetry operation generates two more mirror planes along [21] and [12] (see Figure 3.31). The mirrors are labeled $m_{x\overline{x}}$, m_{x2x}, and m_{2xx}. The matrix for

$$m_{x\overline{x}} = \begin{pmatrix} 0 & \overline{1} \\ \overline{1} & 0 \end{pmatrix}, \quad \text{for } m_{x2x} = \begin{pmatrix} \overline{1} & 1 \\ 0 & 1 \end{pmatrix}, \quad \text{and for } m_{x2x} = \begin{pmatrix} 1 & 0 \\ 1 & \overline{1} \end{pmatrix}$$

See Exercise 3.21.

The point group $3m$ can be represented either by $31m$ or $3m1$, which are related by a rotation of $30°$ and are the same point group although the algebraic form of their matrices is different.

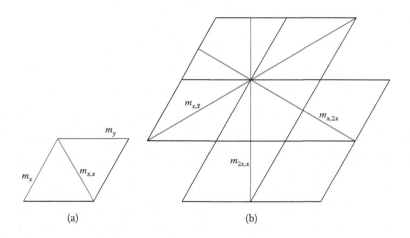

(a)   (b)

FIGURE 3.31   (a) Mirrors in $31m$, (b) mirrors in $3m1$.

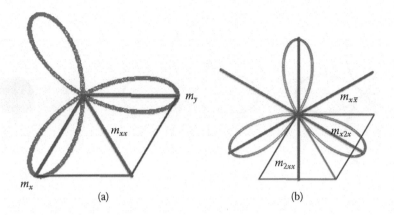

FIGURE 3.32 Motif in (a) $31m$ and (b) $3m1$.

Figure 3.32 illustrates putting the cloverleaf motif in two algebraic forms. The first motif in (a) is rotated by 30° in (b). See Exercise 3.22. The rotation has no effect on the assignment of the point group. These two descriptions of point group $3m$ will be useful in Section 3.6.10.

3.6.9 Point Group 4mm

Point group $4mm$ combines the symmetry operations of point group 4 with the mirror symmetry operation. From Table 3.6 choose

$$m_y = \begin{pmatrix} \bar{1} & 0 \\ 0 & 1 \end{pmatrix}.$$

The description of the mirror with orthogonal coordinates is used. The multiplication

$$m_y \times 4^+ = \begin{pmatrix} \bar{1} & 0 \\ 0 & 1 \end{pmatrix}\begin{pmatrix} 0 & \bar{1} \\ 1 & 0 \end{pmatrix} = \begin{pmatrix} 0 & 1 \\ 1 & 0 \end{pmatrix} = m_{xx}$$

which is the mirror along [11], for it takes x,y into y,x. Point group $4mm$ has a second diagonal mirror, along $[\bar{1}\,\bar{1}]$, that takes x,y into $\bar{y}, \bar{x}$ called $m_{x\bar{x}}$. See Exercises 3.23 through 3.25.

Figure 3.33 shows motifs with point group $4mm$, an abstract motif and the planar molecule xenon tetrafluoride (Ibers and Hamilton, 1963). Note that there are two pairs of perpendicular mirrors. Point group $4mm$ has order eight and has eight symmetry operations, 1, 2, 4^+, 4^-, m_{xx}, $m_{x\bar{x}}$, m_x, and m_y.

3.6.10 Point Group 6mm

Point group $6mm$ combines the symmetry operations of point group 6 with the mirror symmetry operation, m. In agreement with the *International Tables for Crystallography* choose

$$m_{x\bar{x}} = \begin{pmatrix} 0 & \bar{1} \\ \bar{1} & 0 \end{pmatrix}.$$

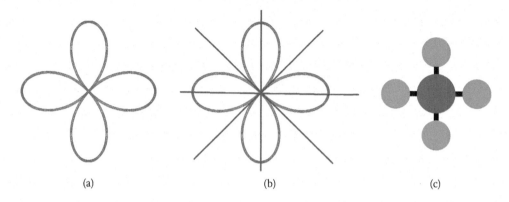

FIGURE 3.33 Motif with point group symmetry 4*mm*. (a) This figure has the equation $r = \cos 2\theta$, (b) mirror planes in purple, the symbol for the fourfold rotation axis is omitted for clarity, (c) xenon tetrafluoride.

FIGURE 3.34 Molecule of benzene, C_6H_6, with point group symmetry 6*mm*.

The notation developed for the mirrors for both 3*m*1 and 31*m* is used. See Exercise 3.26.

Figure 3.34 is an example of a molecule of benzene, C_6H_6 (considered in two dimensions) with point group 6*mm*. A perfect snowflake has point group 6*mm*. Point group 6*mm* has order 12 and has 12 symmetry operations, 1, 3^+, 3^-, 2, 6^-, 6^+, $m_{x\bar{x}}$, m_{x2x}, m_{2xx}, m_{xx}, m_x, and m_y.

3.6.11 Identification of a Crystallographic Point Group

Figure 3.35 shows how to identify the crystallographic point group of a pattern. When examining a pattern, start by looking for the highest possible rotation axis, which is 6. If no sixfold axis is present, then look for a fourfold axis. Keep going following the flow chart until a final identification is made. This systematic approach to the analysis of patterns will be utilized later with more difficult patterns.

3.7 TWO-DIMENSIONAL CRYSTAL SYSTEMS

The operations of the six cyclic two-dimensional point groups—1, 2, 3, 4, 6, and *m*—combine to form 10 point groups, which are divided into four crystal systems. The *crystal systems* classify the point groups as oblique, rectangular, square, or hexagonal (see Table 3.14).

In this section each of the crystal systems is discussed. For each system, each point group of that system is illustrated by two stereographic projections. The first projection, the *point stereographic projection*, indicates what happens to a point under the symmetry operations of the point group. The coordinate axes are indicated by thin solid lines. The *y*-axis runs horizontally from left to right and the *x*-axis points downward.

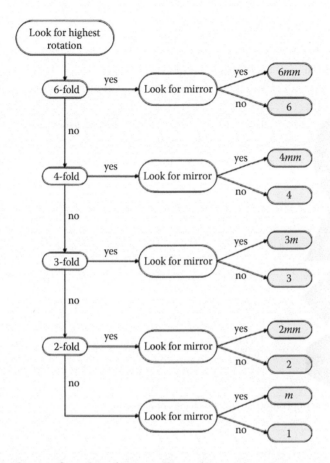

FIGURE 3.35   Identification flow chart for crystallographic point groups.

TABLE 3.14   Two-Dimensional Crystals Systems and Their Point Groups

Crystal System	Point Group
Oblique	1, 2
Rectangular	*m*, 2*mm*
Square	4, 4*mm*
Hexagonal	3, 3*m*, 6, 6*mm*

The dots have two interpretations:

- Miller indices (*hk*) of the equivalent general crystal faces
- General points or centers of atoms *x, y*.

For two dimensions the dot is placed *on* the circumference of the circle, by convention. The matrix representations of the symmetry operations can be obtained from these Miller indices. Equivalent crystallographic planes are related by symmetry operations. It is convenient to include the crystallographic planes in this discussion because they are included in the *International Tables for Crystallography* in the discussion of point groups.

TABLE 3.15   Graphical Symbols for Two Dimensions

	Rotation Axes			Plane	
Printed Symbol	Graphical Symbol	Symmetry Operations	Printed Symbol	Graphical Symbol	Symmetry Operations
1	None	1			
2	(ellipse)	1, 2	m	(bar)	1, m
3	(triangle)	1, 3^+, 3^-			
4	(square)	1, 2, 4^+, 4^-			
6	(hexagon)	1, 2, 3^+, 3^-, 6^-, 6^+			

Note: Proper symmetry operations in red: no change of handedness. Improper symmetry operation in blue: change of handedness.

The second projection, the *symbol stereographic projection*, contains a graphical symbol for the point group. See Exercise 3.27.

Table 3.15 shows the graphical symbols for two dimensions. A *graphical symbol* is a combination of the location of a geometrical object—a point, a line, a plane, or a line with a point—and its related set of symmetry operations. The color red indicates a proper operation, one that does not change handedness. The graphical symbols for the rotation axes are colored red because all their operations are proper. The color blue indicates an improper operation, one that does change handedness. The graphical symbol for a mirror plane is colored purple because half its operations are proper (red), and half its operations are improper (blue).

3.7.1 Oblique Crystal System

The oblique crystal system contains two point groups, 1 and 2 (see Table 3.16). The angle between the x and y axes is unrestrained and is shown here as nonorthogonal. The number of points in the point stereographic projection indicates the order, which is the number of elements in the point group. In point group 1, there is only one point. The (hk) plane is transformed into itself by the identity operation. There is no graphical symbol for point group 1.

In point group 2, the twofold rotation takes the (hk) plane into the ($\overline{hk}$) plane. Specifically the twofold rotation takes the (11) plane into the ($\overline{1}\,\overline{1}$) plane. Thus (11) and

TABLE 3.16   Oblique Crystal System

Point Group	Point Stereographic Projection	Symbol Stereographic Projection	Equivalent General Crystal Faces	Symmetry Matrix Generator
1			(hk)	$\begin{pmatrix} 1 & 0 \\ 0 & 1 \end{pmatrix}$
2			(hk) $(\bar{h}\bar{k})$	$\begin{pmatrix} \bar{1} & 0 \\ 0 & \bar{1} \end{pmatrix}$

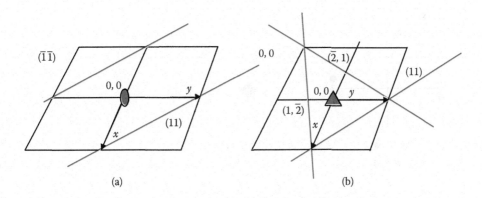

FIGURE 3.36   (a) Equivalent planes in point group 2, (b) equivalent planes in point group 3.

$(\bar{1}\bar{1})$ are equivalent planes (see Figure 3.36a). The twofold symbol in the symbol diagram is shown here in red.

3.7.2 Rectangular Crystal System

The rectangular crystal system contains two point groups, m and $2mm$ (see Table 3.17). In the rectangular system, the x and y axes are orthogonal. In point group m, the mirror in the symbol diagram is shown here as a thick purple horizontal line. In point group $2mm$, the red twofold symbol is shown at the intersection of the two perpendicular purple mirrors. The point group $2mm$ has two independent generators.

TABLE 3.17 Rectangular Crystal System and Square Crystal System

Rectangular System

Point Group	Point Stereographic Projection	Symbol Stereographic Projection	Equivalent General Crystal Faces	Symmetry Matrix Generators
m	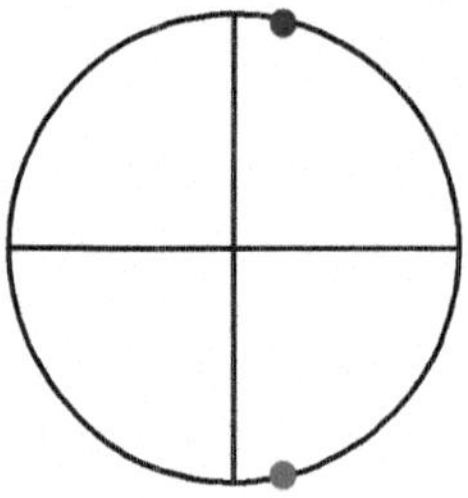	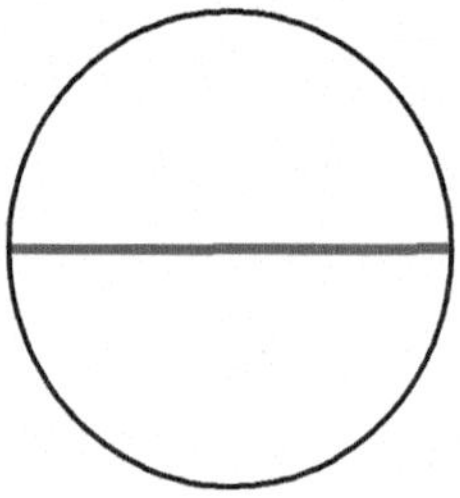	$(hk)\ (\bar{h}k)$	$\begin{pmatrix} \bar{1} & 0 \\ 0 & 1 \end{pmatrix}$
$2mm$	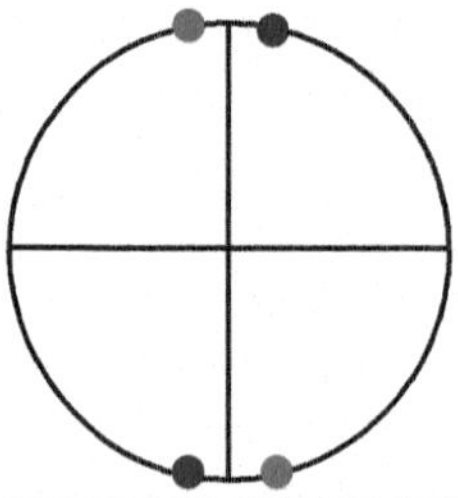	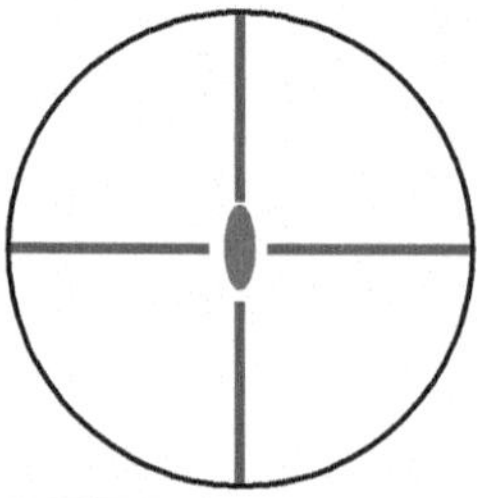	$(hk)(\bar{h}\,\bar{k})$ $(\bar{h}k)(h\bar{k})$	$\begin{pmatrix} \bar{1} & 0 \\ 0 & \bar{1} \end{pmatrix}, \begin{pmatrix} \bar{1} & 0 \\ 0 & 1 \end{pmatrix}$

Square System

Point Group	Point Stereographic Projection	Symbol Stereographic Projection	Equivalent General Crystal Faces	Symmetry Matrix Generators
4	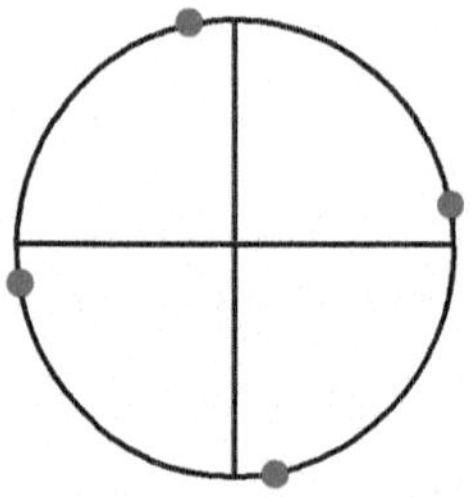	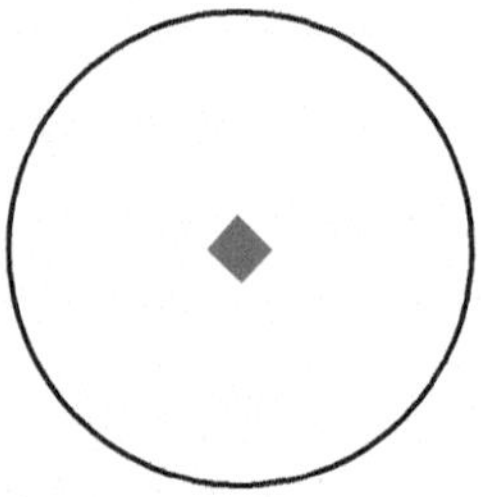	$(hk)\ (\bar{h}\,\bar{k})$ $(\bar{k}h)(k\bar{h})$	$\begin{pmatrix} 0 & \bar{1} \\ 1 & 0 \end{pmatrix}$
$4mm$	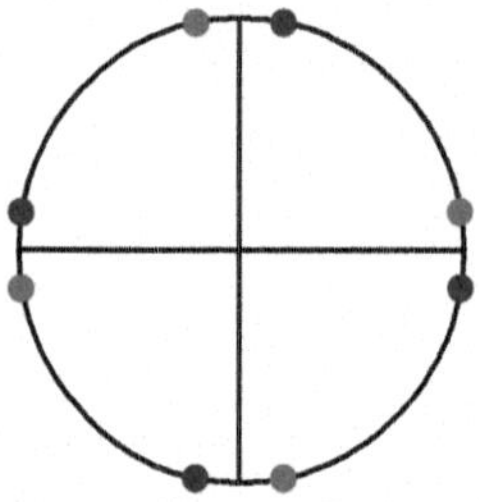	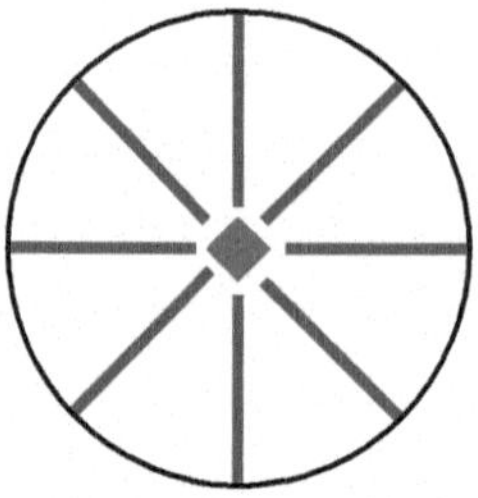	$(hk)\ (\bar{h}\,\bar{k})$ $(\bar{k}h)(k\bar{h})$ $(\bar{h}k)(h\bar{k})$ $(kh)(\bar{k}\,\bar{h})$	$\begin{pmatrix} 0 & \bar{1} \\ 1 & 0 \end{pmatrix}\begin{pmatrix} \bar{1} & 0 \\ 0 & 1 \end{pmatrix}$

3.7.3 Square Crystal System

The square crystal system contains two point groups: 4 and 4mm (see Table 3.17). The square system is characterized by the fourfold rotation. The symbol for the fourfold rotation is a filled-in square with the vertices aligned with the x and y axes. In point group 4mm the appropriate mirrors are added.

TABLE 3.18 Comparison of Three Equivalent Planes in the Hexagonal System

(hk)	(hki)
(11)	$(11\bar{2})$
$(1\bar{2})$	$(1\bar{2}1)$
$(\bar{2}1)$	$(\bar{2}11)$

3.7.4 Hexagonal Crystal System

The hexagonal crystal system contains four point groups: 3, 3m, 6, and 6mm. On the stereographic point projections in addition to the two thin solid lines for the x and y axes, there is a third line for the [11] direction.

The three-index system of Miller indices (hki) is useful to show symmetry, since (hki), (hik), and (ihk) are symmetry equivalent planes. The indices are not independent; the third index, i, is obtained from the Miller indices h and k, with $i = -(h + k)$. An example is $(11\bar{2})$ with equivalent planes $(1\bar{2}1)$ and $(\bar{2}11)$. In the two-index system the equivalent planes are (11), $(1\bar{2})$, and $(\bar{2}1)$ (see Figure 3.36b). The threefold rotation at 0,0 takes the (11) plane into the $(1\bar{2})$ and $(\bar{2}1)$ in red. See Table 3.18 for a comparison of these three equivalent planes.

In Table 3.19 both versions of point group 3m are included. The 3m1 and 31m versions have different forms for their stereographic projections.

Point groups 1, 2, m, 3, 4, and 6 are cyclic groups and thus each can be generated from a single operation. On the other hand, point groups 2mm, 3m, 4mm, 6mm need two generators.

3.8 TWO-DIMENSIONAL POINT GROUP TREE

The 10 point groups are related through supergroups and subgroups. Given point groups 2 and 2mm, point group 2 is a subgroup of point group 2mm because 2 is completely contained in 2mm and 2mm has more elements. This is a proper subgroup. In this book all subgroups are proper subgroups. On the other hand, 2mm is a supergroup of 2 because it contains 2 and has more elements than 2. This is a proper supergroup. In this book all supergroups are proper supergroups.

A *subgroup* is a group wholly contained in a larger group. A *supergroup* is a group that wholly contains a smaller group.

In Figure 3.37a, the point group tree shows the relationship of the point groups to each other. Point group 1 is a subgroup of every point group. The order of each group is shown along the left side of the diagram. For example, point group 4mm has 8 symmetry operations and is therefore order 8. There are no horizontal connecting lines between point groups. The crystal systems—oblique, rectangular, square, and hexagonal—are indicated in color. Each group is linked by a line to its next larger supergroup and by another line, in the opposite direction, to its next smaller subgroup. Increasingly large groups are found near the top of the figure.

Figure 3.37b shows the two-dimensional point group tree with the point stereographic projections. The symmetry equivalent points are shown. In other words, each point represents an operation. If the operation is proper, that is, it does not change handedness, then

TABLE 3.19 Hexagonal Crystal System

Point Group	Point Stereographic Projection	Symbol Stereographic Projection	Equivalent General Crystal Faces	Symmetry Matrix Generators
3			$(hki)(ihk)(kih)$	$\begin{pmatrix} 0 & \bar{1} \\ 1 & \bar{1} \end{pmatrix}$
3m 3m1 version			$(hki)\,(ihk)\,(kih)$ $(\bar{k}\,\bar{h}\,\bar{i})(\bar{i}\bar{k}\,\bar{h})(\bar{h}\,\bar{i}k)$	$\begin{pmatrix} 0 & \bar{1} \\ 1 & \bar{1} \end{pmatrix},\begin{pmatrix} 0 & \bar{1} \\ \bar{1} & 0 \end{pmatrix}$
3m 31m version			$(hki)(ihk)(kih)$ $(khi)(ikh)(hik)$	$\begin{pmatrix} 0 & \bar{1} \\ 1 & \bar{1} \end{pmatrix},\begin{pmatrix} 0 & 1 \\ 1 & 0 \end{pmatrix}$
6			$(hki)\,(ihk)(kih)$ $(\bar{h}\,\bar{k}\,\bar{i})(\bar{i}\bar{h}\,\bar{k})(\bar{k}\bar{i}\bar{h})$	$\begin{pmatrix} 1 & \bar{1} \\ 1 & 0 \end{pmatrix}$
6mm			$(hki)\,(ihk)(kih)$ $(\bar{h}\,\bar{k}\,\bar{i})(\bar{i}\bar{h}\,\bar{k})(\bar{k}\bar{i}\bar{h})$ $(\bar{k}\,\bar{h}\,\bar{i})(\bar{i}\bar{k}\bar{h})(\bar{h}\bar{i}\bar{k})$ $(khi)(ikh)(hik)$	$\begin{pmatrix} 1 & \bar{1} \\ 1 & 0 \end{pmatrix},\begin{pmatrix} 0 & \bar{1} \\ \bar{1} & 0 \end{pmatrix}$

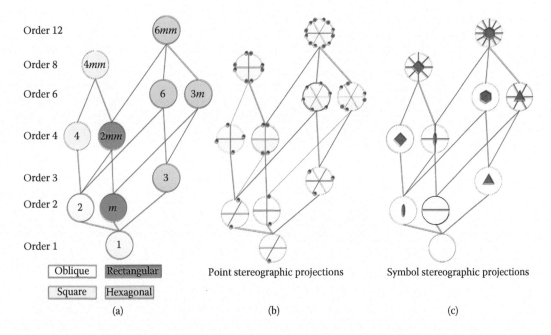

FIGURE 3.37 Two-dimensional point group trees. (a) Crystal systems, (b) point stereographic projections, (c) symbol stereographic projections.

the point is colored red. If the operation is improper, that is, it does change handedness, then the point is colored blue. The identity, represented by a red dot, is by convention in the positive quadrant, which is the lower right-hand quadrant, where the y-axis is horizontal and the x-axis points downward.

Figure 3.37c shows the two-dimensional point group tree using the symbol stereographic projections. These symbols emphasize the subgroups. For example, once the mirror is incorporated, it is always present in its succession of supergroups. The symmetry increases as the order increases. This increasing symmetry is represented by the increasing symmetry of the point group symbol. All the rotation symbols are red and the mirror symbols are purple. See Table 3.15 for the graphical symbols.

EXAMPLE 3.3

Name the two-dimensional point groups that contain point group 2 as a subgroup. Show the relationships between these point groups.

Solution

Consider Figure 3.37a, the point group tree. Point group 2 is contained in point groups 2, 4, 4mm, 2mm, 6, 6mm. Figure 3.38 extracts the information from the complete point group tree and shows the relationships between these point groups.

(continued)

EXAMPLE 3.3 (continued)

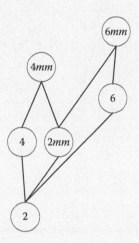

FIGURE 3.38 Subset of two-dimensional point group tree illustrating all the point groups that contain point group 2.

See Exercise 3.28.

The cyclic groups—1, 2, 3, 4, 6, and *m*—are generated from one symmetry operation each and are colored green in Figure 3.39. The remaining four point groups—2*mm*, 3*m*, 4*mm*, and 6*mm*—need two generators each. These groups are colored yellow.

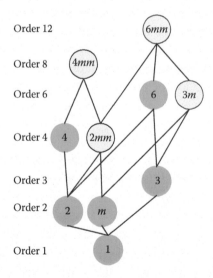

FIGURE 3.39 Two-dimensional point group trees generators. Cyclic groups are in green; groups needing two generators are in yellow.

PART II: THREE DIMENSIONS

3.9 THREE-DIMENSIONAL POINT GROUPS

The definition of a three-dimensional point group is the same as that of a two-dimensional group; namely, a *point group* is a group of symmetry operations that leave at least one point of an object or pattern unmoved. There are 32 point groups in three dimensions. Table 3.20 gives the graphical symbols for three dimensions. The color red indicates a proper operation, one that does not change handedness. The graphical symbols for the rotation axes are colored red because all their operations are proper. The color blue indicates an improper operation, one that does change handedness. The graphical symbols for the rotoinversions are colored purple because half their operations are proper (red), and half their operations are improper (blue). Note that the $\bar{6}$ rotoinversion axis always has an accompanying mirror plane whose normal is along the z-axis. This table parallels Table 3.15, which gives the information for the two-dimensional graphical symbols.

TABLE 3.20 Graphical Symbols in Three Dimensions

Rotation Axes			Rotoinversion Axes		
Printed Symbol	**Graphical Symbol**	**Symmetry Operations**	**Printed Symbol**	**Graphical Symbol**	**Symmetry Operations**
1	None	1	$\bar{1}$	○	$1, \bar{1}$
2		1, 2	m	▬▬▬	$1, m$
3		$1, 3^+, 3^-$	$\bar{3}$		$1, 3^+, 3^-,$ $1, \bar{3}^+, \bar{3}^-$
4		$1, 2, 4^+, 4^-$	$\bar{4}$		$1, 2,$ $\bar{4}^+, \bar{4}^-$
6		$1, 2, 3^+, 3^-, 6^+, 6^-$	$\bar{6}$		$1, 3^+, 3^-,$ $m, \bar{6}^+, \bar{6}^-$

Note: Proper symmetry operations in red: no change of handedness. Improper symmetry operations in blue: change of handedness.

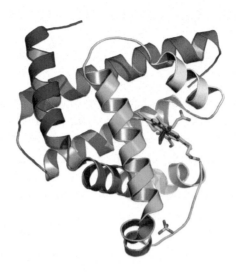

FIGURE 3.40 A molecule of myoglobin protein that has point group 1. (Thanks to David S. Goodsell of the Protein Data Bank, identifier 3RGK.)

3.9.1 n-Fold Rotations

In three dimensions, there are five kinds of n-fold rotations—one-, two-, three-, four-, and sixfold rotations—that are similar to the corresponding rotations in two dimensions. Each rotation is associated with a corresponding cyclic point group that has a single generator. These point groups are 1, 2, 3, 4, and 6. Figure 3.40 shows a molecule of myoglobin protein, which is an example of point group 1. The alpha helices are shown schematically as colored ribbons. The flat porphyrin ring carries the iron.

3.9.1.1 n-Fold Rotation Symmetry Operations

In Table 3.21, the two-, three-, four-, and sixfold rotations are along the z axis. The rotations are all proper operations because the handedness of the figure does not change. In each case the determinant of the symmetry matrix is 1. The first column contains generators of the cyclic groups. These generators are 1, 2, 3^+, 4^+, and 6^+ that generate the cyclic groups 1, 2, 3, 4, and 6, respectively.

See Exercise 3.29.

3.9.1.2 n-Fold Rotation Cyclic Point Groups

As in two dimensions, there is a point group associated with each rotation. In the point stereographic projection, the point—x, y, z—is *interior* to the circle, in contrast with the two-dimensional case, in which the point is *on* the circumference of the circle. Figure 3.41 explains why the points of a three-dimensional point stereographic projection are interior to the circle. A dot in the northern hemisphere (with positive z) is projected onto the equatorial plane through the south pole and an open circle in the southern hemisphere (with negative z) is projected onto the equatorial plane through the north pole.

Table 3.22 summarizes the information about the n-fold rotation point groups. The symmetry operations in the point stereographic projections are indicated by the red dots.

TABLE 3.21   The Three-Dimensional Rotation Symmetry Operations

Operation	Coordinate Transformations	Symmetry Matrix
Identity 1	$x, y, z \rightarrow x, y, z$	$\begin{pmatrix} 1 & 0 & 0 \\ 0 & 1 & 0 \\ 0 & 0 & 1 \end{pmatrix}$
Twofold along z axis $2\ 0, 0, z$	$x, y, z \rightarrow \bar{x}, \bar{y}, z$	$\begin{pmatrix} \bar{1} & 0 & 0 \\ 0 & \bar{1} & 0 \\ 0 & 0 & 1 \end{pmatrix}$
Threefold along z axis $3^+\ 0, 0, z$	$x, y, z \rightarrow \bar{y}, x + \bar{y}, z$	$\begin{pmatrix} 0 & \bar{1} & 0 \\ 1 & \bar{1} & 0 \\ 0 & 0 & 1 \end{pmatrix}$
Fourfold along z axis $4^+\ 0, 0, z$	$x, y, z \rightarrow \bar{y}, x, z$	$\begin{pmatrix} 0 & \bar{1} & 0 \\ 1 & 0 & 0 \\ 0 & 0 & 1 \end{pmatrix}$
Sixfold along z axis $6^+\ 0, 0, z$	$x, y, z \rightarrow x + \bar{y}, x, z$	$\begin{pmatrix} 1 & \bar{1} & 0 \\ 1 & 0 & 0 \\ 0 & 0 & 1 \end{pmatrix}$

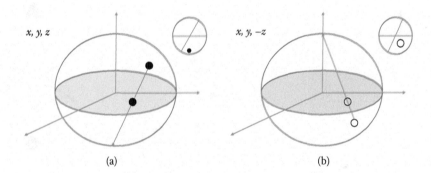

FIGURE 3.41   Three-dimensional stereograms. (a) Positive z, (b) negative z.

The dots are all red because the operations are all proper. The graphical symbols in the symbol stereographic projections for the rotations are also colored red because all their operations are proper.

3.9.2 n-Fold Rotoinversions

In three dimensions, there are five kinds of n-fold rotoinversions—one-, two-, three-, four-, and sixfold rotoinversions. All change the handedness. Each is associated with a corresponding cyclic point group $\bar{1}, \bar{2} = m, \bar{3}, \bar{4}$, and $\bar{6}$, respectively. Table 3.23 gives the rotoinversion symmetry operations. The first column contains generators of the cyclic groups. These generators are $\bar{1}, m, \bar{3}^+, \bar{4}^+$, and $\bar{6}^+$ for the cyclic groups $\bar{1}, m, \bar{3}, \bar{4}$, and $\bar{6}$, respectively.

These operations are developed in the following sections.

TABLE 3.22   Three-Dimensional Point Groups 1, 2, 3, 4, and 6

Crystal System	Point Group	Point Stereographic Projection	Symbol Stereographic Projection
Triclinic	1		
Monoclinic	2		
Trigonal	3		
Tetragonal	4		
Hexagonal	6		

TABLE 3.23 Three-Dimensional Rotoinversion Symmetry Operations

Operation	Coordinate Transformations	Symmetry Matrix
Inversion $\bar{1}\,0,0,0$	$x, y, z \rightarrow \bar{x}, \bar{y}, \bar{z}$	$\begin{pmatrix} \bar{1} & 0 & 0 \\ 0 & \bar{1} & 0 \\ 0 & 0 & \bar{1} \end{pmatrix}$
Mirror m (in xy plane) $m\,x, y, 0$	$x, y, z \rightarrow x, y, \bar{z}$	$\begin{pmatrix} 1 & 0 & 0 \\ 0 & 1 & 0 \\ 0 & 0 & \bar{1} \end{pmatrix}$
$\bar{3}^{+}\,0,0,z$	$x, y, z \rightarrow y, \bar{x}+y, \bar{z}$	$\begin{pmatrix} 0 & 1 & 0 \\ \bar{1} & 1 & 0 \\ 0 & 0 & \bar{1} \end{pmatrix}$
$\bar{4}^{+}\,0,0,z$	$x, y, z \rightarrow y, \bar{x}, \bar{z}$	$\begin{pmatrix} 0 & 1 & 0 \\ \bar{1} & 0 & 0 \\ 0 & 0 & \bar{1} \end{pmatrix}$
$\bar{6}^{+}\,0,0,z$	$x, y, z \rightarrow \bar{x}+y, \bar{x}, \bar{z}$	$\begin{pmatrix} \bar{1} & 1 & 0 \\ \bar{1} & 0 & 0 \\ 0 & 0 & \bar{1} \end{pmatrix}$

3.9.2.1 Inversion Symmetry Operation

The *inversion* symmetry operation $\bar{1}$ involves a point called the inversion point; the operation takes a point at x, y, z into the point at $\bar{x}, \bar{y}, \bar{z}$. The matrix for the inversion is

$$\bar{1} = \begin{pmatrix} \bar{1} & 0 & 0 \\ 0 & \bar{1} & 0 \\ 0 & 0 & \bar{1} \end{pmatrix}$$

Figure 3.42a is an example of a room with an inversion point marked by the circle, o. The inversion operation takes the lamp at x,y,z and inverts it so the lamp is upside down on the ceiling at $\bar{x}, \bar{y}, \bar{z}$. Figure 3.42b shows a molecule, $C_8H_{20}CuN_2O_6$, with the inversion point on the copper atom. Figure 3.42c is a diagram of the molecule.

A *rotoinversion* is a three-dimensional symmetry operation composed of a rotation followed by the inversion operation.

Next, the mirror and the three-, four-, and sixfold rotoinversions are considered.

3.9.2.2 Mirror Operation

Consider the twofold along the y axis. Then the rotoinversion is $\bar{2} = 2 \times \bar{1}$ or, in matrix form,

$$\bar{2} = \begin{pmatrix} \bar{1} & 0 & 0 \\ 0 & 1 & 0 \\ 0 & 0 & \bar{1} \end{pmatrix} \begin{pmatrix} \bar{1} & 0 & 0 \\ 0 & \bar{1} & 0 \\ 0 & 0 & \bar{1} \end{pmatrix} = \begin{pmatrix} 1 & 0 & 0 \\ 0 & \bar{1} & 0 \\ 0 & 0 & 1 \end{pmatrix}$$

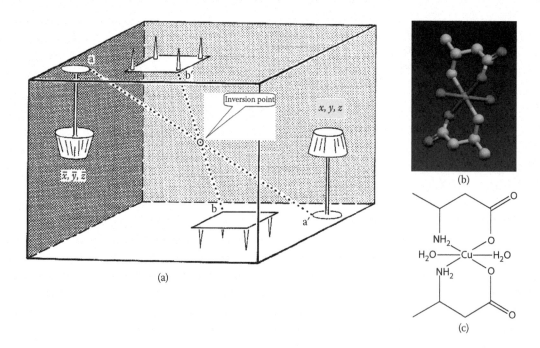

(a)

(b)

(c)

FIGURE 3.42 Inversion point. (a) In room (Adapted from Bloss, F. D. 1971. *Crystallography and Crystal Chemistry: An Introduction.* Holt, Rinehart and Winston, Inc., New York.), (b) molecule with inversion point on copper atom, $C_8H_{20}CuN_2O_6$ (CSD-ABCOPD), (c) diagram.

Now, consider a mirror plane through the dog in Figure 3.43a. The right paw at x, y, z is taken by the mirror in the x,z plane into the left paw at $x, \bar{y}, z$. The matrix for this operation is

$$m_{xz} = \begin{pmatrix} 1 & 0 & 0 \\ 0 & \bar{1} & 0 \\ 0 & 0 & 1 \end{pmatrix}$$

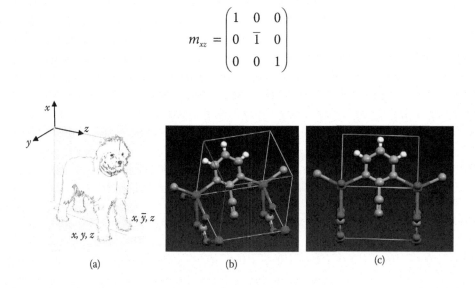

(a)

(b)

(c)

FIGURE 3.43 Mirror plane. (a) On dog (drawn by Kathleen G. Julian, M.D.), (b) perchloric acid, $(C_5H_3AgN_4O_3)_n$ (CSD-QOFBUJ) viewed at angle, (c) viewed along mirror.

which is exactly the matrix for $\overline{2}$. Since $\overline{2} = m$, the symbol $\overline{2}$ will not be used anymore. See Table 3.23 for the mirror in the x,y plane. Figure 3.43b shows perchloric acid $(C_5H_3AgN_4O_3)_n$, viewed at an angle and Figure 3.43c is viewed along the mirror.

Another approach is to multiply a rotation by a mirror with its plane perpendicular to the rotation axis. This operation is called a *rotoreflection*. Multiply each of the rotations by the mirror to get the five cyclic rotoreflection point groups, $\tilde{1}, \tilde{2}, \tilde{3}, \tilde{4}$, and $\tilde{6}$. The corresponding Schoenflies symbols are S_1, S_2, S_3, S_4, and S_6. However, since $\tilde{1} = m, \tilde{2} = \overline{1}, \tilde{3} = \overline{6}, \tilde{4} = \overline{4}, \tilde{6} = \overline{3}$, the rotoreflections do not add any new symmetry operation and are not used in this book. These rotoreflections are important to users of the Schoenflies notation.

See Exercise 3.30.

3.9.2.3 Three-, Four-, and Sixfold Rotoinversions

The threefold rotoinversion $\overline{3}^+$ involves rotating the object by 120° (360/3 = 120), and inverting it through an inversion point. Table 3.23 gives the coordinate transformation and the related symmetry matrix $\overline{3}^+$, which is equal to matrix $3^+ \times$ matrix $\overline{1}$. The fourfold rotoinversion, $\overline{4}^+$, involves rotating the object by 90° (360/4 = 90), and inverting it through an inversion point. Table 3.23 gives the coordinate transformation and the related symmetry matrix $\overline{4}^+$ which is equal to $4^+ \times \overline{1}$. And finally the sixfold rotoinversion, $\overline{6}^+$, involves rotating the object by 60° (360/6 = 60), and inverting it through an inversion point. Table 3.23 gives the coordinate transformation and the related symmetry matrix $\overline{6}^+$ which is equal to $6^+ \times \overline{1}$. See Exercise 3.31.

3.9.2.4 n-Fold Rotoinversion Cyclic Point Groups

There is a point group associated with each rotoinversion. These point groups are $\overline{1}, m, \overline{3}, \overline{4}$, and $\overline{6}$. Since each of these groups can be generated by a single element, they are cyclic groups. Table 3.24 summarizes the information about these point groups. The crystal systems are also given.

First consider point group $\overline{1}$. This is the point group associated with crystalline HMB. In the point stereographic projection, the red dot indicates the identity. The dot shows that the z-coordinate is positive, and it is red to indicate no change of handedness. The blue open circle represents the inversion. The open circle shows that the z-coordinate is negative, and it is blue to indicate the change of handedness. The small circle in the symbol stereographic projection indicates the $\overline{1}$ point group. It is purple because the point group contains equal numbers of proper (red) and improper (blue) operations. The point group $\overline{1}$ consists of the operations 1 and $\overline{1}$. The matrix representing the inversion operation is

$$\overline{1} = \begin{pmatrix} \overline{1} & 0 & 0 \\ 0 & \overline{1} & 0 \\ 0 & 0 & \overline{1} \end{pmatrix}$$

The determinant of this matrix is negative, indicating a change of handedness. The multiplication table for point group $\overline{1}$ is given in Table 3.25.

TABLE 3.24 Three-Dimensional Point Groups $\bar{1}$, m, $\bar{3}$, $\bar{4}$, and $\bar{6}$

Point Group	Crystal System	Point Stereographic Projection	Symbol Stereographic Projection
$\bar{1}$	Triclinic		
m	Monoclinic		
$\bar{3}$	Trigonal		
$\bar{4}$	Tetragonal		
$\bar{6}$	Hexagonal		

In point group m, the mirror is in the plane of the paper. In the point stereographic projection the red dot indicates the identity operation and the blue open circle indicates the mirror in the xy-plane. The solid red dot lies directly above the open blue circle, where the z-coordinate changes. This mirror plane is indicated in the symbol stereographic projection by a solid purple circle.

Figure 3.44 shows a figure with $\bar{3}$ symmetry. The positions for the green and red "atoms" are calculated by choosing $x = 1/3$, $y = 1/4$, $z = 1/5$ for the coordinates in Table 3.26. See Exercise 3.32. Figure 3.44a consists of a green triangle with three unequal sides,

TABLE 3.25 Multiplication Table for Point Group $\bar{1}$

×	1	$\bar{1}$
1	1	$\bar{1}$
$\bar{1}$	$\bar{1}$	1

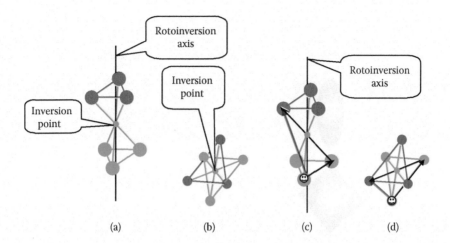

FIGURE 3.44 Figure with $\bar{3}$ symmetry. (a) General view, (b) looking down the symmetry axis, (c) $\bar{3}^+$ operation general view, (d) $\bar{3}^+$ looking down symmetry axis. Balls are colored red and green for clarity.

which is inverted through an inversion point into a red triangle. All the green balls are at the same height (that is the z coordinate) as are the red balls. In Figure 3.44b, the figure is projected down the rotoinversion axis; each ball of the red triangle is related to a corresponding ball of the green triangle by the inversion point. In Figure 3.44c the operation $\bar{3}^+$ is shown. Start with the green ball with a smiley face ☺ on it and follow the black arrow in a counterclockwise rotation of 120°. From this position continue along the black arrow through the inversion point to the red ball. This is a $\bar{3}^+$ operation. The same path could be accomplished by starting with the smiley face and taking the blue arrow directly to the red ball. Figure 3.44d also shows the same $\bar{3}^+$ operation, only this time the view is looking down the rotoinversion axis. These operations are complex and difficult to be seen. The symbol for this point group is a solid triangle with an inversion point in the center of the triangle, as shown in Table 3.24. From the general position diagram the handedness changes for half the coordinates. Table 3.26 gives the symmetry operations and coordinates for point group $\bar{3}$. Similar constructions can be made for the $\bar{4}$ and $\bar{6}$ point groups. See Exercises 3.33 and 3.34.

TABLE 3.26 Symmetry Operations and Coordinates for Point Group $\bar{3}$

Number	Symmetry Operation	Coordinates
(1)	1	x, y, z
(2)	$3^+ \, 0, 0, z$	$-y, x-y, z$
(3)	$3^- \, 0, 0, z$	$-x+y, -x, z$
(4)	$\bar{1}$	$-x, -y, -z$
(5)	$\bar{3}^+ \, 0, 0, z$	$y, -x+y, -z$
(6)	$\bar{3}^- \, 0, 0, z$	$x-y, x, -z$

TABLE 3.27 Graphical Symbols for the Inversion Point with Two-, Four-, and Sixfold Rotation Axes

	Point Group Symbol	Printed Symbol	Symmetry Operations	Minimum Generators
Twofold rotation axis with inversion point		$2/m$	$1, 2, m, \bar{1}$	$2, \bar{1}$
Fourfold rotation axis with inversion point		$4/m$	$1, 2, 4^+, 4^-,$ $\bar{1}, m, \bar{4}^+, \bar{4}^-$	$4^+, \bar{1}$
Sixfold rotation axis with inversion point		$6/m$	$1, 2, 3^+, 3^-, 6^+, 6^-$ $\bar{1}, m, \bar{3}^+, \bar{3}^-, \bar{6}^+, \bar{6}^-$	$6^+, \bar{1}$

3.9.3 The Inversion Point with Two-, Four-, and Sixfold Rotation Axes

In Table 3.20, the purple graphical symbols for $\bar{1}$ and $\bar{3}$ rotoinversion axes both contain the inversion point. On the contrary, m, $\bar{4}$, and $\bar{6}$ do not contain an inversion point. New symbols are needed for combinations of an inversion point with two-, four-, and sixfold rotation axes (see Table 3.27). In each case the purple inversion point is superimposed on the red rotation axis. A mirror plane is generated perpendicular to the rotation axis; so the printed symbols are $2/m$, $4/m$, and $6/m$, respectively for the two-, four-, and sixfold rotation axes combined with the inversion point. Note that in each case the symmetry operations contain the rotation operations, the inversion, and the rotoinversion operations. In each case a minimum of two generators are needed. The choice of the minimal generators is not unique and was made to emphasize the graphical symbol.

Table 3.28 shows how two colors are used for the graphical symbols for the point group $2/m$ in two different orientations. The z-axis is perpendicular to the plane of the paper. In the left diagram the mirror plane is in the xz-plane, represented by a vertical

TABLE 3.28 Point Group $2/m$ in Two Orientations

Mirror in xz-plane, twofold along y axis
Inversion point at origin

Mirror perpendicular to z-axis twofold along z axis
Inversion point at origin

purple line; the twofold rotation axis is along the y-axis, represented by two red ovals connected by a black guide line; and the inversion point is represented by a purple circle at the origin. The only color the twofold can have is red, and the only color the inversion can have is purple. In the right diagram, the mirror plane is perpendicular to the z-axis and is represented by the outer purple circle. The twofold along the z-axis is represented by the red oval at the center and the inversion is colored purple; the two colors are used together.

3.10 THREE-DIMENSIONAL CRYSTAL SYSTEMS

The operations of the 10 cyclic three-dimensional point groups—1, 2, 3, 4, 6, $\bar{1}, m, \bar{3}, \bar{4}$, and $\bar{6}$—combine to form 32 point groups, which are divided into seven crystal systems. In a manner similar to the situation for two dimensions, now the *crystal systems* classify the point groups as triclinic, monoclinic, orthorhombic, tetragonal, trigonal, hexagonal, or cubic (see Table 3.29). The Herman–Mauguin notation is given in column 2 and the Schoenflies notation in column 3.

This book and most crystallographers use the Hermann–Mauguin notation, which has the advantage of directly exhibiting the symmetry of the point group. Each of the point groups 1, 2, 3, 4, 6, $\bar{1}, m, \bar{3}, \bar{4}$, and $\bar{6}$, which have been considered, has a symbol consisting of a single character and can be generated from a single operation. Each of the other 22 point groups has a symbol consisting of multiple characters and needs more than one generator. The point groups $2/m$ and 32 are examples of point groups with multiple generators.

3.10.1 Triclinic Crystal System

The triclinic crystal system contains only the point groups 1 and $\bar{1}$. Both these groups are cyclic. HMB has point group $\bar{1}$. Figure 3.45 shows the point and symbol stereographic projections as well as the minimal generators for the point groups of the triclinic crystal system.

3.10.2 Monoclinic Crystal System

In the monoclinic crystal system 2 and m ($=\bar{2}$) are the cyclic groups. The third point group, $2/m$ (say "2 over m"), has a twofold rotation axis perpendicular to a mirror plane. This

TABLE 3.29 Three-Dimensional Crystal Systems and Their Point Groups

Crystal System	Point Groups (Hermann-Mauguin)	Point Groups (Schoenflies)
Triclinic	$1, \bar{1}$	C_1, C_i
Monoclinic	$2, m, 2/m$	C_2, C_s, C_{2h}
Orthorhombic	$222, mm2, mmm$	D_2, C_{2v}, D_{2h}
Tetragonal	$4, \bar{4}, 4/m, 422, 4mm, \bar{4}2m, 4/mmm$	$C_4, S_4, C_{4h}, D_4, C_{4v}, D_{2d}, D_{4h}$
Trigonal	$3, \bar{3}, 32, 3m, \bar{3}m$	$C_3, C_{3i}, D_3, C_{3v}, D_{3d}$
Hexagonal	$6, \bar{6}, 6/m, 622, 6mm, \bar{6}2m, 6/mmm$	$C_6, C_{3h}, C_{6h}, D_6, C_{6v}, D_{3h}, D_{6h}$
Cubic	$23, m\bar{3}, 432, \bar{4}3m, m\bar{3}m$	T, T_h, O, T_d, O_h

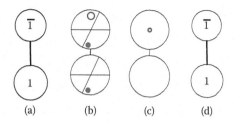

FIGURE 3.45 Triclinic crystal system. (a) Names, (b) point stereographic projections, (c) symbol stereographic projections, (d) minimal generators.

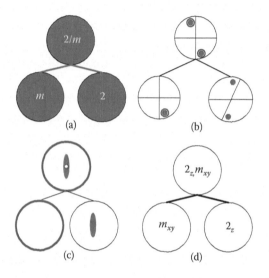

FIGURE 3.46 Monoclinic crystal system. (a) Names, (b) point stereographic projections, (c) symbol stereographic projections, (d) minimal generators.

point group has two generators and is discussed in Section 3.11.1. Figure 3.46 shows the point and symbol stereographic projections as well as the minimal generators for the point groups of the monoclinic crystal system. Note how both the point and symbol stereographic projections for point groups m and 2 are combined in the point group $2/m$. Furthermore, the symbol stereographic projection for $2/m$ contains the generated operation, which is the inversion point. The minimal generators for $2/m$ are not unique and could be chosen as either the pair 2_z and m_{xy} or the pair 2_z and $\bar{1}$. The operations 2_z and m_{xy} are chosen to be a reminder of the symbol $2/m$.

3.10.3 Orthorhombic Crystal System

The orthorhombic crystal system has three point groups—222, $mm2$, and mmm. The structure of this crystal system is similar to that of the monoclinic system. Figure 3.47 shows the point and symbol stereographic projections as well as the minimal generators

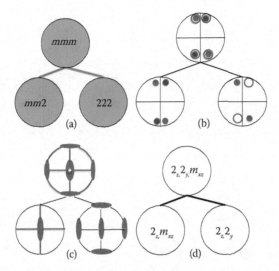

FIGURE 3.47 Orthorhombic crystal system. (a) Names, (b) point stereographic projections, (c) symbol stereographic projections, (d) minimal generators.

for the point groups of this crystal system. Note how both the projections and the generators for the point groups 222 and *mm2* are combined in the point group *mmm*. As shown in the symbol diagram, the point group 222 has three perpendicular red twofolds. Similarly, the point group *mm2* has two perpendicular purple mirror planes and a red twofolds at the intersection of the two mirrors. The combined point group *mmm* has three intersecting purple mirror planes, three perpendicular red twofolds, and the purple inversion point.

The generators, which are not unique, are chosen so as to have a common generator, which is 2_z. The point groups 222 and *mm2* have two generators each. The generators for point group 222 are chosen as 2_z and 2_y. The generators for point group *mm2* are chosen as 2_z and m_{xz}. Finally, the minimal generators for *mmm* are chosen as the triple 2_z, 2_y, and m_{xz}.

3.10.4 Tetragonal Crystal System

The tetragonal crystal system has seven point groups—4, $\bar{4}$, 4/*m*, 422, 4*mm*, $\bar{4}$2*m*, and 4/*mmm*—whose Hermann–Mauguin symbols begin with either 4 or $\bar{4}$. Figure 3.48 shows the point and symbol stereographic projections as well as the minimal generators for the point groups of this crystal system. The point groups 4 and $\bar{4}$ are cyclic and each has order 4. Note how both the projections and the generators for point group 4 are found in the point groups 4/*m*, 4*mm*, and 422. Similar observations can be made for the point group $\bar{4}$ with respect to the point groups $\bar{4}$2*m* and 4/*m*. The point groups 4 and $\bar{4}$ are both subgroups of 4/*m*. The point groups $\bar{4}$2*m*, 4/*m*, 4*mm*, and 422 need a minimum of two generators each; and the point group 4/*mmm* needs a minimum of three generators. Note that for consistency $\bar{4}$2*m* has been chosen in preference to $\bar{4}$*m*2. The symbols $\bar{4}$2*m and $\bar{4}$m2 refer to two orientations of one point group.

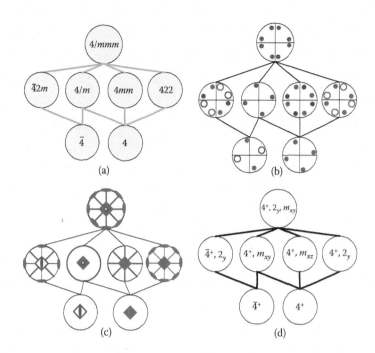

FIGURE 3.48 Tetragonal crystal system. (a) Names, (b) point stereographic projections, (c) symbol stereographic projections, (d) minimal generators.

3.10.5 Trigonal Crystal System

The trigonal crystal system has five point groups—3, $\bar{3}$, 32, 3m, and $\bar{3}m$—whose Hermann–Mauguin symbols begin with either 3 or $\bar{3}$. Figure 3.49 shows the point and symbol stereographic projections as well as the minimal generators for the point groups of this system. The point group 3 is a subgroup of the entire trigonal crystal system. The point groups 3 and $\bar{3}$ with orders 3 and 6, respectively, are cyclic; and the rest of the point groups have a minimum of two generators. The point group 32, the point group of AA, is discussed in Section 3.11.2. Note that the orientations 3m1, 321, and $\bar{3}m1$ have been chosen over the orientations 31m, 312, and $\bar{3}1m$, respectively. For consistency, within this crystal system, either set can be chosen, but not a mixture.

3.10.6 Hexagonal Crystal System

The hexagonal crystal system has seven point groups—6, $\bar{6}$, 6/m, 622, 6mm, $\bar{6}2m$, and 6/mmm—whose Hermann–Mauguin symbols begin with either 6 or $\bar{6}$. The structure of this crystal system is similar to that of the tetragonal crystal system. Figure 3.50 shows the point and symbol stereographic projections as well as the minimal generators for the point groups of this crystal system. The point groups 6 and $\bar{6}$ are cyclic and each has order 6. Note how both the projections and the generators for point group 6 are found in the point groups 6/m, 6mm, and 622. Similar observations can be made for the point group $\bar{6}$ with respect to the point groups $\bar{6}2m$ and 6/m. The point groups 6 and $\bar{6}$ are both subgroups of 6/m. The point groups $\bar{6}2m$, 6/m, 6mm, and 622 need a minimum of two generators each;

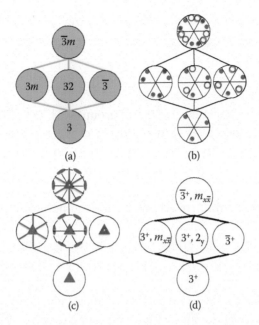

FIGURE 3.49 Trigonal crystal system. (a) Names, (b) point stereographic projections, (c) symbol stereographic projections, (d) minimal generators. Note $3m = 3\,m1$, $32 = 321$, and $\overline{3}m = \overline{3}m1$.

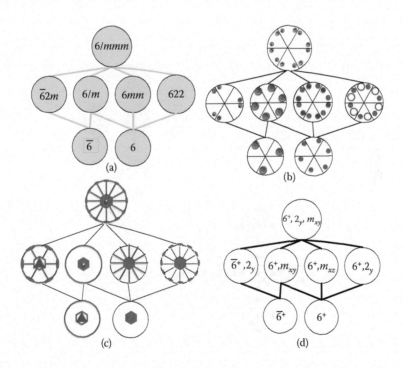

FIGURE 3.50 Hexagonal crystal system. (a) Names, (b) point stereographic projections, (c) symbol stereographic projections, (d) minimal generators.

and the point group $6/mmm$ needs a minimum of three generators. Note that for consistency $\bar{6}2m$ has been chosen in preference to $\bar{6}m2$. The symbols $\bar{6}m2$ and $\bar{6}2m$ refer to two orientations of one point group.

3.10.7 Cubic Crystal System

Finally, the cubic crystal system has five point groups—23, $m\bar{3}$, 432, $\bar{4}3m$, and $m\bar{3}m$. The structure of this crystal system is similar to that of the trigonal crystal system. In the Hermann–Mauguin notation the cubic crystal system is indicated by a 3 or $\bar{3}$ in the *second* place of the symbol. All cubic point groups have threefold rotations down the four diagonals, for example, down the [111]. Figure 3.51 shows the point and symbol stereographic projections as well as the minimal generators for the point groups of the cubic crystal system.

Surprisingly, each cubic point group needs only two generators. Examine Figure 3.51. Because the point group 23 is a subgroup of every other cubic point group and because generators of the point group 23 are 2_z and 3^+_{xxx} one would expect that those generators, 2_z and 3^+_{xxx} would be among the generators of every other cubic point group. In fact they are, although at first glance they do not appear to be. For the point group $\bar{4}3m$ the operation 2_z is present because it is included in the cyclic subgroup generated by $\bar{4}^+_z$. Note $2_z = (\bar{4}^+_z)^2$. For the point group 432, the operation 2_z is present because it is included in the cyclic subgroup generated by 4^+_z. Note $2_z = (4^+_z)^2$. For the point group $m\bar{3}$, the operations 3^+_{xxx} present because it is included in the cyclic subgroup generated by $\bar{3}^+_{xxx}$. Note $3^+_{xxx} = (\bar{3}^+_{xxx})^4$. Similarly, the operations 2_z and 3^+_{xxx} are included in the cyclic subgroups of the generators of the point group $m\bar{3}m$.

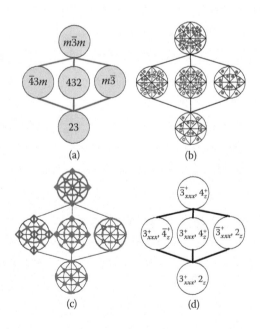

FIGURE 3.51 Cubic crystal system. (a) Names, (b) point stereographic projections, (c) symbol stereographic projections, (d) minimal generators.

3.10.8 Overview

Now that the crystal systems have been introduced, some of their properties are discussed. In each point group every symmetry operation can be represented as a product of generators. In fact, at most three generators are needed to produce any of the 32 three-dimensional point groups. This section gives examples to support this idea.

3.10.8.1 Examples of Multiplication Tables

Multiplication tables show the relationships among the symmetry operations. Examples of the multiplication tables for a point group of each of the seven crystal systems are found in Chapters 9 through 15. They were all produced with the Starter Program for this chapter.

- Triclinic crystal system; example, point group $\bar{1}$

  Its 2×2 multiplication table is in Table 9.2 in Chapter 9.

- Monoclinic crystal system; example, point group 2

  Its 2×2 multiplication table is in Table 10.3 in Chapter 10.

- Orthorhombic crystal system; example, point group mmm

  Its 8×8 multiplication table is in Table 11.3 in Chapter 11.

- Tetrahedral crystal system; example, point group 422

  Its 12×12 multiplication table is in Table 12.3 in Chapter 12.

- Trigonal crystal system; example, point group $\bar{3}m$

  Its 12×12 multiplication table is in Table 13.2 in Chapter 13.

- Hexagonal crystal system; example, point group 6/mmm

  Its 24×24 multiplication table is in Table 14.2 in Chapter 14.

- Cubic crystal system; example, point group $m\bar{3}$

  Its 24×24 multiplication table is in Table 15.5 in Chapter 15.

See Exercises 3.38 through 3.40.

3.10.8.2 Point Groups n and $\bar{n}$

Next, the discussion is about the extraordinary relationship of point group n to point group $\bar{n}$ that depends on whether the integer n is even or odd.

As can be seen from Figures 3.45, 3.49, and 3.51, the triclinic, trigonal, and cubic crystal systems all have similar structures. This is because the rotations with odd numbers and their corresponding rotoinversions all exhibit the characteristic that for n odd, the order of $\bar{n}$ is twice the order of n. This characteristic applies to point groups 1 and $\bar{1}$ and to 3 and $\bar{3}$. (The noncrystallographic groups such as 5 and $\bar{5}$, 7 and $\bar{7}$, etc. also have this characteristic.)

As can be seen from Figures 3.46 through 3.48, and Figure 3.50, the monoclinic, orthorhombic, tetragonal, and hexagonal crystal systems also have similar structures. This is because the rotations with even numbers and their corresponding rotoinversions all exhibit the characteristic that for n even, the order of $\bar{n}$ is equal to the order of n. This condition applies to point groups 2 and m, to 4 and $\bar{4}$ and to 6 and $\bar{6}$ (The noncrystallographic groups such as 8 and $\bar{8}$, 10 and $\overline{10}$, etc. also have this characteristic.)

3.10.8.3 Minimal Generators for Three-Dimensional Point Groups

This section proceeds from the specific multiplication table for the point group 23 to a presentation of the breakdown of its operations in terms of the minimal generators of that point group, and finally to Table 3.32 containing the minimal generators of all 32 three-dimensional point groups.

See Appendix 3 for the symmetry operations for the cubic point group 23. Table 3.30 gives the multiplication table, which was produced with the Starter Program for this chapter. The 12 symmetry operations produce a table with 144 entries. Note how subgroups can be seen in the multiplication table.

Table 3.31 is a remarkable table that shows how the symmetry operations of a point group can be expressed in terms of the minimal generators of that point group. In this case, for the cubic point group 23 the generators $A = 2_z$ and $B = 3^+_{xxx}$ are combined as products to produce all the symmetry operations in the point group. These products can be read from the multiplication table. The *International Tables for Crystallography Vol. A* chooses to give generators but not the minimal number of generators. For example, for the point group 23, the *International Tables for Crystallography* give 1, 2_z, 2_y, and 3^+_{xxx}. This selection of generators has a different purpose, which is to emphasize the subgroups 1, 2, and 222 and finally to emphasize the whole group 23.

TABLE 3.30 Multiplication Table for Cubic Point Group 23

$\times$	1	2_z	2_y	2_x	3^+_{xxx}	$3^+_{\bar{x}\bar{x}x}$	$3^+_{x\bar{x}\bar{x}}$	$3^+_{\bar{x}x\bar{x}}$	3^-_{xxx}	$3^-_{x\bar{x}\bar{x}}$	$3^-_{\bar{x}\bar{x}x}$	$3^-_{\bar{x}x\bar{x}}$
1	1	2_z	2_y	2_x	3^+_{xxx}	$3^+_{\bar{x}\bar{x}x}$	$3^+_{x\bar{x}\bar{x}}$	$3^+_{\bar{x}x\bar{x}}$	3^-_{xxx}	$3^-_{x\bar{x}\bar{x}}$	$3^-_{\bar{x}\bar{x}x}$	$3^-_{\bar{x}x\bar{x}}$
2_z	2_z	1	2_x	2_y	$3^+_{x\bar{x}\bar{x}}$	$3^+_{\bar{x}x\bar{x}}$	3^+_{xxx}	$3^+_{\bar{x}\bar{x}x}$	$3^-_{\bar{x}x\bar{x}}$	$3^-_{\bar{x}\bar{x}x}$	$3^-_{x\bar{x}\bar{x}}$	3^-_{xxx}
2_y	2_y	2_x	1	2_z	$3^+_{\bar{x}\bar{x}x}$	3^+_{xxx}	$3^+_{\bar{x}x\bar{x}}$	$3^+_{x\bar{x}\bar{x}}$	$3^-_{x\bar{x}\bar{x}}$	3^-_{xxx}	$3^-_{\bar{x}x\bar{x}}$	$3^-_{\bar{x}\bar{x}x}$
2_x	2_x	2_y	2_z	1	$3^+_{\bar{x}x\bar{x}}$	$3^+_{x\bar{x}\bar{x}}$	$3^+_{\bar{x}\bar{x}x}$	3^+_{xxx}	$3^-_{\bar{x}\bar{x}x}$	$3^-_{\bar{x}x\bar{x}}$	3^-_{xxx}	$3^-_{x\bar{x}\bar{x}}$
3^+_{xxx}	3^+_{xxx}	$3^+_{\bar{x}x\bar{x}}$	$3^+_{x\bar{x}\bar{x}}$	$3^+_{\bar{x}\bar{x}x}$	3^-_{xxx}	$3^-_{\bar{x}x\bar{x}}$	$3^-_{\bar{x}\bar{x}x}$	$3^-_{x\bar{x}\bar{x}}$	1	2_z	2_y	2_x
$3^+_{\bar{x}\bar{x}x}$	$3^+_{\bar{x}\bar{x}x}$	$3^+_{x\bar{x}\bar{x}}$	$3^+_{\bar{x}x\bar{x}}$	3^+_{xxx}	$3^-_{x\bar{x}\bar{x}}$	$3^-_{\bar{x}\bar{x}x}$	$3^-_{\bar{x}x\bar{x}}$	3^-_{xxx}	2_y	2_x	1	2_z
$3^+_{x\bar{x}\bar{x}}$	$3^+_{x\bar{x}\bar{x}}$	$3^+_{\bar{x}\bar{x}x}$	3^+_{xxx}	$3^+_{\bar{x}x\bar{x}}$	$3^-_{\bar{x}x\bar{x}}$	3^-_{xxx}	$3^-_{x\bar{x}\bar{x}}$	$3^-_{\bar{x}\bar{x}x}$	2_z	1	2_x	2_y
$3^+_{\bar{x}x\bar{x}}$	$3^+_{\bar{x}x\bar{x}}$	3^+_{xxx}	$3^+_{\bar{x}\bar{x}x}$	$3^+_{x\bar{x}\bar{x}}$	$3^-_{\bar{x}\bar{x}x}$	$3^-_{x\bar{x}\bar{x}}$	3^-_{xxx}	$3^-_{\bar{x}x\bar{x}}$	2_x	2_y	2_z	1
3^-_{xxx}	3^-_{xxx}	$3^-_{x\bar{x}\bar{x}}$	$3^-_{\bar{x}\bar{x}x}$	$3^-_{\bar{x}x\bar{x}}$	1	2_x	2_y	2_z	3^+_{xxx}	$3^+_{x\bar{x}\bar{x}}$	$3^+_{\bar{x}x\bar{x}}$	$3^+_{\bar{x}\bar{x}x}$
$3^-_{x\bar{x}\bar{x}}$	$3^-_{x\bar{x}\bar{x}}$	3^-_{xxx}	$3^-_{\bar{x}x\bar{x}}$	$3^-_{\bar{x}\bar{x}x}$	2_y	2_z	1	2_x	$3^+_{\bar{x}\bar{x}x}$	$3^+_{x\bar{x}\bar{x}}$	$3^+_{\bar{x}x\bar{x}}$	3^+_{xxx}
$3^-_{\bar{x}\bar{x}x}$	$3^-_{\bar{x}\bar{x}x}$	$3^-_{\bar{x}x\bar{x}}$	3^-_{xxx}	$3^-_{x\bar{x}\bar{x}}$	2_x	1	2_z	2_y	$3^+_{\bar{x}x\bar{x}}$	3^+_{xxx}	$3^+_{\bar{x}\bar{x}x}$	$3^+_{x\bar{x}\bar{x}}$
$3^-_{\bar{x}x\bar{x}}$	$3^-_{\bar{x}x\bar{x}}$	$3^-_{\bar{x}\bar{x}x}$	$3^-_{x\bar{x}\bar{x}}$	3^-_{xxx}	2_z	2_y	2_x	1	$3^+_{x\bar{x}\bar{x}}$	$3^+_{\bar{x}\bar{x}x}$	3^+_{xxx}	$3^+_{\bar{x}x\bar{x}}$

TABLE 3.31 Symmetry Operations for Point Group 23 as Products of Minimal Generators

Symmetry Operation	Products of Symmetry Operations, $A = 2_z$ and $B = 3^+_{xxx}$	Products of Symmetry Operations
2_z	A	2_z
3^+_{xxx}	B	3^+_{xxx}
1	A^2	$(2_z)^2$
$3^+_{\overline{xx}x}$	AB	$(2_z)(3^+_{xxx})$
3^-_{xxx}	B^2	$(3^+_{xxx})^2$
$3^+_{\overline{xx}\overline{x}}$	BA	$(3^+_{xxx})(2_z)$
$3^+_{\overline{xx}x}$	ABA	$(2_z)(3^+_{xxx})(2_z)$
$3^-_{\overline{xx}x}$	BAB	$(3^+_{xxx})(2_z)(3^+_{xxx})$
$3^-_{\overline{x}x\overline{x}}$	AB^2	$(2_z)(3^+_{xxx})^2$
$3^-_{\overline{xx}x}$	B^2A	$(3^+_{xxx})^2(2_z)$
2_x	BAB^2	$(3^+_{xxx})(2_z)(3^+_{xxx})^2$
2_y	B^2AB	$(3^+_{xxx})^2(2_z)(3^+_{xxx})$

TABLE 3.32 Three-Dimensional Point Groups with Minimal Generators

Number of Minimal Generators	Point Group (Minimal Generators)
10 Cyclic groups	$1(1)$; $\overline{1}(\overline{1})$; $2(2)$; $m(m)$; $3(3^+)$; $\overline{3}(\overline{3}^+)$; $4(4^+)$; $\overline{4}(\overline{4}^+)$; $6(6^+)$; $\overline{6}(\overline{6}^+)$
19 Two-generator groups	$2/m$ $(2_z, m_{xy})$; 222 $(2_z, 2_y)$; $mm2$ $(2_z, m_{xz})$; $4/m$ $(4^+, m_{xy})$; 422 $(4^+, 2_y)$; $4mm(4^+, m_{xz})$; $\overline{4}2m(\overline{4}^+, 2_y)$; 32 $(3^+, 2_y)$; $3m$ $(3^+, m_{x\overline{x}})$; $\overline{3}m(3^+, m_{x\overline{x}})$; $6/m$ $(6^+, \overline{4}m_{xy})$; 622 $(6^+, 2_y)$; $6mm$ $(6^+, m_{xz})$; $\overline{6}2m(\overline{6}^+, 2_y)$; $23(3^+_{xxx}, 2z)$; $m\overline{3}(3^+_{xxx}, 2_z)$,; $432(3^+_{xxx}, 4^+_z)$; $\overline{4}3m(3^+_{xxx}, 4^+_z)$; $m\overline{3}m(3^+_{xxx}, 4^+_z)$
3 Three-generator groups	mmm $(2_z, 2_y, m_{xz})$; $4/mmm$ $(4^+, 2_y, m_{xy})$; $6/mmm$ $(6^+, 2_y, m_{xy})$

Table 3.32 gives minimal generators for all 32 point groups. There are 10 cyclic point groups, 19 point groups that need at least two generators, and finally three groups that need at least three generators.

The Schoenflies notation, which is not used for two dimensions, is applied, for example, in spectroscopy and is often included in textbooks of physical chemistry. The symbols C, D, T, and O stand for cyclic, dihedral, tetrahedral, and octahedral. S is a rotoreflection. In the subscripts, i is an inversion point, v is a vertical mirror, h is a horizontal mirror, d is for mirrors that bisect the horizontal axes, and $n = 1, 2, 3, 4$, and 6 are the rotations.

3.11 EXAMPLES OF THREE-DIMENSIONAL POINT GROUPS WITH MULTIPLE GENERATORS

Two point groups with multiple generators are considered: the monoclinic point group $2/m$ and the trigonal point group 32, which is the associated point group of crystalline anhydrous alum (AA).

3.11.1 Point Group 2/m

Figure 3.52 shows a graph generated by the equation

$$z = f(x, y) = xe^{-(x^2+y^2)}$$

Inspection of the graph shows an inversion point, a mirror and a twofold rotation axis. The corresponding point and symbol stereographic diagrams are shown in Table 3.33. These features are now examined mathematically. The point group for this figure is 2/m.

First, the inversion operation takes x, y, z into $\bar{x}, \bar{y}, \bar{z}$. Consider

$$f(\bar{x}, \bar{y}) = -xe^{-(x^2+y^2)} = -f(x, y) = -z$$

Thus, the function has an inversion point which is at position (0,0,0).

Second, a twofold along the y axis takes x, y, z into $\bar{x}, y, \bar{z}$. Consider

$$f(\bar{x}, y) = -xe^{-(x^2+y^2)} = -f(x, y) = -z$$

Thus, the function has a twofold rotation axis along the y-axis.

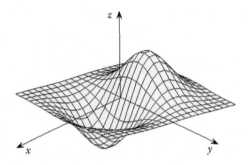

FIGURE 3.52   Figure with point group 2/m.

TABLE 3.33   Group Properties for 2/m

Point Group	Matrix Generators	Point Stereographic Projection	Symbol Stereographic Projection	Crystal System Monoclinic
2/m	$\bar{1} = \begin{pmatrix} \bar{1} & 0 & 0 \\ 0 & \bar{1} & 0 \\ 0 & 0 & \bar{1} \end{pmatrix}$, $2_y = \begin{pmatrix} \bar{1} & 0 & 0 \\ 0 & 1 & 0 \\ 0 & 0 & \bar{1} \end{pmatrix}$			

TABLE 3.34   Multiplication Table for Point Group 2/m

×	1	2	$\bar{1}$	m
1	1	2	$\bar{1}$	m
2	2	1	m	$\bar{1}$
$\bar{1}$	$\bar{1}$	m	1	2
m	m	$\bar{1}$	2	1

Third, a mirror in the xz plane takes x, y, z into $x, \bar{y}, z$. Consider

$$f(x, \bar{y}) = xe^{-(x^2 + y^2)} = f(x, y) = z$$

Thus, the function has a mirror plane in the xz plane.
The matrix

$$m_{xz} = \begin{pmatrix} 1 & 0 & 0 \\ 0 & \bar{1} & 0 \\ 0 & 0 & 1 \end{pmatrix}.$$

The multiplication table for point group $2/m$ is shown in Table 3.34. See Exercises 3.35 and 3.36.

3.11.2 Point Group 32

Point group 32, which is in the trigonal system, is the associated point group of crystalline AA. Two symmetry operations are needed to generate this point group. These operations are threefold and twofold with its axis perpendicular to the threefold axis (see Table 3.35). The *International Tables for Crystallography* chooses the threefold axis to be the z-axis and

TABLE 3.35   Point Group Properties for 32

Point Group	Matrix Generators	Point Stereographic Projection	Symbol Stereographic Projection	Crystal System
32	$3^+ = \begin{pmatrix} 0 & \bar{1} & 0 \\ 1 & \bar{1} & 0 \\ 0 & 0 & 1 \end{pmatrix}$, $2_{xx} = \begin{pmatrix} 0 & 1 & 0 \\ 1 & 0 & 0 \\ 0 & 0 & \bar{1} \end{pmatrix}$			Trigonal

places a twofold axis along the [110] crystallographic direction with the origin placed at the intersection of the two axes. The twofold along the [110] crystallographic direction takes x, y, z into $y, x, \bar{z}$.

From Table 3.35, the matrix

$$2_{xx} = \begin{pmatrix} 0 & 1 & 0 \\ 1 & 0 & 0 \\ 0 & 0 & \bar{1} \end{pmatrix}$$

and the matrix

$$3^+ = \begin{pmatrix} 0 & \bar{1} & 0 \\ 1 & \bar{1} & 0 \\ 0 & 0 & 1 \end{pmatrix}.$$

The multiplication is

$$2_{xx} \times 3^+ = \begin{pmatrix} 0 & 1 & 0 \\ 1 & 0 & 0 \\ 0 & 0 & \bar{1} \end{pmatrix} \begin{pmatrix} 0 & \bar{1} & 0 \\ 1 & \bar{1} & 0 \\ 0 & 0 & 1 \end{pmatrix} = \begin{pmatrix} 0 & \bar{1} & 0 \\ 1 & \bar{1} & 0 \\ 0 & 0 & \bar{1} \end{pmatrix} = 2_x.$$

Continue with

$$2_x \times 3^+ = \begin{pmatrix} 1 & \bar{1} & 0 \\ 0 & \bar{1} & 0 \\ 0 & 0 & \bar{1} \end{pmatrix} \begin{pmatrix} 0 & \bar{1} & 0 \\ 1 & \bar{1} & 0 \\ 0 & 0 & 1 \end{pmatrix} = \begin{pmatrix} \bar{1} & 0 & 0 \\ \bar{1} & 1 & 0 \\ 0 & 0 & \bar{1} \end{pmatrix} = 2_y.$$

Table 3.36 gives the multiplication table for point group 32. See Exercise 3.37. In Figure 3.53, the airplane propeller with three blades all tipped by the same angle is an example of an object with point group 32.

TABLE 3.36   Multiplication Table for Point Group 32

×	1	3⁺	3⁻	2_{xx}	2_x	2_y
1	1	3⁺	3⁻	2_{xx}	2_x	2_y
3⁺	3⁺	3⁻	1	2_y	2_{xx}	2_x
3⁻	3⁻	1	3⁺	2_x	2_y	2_{xx}
2_{xx}	2_{xx}	2_x	2_y	1	3⁺	3⁻
2_x	2_x	2_y	2_{xx}	3⁻	1	3⁺
2_y	2_y	2_{xx}	2_x	3⁺	3⁻	1

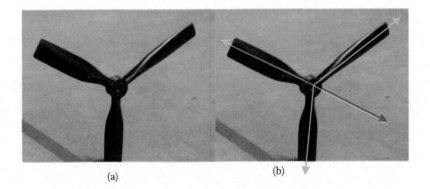

FIGURE 3.53 Propeller. (a) With point group 32 symmetry, (b) a single red threefold axis and three green twofold axes superimposed on the propeller.

3.12 THREE-DIMENSIONAL POINT GROUP TREES

The 32 three-dimensional point groups are related to each other through supergroups and subgroups in a manner similar to the two-dimensional case. Figures 3.54 through 3.56 collectively show how all the crystal systems are assembled into the complete

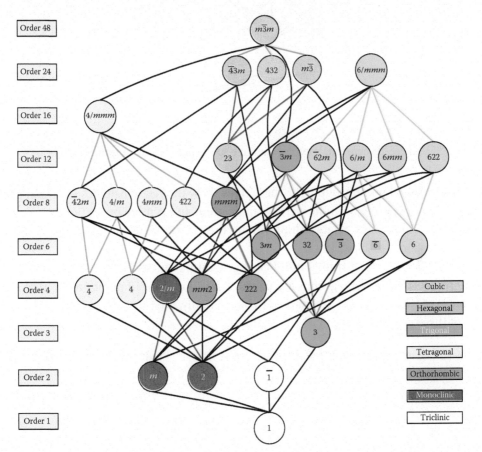

FIGURE 3.54 Three-dimensional point group tree with crystal systems connected in color.

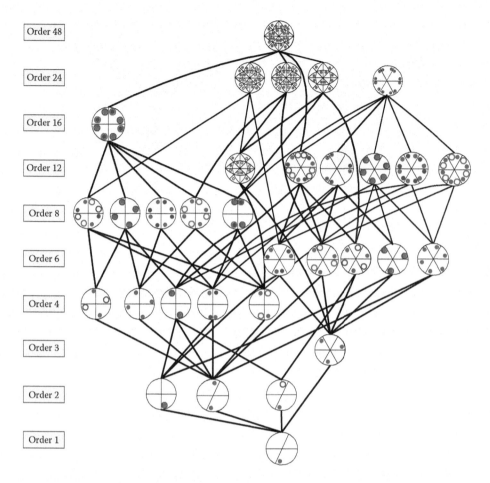

FIGURE 3.55 Three-dimensional point group tree of the point stereographic projections.

three-dimensional point group tree. These trees show the relationship of the point groups to each other, including both the stereographic point and symbol projections.

Figure 3.57 shows the relationships among the crystal systems. It summarizes Figure 3.54, emphasizing the black lines.

Figure 3.58 gives the minimal number of generators for each of the three-dimensional point groups. The 10 cyclic groups are marked in yellow. The 19 point groups that need a minimum of two generators are marked in blue. Finally, the three groups that need a minimum of three generators are marked in pink.

3.13 POINT GROUP SYMMETRY AND SOME PHYSICAL PROPERTIES OF CRYSTALS

3.13.1 Piezoelectricity

Piezoelectric materials produce electricity when squeezed or placed under mechanical stress. The prefix piezo- is from the Greek word *piezein* which means to squeeze or press. In a piezoelectric material, an electric dipole moment arises upon compression, generating

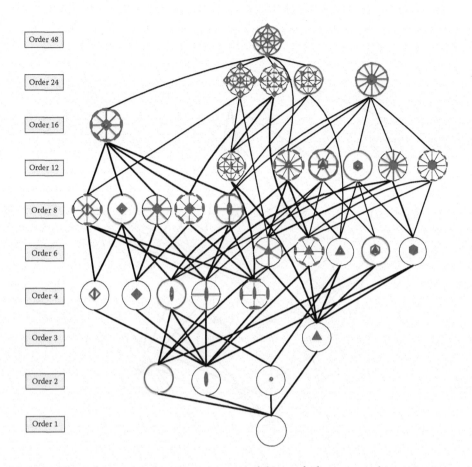

FIGURE 3.56 Three-dimensional point group tree of the symbol stereographic projections.

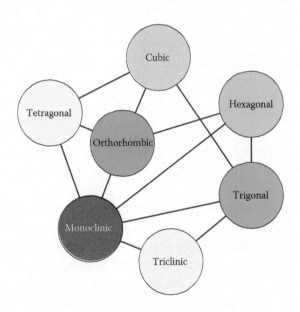

FIGURE 3.57 Relationships among crystal systems.

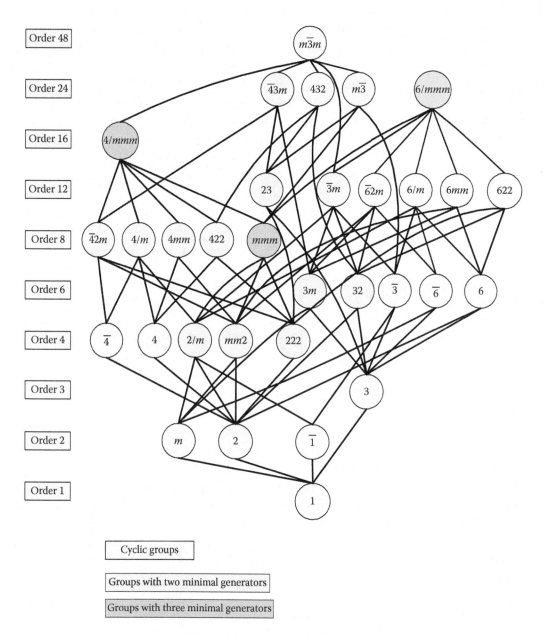

FIGURE 3.58 Point group tree with minimal generators.

a voltage. Reversing the direction of stress, under tension, reverses the sign of the voltage. There is a great variation in the amount of voltage generated under stress. Even when the piezoelectric effect may be theoretically possible, the effect is often so small that it is difficult to detect. Currently, piezoelectric particles are introduced into composites to control the ability of the material to dissipate mechanical or acoustical energy (Goff et al., 2004). The addition of such 'functionally active' components to composites produces a whole new level of material design.

Centrosymmetric crystals are crystals which contain an inversion point. They are never piezoelectric. The inversion point implies the impossibility of an induced dipole moment. Additionally, the cubic point group 432, though lacking an inversion point, cannot exhibit the piezoelectric effect. In this point group the symmetries cooperate to prevent an induced dipole moment.

On the point group tree in Figure 3.59, begin with point group $\bar{1}$, which is the lowest-order point group that contains an inversion point. Now look at the next level up. From the point group tree, $\bar{1}$ has only two immediate supergroups, $2/m$ and $\bar{3}$. Both these groups contain an inversion point. Continue up the tree until all the point groups containing an inversion point are located. All 11 centrosymmetric point groups are in yellow. *Noncentrosymmetric crystals* are crystals which do not contain an inversion point. If a crystal is to be piezoelectric, it must form in one of the noncentrosymmetric point groups,

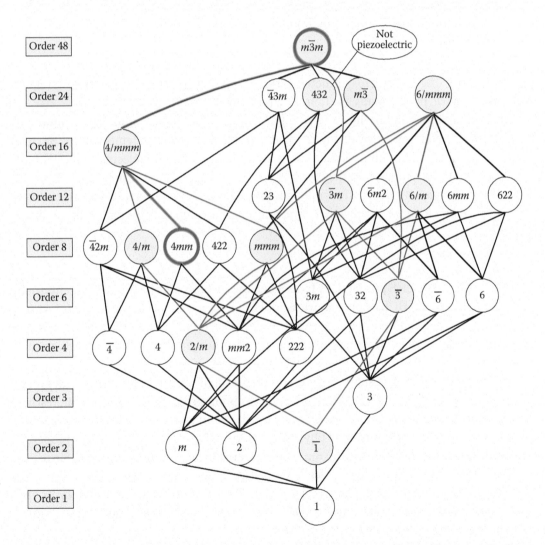

FIGURE 3.59 Point group tree with centrosymmetric point groups in yellow with connecting lines in red. Point group 432 is labeled. Point groups of barium titanate, $4mm$ and $m\bar{3}m$, are circled in blue.

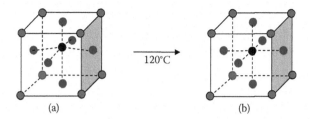

FIGURE 3.60   Barium titanate (a) below and (b) above the Curie temperature of 120°C.

TABLE 3.37   Properties of Barium Titanate below and above the Curie Point

	BaTiO$_4$ below 120°C	BaTiO$_4$ above 120°C
Crystal system	Tetragonal	Cubic
Point group	4mm	$m\bar{3}m$
Centrosymmetric	No	Yes
Piezoelectric	Yes	No

excluding 432. The noncentrosymmetric point groups are not colored. Examples of non-centrosymmetric point groups are 1, *m*, 2, 3, and *4mm*.

At room temperature, barium titanate, BaTiO$_4$, has a permanent ionic dipole moment associated with its unit cell (see Figure 3.60). At 120°C, the Curie temperature, BaTiO$_4$ transforms from a strongly piezoelectric material into a nonpiezoelectric material (see Table 3.37). At this temperature the crystal structure changes from the noncentrosymmetric point group *4mm* to the centrosymmetric point group $m\bar{3}m$. In Figure 3.59, trace *4mm* through *4/mmm* to $m\bar{3}m$ to show that *4mm* is a subgroup of $m\bar{3}m$. This path is highlighted in blue.

3.13.2 Etch Figures

Some crystals reveal etch pits on their faces when the surfaces are attacked by reagents such as hydrochloric acid, HCl, or phosphoric acid, H$_3$PO$_4$. The etch pits are sometimes formed by surface evaporation upon heating the crystal in a vacuum. Usually the pits form at the ends of dislocation lines. The shape of the pits reveals the symmetry of the face which corresponds to one of the 10 two-dimensional point groups.

For example, Figure 3.61 shows etch figures on the (0001) face of apatite. The point group of this face is point group 6. The crystal belongs to point group *6/m*. This is an interesting example of *Neumann's principle*, which says that the point group of a crystal is a *subgroup* of the symmetry group of any of its physical properties. In this case the physical property, namely the etching, has point group *6mm* and the crystal face has point group 6. Since 6 is a subgroup of *6mm*, Neumann's principle is obeyed. See Honess (1927).

The investigation of etch pits can be used to study twinning, for example in the perovskite-like crystals YAlO$_3$ and Nd GaO$_3$. Both of these crystals undergo twinning during the growth process (Savytskii et al., 2000).

In noncentrosymmetric crystals the etch pits on parallel but opposite faces may be different in size or shape. This physical property can distinguish centrosymmetric from

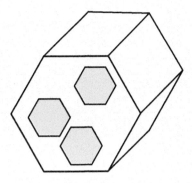

FIGURE 3.61 Apatite etched by dilute hydrochloric acid.

noncentrosymmetric crystals. Studying etch pits applies to natural growth planes, cleaved planes, or arbitrarily cut planes.

See the *International Tables for Crystallography, Vol. D, Physical Properties of Crystals*, edited by A. Authier (2003), *Physical Properties of Crystals* by J. F. Nye (2010), and Ma and Gu (2004) for more advanced treatments. Also see Julian (1999) for a comparison of stretching force constants in symmetry coordinates. Optical properties are considered in Julian and Bloss (1987).

DEFINITIONS

Associative law
Centrosymmetric crystals
Clathrate
Closure
Commutative group
Crystal systems
Crystallographic point group
Crystallographic rotations
Cyclic group
Fourfold rotation
Generator
Graphical symbol
Group
Identity operation
Identity, E, of a group
Improper operation
Invariant
Inverse in a group
Isomorphic group
Mirror
Multiplication
Multiplication table

Neumann's principle
n-fold rotation
Noncommutative group
Noncrystallographic point group
Order of a group
Piezoelectric material
Point group
Point stereographic projection
Proper operation
Rotoinversion
Rotoreflection
Set
Sixfold rotation
Subgroup
Supergroup
Symbol stereographic projection
Symmetry operation
Threefold rotation
Trace
Twofold rotation

EXERCISES

3.1 Demonstrate that the identity $E = 0$ is true for Example 3.1.

3.2 Show that the associative law holds in Example 3.1.

3.3 What are the inverses of the elements in Example 3.2. Demonstrate by two examples how the associative law holds.

3.4 Show that the elements in Table 3.4 form a group. Include only one example for associative law.

3.5 List the letters of the alphabet that have a single mirror plane. Use upper case Ariel font.

3.6 Plot Figure 3.9a the transcendental butterfly, using the parametric equations: $x = \sin t\ (\exp(\cos t) - 2\cos 4t - \sin^5 t/12)$ and $y = \cos t\ (\exp(\cos t) - 2\cos 4t - \sin^5 t/12)$. Include computer code.

3.7 Construct figures with ninefold and 13-fold rotation symmetry.

3.8 Show the construction as in Figure 3.18a for $n = 2, 3,$ and 4.

3.9 Show that the construction as in Figure 3.18b fails for $n = 7$ and $n = 8$.

3.10 Calculate the determinant and the trace for each matrix in Table 3.6.

3.11 List the letters of the alphabet that have point group 1. Use upper case Arial font.

⌨ 3.12 Verify the multiplication table for point group 3 given in Table 3.10. Use the Starter Program to multiply the matrices.

⌨ 3.13 Verify the multiplication table for point group 4 given in Table 3.11. Use the Starter Program to multiply the matrices.

⌨ 3.14 What is the matrix for symmetry operation 6⁻, the matrix needed for closure. Create the multiplication table for point group 6 using the Starter Program. Find the identity and the inverses of all the group elements. Use Appendix 2.

⌨ 3.15 Calculate the multiplication table for point group 6 starting with group elements 3^+ and 2. Use Appendix 2.

3.16 List the letters of the alphabet that have the symmetry of point group 2mm. Use upper case Arial font.

3.17 Point groups 4 and 2mm both have the same order. Are they isomorphic point groups? Name the subgroups in 2mm. What is the identity? Give the inverse of every group element in 2mm.

⌨ 3.18 Plot the motif from Figure 3.28a, $r = \cos 3\theta$, with 3m symmetry. Hint $x = \cos 3\theta \cos \theta$ and $y = \cos 3\theta \sin \theta$.

3.19 In point group 3m, construct the matrix for m_y which is a counterclockwise rotation of m_x for 120°. Label each operation in 3m as proper or improper.

⌨ 3.20 Complete the multiplication table for 3m as in Table 3.13. Use Starter Program and use Appendix 2.

⌨ 3.21 For the point group 31m calculate the six group elements and construct a multiplication table. Is this table isomorphic to the multiplication table for point group 3m1? Use Appendix 2.

⌨ 3.22 Rotate the figure from Exercise 3.18 by 30° and draw Figure 3.32a (omit axes.) Hint: To rotate a matrix by θ in orthonormal coordinates, multiply by $\begin{pmatrix} \cos\theta & -\sin\theta \\ \sin\theta & \cos\theta \end{pmatrix}$.

⌨ 3.23 For point group 4mm construct matrices for the eight symmetry operations. Use matrix multiplication to construct the multiplication table. What is the identity? What are the inverses of each of the elements? Is this a commutative group? Use Appendix 2.

⌨ 3.24 Plot the motif in Figure 3.33a, $r = \cos 2\theta$, with 4mm symmetry. Hint: $x = \cos 2\theta \cos \theta$ and $y = \cos 2\theta \sin \theta$.

⌨ 3.25 Plot $r = \cos n\theta$ where n is an integer. Experiment with several integers and find a rule. Hint: $x = \cos n\theta \cos \theta$ and $y = \cos n\theta \sin \theta$.

⌨ 3.26. Create a multiplication table for point group 6mm. What is the identity and what are the inverse elements for each of the group elements. Is this point group a commutative group? Use Appendix 2.

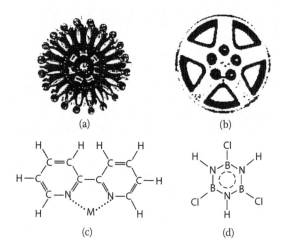

FIGURE 3.62   Assign a point group to these objects.

3.27   What is the two-dimensional point group of each object in Figure 3.62? Label crystallographic groups.

3.28   Name two-dimensional point groups that contain point group $2mm$ as a subgroup. Show the relationship among these point groups with a point group tree similar to Figure 3.38.

3.29   In Table 3.23 determine if the symmetry matrices represent proper or improper operations. Compare this result to the rotation operations in two dimensions.

3.30   Using matrix multiplication show that $\tilde{1} = m, \tilde{2} = \bar{1}, \tilde{3} = \bar{6}, \tilde{4} = \bar{4}, \tilde{6} = \bar{3}$.

3.31   In Table 3.23 determine if the symmetry matrices represent proper or improper operations. What operation in two dimensions is improper.

3.32   Construct Figure 3.44 using the coordinates in Table 3.6 with $x = 2/3$, $y = 4/5$, and $z = 1/2$.

3.33   Construct an abstract figure illustrating point group $\bar{4}$ using the coordinates in Table 3.38. Use $x = 1/3$, $y = 1/4$, and $z = 1/5$. Does this figure have an inversion symmetry operation?

TABLE 3.38   Symmetry Operations and Coordinates for Point Group $\bar{4}$

Number	Symmetry Operation	Coordinates
(1)	1	x, y, z
(2)	$2\,0, 0, z$	$\bar{x}, \bar{y}, z$
(3)	$\bar{4}^{+}0, 0, z$	$y, \bar{x}, \bar{z}$
(4)	$\bar{4}^{-}0, 0, z$	$\bar{y}, x, \bar{z}$

🖳 3.34 Construct the multiplication table for three-dimensional point group $\bar{6}$. What is the identity and what are the inverse elements for each of the group elements? Is this point group a commutative group? Use Appendix 3.

🖳 3.35 Construct the multiplication table for $2/m$. Adapt the Starter Program to three dimensions. Use Appendix 3.

🖳 3.36 Construct Figure 3.52 which is a three-dimensional plot with

$$z = x^*\exp(-x^\wedge 2 - y^\wedge 2).$$

Use a surf function. Hint: use $[x, y] = \text{meshgrid}(-2:.2:2, -2:.2:2)$. Now plot $z = y^*\exp(-x^\wedge 2 - y^\wedge 2)$ and compare the symmetries of the two functions.

🖳 3.37 Construct the multiplication table for three-dimensional point group 32. Check with Table 3.36. What is the identity and what are the inverse elements for each of the group elements. Is this point group a commutative group? Use the Starter Program and information from Appendix 3.

🖳 3.38 Construct the multiplication table for three-dimensional point group 622. What is the identity and what are the inverse elements for each of the group elements? Is this point group a commutative group? Use the Starter Program and information from Appendix 3.

🖳 3.39 Construct the multiplication table for three-dimensional point group $6/mmm$. What is the identity and what are the inverse elements for each of the group elements? Is this point group a commutative group? List the proper symmetry operations and the improper symmetry operations. Use the Starter Program and information from Appendix 3.

3.40 Show that 3^+_{xxx} and $3^+_{\bar{x}x\bar{x}}$ are a set of minimal generators for point group 23 by listing all the symmetry operations as powers of these minimal generators. Use Table 3.30.

MATLAB CODE: STARTER PROGRAM FOR CHAPTER 3: POINT GROUP MULTIPLICATION TABLE

This program produces the multiplication table for both the two- and three-dimensional point groups. The elements that are multiplied are the matrices which represent the symmetry operations. The matrices are found in Appendix 2 for the two-dimensional point groups and in Appendix 3 for the three-dimensional point groups. This program produces the products. See the book's web site for instructions on how to use Word to convert the results of this program into the format found, for example, in Table 3.30.

```
function groupmult
clc
```

```
close all
clear

% Purpose: group multiplication tables
% Record of revisions:
% Date        Programmer  Description of change
% 1/10/       M. Julian    original code
%             your name here
% define variables:
% m1 2x2 identity matrix
% m2 2x2 twofold rotation matrix
% M matrix formed from m1 and m2
% k, kk, kkk running indices
% nn,mm matrices selected for multiplication
% garlic product of nn and mm
% multiplication_table  matrix multiplication table  THE ANSWER
% order is number of elements
% dimension  either 2D or 3D
% SymOper  char which hold names
% multtb  three dimensional array that is THE ANSWER
m1 =[1 0
     0 1];%two-dimensional identity matrix
m2 = [-1 0
     0 -1];%twofold matrix
%set of individual matrices—add more matrices here
M = [m1;m2];
[row,dimension] = size(M);
order = row/dimension;
SymOper = char('  *1','  *2');%each entry has exactly four spaces
%now we will multiply the matices together
for k = 1:order
    for kk = 1:order
        nn = matrixx(k,M,dimension);%calls function that selects
matrix k = 1 identity, k = 2 twofold
        mm = matrixx(kk,M,dimension);%calls function that selects
matrix k = 1 identity, k = 2 twofold
        garlic = nn*mm;%multipy matrices
        % for value of k, name is identified
        %now we want to find out the name of garlic
        for kkk = 1:order
            if garlic==matrixx(kkk,M,dimension);%find out what the
name of garlic is
                multtb(k,kk,:)=SymOper(kkk,:);
                %fprintf(multtb(k,kk,:))
            end
        end
    end
end
```

```matlab
end
%now print out the answer
%%% multiplication_table%old way
for k = 1:order
    for kk = 1:order
        fprintf(multtb(k,kk,:))
        if kk==order
            fprintf('_n')
        end
    end
end
%************************************************************
function n3 = matrixx(k,M,dimension)
if dimension == 2 %select 2x2 symmetry matrix from source matrix M
i1 = 1 + 2*(k-1);
i2 = 2 + 2*(k-1);
n3 = M(i1:i2,1:2);
else  % selects 3x3 symmetry matrix from source matrix M
i1 = 1 + 3*(k-1);
i2 = 2 + 3*(k-1);
i3 = 3 + 3*(k-1);
n3 = M(i1:i3,1:3);
end
```

## CHAPTER 4

Space Groups

This chapter adds the translations of Chapter 1 to the symmetry operations of Chapter 3. The diagram above shows the unit cell of hexamethylbenzene, HMB, translated 12 times. Ideally, these unit cells are repeated forever, covering the entire plane. In this chapter the concept of space group is introduced to study the organization of the symmetry operations when translations are included. In two dimensions these space groups may be informally called "wallpaper groups." This chapter teaches the tools necessary to use the *International Tables for Crystallography* to interpret crystal structures.

CONTENTS

CHAPTER OBJECTIVES

PART I: TWO DIMENSIONS

- Examine the five two-dimensional Bravais lattices.

- Combine a Bravais lattice with a point group to get a symmorphic space group.

- Recognize nonsymmorphic space groups.

- Introduce glides with color codes.

- Interpret the space group tables in the *International Tables for Crystallography*.

- Assign a space group to a periodic pattern.

- Construct general position and symbol diagrams.

- Color-code all 17 two-dimensional space groups.

- Understand the asymmetric unit for each of the 17 two-dimensional space groups.

- Construct the space group tree.

PART II: THREE DIMENSIONS

- Expand concepts of Bravais lattices and space groups to three dimensions.

- Classify the three-dimensional space groups by point groups or by lattices.

- Introduce three-dimensional glides and screws with color codes.

- Recognize symmorphic and nonsymmorphic space groups.

- Compare the Hermann–Mauguin and the Schönflies notations.

- Understand how to combine the space group information with the crystallographic data for HMB, AA, and caffeine.

- Interpret the CIF, Crystallographic Information File.

- Use the Starter Program and the CIF to populate the unit cell of caffeine monohydrate.

- Examine the special projections for HMB, AA, and caffeine to recognize their space group symmetries and construct the asymmetric unit.

- Use the *International Tables for Crystallography* to understand and apply concepts related to symmetry operations, generators, general position, special positions, Wyckoff letter, multiplicity, and site symmetry.

- To construct part of the three-dimensional space group tree with the maximal type I subgroups and the minimal type I supergroups found in the *International Tables for Crystallography*.

- Examine polyhedrons in crystal structures and interpret deviations from ideal polyhedrons.

- Compare isotypic crystal structures.

- Obtain Z and Z'.

- Apply the Starter Program to a unit cell with a volume of around 10^3 Å^3.

PART I: TWO DIMENSIONS

4.1 INTRODUCTION

The last chapter introduced point groups to classify the symmetries of individual objects, including molecules. In a point group, at least one point remains stationary under the symmetry operations. There are 10 two-dimensional crystallographic point groups and 32 three-dimensional crystallographic point groups.

This chapter adds the translations of Chapter 1 to the symmetry operations of Chapter 3. The effects of global symmetries are analyzed. Global symmetries are symmetries that apply to the whole pattern. Each space group has an infinite order because there is an infinite number of translation operations. The 17 two-dimensional space groups, or wallpaper groups, are introduced. In three dimensions there are 230 space groups. In this book the notation of the *International Tables for Crystallography* is carefully followed in order to facilitate interpreting the crystallographic literature.

4.2 TWO-DIMENSIONAL BRAVAIS LATTICES

As in Chapter 1, lattices are associated with patterns that repeat in space. The lattice of a pattern is an array of points with identical neighborhoods. The lattice is unique to the pattern. The lattice points, in two dimensions, are described by $t(u,v) = u\,\mathbf{a} + v\,\mathbf{b}$ where u and v are integers and $\mathbf{a}$ and $\mathbf{b}$ are basis vectors. The integers u and v each range from $-\infty$ to $+\infty$, which makes the lattice infinite. The lattice is chosen to be infinite, as an idealization, to avoid dealing with the edges of the pattern. The basis vectors are not unique to the lattice. Lattices are categorized according to symmetry. Most point-group symmetries place restrictions on a, b, and/or γ. *Bravais lattices* are a classification of the five two-dimensional lattices based on unit cells that are parallelograms. These lattices are named in honor of the French physicist Auguste Bravais (1811–1863). Each Bravais lattice has a point group that describes its symmetry. Ultimately the descriptive classification of a lattice is determined by symmetry. The five two-dimensional Bravais lattices are oblique, rectangular primitive, rectangular centered, square, and hexagonal. These are listed according to the increasing symmetry of the associated point groups.

For the illustrations in this book, the origin is in the upper left-hand corner, the y axis is drawn horizontally pointing to the right and the x axis pointing generally downward, following the notation in the *International Tables for Crystallography*. Conventional unit cells are used.

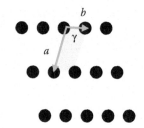

FIGURE 4.1   Primitive oblique lattice, *mp.*

4.2.1 Oblique

Figure 4.1 shows part of an oblique lattice. An *oblique lattice* is a primitive lattice that has the symmetry of point group 2, which is the least symmetry that a two-dimensional lattice can have. Only the identity and the twofold operations are present. Lattice parameters *a*, *b*, and γ must be measured.

An oblique lattice has a twofold axis at each lattice point. To demonstrate this, trace the lattice onto a separate piece of paper. Rotate the paper 180° about any lattice point. The resulting pattern is indistinguishable from the unrotated lattice. All two-dimensional lattices have a twofold axis. The symbol for the oblique lattice from the *International Tables for Crystallography* is *mp*, where *m* stands for monoclinic and *p* for primitive.

4.2.2 Rectangular Primitive

Figure 4.2 shows part of a rectangular primitive lattice. A *rectangular primitive lattice* is a primitive lattice that has the symmetry of point group 2*mm*. Consider any single lattice point. This point has a twofold axis and two perpendicular mirror planes. These symmetry operations restrict γ to 90°. Lattice parameters *a* and *b* must be measured. The symbol for the rectangular primitive lattice from the *International Tables for Crystallography* is *op*, where *o* stands for orthorhombic and *p* for primitive.

4.2.3 Rectangular Centered

Figure 4.3 shows part of a rectangular centered lattice. A *rectangular centered lattice* is a centered lattice that has the symmetry of point group 2*mm*. There are two conventional unit cells. In the most commonly used cell, the rectangular cell, γ = 90° and *a* and *b* must be measured. Alternatively, a primitive rhombic unit cell may be chosen. In the rhombic unit cell *a′* = *b′* and γ′ must be measured. A rhombus is a parallelogram with two equal

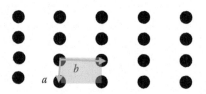

FIGURE 4.2   Rectangular primitive lattice, *op.*

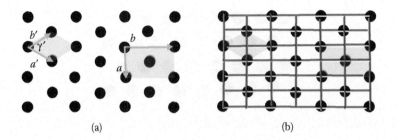

(a)   (b)

FIGURE 4.3 Rectangular centered lattice, *oc*. (a) Primitive and centered unit cells, (b) with mirror planes.

adjacent sides. The rectangular unit cell emphasizes the symmetry. In Figure 4.3b the horizontal mirrors are red and the vertical mirrors are blue. (The symbols for the twofold axes are omitted for clarity.) The **a** and **b** basis vectors lie along the mirrors. The transformation between the two sets of basis vectors is

$$a = 2\,a'\sin(\gamma'/2)$$

$$b = 2\,a'\cos(\gamma'/2).$$

The rectangular cell has two lattice points in the unit cell, while the rhombic cell is primitive. The point group of the lattice is independent of the choice of the basis vectors. The symbol for the rectangular centered lattice from the *International Tables for Crystallography* is *oc*, where *o* stands for orthorhombic and *c* for centered.

4.2.4 Square

Figure 4.4 shows part of a square lattice. A *square lattice* is a primitive lattice that has the symmetry of point group 4*mm*. Consider any single lattice point. This point has a fourfold axis and two sets of perpendicular mirror planes. These symmetries require $a = b$ and $\gamma = 90°$. Lattice parameter a must be measured. The symbol for the square lattice from the *International Tables for Crystallography* is *tp*, where *t* stands for tetragonal and *p* for primitive.

4.2.5 Hexagonal

Finally, Figure 4.5 shows part of a hexagonal lattice. A *hexagonal lattice* is a primitive lattice that has the symmetry of point group 6*mm*. Consider any single lattice point. This point has a sixfold axis and three sets of perpendicular mirror planes. These symmetries require $a = b$ and $\gamma = 120°$. Lattice parameter a must be measured. Although this lattice also has

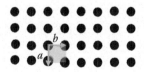

FIGURE 4.4 Square lattice, *tp*.

FIGURE 4.5 Hexagonal lattice, *hp*, primitive and centered cells.

TABLE 4.1 Point Groups of the Five Two-Dimensional Bravais Lattices

Bravais Lattice	Symbol for Bravais Lattice	Point Group of the Bravais Lattice
Oblique, *p*	*mp*	2
Rectangular, *p*	*op*	2*mm*
Rectangular, *c*	*oc*	2*mm*
Square, *p*	*tp*	4*mm*
Hexagonal, *p*	*hp*	6*mm*

an orthogonal set of basis vectors **a′**, **b′** that forms a rectangular centered cell, the primitive hexagonal unit cell is used. The symbol for the hexagonal lattice from the *International Tables for Crystallography* is *hp*, where *h* stands for hexagonal and *p* for primitive. Table 4.1 summarizes the point groups of the five two-dimensional Bravais lattices.

4.3 CRYSTAL SYSTEMS AND THE **G** MATRICES

In Chapter 3, Table 3.14, the 10 point groups were classified according to their crystal systems. Consider the oblique Bravais lattice *mp*, defined by point group 2. This lattice is associated with the oblique system. The resulting **G** matrix has the most general form and is a function of *a*, *b*, and γ. Table 4.2 shows the relationships among the crystal systems, the point groups, the Bravais lattices, and the **G** matrices.

TABLE 4.2 Two-Dimensional Crystal Systems with the Point Groups of the Bravais Lattices in Red

Crystal System	Point Groups	Bravais Lattices	G-matrix
Oblique	1,2	Oblique, *mp*	$\mathbf{G} = \begin{pmatrix} a^2 & ab\cos\gamma \\ ab\cos\gamma & b^2 \end{pmatrix}$
Rectangular	*m*,2*mm*	Rectangular, *op* Rectangular, *oc*	$\mathbf{G} = \begin{pmatrix} a^2 & 0 \\ 0 & b^2 \end{pmatrix}$
Square	4,4*mm*	Square, *tp*	$\mathbf{G} = \begin{pmatrix} a^2 & 0 \\ 0 & a^2 \end{pmatrix}$
Hexagonal	3,3*m*, 6,6*mm*	Hexagonal, *hp*	$\mathbf{G} = \begin{pmatrix} a^2 & -\dfrac{a^2}{2} \\ -\dfrac{a^2}{2} & a^2 \end{pmatrix}$

Both the rectangular primitive Bravais lattice *op* and the rectangular centered Bravais lattice *oc* have the point group 2*mm*. These lattices are associated with the rectangular system. The **G** matrix in column 4 is a function of *a* and *b*. In the rectangular lattice γ = 90° and cos γ = 0.

The square Bravais lattice *tp* and the hexagonal lattice *hp* are defined by the point groups 4*mm* and 6*mm*, respectively. These lattices are associated with the square and hexagonal systems, respectively. Their **G** matrices are given in column 4. In the square lattice γ = 90° and cos γ = 0. In the hexagonal lattice γ = 120° and cos γ = −1/2 (see Exercise 4.1).

For a given crystal system, the associated Bravais lattice has the point group with the highest symmetry.

4.4 TWO-DIMENSIONAL SPACE GROUPS

A *space group* is the symmetry group of a regularly repeating infinite pattern. The diagram of the sheet of HMB unit cells at the beginning of this chapter suggests the idea of symmetry operations that pave space. The space group is the mathematical description of the symmetry of a crystal structure. Since it is assumed that the crystal never ends, edge effects are not considered. A two-dimensional space group is also called a plane group or, informally, a wallpaper group. In this book the term "space group" is used for both two and three dimensions.

The operation that paves space is called translation. *Translation* is an operation in which every point of an object is displaced so that its position after the displacement minus the original position equals

$$\mathbf{t}(u,v) = u\,\mathbf{a} + v\,\mathbf{b}.$$

In general, *u* and *v* are not necessarily integers. Figure 4.6 shows a green plant translated by the displacement vector

$$\mathbf{t}(-2,2) = -2\mathbf{a} + 2\mathbf{b}.$$

(see Exercise 4.2).

First, an overview is given. Then the space groups are considered individually.

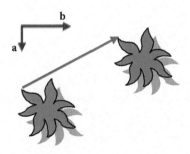

FIGURE 4.6   A green plant translated by the displacement vector **t**(−2,2) = −2**a** + 2**b.**

4.4.1 Overview

There are 17 two-dimensional space groups. Every space group has a unique point group associated with it. The first 13 space groups are generated by combining each point group, which has a finite order, with an appropriate Bravais lattice, which has infinite order. These are the symmorphic space groups. A *symmorphic space group* has at least one point in each unit cell with the symmetry of its associated point group. Table 4.3 uses the Hermann–Mauguin notation. For example, in this notation the symbol *p2mm* means that the unit cell is primitive (Note the lower case italic *p*.) and that the associated point group is 2*mm*. The lower case *c* in *c2mm* means that the Bravais lattice is centered.

In addition to the rotation axes and the mirror plane, another symmetry which incorporates translation is needed. This symmetry is called a glide. A glide is the product of a mirror and a translation to be explained later. The symbol for the glide is *g*.

In addition to the symmorphic space groups, there are four more space groups. These are the nonsymmorphic space groups. A *nonsymmorphic space group* does not have any point in the unit cell with the symmetry of its associated point group. The nonsymmorphic groups are the logical complement of the symmorphic groups. In two dimensions, these space groups contain a glide as a generator of the space group. The glide is indicated in the notation. See Table 4.3, column four.

The Herman–Mauguin notation is designed to make the associated point group obvious from the symbol for the space group.

For example, from the notation, space group *p2mg* is primitive and nonsymmorphic. To get the associated point group remove the translation component of the symbol, which in this case is the *p*, and replace the *g* with an *m*. Thus space group *p2mg* is associated with point group 2*mm*. The use of the lower case *p* and *c* in the space groups indicates two dimensions. These space groups and their symmetry operations are developed in detail later in this chapter.

TABLE 4.3 The Two-Dimensional Space Groups

Bravais Lattice	Point Groups	Symmorphic Space Groups	Nonsymmorphic Space Groups
Oblique, *mp*	1	*p*1	
	2	*p*2	
Rectangular, *op*	*m*	*pm*	*pg*
	2*mm*	*p2mm*	*p2mg, p2gg*
Rectangular, *oc*	*m*	*cm*	
	2*mm*	*c2mm*	
Square, *tp*	4	*p*4	
	4*mm*	*p4mm*	*p4gm*
Hexagonal, *hp*	3	*p*3	
	3*m*	*p31m, p3m1*	
	6	*p*6	
	6*mm*	*p6mm*	

There are two ways of inserting an object with $3m$ symmetry into the hexagonal unit cell. The mirror can be along the long diagonal or along the short diagonal. This results in two distinct space groups, $p31m$ and $p3m1$. This example is discussed later in this chapter (see Exercise 4.3).

4.4.2 Space Group, No. 1, $p1$

The first space group, No. 1, in the *International Tables for Crystallography*, is $p1$. In Chapter 3, the horse is given as an example of a motif with point group 1. Insert this motif into a unit cell and apply translations both horizontally and vertically. These translations are infinite in number and pave the entire plane. In two dimensions the translation generators are indicated by $\mathbf{t}(1,0)$ and $\mathbf{t}(0,1)$. Figure 4.7 shows six unit cells of the horses and of the $C_9H_4F_4O$ molecules. Since the displacement vector $\mathbf{t}(u,v) = u\,\mathbf{a} + v\,\mathbf{b}$, the displacement vector $\mathbf{t}(1,0) = \mathbf{a}$; and the displacement vector $\mathbf{t}(0,1) = \mathbf{b}$. The generators for this space group are the symmetry operation 1 combined with the translation generators. In Figure 4.7 the integer u runs from 1 to 2 and v runs from 1 to 3. Figure 4.8 shows the placement of the origin and the orientations of the axes from the *International Tables for Crystallography*. The origin is in the upper left-hand corner. In this space group the $\mathbf{a}$ and $\mathbf{b}$ vectors are not orthogonal; the $\mathbf{b}$ axis is horizontal pointing to the right and the $\mathbf{a}$ axis points downward. This arrangement is used for all space groups.

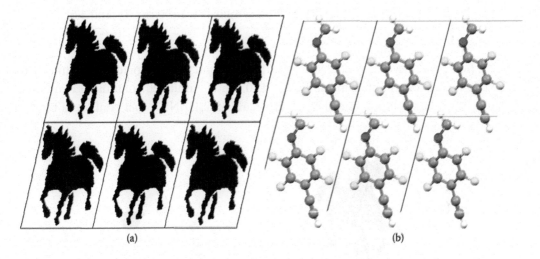

(a)   (b)

FIGURE 4.7   Examples of $p1$. (a) Horses, (b) $C_9H_4F_4O$. (CSD: AYOHUR.)

FIGURE 4.8   Origin and orientation of the basis vectors $\mathbf{a}$ and $\mathbf{b}$.

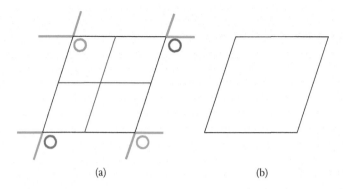

(a)   (b)

FIGURE 4.9   Space group $p1$. (a) General position diagram, (b) symmetry diagram.

Each space group is illustrated with both a general position diagram and a symbol diagram. They have the same information. If the symbol diagram is known, the general position diagram can be drawn and vice versa. Figure 4.9a shows the general position diagram for the space group $p1$. The unit cell is drawn in black. The green extensions and the blue lines inside the unit cell serve as guides for the general position diagram. The green circle, representing a typical point, has fractional coordinates x,y; the red circle has fractional coordinates $x + 1, y$; the blue circle has fractional coordinates $x, y + 1$; and the purple circle has fractional coordinates $x + 1, y + 1$. In Figure 4.9b the only symmetry operation is the identity, which is indicated by the unit cell without any symmetry symbol (see Exercises 4.4 and 4.5).

4.4.3 Space Group, No. 2, $p2$

Take the horse and insert a twofold axis, creating a second rotated horse. The combination of the two horses forms a new motif. Figure 4.10 shows a unit cell containing two horses related by a twofold rotation and shows a molecule of $C_6H_{12}N_2O_2$ that contains a twofold

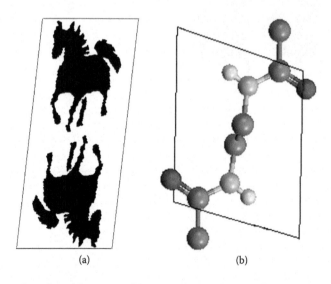

(a)   (b)

FIGURE 4.10   Examples of $p2$. (a) Horses, (b) $C_6H_{12}N_2O_2$ molecule. (CSD-ABAWEG.)

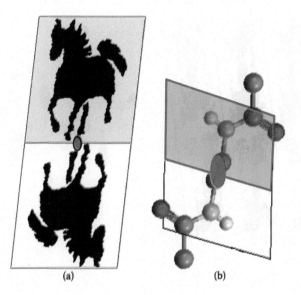

FIGURE 4.11 Space group *p2* with the asymmetric unit shown in pink and the rotation axis in red for (a) horses, (b) molecule.

rotation. Each cell will be translated to create a two-dimensional paving of the plane. Each cell is primitive because there is only one motif or translational component per unit cell. However, there is a relationship between the two horses. Starting with the upper horse, the rotation generates the bottom half of the unit cell. In Figure 4.11 the part of each unit cell in pink is the asymmetric unit. The *asymmetric unit* is the smallest part of the unit cell that, when operated on by the symmetry operations, produces the whole unit cell. The asymmetric unit is not unique. For example, the bottom part of the unit cell could be chosen. This book adheres to the *International Tables for Crystallography*, which gives the asymmetric unit for *p2* as $0 \leq x \leq 1/2$; $0 \leq y \leq 1$. Begin with the unit cell in Figure 4.10 and apply translations horizontally and vertically. Figure 4.12 shows six unit cells (see Exercise 4.6).

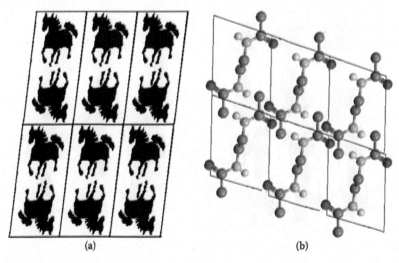

FIGURE 4.12 Six unit cells of a pattern with space group *p2* for (a) horses, (b) molecules.

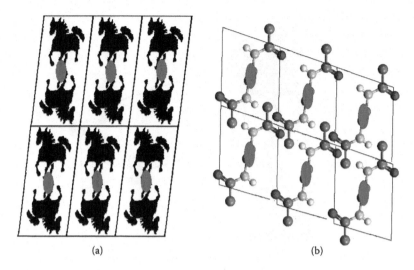

(a)　　　　(b)

FIGURE 4.13　The symbol for a twofold rotation axis, colored in red, is inserted into each unit cell for (a) horses, (b) molecules.

In Figure 4.13 the symbol for a twofold rotation axis, colored in red, is inserted in each unit cell. This rotation was used to generate the original motif. Careful inspection reveals other rotations. For example, in Figure 4.14 the twofold rotations with axes colored green take the head of one horse and rotate it into the head of the horse below it. The molecular version is similar. Now follow the other rotations with axes colored blue and yellow. These other rotations, all different, because their axes have different neighborhoods, are generated by the first rotation *plus* the translation operations. The axes of all the green rotations have the same neighborhood. These axes form a lattice. The axes of the blue rotations form a lattice labeled with a different origin. The lattice points of the same origin have the same color. There is nothing special about the red rotations. The entire figure could have been generated by the blue rotations plus translations.

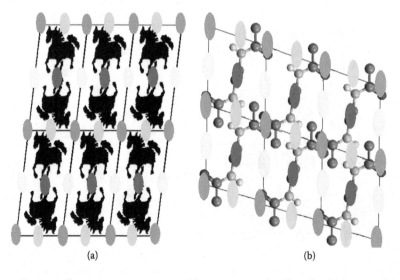

(a)　　　　(b)

FIGURE 4.14　Generated rotation operations in blue, green, and yellow for (a) horses, (b) molecules.

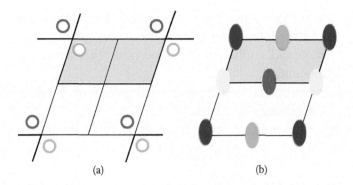

FIGURE 4.15 Space group $p2$. (a) General position diagram, (b) symmetry diagram. Asymmetric unit in pink.

Figure 4.15 shows the general position diagram and the symbol diagram. The green circle, representing a typical point, has fractional coordinates x,y. All the green circles are related by unit lattice translations. The red circle is related to the green circle by a rotation indicated in blue in the symbol diagram. The other rotations are green, red, and yellow, following the notation in Figure 4.14. Any of the rotations can be considered to be inserted, and the rest then are generated. Thus, in space group $p2$, the single twofold rotation with its axis at the origin and the translations generate three more rotation axes. A *position* is a set of symmetry-equivalent points. In $p2$ the green and red circles taken together represent the set of symmetry equivalent points that form the general position. The asymmetric unit is shown in pink. The lower half of the cell is generated from the asymmetric unit by the red rotation. The area of the asymmetric unit is one-half the area of the unit cell (see Exercise 4.7).

Figure 4.16 shows part of the entry in the *International Tables for Crystallography* for the two-dimensional space group $p2$. The word "Oblique" in the upper left-hand corner gives the crystal system. The "2" in the middle of the heading gives the associated point group

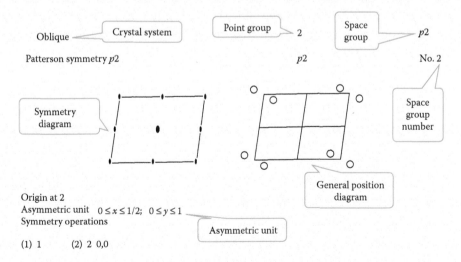

FIGURE 4.16 Part of the entry for space group $p2$. (From the *International Tables for Crystallography*, A.)

and the "*p2*" on the right gives the space group. The "No. 2" refers to the number of the space group chosen for the *International Tables for Crystallography*.

The general position and symbol diagrams are next. They are not in color; so care is needed in their interpretation.

If there is a symmetry operation at the origin, it is given next. In this case, a twofold axis is placed at the origin indicated by "Origin at 2."

The asymmetric unit is indicated by inequalities. Here $0 \leq x \leq 1/2$; $0 \leq y \leq 1$. This is the area in pink in Figure 4.15. For space groups in which the inequalities are difficult to interpret, the vertices of the asymmetric unit are given.

The symmetry operations are labeled in sequence and are indicated by their symbols. In this example the entries "(1) 1" and "(2) 2 0,0" mean that there are two symmetry operations and that the first is the identity, 1, and the second is a twofold rotation with its axis placed at the origin, 2 0,0.

Other examples of the space group *p2* are found in

- Section 9.6.1 for DL-leucine along [001], [100], [010]

- Section 10.6.1 for sucrose along [010]

- Section 13.6.1 for the boron compound $H_{12}B_{12}{}^{2-}$, $3K^+$, Br^- along $[1\,\overline{1}0]$

4.4.4 Space Group, No. 3, *pm*

In Chapter 3, the letter **B** was given as an example of a motif with point group *m*. Insert this motif into the rectangular primitive Bravais lattice to obtain the space group *pm* (see Table 4.3). A Bravais lattice is chosen that allows the entire pattern to have a mirror. Therefore, the mirror is a global symmetry. A mirror plane m_y operates on the top half of the letter **B** and produces the whole letter. Figure 4.17a shows the motif **B** in the unit cell with the mirror m_y drawn at 1/2, *y*; Figure 4.17b shows the asymmetric unit in pink, and Figure 4.17c shows two molecules of $C_{14}H_{12}FNO_2$ in a unit cell. In Figure 4.18, the unit cells in Figure 4.17 are translated nine times for the letter B and six times for the molecular figure. The mirror planes are emphasized by the color red. There are three red mirrors.

In space group *pm*, the red mirror at 1/2,*y* plus the translation operations generates another distinct mirror at 0, *y*. The generated mirrors are in green. Visualize the lattice continuing along **a** to understand why the top and bottom green mirrors are included. With a first mirror at 1/2, *y* the operation of translation along **a** generates a second mirror at 0,*y*. The two mirrors are distinct: the red mirrors go through the center of the letter **B** and the green mirrors go between the rows of **B**s. In other words, the neighborhood of a red mirror plane is different from the neighborhood of a green mirror plane. The molecular version is similar. Start with the red mirrors and then greens are generated by the translation. Likewise, start with the green mirrors and the reds are generated. It does not matter which mirror is selected first, the second mirror is generated by the translation.

Figure 4.19a shows the general position diagram for the space group *pm*. The unit cell has the two symmetry operations, the identity and a mirror. In Chapter 3, Table 3.6, a

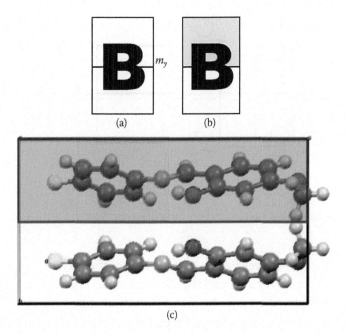

FIGURE 4.17 Space group *pm*. (a) Unit cell with mirror m_y drawn at ½, y, (b) asymmetric unit in pink, (c) $C_{14}H_{12}FNO_2$ molecule. (CSD-DUMXUF.)

mirror at 0, y takes a point x, y into $\bar{x}, y$. The mirror, which is an improper operation, changes the handedness of the system.

A change of handedness is indicated by a comma inside the circle.

Operating the mirror *m* at 0, y in Figure 4.19b on the red open circle in Figure 4.19a produces the red circle with the comma in it. This symbol is not in the unit cell. Operating

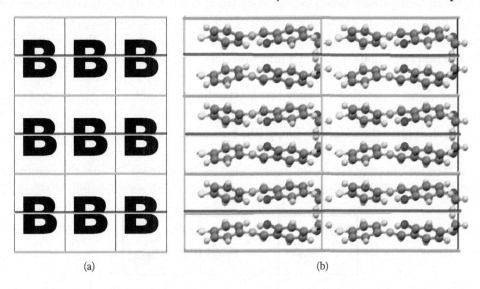

FIGURE 4.18 Space group *pm* with both green and red mirrors for (a) letter B, (b) molecules.

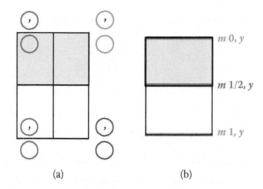

(a)   (b)

FIGURE 4.19 Space group *pm*. (a) General position diagram, (b) symbol diagram. The asymmetric unit is in pink.

the mirror *m* at 0, *y* on the green open circle produces the green circle with the comma. Similarly operating the mirror at 1, *y* on the purple and blue open circles produce their mirror images. In the symbol diagram, the mirror is indicated by a heavy solid line. The two distinct mirrors are in red and green, following the notation in Figure 4.18. Look at space group No. 3, *pm* in the *International Tables for Crystallography* (see Exercises 4.8 and 4.9).

4.4.5 Space Group, No. 4, *pg*

In addition to the rotation axes and mirror plane a new symmetry, incorporating translation, is needed. In this book the symmetry, for either two or three dimensions, is called a glide. Figure 4.20 shows three unit cells, each containing a set of left and right footprints. The left foot print is first; the second footprint, the right, is a mirror image of the first and is translated by one-half unit cell. A *glide* symmetry operation is a combination of a mirror and a translation of a half unit cell parallel to the mirror plane. In this case the mirror plane is along 1/2, *y* and the glide is in the *y* direction. The designation is *b* 1/2, *y*. The glide is represented by a dashed line, as shown in green. Space group *pg* is a nonsymmorphic space group.

Figure 4.21 shows two molecules of $C_4H_3F_3$ related by a red glide. There are two molecules in a unit cell. Figure 4.22 shows nine unit cells. By inserting a glide *b* at 1/2, *y* a second glide along *b* is generated at 0, *y*. The second glide is marked by a green dashed line. The two glides are distinct. Starting with the red glides, the green glides are generated. Likewise starting with the green glides, the red glides are generated.

Figure 4.23 gives the diagrams for *pg*. The two distinct glides are in red and green, following the notation in Figure 4.22. Point *x*, *y*, represented by the red circle inside the unit cell, is taken into point $\bar{x}$, *y* + 1/2, represented by the blue circle with the comma in it outside the unit cell, by a mirror across 0, *y* combined with a translation of 1/2 along **b**.

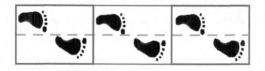

FIGURE 4.20 Glide operation illustrated by footprints.

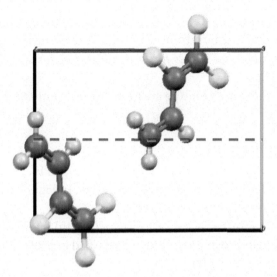

FIGURE 4.21   Glide operation illustrated by molecules of $C_4H_3F_3$. (CSD-AMECAW.)

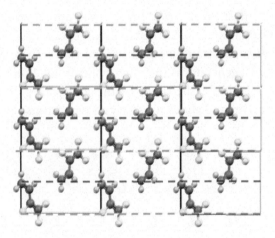

FIGURE 4.22   Nine unit cells with glides.

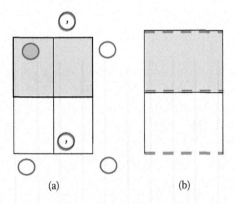

FIGURE 4.23   Space group *pg*. (a) General position diagram, (b) symmetry diagram. The asymmetric unit is in pink.

This operation is represented by b 0, y. The blue circle with the comma outside the unit cell is brought inside the unit cell by a translation of **a**. The asymmetric unit has $0 \leq x \leq 1/2$; $0 \leq y \leq 1$. See the *International Tables for Crystallography* for space group No. 4, *pg* (see Exercises 4.10 and 4.11).

Other examples of the space group *pg* are found in Section 10.6.1 for sucrose along [001] and [100].

4.4.6 Space Group No. 5, *cm*

In this symmorphic space group, a motif with point group *m* is combined with the rectangular centered Bravais lattice to give the space group *cm*. In Figure 4.24 the butterfly, which has point group *m*, is inserted into a centered cell.

In Figure 4.25 the red mirror planes are drawn. Careful inspection reveals other symmetries in Figure 4.25. Glide planes, indicated by dashed green lines, are half-way between the red mirrors, as shown in Figure 4.26.

Figure 4.27 shows the unit cell extracted from Figure 4.26 and rotated by 90°. The reason for this rotation is to have the same conformation as space group *cm* in the *International Tables for Crystallography*. The origin is chosen on the mirror *m* 0, *y*. The motif is translated by fractional coordinates, 1/2, 1/2. There are five generators for this space group. They include (1) 1, the identity, and translations **t**(1,0) and **t**(0,1). To get the centering there is **t**(1/2,1/2). Finally, the fifth generator is the mirror operation (2) *m* 0, *y*. The red solid lines are the mirror planes, the green dashed lines are the glide planes. The unit cell boundaries, perpendicular to the mirrors and glides, are freely chosen because no symmetry constrains

FIGURE 4.24   Example for the space group *cm*.

FIGURE 4.25   Space group *cm* with red mirrors.

FIGURE 4.26   Space group *cm* with red mirrors and green glides.

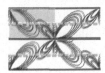

FIGURE 4.27   Unit cell for *cm*. The asymmetric unit is in pink.

their position. The asymmetric unit is in pink. The glide plane goes through the asymmetric unit, but the mirror planes are on the boundaries.

Mirror planes never go through the interior of an asymmetric unit.

Figure 4.28 shows the general position diagram and the symbol diagram for *cm*. The point x, y is represented by the red circle inside the unit cell. The red mirror $m\ 0, y$ operates on the red circle, generating the purple circle with the comma. The translation $\mathbf{t}(1/2,1/2)$ takes the point x, y, the red circle, and translates it to $x + 1/2, y + 1/2$, which is represented by the green circle. The mirror at $m\ 1/2, y$ takes the green circle into the blue circle with the comma. The neighborhood of the origin and the neighborhood of 1/2, 1/2 are identical. The Bravais lattice is rectangular centered.

The relationship between the red circle and the blue circle with a comma is a glide. This glide is generated. Here is an example of a symmorphic group with a glide! On the symbol

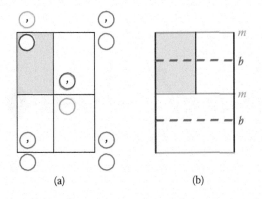

FIGURE 4.28   Space group *cm*. (a) General position diagram, (b) symmetry diagram.

diagram the glides are indicated by green dashed lines. In this case the centering plus the mirror produces a glide along the *b* direction. The glide is indicated as *b* 1/4, *y*. A second glide is generated at *b* 3/4, *y*. The asymmetric unit has $0 \leq x \leq 1/2$; $0 \leq y \leq 1/2$. Again note that the mirror planes form boundaries of the asymmetric unit, and a glide plane goes through the interior of the asymmetric unit (see Exercises 4.12 and 4.13).

4.4.7 Space Group No. 6, *p2mm*; Space Group No. 7, *p2mg*

Space groups, *p2mm* and *p2mg* both have Bravais lattices with point group *2mm*. Space group *p2mm* is symmorphic while *p2mg* is not. In the symmorphic space group, *p2mm*, a motif with point group *2mm* is combined with the rectangular primitive Bravais lattice to give the space group. For example, consider the molecule in Figure 4.29a. The lattice is a rectangular primitive lattice. Figure 4.29b shows the crystallographic symbols superimposed on the *2mm* pattern. The molecular pattern is repeated to form the periodic pattern in Figure 4.30a. Figure 4.30b shows the perpendicular purple mirror planes with the red twofold rotation axes on the intersecting mirror planes. Figure 4.31 shows the unit cell for the example of *p2mm*. The origin of the unit cell is chosen at the upper left hand on the twofold axis. The asymmetric unit, in pink is $0 \leq x \leq 1/2$; $0 \leq y \leq 1/2$. The symmetries present are red twofolds, purple mirrors, and the translations. Compare the symmetries for a single object with point group *2mm* symmetry in Figure 4.29 with the unit cell of a pattern with *p2mm* symmetry in Figure 4.31.

Figure 4.32 shows the general position diagram and the symbol diagram for *p2mm*. There are four circles in the unit cell of the general position diagram and thus the asymmetric unit is one-fourth of the unit cell (see Exercises 4.14 and 4.15).

The asymmetric unit always has only one circle.

By contrast space group No. 7, *p2mg*, is a nonsymmorphic space group.

In contrast with the previous examples in which the motif is inserted into the appropriate Bravais lattice, the following example begins with the generators of the space group and constructs the general position diagram and the symbol diagram.

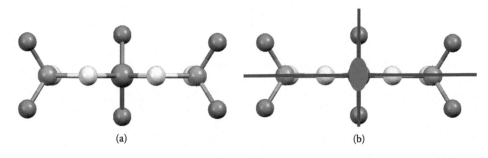

(a)   (b)

FIGURE 4.29   (a) Molecule of $C_6H_{18}As_2S_4Sn$ with point group *2mm*. (b) Crystallographic symbols superimposed on molecule. (CSD-CERJEO.)

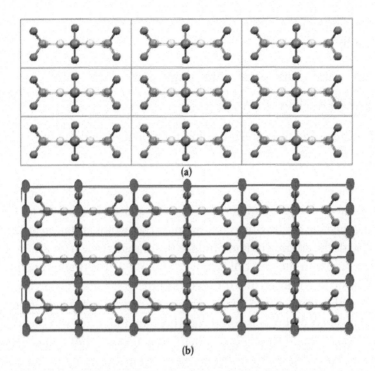

FIGURE 4.30 Example of space group *p2mm*. (a) Nine unit cells, (b) with graphical symbols.

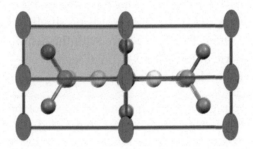

FIGURE 4.31 Unit cell for *p2mm*.

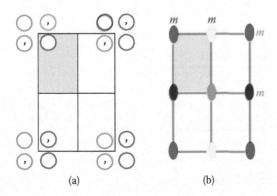

FIGURE 4.32 Space group *p2mm*. (a) General position diagram, (b) symbol diagram. The asymmetric unit is shown in pink.

The generators for space group *p2mg* are (in addition to the identity and the translations along the **a** and **b**)

- A twofold rotation with its axis at the origin and

- A mirror with its plane at 1/4, *y*

Figure 4.33a shows the nine green twofold rotation axes and the two red mirror planes at 1/4*y* and 3/4*y*. Figure 4.33b starts out with the green circle at *x,y* and translates it to four nearest cells. These circles are green to show that they are related by translation. The twofold rotation at the origin operates on the green circle at *x,y* and generates the red circle at $\bar{x}, \bar{y}$; similar pairs are at each of the four corners. The green circle is operated on by the mirror at 1/4,*y* producing the blue circle with a comma in it. The twofold rotation at 1/2,0 operates on the blue circle with the comma in it to produce the orange circle with a comma. The orange circle with the comma in the unit cell is at $x + (1/2), \bar{y} + 1$. The orange and blue circles have commas to indicate that they are mirror images of the green and red circles. The black lines are the outline of the unit cell and do not represent any symmetry operation.

The general position diagram is complete. There are four circles in the unit cell. Careful consideration reveals a glide along **a**. The symbol diagram is not complete until the glide is inserted, as in Figure 4.34. This nonsymmorphic space group has an associated point group of 2*mm* (see Table 4.3). Point group 2*mm* has a twofold axis at the intersection of two perpendicular mirrors. This space group *p2mg* has twofold axes along the glide plane. No twofold axis lies on a mirror plane. In other words no single point in *p2mg* has the symmetry of the point group 2*mm*. This is in accord with the definition of a nonsymmorphic space group. The asymmetric unit has $0 \le x \le 1/4; 0 \le y \le 1$.

Figure 4.35 shows a urea molecule projection with *p2mg* symmetry.

In contrast with the previous example in which the generators of the space group were given and the general position and symbol diagrams were constructed, the next example begins with a periodic pattern and determines the general position and symbol diagrams. Later in the chapter, this process of assignment will be generalized with a flow chart.

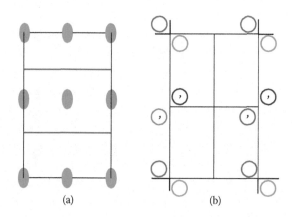

(a)   (b)

FIGURE 4.33   Space group *p2mg*. (a) INCOMPLETE symbol diagram, (b) general position diagram.

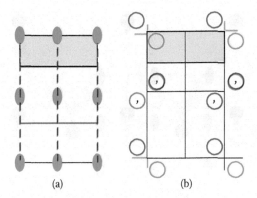

FIGURE 4.34   Space group *p2mg*. (a) Symbol diagram (b) general position diagram. The asymmetric unit is in pink.

Figure 4.36 shows the unit cell from Figure 4.35. In analyzing this pattern to construct the symbol diagram, first look for periodicity and then look for the highest-order rotation axis. Here it is a twofold. Mark the nine twofolds and then look for mirror planes. Now mark the mirror planes. In this case there are two vertical mirror planes. Next, look for glide planes. In this case there are three horizontal glide planes. The asymmetric unit is colored pink. Use a mirror for a boundary of the asymmetric unit. The asymmetric unit is one-quarter of the unit cell. This means that the space group has four symmetry operations. Next, analyze the pattern to construct the general position diagram. Put a circle in the asymmetric unit representing the identity operation. Use the first vertical mirror to construct the circle with the comma in it. The rotation axis at 1/2, 1/2 creates the next circle with a comma in it. Finally, the second mirror is used to create the final circle in the upper right-hand corner. Now there are four circles, one for each symmetry operation. As a final check, match the diagrams with those in the *International Tables for Crystallography*. In this case the symbol diagram and the general position diagram are drawn at 90° to the diagrams in the *International Tables for Crystallography* (see Exercises 4.16 and 4.17).

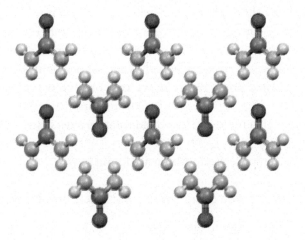

FIGURE 4.35   Urea, CH_4N_2O, projection with *p2mg* symmetry. (CSD UREAXX23.)

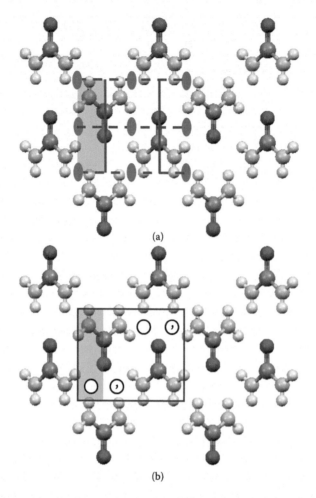

FIGURE 4.36 Urea, CH_4N_2O, projection with *p2mg* symmetry. (a) Symbol diagram, (b) general position diagram.

Figure 4.37 compares the symmorphic space group *p2mm* with the nonsymmorphic space group *p2mg*. Both space groups have the associated point group *2mm*, whose symbol stereographic projection is shown in the upper right-hand corner. Remember that the definition of a symmorphic space group is that it has at least one point in the unit cell with the symmetry of the associated point group. In the case of *p2mm every* twofold rotation axis sits on the intersection of a pair of perpendicular mirrors. This contrasts with the symbol diagram for *p2mg*, in which none of the twofold rotation axes sits at an intersection of perpendicular mirror planes.

Other examples of the space group *p2mm* are found in

- Section 13.7.1 for the boron compound $H_{12}B_{12}{}^{2-}, 3K^+, Br^-$ along $[2\,\overline{1}\,\overline{1}]$
- Section 14.6.4 for magnesium along [210]
- Section 15.6.1 for acetylene along [210]

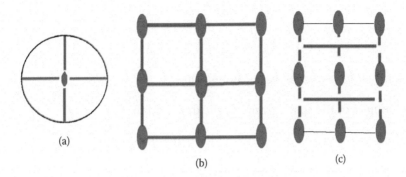

FIGURE 4.37   Symbol diagrams. (a) Point group 2*mm*, (b) symmorphic space group *p*2*mm*, (c) nonsymmorphic space group *p*2*mg*.

Other examples of the space group *p*2*gm* are found in

- Section 11.6.1 for polyethylene along [001]
- Section 12.6.1 for α-cristobalite along [110]
- Section 14.6.4 for magnesium along [100]
- Section 15.6.1 for acetylene along [100]

4.4.8 Space Group No. 8, *p*2*gg*; Space Group No. 9, *c*2*mm*; Space Group No. 10, *p*4; Space Group No. 11, *p*4*mm*; Space Group No. 12, *p*4*gm*

These space groups are left as exercises since the concepts needed to construct and analyze them have already been given. In Chapter 3 there is an example of *p*4 in Figure 3.15, and in this chapter Figure 4.38 is an example of *p*4*gm*. Start the analysis by searching for the

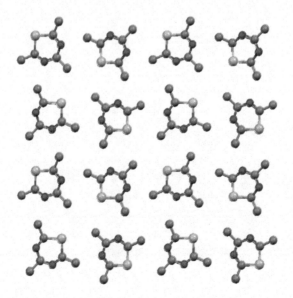

FIGURE 4.38   Space group *p*4*gm*: projection of molecules of $C_6H_{12}O_2Se$. (CSD-MSEPRL20.)

highest-order rotation axis. Begin by looking for a sixfold rotation and, if that is not there, look for the next lowest rotation, which is a fourfold. A fourfold can be seen in Figure 4.38.

Another example of the space group *p2gg* is found in

- Section 12.6.1 for α-cristobalite along [100]

Other examples of the space group *c2mm* are found in

- Section 11.6.1 for polyethylene along [100] and [010]

An example of the space group *p4gm* is found in

- Section 12.6.1 for α-cristobalite along [001]

For more assistance consult the flow chart given later in the chapter (see Exercises 4.18 through 4.26).

4.4.9 Space Group No. 13, *p3*

Space group *p3* is symmorphic. An example of a pattern with this space group is analyzed next. Consider the projection of molecules of $C_2H_5N_5$ in Figure 4.39. To analyze a pattern first find the highest-order rotation axis. The symmetry must be global. Consider the center of the three green hydrogen atoms. Place a red triangle indicating a threefold axis. This is the origin of the lattice. Now put red triangles on the other lattice points. The unit cell is created by joining the lattice points in Figure 4.40. Now look inside the unit cell. Three pink hydrogen atoms have a threefold axis at their center. Mark this with a green triangle. Note that the environment of a red triangle is different from that of the green triangle. A third crystallographically distinct threefold is marked with a blue triangle. There are no mirrors or glides (see Figure 4.41).

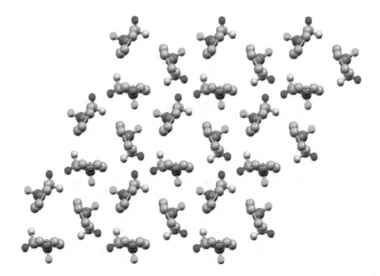

FIGURE 4.39 Projection of molecules of $C_2H_5N_5$, giving space group *p3*. The three crystallographically distinct hydrogen atoms are colored green, pink, and white. (CSD-KUQCAB.)

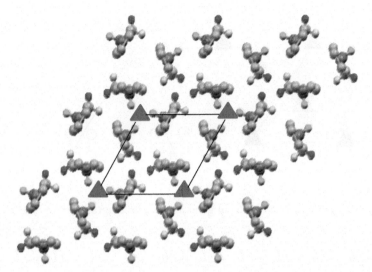

FIGURE 4.40   The preceding pattern, showing the threefold axes at the vertices of the unit cell.

The symbol diagram, Figure 4.42a, is taken directly from Figure 4.41. The next step is to draw the general position diagram. In Figure 4.42b, the outline of the unit cell and the guide lines are in black. The threefold axis at the origin acts on the blue circle at x,y, which is chosen to represent the identity, producing circles at x, y; $\bar{y}, x - y$; and $\bar{x} + y, \bar{x}$—as in Chapter 3. The generators for $p3$ are the identity, the translations along **a** and **b**, and the 3^+ with its axis at the origin. The asymmetric unit is $0 \leq x \leq 2/3$; $0 \leq y \leq 2/3$; $y \leq \min (1 - x, (1 + x)/2)$. Alternatively the *International Tables for Crystallography* gives the vertices; 0,0; 1/2,0; 2/3,1/3; 1/3,2/3; 0,1/2. The vertices are usually easier to understand than the inequalities. The asymmetric unit is in pink. Three of the vertices of the asymmetric unit are independent threefold axes as indicated by the three different colors. As always, there is only

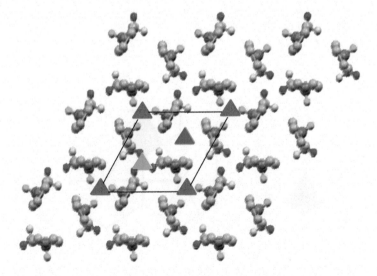

FIGURE 4.41   The preceding pattern with threefold axes in red, green, and blue.

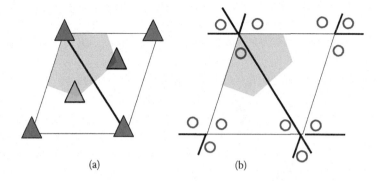

(a)   (b)

FIGURE 4.42   (a) Symbol diagram and (b) general position diagram for $p3$. The asymmetric unit is in pink.

one circle in the asymmetric unit. For a primitive cell, such as this one, the area of the asymmetric unit is the area of the unit cell divided by the number of circles in the unit cell. As always, each circle represents a symmetry operation. In this case there are three circles within the unit cell; therefore, the area of the asymmetric unit for $p3$ is 1/3 of the total area of the unit cell (see Exercises 4.27 through 4.29).

The relationship of the asymmetric unit to the unit cell for the hexagonal space groups is clarified by considering Figure 4.43. This construction applies not only to the space group $p3$, which is presently under consideration, but also to the space groups $p3m1$, $p31m$, $p6$, and $p6mm$, which will be discussed next. Figure 4.43a shows a hexagon partitioned into six equal parts with one of the slices colored red. In Figure 4.43b four hexagons are assembled, and the hexagonal unit cell is outlined in blue. The four sides of the unit cell connect the centers of the four hexagons. Note that the unit cell consists of six slices, one of which is colored red. This slice is the asymmetric unit used for space groups $p31m$ and $p6$. In each case the asymmetric unit is 1/6 of the unit cell. Two of the slices taken together are colored pink. This is the asymmetric unit for the space group $p3$ and has an area 1/3 of the unit cell. The light blue section, a triangle, shows one of the slices cut in half; and this is the asymmetric unit for $p6mm$. This asymmetric unit has an area of 1/12 of the unit cell.

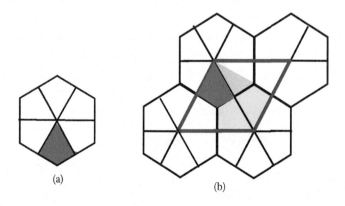

(a)   (b)

FIGURE 4.43   (a) Hexagon divided into six slices, (b) relationship of the hexagonal unit cell, with blue boundaries, to hexagons paving space.

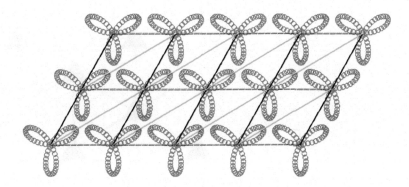

FIGURE 4.44 Example of *p3m1* with green mirror planes along the long diagonals of the unit cells.

FIGURE 4.45 Example of *p31m* with mirror, in green, along the short diagonal.

4.4.10 Space Group No. 14, *p3m1* and Space Group No. 15, *p31m*

The next two space groups involve inserting a mirror plane into the unit cell of *p3*. There are two possible ways to do this, either along the long diagonal or along the short diagonal. The space group *p3m1* is produced if the mirror is inserted along the long diagonal and the space group *p31m* is produced if the mirror is inserted along the short diagonal. Note the difference between the symbols for these space groups. These are two distinct symmorphic space groups. Consider the motif for the point group 3*m* from Section 3.6.8 and insert it into a hexagonal lattice.

In the first case, the space group *p3m1*, the mirror plane is along the long diagonal of the unit cell (see Figure 4.44).

For the second case, the space group *p31m*, rotate the motif until its mirror plane is along the short diagonal of the unit cell (see Figure 4.45). The angle for the rotation is 30°.

Close examination of Figures 4.46 and 4.47 reveals distinct general position and symbol diagrams. A striking difference is in the asymmetric unit. Thus *p3m1* and *p31m* are distinct space groups. Both symmorphic space groups generate glide planes (see Exercise 4.30).

4.4.11 Space Group No. 16, *p6*; Space Group No. 17, *p6mm*

These space groups are left as exercises since the concepts needed to construct and analyze them have already been given. Begin the analysis with a search for the highest-order rotation

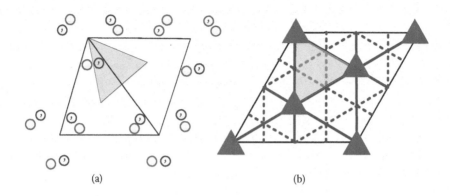

(a)

(b)

FIGURE 4.46 Space group *p3m1*. (a) General position diagram, (b) symbol diagram. The asymmetric unit is in green.

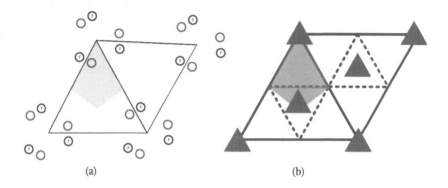

(a)

(b)

FIGURE 4.47 Space group *p31m*. (a) General position diagram, (b) symbol diagram. The asymmetric unit is in green in (a) and in pink in (b).

(a)

(b)

FIGURE 4.48 Figures illustrating. (a) Space group *p6*, (b) space group *p6mm*. (Courtesy of Prof. Slavik V. Jablan.)

axis. First look for a sixfold rotation. Sixfold axes are in both Figures 4.48a and b. For more assistance consult the flow chart given later in the chapter to distinguish the two figures.

Other examples of the space group *p6mm* are found in

- Section 13.7.1 for the boron compound $H_{12}B_{12}^{2-}$, $3K^+$, Br^- along [111]

- Section 14.6.4 for magnesium along [001]
- Section 15.6.1 for acetylene along [001]

(see Exercises 4.31 and 4.32).

4.5 COLOR CODING AND OVERVIEW OF THE TWO-DIMENSIONAL SYMBOL DIAGRAMS

Color coding of the symbol diagrams according to change of handedness gives insights into their relationships. This color coding is then applied to all 17 of the two-dimensional space groups in one figure. An overview of the asymmetric unit is given.

4.5.1 Color Coding

Figure 4.49 gives the color scheme for the two-dimensional rotation axes, mirror planes, and glide planes. This figure is an extension of the color codes introduced for the point groups in Table 3.15. The symmetry operations in red have no change of handedness. Note that all the rotation axes have only red symmetry operations and therefore are colored red. The symmetry operations in blue have a change of handedness. The mirrors and the glides have both red and blue symmetry operations; thus the mirrors and glides are colored purple.

4.5.2 Symbol Diagrams with Their Asymmetric Units

Next all 17 space groups are compared and contrasted. Space is paved when all symmetry operations of the space group act on the asymmetric unit. The symmetry operations are rotations, mirrors, glides, and translations. By contrast, the *unit cells* pave space with *translation* operations only. Figure 4.50 shows the relationship between the unit cell and

2-D Rotation axes			2-D Planes		
Printed symbol	Graphical symbol	Symmetry operations	Printed symbol	Graphical symbol	Symmetry operations
1	None	1			
2	⬮	1, 2	m g	▬▬▬ ▬ ▬	1, m 1, g
3	 ▲	1, 3⁺, 3⁻			
4	 ◆	1, 2, 4⁺, 4⁻			
6	 ⬡	1, 2, 3⁺, 3⁻, 6⁺, 6⁻			

Symmetry operations in red: No change of handedness
Symmetry operations in blue: Change of handedness

FIGURE 4.49 Two-dimensional rotation axes, mirror planes, and glide planes.

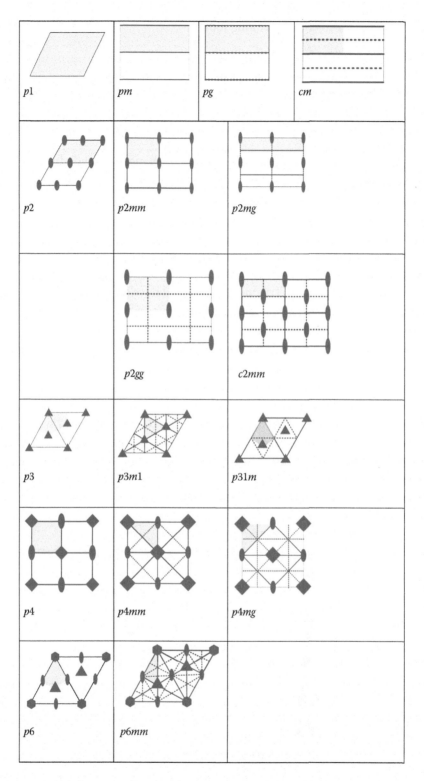

FIGURE 4.50 Symbol diagrams showing the relationship between the unit cell and the asymmetric unit (in green) for the 17 two-dimensional space groups.

the asymmetric unit for all 17 two-dimensional space groups. This diagram is one of the most important diagrams in the book as it summarizes two-dimensional crystallography.

If there is a mirror plane in the unit cell, then at least one edge of the asymmetric unit must lie along a mirror plane. Look, for example, at space group *pm*. The horizontal mirror takes the asymmetric unit into the bottom of the unit cell. A mirror plane can *never* be in the interior of the asymmetric unit. If it were, then applying that mirror to the asymmetric unit would duplicate material within the asymmetric unit. On the other hand, a glide plane may be found in the interior of the asymmetric unit, as seen in *p2mg*, because the translation part of the glide takes the material outside the asymmetric unit.

If there is a rotation axis, then it must lie on a vertex or a side of the asymmetric unit. A rotation axis is *never* inside the asymmetric unit. If it were, then applying that rotation to the asymmetric unit would duplicate material within the asymmetric unit. In space group *p2* the rotation with its axis at the center of one side of the asymmetric unit takes the asymmetric unit into the bottom half of the unit cell. If a higher-order rotation axis—3, 4, or 6—is present, then it lies on at least one of the vertices of the asymmetric unit. For example, the asymmetric unit in the space group *p6* has four vertices; and each vertex has a rotation axis. Specifically, there are a sixfold, a threefold, and two twofolds.

The area of the asymmetric unit is related to the area of the unit cell, A_{uc}, by the number of lattice points per unit cell, n, and by the number of operations in the associated point group, h.

$$A_{au} = \frac{A_{uc}}{n\,h}$$

Note also that the product nh is equal to the number of symmetry equivalent points found in the CIF.

There is one lattice point in a primitive unit cell and there are two lattice points in a rectangular centered unit cell. The area of the asymmetric unit is either equal to or less than that of the unit cell. For example, the unit cell and the asymmetric unit are identical in the space group *p1*. This is the only space group for which this is true. In the space group *c2mm*, there are two lattice points in the unit cell, $n = 2$; and there are four symmetry operations in the associated point group, which is *2mm*; thus $h = 4$. Therefore, the area of the asymmetric unit is one-eighth of the area of the unit cell.

Consider the general position diagrams; for example, for the space group *p2mg* in Figure 4.36b. Note that there is exactly one circle in the asymmetric unit. This is true for all asymmetric units.

Populated asymmetric units are found for

- Space group *p2* for DL-leucine in Section 9.6.1, for sucrose in Section 10.6.1, for the boron compound $H_{12}B_{12}{}^{2-}$,3 K^+, Br^- in Section 13.7.1

- Space group *pg* for sucrose in Section 10.6.1

- Space group *p2mm* for the boron compound $H_{12}B_{12}{}^{2-}$, 3 K^+, Br^- in Section 13.7.1 and for magnesium in Section 14.6.4

- Space group *p2mg* for polyethylene in Section 11.6.1, for α-cristobalite in Section 12.6.1, for magnesium in Section 14.6.4, and for acetylene in Section 15.6.1

- Space group *p2gg* for polyethylene in Section 11.6.1, α-cristobalite in Section 12.6.1, and for acetylene in Section 15.6.1

- Space group *c2mm* for polyethylene in Section 11.6.1

- Space group *p4gm* for α-cristobalite in Section 12.6.1

- Space group *p6* for acetylene in Section 15.6.1

- Space group *p6mm* for the boron compound in Section 13.7.1 and for magnesium in Section 14.6.4

(see Exercises 4.33 and 4.34).

4.6 RECIPE FOR ANALYZING A PERIODIC PATTERN

Earlier a pattern with space group *p3* was analyzed and the symbol and general position diagrams were drawn. A flowchart is given for the general procedure in Figure 4.51.

In the chart there are two places where the question "Is cell centered" appears. This question is asked in order to distinguish between *p2mm* and *c2mm*, and also between *pm* and *cm*. In the case of distinguishing between *p2mm* and *c2mm*, the question could have been "In addition to a mirror through the twofold is there a glide?" In the case of distinguishing between *pm* and *cm*, the question could have been "Are there both a mirror and a glide?" For simplicity, the question "Is cell centered?" has been chosen for the chart.

The assignment of the space group is done entirely through symmetries.

4.7 PRIMITIVE CELLS FOR *cm* AND *c2mm*

This section prepares for the space group tree given in the next section. Of the 17 two-dimensional space groups, 15 are primitive and 2 are centered. The two centered space groups are *cm* and *c2mm*. For these two space groups, a *multiple* cell is chosen by the *International Tables for Crystallography* because in both cases at least one basis vector then coincides with a mirror plane. In this section the general position and symbol diagrams are described for the corresponding *primitive* unit cells.

4.7.1 Primitive Cell for *cm*

Consider the case of *cm* which is illustrated in Figure 4.52. Two unit cells are indicated. The pink unit cell has two motifs per unit cell and is a centered cell. The green unit cell has only one motif per cell and is a primitive cell. In both cases, the pattern inside the unit cell is characterized by alternating mirrors and glides.

Figure 4.53 shows the corresponding general position diagrams for *cm* with their corresponding asymmetric units. Both unit cells in Figure 4.53 from Figure 4.52 have been rotated to conform to the *International Tables for Crystallography*. Careful examination of

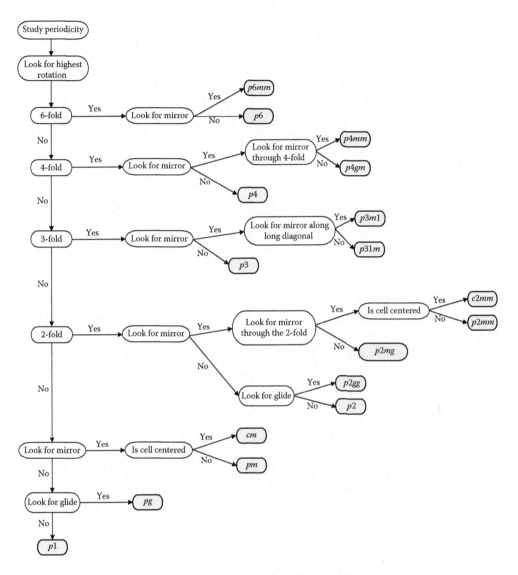

FIGURE 4.51 Flow chart to assign two-dimensional space groups.

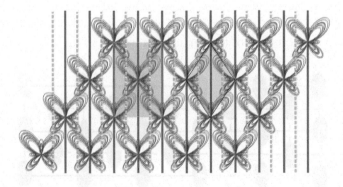

FIGURE 4.52 Pattern showing centered (pink) and primitive (green) unit cells for space group No. 5, *cm*.

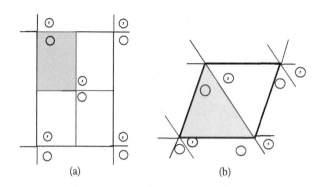

FIGURE 4.53   Centered (a) and primitive (b) general position diagrams for the space group No. 5, *cm*.

the two cells shows that the pattern of the general position in the centered cell is related to that of the primitive cell by a rotation, as seen in Figure 4.52. The area of the asymmetric unit of the rectangular centered cell is one-quarter of the area of the unit cell. In contrast, the area of the asymmetric unit of the primitive cell is one-half of the area of *its* unit cell. The primitive cell has the shape of a rhombus. Since the mirror of the primitive cell is along the diagonal of the unit cell, then the edge of the asymmetric unit must follow along the mirror. The area of the asymmetric unit is the same in either case, $ab/4$, where a and b are the lattice parameters for the rectangular centered unit cell.

4.7.2 Primitive Cell for *c2mm*

Next consider the space group *c2mm*. The space group, as always, is determined by the symmetry of the pattern. In Figure 4.54, the pattern is the symbol diagram for *c2mm* continued for more than one cell. The pattern consists of two sets of twofolds: one set has perpendicular mirror planes and the other set has perpendicular glide planes. Both centered (blue) and primitive (green) unit cells are overlaid on the pattern. The associated point group is *2mm*.

Figure 4.55 shows the isolated symbol diagrams for the centered and primitive unit cells for space group *c2mm*. The asymmetric unit for the centered cell has mirror planes on three sides and a glide plane on the fourth side. Two of the vertices have rotation axes and there is another rotation axis in the middle of the side that contains a glide plane. The primitive unit cell is a rhombus. Its asymmetric unit is a triangle with mirror planes on two sides and rotation axes on each of the vertices, as well as in the center of the non-mirror

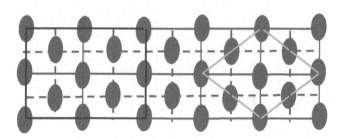

FIGURE 4.54   Unit cells for *c2mm*: centered (blue) and primitive (green).

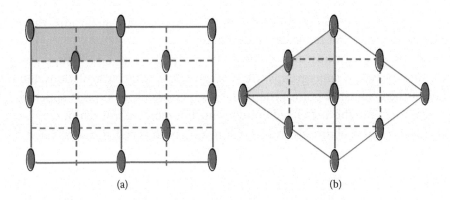

FIGURE 4.55 Symbol diagrams for centered (a) and primitive (b) unit cells for space group *c2mm*.

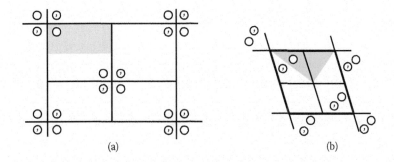

FIGURE 4.56 General position diagrams for centered (a) and primitive (b) unit cells for space group *c2mm*.

side. The glide planes are internal. The areas of both asymmetric units are equal, as they must be. The area is $ab/8$ where a and b are the lattice parameters of the centered cell.

Now consider the general position diagrams in Figure 4.56. In both the centered and primitive cells the asymmetric unit has only one circle at x,y, as must be so.

The information about the relationships between the general position and symbol diagrams of the centered and primitive unit cells is necessary to construct the space group tree.

4.8 TWO-DIMENSIONAL SPACE GROUP TREE

Relationships between crystal structures imply relationships between their space groups. Just as the point groups are related through supergroups and subgroups, so are the space groups. See Sections 3.8 and 3.12. The space group tree is more intricate than the point group tree. The order of a space group is infinite because the corresponding idealized crystal structure is infinite. However, since each space group has an associated point group, the order of that point group can be associated with the space group.

There are two ways to build a space group tree. The first way uses *t*-subgroups (for *translationengleiche*). The *t-subgroups*, or type I subgroups, are subgroups that keep translations, but reduce the order of the associated point group. The second way uses *k-subgroups* (for *klassengleiche*). The *k-subgroups*, or type II subgroups, are subgroups that keep the associated point group, but reduce the translations.

Similarly the supergroups are divided into *t*-supergroups and *k*-supergroups. The *t*-supergroups, or type II supergroups, are supergroups that keep translations, but increase the order of the associated point group. The *k*-supergroups, or type II supergroups, are supergroups that keep the associated point group, but increase translations.

In this book, only *t*-subgroups and *t*-supergroups, both type I, are considered.

EXAMPLE 4.1

Determine the maximal subgroups and minimal supergroups for space group *p2*.

Solution

Consider space group *p2*. The associated point group is the point group 2. From the point group tree in Chapter 3, the point group 2 has one maximal subgroup—the point group 1—and three minimal supergroups—the point groups 4, 2*mm*, and 6 (see Figure 4.57).

First, consider the maximal, that is to say, largest subgroups. The point group 1 is a subgroup of the point group 2. Keep the primitive cell and consider the largest wholly contained subgroups. In this case there is only one, the point group 1. Thus *p1* is the maximal subgroup of *p2*. Second, consider the minimal supergroups of the point group 2. These point groups are 2*mm*, 4, and 6. The new space groups are generated by adding a mirror, a glide, or an *n*-fold rotation with *n* greater than 2. The translational part is kept; thus the new cell is primitive. This example is guided by the symbol diagrams for the 17 space groups in the *International Tables for Crystallography*.

Case 1: In Figure 4.58, begin with *p2* and place a blue mirror plane along [0,1] through the twofold axis at the origin. The symmetries generate the other blue mirrors and cause the crystal system to go from oblique to rectangular. The result is space group *p2mm*, which is symmorphic and is associated with point group 2*mm*.

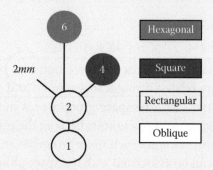

FIGURE 4.57 The maximal subgroup and the three minimal supergroups for the point group 2.

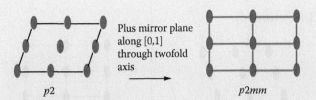

FIGURE 4.58 Path from *p2* to *p2mm*.

Case 2: In Figure 4.59, again begin with the parallelogram for *p2* and place a blue mirror plane along [11] through the twofold axis at the origin. The symmetries generate the other blue mirrors and glides. The symmetries restrict the shape of the unit cell to a rhombus. Examination of Figure 4.55 shows that this symbol diagram represents the primitive cell for the symmorphic space group *c2mm*. The crystal system has gone from oblique to rectangular. The associated point group is *2mm*.

Case 3: In Figure 4.60, begin with *p2* and place a blue glide plane along [10] through the twofold axis at the origin. The symmetries generate more blue glide planes and blue mirror planes at 1/4, *y* and 3/4, *y*. The crystal system goes from oblique to rectangular. The result is the nonsymmorphic space group *p2mg* with the associated point group 2*mm*.

Case 4: In Figure 4.61, begin with *p2* and insert the blue glide plane *b* 1/4*y*. The symmetries generate another blue glide plane *a x* 1/4. The crystal system goes from oblique to rectangular. The result is the nonsymmorphic space group *p2gg* with the associated point group 2*mm*.

Case 5: In Figure 4.62, begin with *p2* and combine a threefold rotation with the twofold rotation at the origin. Both point group 2 and point group 3 are subgroups of point group 6. Thus, a sixfold rotation is generated at the origin. The crystal system

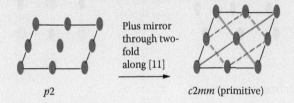

FIGURE 4.59 Path from *p2* to *c2mm* (primitive).

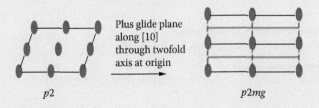

FIGURE 4.60 Path from *p2* to *p2mg*.

(continued)

EXAMPLE 4.1 (continued)

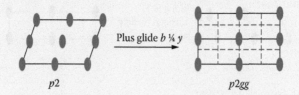

FIGURE 4.61   Path from *p*2 to *p*2*gg*.

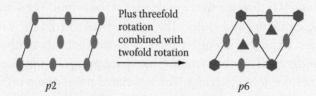

FIGURE 4.62   Path from *p*2 to *p*6.

goes from oblique to hexagonal. The result is the symmorphic space group *p*6 with the associated point group 6.

Case 6: In Figure 4.63, begin with *p*2 and replace the twofold rotation at the origin with a fourfold rotation. The point group 2 is a subgroup of the point group 4. The crystal system goes from oblique to square. The result is the symmorphic space group *p*4 with the associated point group 4.

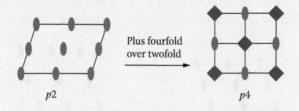

FIGURE 4.63   Path from *p*2 to *p*4.

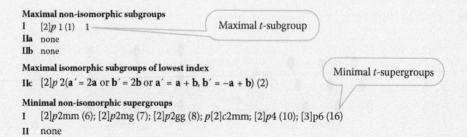

Maximal non-isomorphic subgroups
I   [2]*p* 1 (1)   1 ——————— Maximal *t*-subgroup
IIa   none
IIb   none

Maximal isomorphic subgroups of lowest index
IIc   [2]*p* 2(**a**′ = 2**a** or **b**′ = 2**b** or **a**′ = **a** + **b**, **b**′ = −**a** + **b**) (2)   Minimal *t*-supergroups

Minimal non-isomorphic supergroups
I   [2]*p*2mm (6); [2]*p*2mg (7); [2]*p*2gg (8); *p*[2]*c*2mm; [2]*p*4 (10); [3]p6 (16)
II   none

FIGURE 4.64   Maximal *t*-subgroup and minimal *t*-supergroups for space group *p*2. (From the *International Tables for Crystallography, A.*)

Summary: Space group *p2* has one maximal *t*-subgroup namely *p*1. Space group *p2* has six minimal *t*-supergroups namely *p2mm*, *p2mg*, *p2gg*, *c2mm*, *p4*, and *p6*. This information is included in the *International Tables for Crystallography* for each space group under type I. (see Figure 4.64). In Figure 4.65 part of the space group tree is given for *p2*.

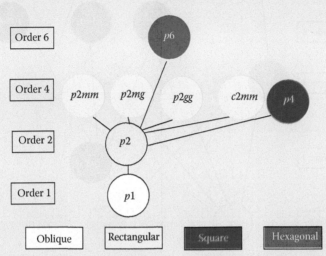

FIGURE 4.65 Maximal subgroup and minimal supergroups of space group *p2*. The order is the order of the point group associated with a particular space group.

The procedure used on space group *p2* can be expanded similarly to include all the relationships among the two-dimensional space groups. A completed tree is in Figure 4.66. The order in the figure refers to the order of the point group associated with each space group (see Exercises 4.35 and 4.36).

4.9 SUMMARY OF TWO-DIMENSIONAL SPACE GROUPS

The flow chart in Figure 4.51 identifies the space group of a pattern. Both Figure 4.66 and Table 4.4 classify the space group in several different ways. The figure shows the space group tree and the table summarizes the relationships among the two-dimensional space groups, the crystal systems, the point groups, and the Bravais lattices.

The space groups can be either symmorphic or nonsymmorphic. Each space group has a point group associated with it. Some point groups have more than one space group associated with them. For example, the point group *m* is associated with space groups *pm*, *pg*, and *cm*. Space groups may be classified according to their Bravais lattices. Each space group has a definite Bravais lattice. For example, the space group *pm* has the Bravais lattice *op*. On the other hand, the Bravais lattice *op* has five space groups: *pm*, *p2mm*, *pg*, *p2mg*, and *p2gg*. Finally, the space groups are classified by four crystal systems. Again each space group has a definite crystal system. For example, *p2* is in the oblique crystal system. The oblique crystal system has two space groups, *p*1 and *p2*.

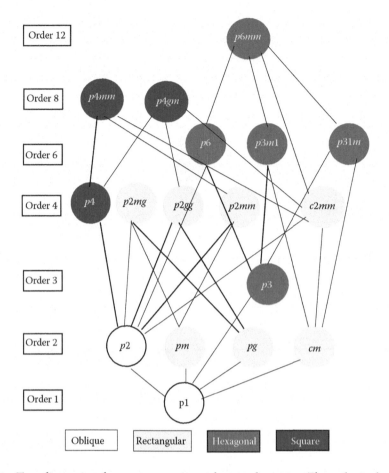

FIGURE 4.66 Two-dimensional space group tree with crystal systems. The order is the order of the point group associated with a given space group.

TABLE 4.4 Review of the Relationships among the Two-Dimensional Space Groups

| Crystal System | Point Group | Space Group | | Bravais Lattice |
		Symmorphic	Nonsymmorphic	
Oblique	1, 2	p1, p2		mp
Rectangular	m, 2mm	pm, p2mm	pg, p2mg, p2gg	op
		cm, c2mm		oc
Square	4, 4mm	p4, p4mm	p4gm	tp
Hexagonal	3, 6, 3m, 6mm	p3, p31m, p3m1, p6, p6mm		hp
4 crystal systems	**10 point groups**	**17 space groups**		**5 Bravais lattices**

PART II: THREE DIMENSIONS

4.10 THREE-DIMENSIONAL BRAVAIS LATTICES

In three dimensions, there are 14 Bravais lattices. As presented in Chapter 1, the lattice of a crystal is an array of points with identical neighborhoods. The lattice points are described by $\mathbf{t}(u,v,w) = u\,\mathbf{a} + v\,\mathbf{b} + w\,\mathbf{c}$ where $u,v,$ and w are integers and $\mathbf{a}, \mathbf{b},$ and $\mathbf{c}$ are basis vectors.

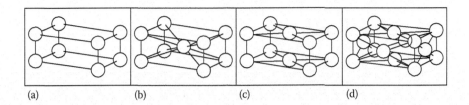

(a)   (b)   (c)   (d)

FIGURE 4.67 (a) Primitive, (b) body-centered, (c) C-centered, and (d) face-centered unit cells. (From the *International Tables for Crystallography, A.*)

The integers u, v, and w each range from $-\infty$ to $+\infty$. This infinite range makes the lattice infinite. The lattice is chosen to be infinite, as an idealization, to avoid concern about what happens at the surface of a crystal.

As was seen earlier in this chapter, lattices are categorized according to symmetry. Most point group symmetries place restrictions on one or more of the lattice parameters a, b, c, α, β, and/or γ. Each lattice has its own point group. Some Bravais lattices share the same point group. For example, the three cubic lattices share the point group $m\overline{3}m$. The Bravais lattices are distinct, but the subtlety of the distinctions between them is greater than for two dimensions.

The 14 *Bravais lattices* are classified into seven *lattice systems*—triclinic, monoclinic, orthorhombic, tetragonal, hexagonal, rhombohedral, and cubic. Each Bravais lattice is characterized by its unit cell.

The 14 conventional unit cells for the Bravais lattices, as defined by the *International Tables for Crystallography*, are chosen for this book and described in detail in the following subsections.

See Figure 4.67. A *primitive unit cell*, *P*, has just one lattice point with fractional coordinates 0,0,0. As in Section 1.3.3, a unit cell can be chosen to have multiple lattice points. A lattice point may be centered on a face. To be specific, choose the *C face-centered unit cell*, *C*, which has two lattice points, one at the origin and another at 1/2, 1/2, 0. A *body-centered unit cell*, *I*, has two lattice points, whose fractional coordinates are 0,0,0 and 1/2,1/2,1/2. A *face-centered unit cell*, *F*, has four lattice points, whose fractional coordinates are 0,0,0; 0,1/2,1/2; 1/2,0,1/2; and 1/2,1/2,0. There is also a hexagonal triple cell that is dealt with later in this section.

Figure 4.68 shows the conventional unit cell for each of the 14 Bravais lattices. These lattices can be further classified into seven Bravais lattice systems, which are next described individually from the least symmetric to the most symmetric.

4.10.1 Triclinic

A *triclinic P lattice* is the Bravais lattice that has point group $\overline{1}$. From Chapter 3, this point group has two symmetry operations: the identity and the inversion. There is one lattice point per unit cell at 0,0,0. All of the lattice parameters a, b, c, α, β, and γ are measured. The symbol for the primitive Bravais lattice is aP, where a stands for "anorthic."

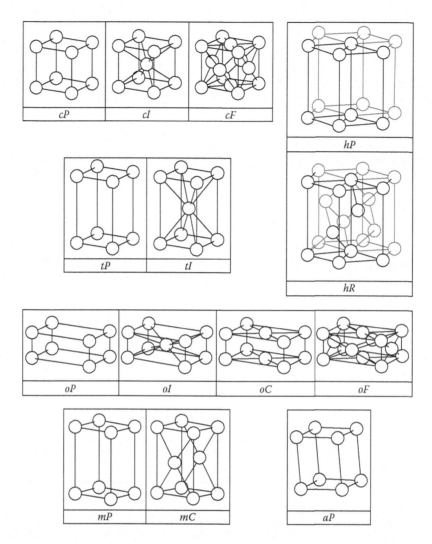

FIGURE 4.68 Conventional unit cells for the fourteen three-dimensional Bravais lattices. (From the *International Tables for Crystallography, A.*)

4.10.2 Monoclinic

There are two distinct monoclinic Bravais lattices, each with point group 2/*m*. From Chapter 3, a twofold axis is perpendicular to a mirror plane. This point group has four symmetry operations. If **b** is chosen along the twofold axis, with **a** and **c** in the mirror plane, then only the lattice parameters *a*, *b*, *c*, and β are measured. Lattice parameters α and γ are exactly 90°. The unit cell may be either primitive or centered on one face (see Figure 4.68). The *monoclinic P lattice* is a primitive Bravais lattice that has point group 2/*m*. There is one lattice point in the unit cell at 0,0,0. The *monoclinic C lattice* is a *C*-centered Bravais lattice that has point group 2/*m*. There are two lattice points in the unit cell at 0,0,0 and 1/2, 1/2, 0. To be specific the *C*-face has been chosen. The symbols for the monoclinic Bravais lattices are *mP* and *mC*, respectively.

4.10.3 Orthorhombic

There are four distinct orthorhombic Bravais lattices, each with point group *mmm*. From Chapter 3, there are three perpendicular mirrors that generate the point group. The eight symmetry operations include three mutually perpendicular twofold axes at the intersection of the mirrors. Choose **a**, **b**, and **c** along these perpendicular twofold axes. Lattice parameters *a*, *b*, and *c* are measured. Lattice parameters α, β, and γ are exactly 90°. This unit cell may be primitive, centered on one face, body-centered, or face-centered. See Figure 4.68. The *orthorhombic P lattice* is a primitive Bravais lattice that has the point group *mmm*. There is one lattice point in the unit cell at 0,0,0. The *orthorhombic C lattice* is a *C*-centered Bravais lattice that has the point group *mmm*. There are two lattice points in the unit cell at 0,0,0 and 1/2, 1/2, 0. To be specific the C-face has been chosen. The *orthorhombic I lattice* is a body-centered Bravais lattice that has the point group *mmm*. There are two lattice points in the unit cell at 0,0,0 and 1/2, 1/2, 1/2. (To remember that the symbol "*I*" refers to a body-centered cell, think "inner.") Finally, the *orthorhombic F lattice* is a face-centered Bravais lattice that has the point group *mmm*. There are four lattice points in the unit cell at 0,0,0; 0,1/2,1/2; 1/2,0,1/2; and 1/2,1/2,0. The symbols for the orthorhombic Bravais lattices are *oP*, *oC*, *oI*, and *oF*, respectively.

4.10.4 Tetragonal

There are two distinct tetragonal Bravais lattices, each with the point group 4/*mmm*. The point group has a fourfold axis and three perpendicular mirrors. There are 16 symmetry operations including three mutually perpendicular axes—the fourfold axis and two twofold axes. Choose **c** along the fourfold axis, and **a** and **b** along the twofold axes. Lattice parameters *a* and *c* are measured. Lattice parameters *a* and *b* are exactly equal and lattice parameters α, β, and γ are exactly 90°. The unit cell may be primitive or body centered (see Figure 4.68). The *tetragonal P lattice* is a primitive Bravais lattice that has the point group 4/*mmm*. There is one lattice point in the unit cell at 0,0,0. The *tetragonal I lattice* is a body-centered Bravais lattice that has the point group 4/*mmm*. There are two lattice points in the unit cell at 0,0,0 and 1/2, 1/2, 1/2. The symbols for the tetragonal Bravais lattices are *tP* and *tI*, respectively.

4.10.5 Hexagonal

The *hexagonal P lattice* is a Bravais lattice with the point group 6/*mmm*. There is a mirror plane perpendicular to the sixfold axis. This point group has 24 symmetry operations including two twofold axes perpendicular to the sixfold axis with an angle of 120° between them. Choose the sixfold axis to be along **c** with **a** and **b** along these twofold axes. Lattice parameters *a* and *c* are measured. Lattice parameters *a* and *b* are equal, α and β are exactly 90°, and γ is exactly 120°. There is one lattice point in the unit cell at 0,0,0 (see Figure 4.68). The symbol for the hexagonal *P* Bravais lattice is *hP*.

4.10.6 Rhombohedral

The *hexagonal R lattice*, also known as the *rhombohedral lattice*, is a Bravais lattice that has point group $\bar{3}m$. The symbol "*R*" refers to the word "rhombohedral." There are 12

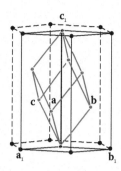

FIGURE 4.69   Bravais lattice hexagonal R showing both hexagonal and rhombohedral (red) reference axes. (From the *International Tables for Crystallography, A.*)

symmetry operations in this point group, including two twofold axes perpendicular to the threefold axis with an angle of 120° between them. Figure 4.69 shows two choices for the unit cell. In both choices there are two lattice parameters to be measured. One possibility, labeled "Hexagonal Axes" in the *International Tables for Crystallography*, is to have the threefold rotation axis along c_1 with a_1 and b_1 along the twofold axes. Lattice parameters a_1 and c_1 are measured. Lattice parameters a_1 and b_1 are equal, α_1 and β_1 are exactly 90°, and γ_1 is exactly 120°. This unit cell is like the unit cell for the hexagonal P lattice. However, this unit cell for the hexagonal R lattice is a multiple cell with three lattice points—0,0,0; 1/3, 2/3, 2/3; and 2/3, 1/3, 1/3. There is a second choice, labeled "Rhombohedral Axes" in the *International Tables for Crystallography*, to describe this lattice, in which the threefold rotation axis is along [111] and the twofold rotation axes are along $[1\bar{1}0]$ and $[01\bar{1}]$. This unit cell is primitive with one lattice point at 0,0,0. The lattice parameters a and α are measured with $a = b = c$ and $\alpha = \beta = \gamma$. The figure shows the obverse setting for the rhombohedral reference axes.

There are two ways to fit the rhombohedron into the hexagonal cell called obverse and reverse. The *International Tables for Crystallography* uses the obverse setting. The transformation from the primitive rhombohedral cell, $\mathbf{a}, \mathbf{b}, \mathbf{c}$ with rhombohedral axes of reference to the triple hexagonal cell, $\mathbf{a}_1, \mathbf{b}_1, \mathbf{c}_1$ with hexagonal axes of reference using the obverse setting is

$$(\mathbf{a}_1 \mathbf{b}_1 \mathbf{c}_1)_H = (\mathbf{a} \; \mathbf{b} \; \mathbf{c})_R \begin{pmatrix} 1 & 0 & 1 \\ \bar{1} & 1 & 1 \\ 0 & \bar{1} & 1 \end{pmatrix}$$

The symbol for this hexagonal rhombohedral Bravais lattice is hR. See Chapter 13 for a more detailed discussion of hexagonal and rhombohedral axes.

4.10.7 Cubic

There are three distinct cubic Bravais lattices, each with point group $m\bar{3}m$. This point group is generated by a threefold axis along the main diagonal of the cube, a twofold axis

TABLE 4.5   Three-Dimensional Bravais Lattices (hR Uses Hexagonal Axes)

Bravais Lattice	Number of Lattice Points in the Unit Cell	Point Group of the Bravais Lattice
Triclinic, aP	1	$\bar{1}$
Monoclinic, mP	1	$2/m$
Monoclinic, mC	2	$2/m$
Orthorhombic, oP	1	mmm
Orthorhombic, oC	2	mmm
Orthorhombic, oI	2	mmm
Orthorhombic, oF	4	mmm
Tetragonal, tP	1	$4/mmm$
Tetragonal, tI	2	$4/mmm$
Hexagonal, hP	1	$6/mmm$
Hexagonal, hR (R = rhombohedral)	3	$\bar{3}m$
Cubic, cP	1	$m\bar{3}m$
Cubic, cI	2	$m\bar{3}m$
Cubic, cF	4	$m\bar{3}m$

along a face diagonal, and an inversion point. This point group has 48 symmetry operations including three mutually perpendicular fourfold rotation axes. Choose **a**, **b**, and **c** along these fourfold axes. Only lattice parameter a must be measured. Lattice parameters $a = b = c$, and lattice parameters α, β, and γ are exactly 90°. The unit cells are primitive, body-centered, or face-centered (see Figure 4.68). The *cubic P lattice* is a primitive Bravais lattice that has the point group $m\bar{3}m$. There is one lattice point in the unit cell at 0,0,0. The *cubic I lattice* is a body-centered Bravais lattice that has the point group $m\bar{3}m$. There are two lattice points in the unit cell at 0, 0, 0 and 1/2, 1/2, 1/2. Finally, the *cubic F lattice* is a face-centered Bravais lattice that has the point group $m\bar{3}m$. There are four lattice points in the unit cell at 0, 0, 0; 0, 1/2, 1/2; 1/2, 0, 1/2; and 1/2, 1/2, 0. The symbols for the cubic Bravais lattices are cP, cI, and cF, respectively.

4.10.8 Summary of Bravais Lattices

Each Bravais lattice has a point group shown in Table 4.5. The table gives the number of lattice points in the conventional unit cell. Hexagonal axes are used for hR.

4.11 THREE-DIMENSIONAL SPACE GROUPS

Earlier in this chapter, the 17 two-dimensional space groups were introduced. Similarly, there are 230 three-dimensional space groups. The classification of these space groups is more difficult than that of the two-dimensional space groups.

4.11.1 Classification of the Space Groups

Figure 4.70 shows two different classifications of the 230 space groups. On the left side of the diagram the classification is by point groups. On the right side the classification is by

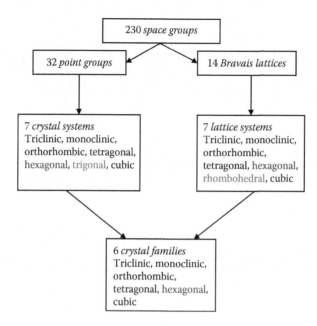

FIGURE 4.70 Classification of the three-dimensional space groups.

Bravais lattices. The *International Tables for Crystallography* uses both. These classifications arose for historical reasons.

In Chapter 3, the 32 crystallographic point groups were classified into seven crystal systems, including trigonal and hexagonal. The trigonal system consists of the point groups that include either a *single* threefold rotation axis or a *single* threefold rotoinversion axis. The hexagonal system consists of the point groups that include either a sixfold rotation axis or a sixfold rotoinversion axis.

On the other hand, the 14 Bravais lattices were developed in this chapter. A lattice is classified according to its own *point group*. There are seven point groups represented among the lattices. As was seen in the last section, the rhombohedral lattice has the point group $\bar{3}m$ and the hexagonal lattice has the point group $6/mmm$.

The bottom of the diagram shows a more general way of classifying the space groups by *crystal family*, as seen in Table 4.6. The hexagonal family includes both the trigonal and hexagonal crystal systems and also both Bravais lattices hP and hR. The six crystal families may be represented by six algebraic forms of the **G** matrix, as will be demonstrated in the next section. Since the **G** matrix depends on the particular basis vectors chosen, the conventional cells are selected.

The usage of the terms "triclinic," "monoclinic," "orthorhombic," "tetragonal," and "cubic" is consistent for crystal systems, Bravais lattices, and crystal families. Unfortunately the word "hexagonal" takes on three different meanings, depending on whether it is applied to crystal systems, Bravais lattices, or crystal families. The word "hexagonal" is found throughout the crystallographic literature, and caution must be used in interpreting it.

TABLE 4.6 Crystal Families with the Point Groups of Their Bravais Lattices Highlighted in Red

Crystal Family	Crystal System	Symmetry Requirements for Crystal Systems	Point Groups	Symmorphic Space Groups	Nonsymmorphic Space Groups	Number of Space Groups	Bravais Lattice
Triclinic (anorthic)	Triclinic	1	$1\ \bar{1}$	2	—	2	aP
Monoclinic	Monoclinic	2 or m	$2, m, 2/m$	6	7	13	mP mC
Orthorhombic	Orthorhombic	Two $\perp 2$ or two $\perp m$	$222, mm2, mmm$	13	46	59	oP oC oI oF
Tetragonal	Tetragonal	4 or $\bar{4}$	$4, \bar{4}, 4m, 422, 4mm, \bar{4}2m, 4/mmm$	16	52	68	tP tI
Hexagonal	Trigonal	Single 3	$3, \bar{3}, 32, 3m, \bar{3}m$	8	10	18	hP hR
	Hexagonal	6 or $\bar{6}$	$6, \bar{6}, 6/m, 622, 6mm, \bar{6}2m, 6/mmm$	8	19	27	hP
Cubic	Cubic	Two 3s	$23, m\bar{3}, 432, \bar{4}3m, m\bar{3}m$	15	21	36	cP cI cF
6 crystal families	7 crystal systems		32 point groups	73 symmorphic space groups	157 non symmorphic space groups	230 space groups	14 Bravais lattices

4.11.2 Minimal Symmetries for Crystal Systems

As in Figure 4.70, the crystal systems classify the point groups. The intention of the systems, embodied in the terminology, is to indicate the geometry of the unit cell that is useful for the related space groups. Now the minimal symmetries needed for each crystal system are examined. The point group tree from Chapter 3 may be used in finding the minimal symmetries (see Table 4.6).

When a lattice becomes populated with atoms such as the atoms of K, Al, S, and O in anhydrous alum, the space group often has an associated point group that is less symmetric than the point group of the lattice. For example, consider Figure 1.1 in Chapter 1 which shows cats and mice. The repeat unit consists of one mouse and one cat and the space group has the associated point group 1. Similarly, in a three-dimensional case, a populated triclinic lattice could have associated point group 1. The *triclinic crystal system* has 1 as its minimum symmetry.

For the *monoclinic crystal system*, the minimum symmetry is either 2 or *m*. If either of these symmetries exists, then two of the lattice angles are 90°.

For the *orthorhombic crystal system*, the minimum symmetry is the presence of two perpendicular twofold axes, or two perpendicular mirror planes. With these symmetries all three of the lattice angles are 90°. Note that two perpendicular twofold axes generate a third twofold axis perpendicular to the first two. Also two perpendicular mirror planes generate a twofold axis at the intersection of the two mirror planes.

For the *tetragonal crystal system*, the minimum symmetry is the presence of either a single fourfold rotation axis or a single fourfold rotoinversion axis. The presence of either requires that two of the lattice repeat distances are exactly equal, for example $a = b$, and that all three of the lattice angles are 90°.

For the *trigonal crystal system*, the minimum symmetry is the presence of a single threefold rotation axis. Its presence requires that two of the lattice repeat distances are exactly equal, for example, $a = b$, and that two of the lattice angles are 90° and one of them is 120°; for example, $\alpha = \beta = 90°$ and $\gamma = 120°$. Note that the presence of a threefold rotoinversion axis also requires the simultaneous presence of the threefold rotation axis. This can be seen in the point group tree, where point group 3 is a subgroup of $\bar{3}$.

For the *hexagonal crystal system*, the minimum symmetry is the presence of either a sixfold rotation axis or a sixfold rotoinversion axis. The presence of either of these requires that two of the lattice repeat distances be equal and that two of the lattice angles are 90°and one of them is 120°.

Finally for the *cubic crystal system*, the minimum symmetry is the presence *two* threefold rotation axes. As discussed in Chapter 3, if the point group contains two threefold rotation axes, then two more are generated. Thus a point group in the cubic crystal system contains four threefold rotation axes. The angle between any two threefold axes is the tetrahedral angle, 109°28'. These four threefold axes are along the four body diagonals of the unit cell. In the Hermann–Mauguin symbol for a point group in the cubic crystal system, either a 3 or $\bar{3}$, is placed in the second place in the symbol. It may come as a surprise that it is not necessary for a fourfold rotation axis to be present in the cubic system. However fourfold rotation axes or fourfold rotoinversion axes are present in three of the cubic point groups; namely—432, $\bar{4}3m$, and $m\bar{3}m$. If two threefold rotation axes are present, then all three of the lattice repeat distances are equal and all three of the lattice angles are 90°.

4.11.3 G Matrices

In Chapter 2, the G matrix was introduced to calculate bond distances, bond angles, and volumes for HMB, which is triclinic. The **G** matrix for the triclinic crystal family is in fact always correct.

The lattice parameters in three dimensions are a, b, c, α, β, and γ. As shown earlier in this chapter, symmetries put restrictions on the lattice parameters. Therefore, symmetries influence the G-matrix. Table 4.7 gives the **G** matrices for the six crystal families. The **G** matrices shown for the other crystal families are special versions of the triclinic case. The

TABLE 4.7 Three-Dimensional Crystal Families, **G**-Matrices, and Restrictions

Crystal Family	G-Matrix	Lattice Parameters That Are Measured in the Conventional Cell
Triclinic	$G = \begin{pmatrix} a^2 & ab\cos\gamma & ac\cos\beta \\ ab\cos\gamma & b^2 & bc\cos\alpha \\ ac\cos\beta & bc\cos\alpha & c^2 \end{pmatrix}$	$a, b, c, \alpha, \beta, \gamma$
Monoclinic	$G = \begin{pmatrix} a^2 & ab\cos\gamma & 0 \\ ab\cos\gamma & b^2 & 0 \\ 0 & 0 & c^2 \end{pmatrix}$	a, b, c, γ
Orthorhombic	$G = \begin{pmatrix} a^2 & 0 & 0 \\ 0 & b^2 & 0 \\ 0 & 0 & c^2 \end{pmatrix}$	a, b, c
Tetragonal	$G = \begin{pmatrix} a^2 & 0 & 0 \\ 0 & a^2 & 0 \\ 0 & 0 & c^2 \end{pmatrix}$	a, c
Hexagonal	$G = \begin{pmatrix} a^2 & -\dfrac{a^2}{2} & 0 \\ -\dfrac{a^2}{2} & a^2 & 0 \\ 0 & 0 & c^2 \end{pmatrix}$	a, c
Cubic	$G = \begin{pmatrix} a^2 & 0 & 0 \\ 0 & a^2 & 0 \\ 0 & 0 & a^2 \end{pmatrix}$	a

table uses the conventional cells. The Bravais lattice hexagonal R, also known as the rhombohedral lattice, is given with respect to hexagonal reference axes.

As was explained in detail in the previous section, in the case of the Bravais lattice hexagonal R, an alternative basis with rhombohedral axes can be chosen where the restrictions are $a = b = c$ and $\alpha = \beta = \gamma$. This unit cell is primitive. In this case the **G** matrix becomes

$$\begin{pmatrix} a^2 & a^2\cos\alpha & a^2\cos\alpha \\ a^2\cos\alpha & a^2 & a^2\cos\alpha \\ a^2\cos\alpha & a^2\cos\alpha & a^2 \end{pmatrix}$$

If hexagonal axes are used instead of rhombohedral axes, then all members of the hexagonal crystal family have the same form of the **G** matrix. On the other hand, it is often convenient to use rhombohedral axes as used in Chapter 13 for the structure of $H_{12}B_{12}^{2-}$, $3K^+$, Br^- (see Exercise 4.37).

4.11.4 Symmetry Operations

In addition to the rotation axes, rotoinversion axes and the mirror plane, more symmetries which incorporate translation are needed. These are glide planes and screw axes. The terms "glide" and "screw" are intentionally chosen to describe vividly these operations.

4.11.4.1 Glides

The glide in three dimensions resembles the glide in two dimensions. A *lattice vector* connects any two lattice points, provided there is no intermediate lattice point between the two end points. A *glide* is a combination of a mirror and a translation that is a fraction of a lattice vector in the plane of the mirror. In three dimensions the glide is more general than the two-dimensional glide. The glides include axial glides, double glides, diagonal glides, and diamond glides (see Table 4.8). There is also a transverse glide k for which there is no graphical symbol. The glides are colored purple as in the two-dimensional case in Section 4.5.1.

4.11.4.2 Screws

A screw operation is found only in three dimensions and is the combination of a rotation and a translation that is a fraction of the lattice vector along the rotation axis. An *n-fold screw operation*, indicated by the symbol n_p, is a rotation of $360°/n$ around an axis parallel to a lattice vector, **L**, combined with a translation of a distance $(p/n)L$, where p is a positive integer less than n, in the direction specified by the right-hand rule for vector products. There are 11 screw axes—2_1, 3_1, 3_2, 4_1, 4_2, 4_3, 6_1, 6_2, 6_3, 6_4, and 6_5. Table 4.9 shows the printed symbol, the graphical symbol, and the symmetry operations for all the screw axes. The graphical symbols of the screw axes are colored red. There is no change of handedness, and thus all the symbols and operations are in red as in the two-dimensional case for rotations discussed in Section 4.5.1.

TABLE 4.8 Glide Planes in Three Dimensions

Glide Plane	Letter Symbol	Graphical Symbol for Planes Normal to the Plane of Projection	Graphical Symbol for Planes Parallel to the Plane of Projection	Description
Axial glide	a b c			1/2 lattice vector normal to plane
Double glide, found only in centered cells	e			2 glide vectors 1/2 parallel to plane and 1/2 normal to plane
Diagonal glide	n			1 glide vector 1/2 parallel to plane and 1/2 normal to plane
Diamond glide, pairs of planes found only in centered cells	d		3/8 1/8	1/4 parallel to plane and 1/4 normal to plane

TABLE 4.9 Screw Axes

Printed Symbol	Graphical Symbol	Symmetry Operations	Printed Symbol	Graphical Symbol	Symmetry Operations
2_1		$1, 2_1$	6_1		$1, 3_1^+, 3_1^-, 2_1, 6_1^-, 6_1^+$
3_1		$1, 3_1^+, 3_1^-$	6_2		$1, 3_2^+, 3_2^-, 2, 6_2^-, 6_2^+$
3_2		$1, 3_2^+, 3_2^-$	6_3		$1, 3^+, 3^-, 2_1, 6_3^-, 6_3^+$
4_1		$1, 2_1, 4_1^+, 4_1^-$	6_4		$1, 3_1^+, 3_1^-, 2, 6_4^-, 6_4^+$
4_2		$1, 2, 4_2^+, 4_2^-$	6_5		$1, 3_2^+, 3_2^-, 2_1, 6_5^-, 6_5^+$
4_3		$1, 2_1, 4_3^+, 4_3^-$			

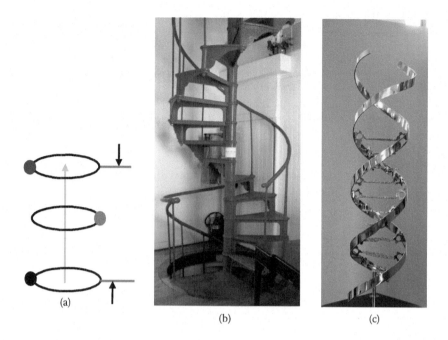

FIGURE 4.71   (a) Screw axis 2_1, (b) spiral staircase (photograph by Maureen M. Julian), (c) Watson-Crick double helix (Sculpture by Byron Rubin; photograph by Maureen M. Julian).

Figure 4.71a shows a twofold screw axis, 2_1. The black disk is at the starting place. If a 2_1 is applied, there is a rotation of 180° followed by a translation of 1/2 lattice vector along the green rotation axis to get to the red disk. Finally, a further rotation of 180°, followed by a translation of 1/2 lattice vector along the green rotation axis, gets to the blue disk, which is one lattice vector from the black disk. The green arrow shows the length of the repeat unit. Figure 4.71b shows a spiral staircase. Note the discrete steps. Finally in Figure 4.71c is the Watson–Crick double helix of DNA. The screw axis is evident in both the spiral staircase and the DNA molecule. Many materials of biological interest precipitate in crystals that have screw axes. See Section 12.1.2 on fourfold axes and Section 14.3.1 on sixfold axes.

4.11.4.3 The Inversion Point and Various Axes

The odd rotoinversions, $\bar{1}$ and $\bar{3}$, both contain the inversion operation, while the even rotoinversions, $\bar{2} = m, \bar{4}$, and $\bar{6}$, do not contain the inversion operation. When a threefold rotation axis and an inversion point located on that axis are present the result is a threefold rotoinversion axis for which there is already a symbol in Table 3.20.

Table 4.10 is now interpreted. Consider the first two columns. A graphical symbol is needed when both an even-numbered rotation axis and an inversion point on that axis are present. The *International Tables for Crystallography* has a graphical symbol that emphasizes this combination. This symbol is only used when the rotation axis is perpendicular to the plane of the figure. This book chooses colors to highlight the component graphical symbols. The 2, 4, and 6 rotation axes are red and the inversion point is purple. The example of

TABLE 4.10   Combination of Inversion Point with Various Axes

Description	Graphical Symbol	Description	Graphical Symbol
Twofold rotation axis and an inversion point		Twofold screw axis and an inversion point	
Fourfold rotation axis and an inversion point		4_2 screw axis and an inversion point	
Sixfold rotation axis and an inversion point		6_3 screw axis and an inversion point	

a rotation axis combined with an inversion point introduced in Section 3.9.3 shows that two colors are needed; note that when the rotation axis lies in the plane of the figure the components have completely separate graphical symbols. See Section 8.4 for more discussion.

Consider the last two columns of the table. Here is a similar situation when both a screw axis—2_1, 4_2, or 6_3—and an inversion point located on that axis are present. The corresponding two-color symbols are given.

4.11.5 Symmorphic and Nonsymmorphic Space Groups

There are 73 symmorphic space groups generated by combining the 32 point groups with the 14 Bravais lattices in a manner similar to that described earlier in this chapter for the two-dimensional space groups. There are 157 nonsymmorphic space groups, for a total of 230 space groups. Table 4.11 selects 4 of the 14 Bravais lattices and indicates the symmorphic and nonsymmorphic space groups.

The Hermann–Mauguin notation conveniently shows the associated point group for each space group.

The primitive, C-centered, body-centered, and face-centered space groups are indicated by the capital letters P, C, I, and F, respectively. The nonsymmorphic space groups show glides or screws in the symbols. The glides are represented by a, b, c, e, n, and d. As in the proceeding section, the screws are 2_1, 3_1, 3_2, 4_1, 4_2, 4_3, 6_1, 6_2, 6_3, 6_4, and 6_5.

TABLE 4.11   Hermann–Mauguin Notation for the Space Groups of
Four Bravais Lattices, with HMB and AA in Red

Bravais Lattice	Point Group	Symmorphic Space Group(s)	Nonsymmorphic Space Group(s)
Triclinic P	1	$P1$	
	$\bar{1}$	$P\bar{1}$	
Monoclinic P	2	$P2$	$P2_1$
	m	Pm	Pc
	$2/m$	$P2/m$	$P2_1/m, P2/c, P2_1/c$
Monoclinic C	2	$C2$	
	m	Cm	Cc
	$2/m$	$C2/m$	$C2/c$
Trigonal P	3	$P3$	$P3_1, P3_2$
	$\bar{3}$	$P\bar{3}$	
	32	$P312$	$P3_112, P3_212$
		$P321$	$P3_121, P3_221$
	$3m$	$P3m1$	$P3c1$
		$P31m$	$P31c$
	$\bar{3}m$	$P\bar{3}1m$	$P\bar{3}1c$
		$P\bar{3}m1$	$P\bar{3}c1$

The associated point group of every space group can be read from the Hermann–Mauguin symbol in the following way. The procedure is an extension of the procedure explained for two dimensions earlier in this chapter.

For the symmorphic space groups, remove the symbol for the lattice to find the associated point group. For example, for space group $C2/m$, remove the symbol for the lattice, "C," and read the associated point group, $2/m$.

For the nonsymmorphic space groups, first remove the symbol for the lattice. Then replace the symbol for a glide by the symbol for a mirror and/or the symbol for a screw by the symbol for the corresponding rotation. Specifically, replace the symbol for a glide—a, b, c, e, n, or d—by the symbol for the mirror, m. And replace the symbol for a screw—2_1, 3_1, 3_2, 4_1, 4_2, 4_3, 6_1, 6_2, 6_3, 6_4, or 6_5—by the symbol for the corresponding rotation—2, 3, 4, or 6. Thus space group $P2_1/c$ has the associated point group $2/m$.

An object with point group 32 can be inserted into a hexagonal Bravais lattice in two ways. The twofold rotation axis (that is perpendicular to the threefold rotation axis) can be along the long diagonal, as in $P312$, or along the short diagonal, as in $P321$. Each of those two symmorphic groups yields two more nonsymmorphic groups (see Table 4.11). Thus $P321$ is primitive, symmorphic, and associated with point group 32. Anhydrous alum crystallizes in $P321$.

Figure 4.68 shows the monoclinic Bravais lattice mC with the C face containing a lattice point at 1/2 1/2 0. Combining this Bravais lattice with point group 2 gives space group $C2$ (see Table 4.11). On the other hand, the Bravais lattice could have its second lattice point located at 0 1/2 1/2. Combining this lattice with point group 2 gives space group $A2$, since the A face is centered. In this way some of the Hermann–Mauguin space group notation depends on the orientation of the Bravais lattice. Space groups $C2$ and $A2$ are both No. 5 in the *International Tables for Crystallography*.

Another system that labels the space groups is the Schönflies system. Table 4.12 gives the Schönflies symbols corresponding to the Hermann–Mauguin symbols in Table 4.11. The Schoenflies symbol for a space group takes the symbol for the associated point group and adds a right-hand superscript. The relationships between the symmorphic and nonsymmorphic space groups are lost in this system. However, the Schoenflies symbol does not depend on the choice of the basis vectors. In many cases the Hermann–Mauguin symbols do depend on the choice of the basis vectors, as discussed in the preceding paragraph.

HMB is triclinic with point group $\bar{1}$ (or C_i) and space group $P\bar{1}$ (or C_i^1). Anhydrous alum is trigonal with point group 32 (or D_3) and space group $P321$ (or D_3^2). To show how both crystal structures fit into the overall picture, they are highlighted in red in Tables 4.11 and 4.12. These space groups are discussed in this chapter.

The space groups in Chapters 9 through 15 have been chosen to illustrate

- Symmorphic

  - Triclinic, #2, $P\bar{1}$ in Chapter 9

  - Trigonal, #166, $R\bar{3}m$ in Chapter 13

TABLE 4.12 Schönflies Notation for the Space Groups of Four Bravais Lattices, with HMB and AA in Red

Bravais Lattice	Point Group	Symmorphic Space Group(s)	Nonsymmorphic Space Group(s)
Triclinic P	C_1	C_1^1	
	C_i	C_i^1	
Monoclinic P	C_2	C_2^1	C_2^2
	C_s	C_s^1	C_s^2
	C_{2h}	C_{2h}^1	$C_{2h}^2, C_{2h}^4, C_{2h}^5$
Monoclinic C	C_2	C_2^3	
	C_s	C_s^3	C_s^4
	C_{2h}	C_{2h}^3	C_{2h}^6
Trigonal P	C_3	C_3^1	C_3^2, C_3^3
	C_{3i}	C_{3i}^1	
	D_3	D_3^1	D_3^3, D_3^5
		D_3^2	D_3^4, D_3^6
	C_{3v}	C_{3v}^1	C_{3v}^3
		C_{3v}^2	C_{3v}^4
	D_{3d}	D_{3d}^1	D_{3d}^2
		D_{3d}^3	D_{3d}^4

- Nonsymmorphic

  - Monoclinic, #4, $P2_1$ in Chapter 10

  - Orthorhombic, #62, $Pnma$ ($Pnam$) in Chapter 11

  - Tetragonal, #92, $P4_12_12$ in Chapter 12

  - Hexagonal, #194, $P6_3/mmc$ in Chapter 14

  - Cubic, #205, $Pa\overline{3}$ in Chapter 15

(see Exercises 4.38 through 4.48).

4.11.6 Asymmetric Unit

As in two dimensions, the *asymmetric unit* in three dimensions is the smallest part of the unit cell that, when operated on by the symmetry operations, produces the whole unit cell. Examples of an asymmetric unit for a space group of each of the seven crystal systems are found in Chapters 9 through 15 as follows:

- Triclinic system: space group #2, $P\overline{1}$ has an asymmetric unit with a volume of 1/2 of the unit cell in Section 9.3.3 in the DL leucine chapter.

- Monoclinic system: space group #4, $P2_1$ has an asymmetric unit with a volume of 1/2 of the unit cell in Section 10.3.3 in the sucrose chapter.

- Orthorhombic system: space group #62, $Pnam$ has an asymmetric unit with a volume of 1/8 of the unit cell in Section 11.3.3 in the polyethylene chapter.

- Tetrahedral system: space group #92, $P4_12_12$ has an asymmetric unit with a volume of 1/8 of the unit cell in Section 12.3.3 in the alpha-cristobalite chapter.

- Trigonal system: space group #166, $R\overline{3}m$ has an asymmetric unit with a volume of 1/12 of the unit cell in Section 13.3.3 in the $H_{12}B_{12}{}^{2-},3K^+,Br^-$ chapter.

- Hexagonal system: space group #194, $P6_3/mmc$ has an asymmetric unit with a volume of 1/24 of the unit cell in Section 14.3.4 in the magnesium chapter.

- Cubic crystal system: space group #205, $Pa3$ has an asymmetric unit with a volume of 1/24 of the unit cell in Section 15.3.4 in the acetylene chapter.

4.12 HMB AND SPACE GROUP NO. 2, $P\overline{1}$

Up to now an overview of the three-dimensional space groups has been given. Now three space groups are considered in detail. These space groups are used as vehicles to introduce ideas and information that are applicable much more generally. These three space groups are those of HMB, AA, and caffeine. The general properties of these three space groups are considered and then their specific applications to HMB, AA, and caffeine are given. The three-dimensional cells are populated with atoms. The three two-dimensional special projections are analyzed using their two-dimensional space group symbol diagrams. Finally some new topics, such as isotypic series, are introduced.

HMB crystallizes in space group No. 2, $P\bar{1}$. First some general properties of this space group are discussed and then the data for the HMB crystal are applied to its space group.

4.12.1 Interpretation of the Space Group Symmetries from the *International Tables for Crystallography*

Figure 4.72 shows the first page for space group $P\bar{1}$ from the *International Tables for Crystallography* with the crystal system, the point group, the space group, the general position diagram, and the symbol diagrams appropriately labeled. The Bravais lattice is triclinic P and the point group is $\bar{1}$, as is indicated by the symbol for the space group. There are three symbol diagrams, each a projection along [001], [010], and [100]. In the general position diagram the "+" next to a circle means that the z coordinate is positive or zero (and the point lies on or above the xy plane) and similarly the "−" indicates that the z coordinate is negative (and the point lies below the xy plane).

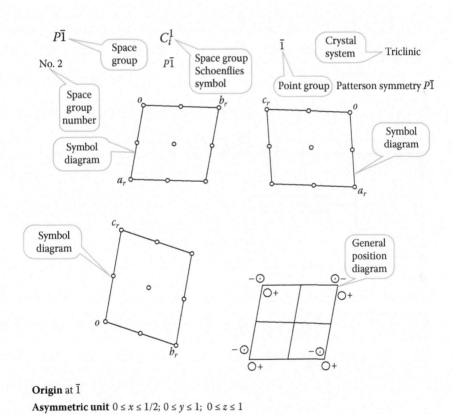

Origin at $\bar{1}$

Asymmetric unit $0 \le x \le 1/2; \ 0 \le y \le 1; \ 0 \le z \le 1$

Symmetry operations

(1) 1   (2) $\bar{1}$ 0, 0, 0

FIGURE 4.72 Space group No. 2, $P\bar{1}$, first page. (From the *International Tables for Crystallography, A.*)

The next term, "Origin at $\bar{1}$," gives the location of the origin. In this case the origin is placed at the inversion point.

The asymmetric unit is $0 \leq x \leq 1/2$; $0 \leq y \leq 1$; $0 \leq z \leq 1$. Except for $P1$, in which the asymmetric unit is the entire unit cell, the asymmetric unit is always smaller than the unit cell. Except for $P1$, the choice of the asymmetric unit is not unique. This book follows the choice made by the *International Tables for Crystallography*.

Under "Symmetry operations," there are two—the first, labeled "(1)" is the identity 1, and the second, labeled "(2)" is the inversion at the origin, $\bar{1}$ 0, 0, 0. The number in parentheses is the numerical label of the symmetry operation.

Figure 4.73 shows the second page. The line "Generators selected" gives the generators, both symmetry operations and translation operations. The first generator is always (1), which is always the identity. The translations $t(1,0,0)$, $t(0,1,0)$, $t(0,0,1)$ generate the infinite three-dimensional lattice. Finally, here, there is another symmetry operation, (2), which, in this case, is the inversion. In more symmetric space groups there are more generators.

The "Positions" are next. As in two dimensions, a *position* (singular) is a set of points that are equivalent by symmetry. The first column heading is "Multiplicity" followed by "Wyckoff letter," and "Site symmetry." The second heading is "Coordinates" and the third is "Reflection conditions." The reflection conditions are considered later in Chapter 7.

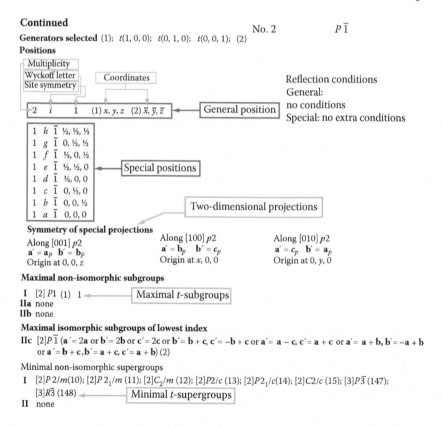

FIGURE 4.73 Space group No. 2, $P\bar{1}$, second page. (From the *International Tables for Crystallography, A*.)

The first line after the column headings under "Positions" gives information about the general position. The *multiplicity* is the number of symmetry-equivalent points in the unit cell. In this case the multiplicity is 2; and there are two coordinates—(1) x,y,z and (2) $\bar{x},\bar{y},\bar{z}$.

The labels "(1)" and "(2)" connect each coordinate with its corresponding symmetry operation. This is an important feature of the *International Tables for Crystallography*.

The *site symmetry of a point* is the set of all operations in a space group that leave that point fixed. Screws, glides, and translations cannot be included. The *general position* of a space group is the position with site symmetry 1 and therefore has the greatest multiplicity. Every space group has exactly one general position. The coordinates of a point in the general position are displayed with the number, in parentheses, that is used to label the corresponding symmetry operation. For example, the point in the general position corresponding to the symmetry operation labeled (1), the identity, has fractional coordinates x, y, z. The point in the general position corresponding to the symmetry operation labeled (2), the inversion, has fractional coordinates $\bar{x},\bar{y},\bar{z}$.

The multiplicity of the general position is equal to the order of the associated point group multiplied by the number of lattice points per unit cell.

The next eight lines list the special positions for this space group. A *special position* of a space group is a position with multiplicity less than the multiplicity of the general position. This position is associated with a site symmetry other than the identity. In this case, all the special positions are associated with the inversion, $\bar{1}$.

Consider the inversion point at the origin 0,0,0 labeled by Wyckoff letter a. Substitute the coordinates of the origin into the general position coordinates x,y,z and $\bar{x},\bar{y},\bar{z}$. They both become 0,0,0, giving a multiplicity of 1. For another example, consider the inversion point at 1/2, 1/2, 0 labeled by Wyckoff letter e. Substitute these coordinates into the general position coordinates x,y,z and $\bar{x},\bar{y},\bar{z}$. Remember that the fractional coordinate 1/2 is equivalent to −1/2. Then 1/2, 1/2, 0 becomes −1/2, −1/2, 0, which is equivalent to 1/2, 1/2, 0, again giving a multiplicity of 1.

The *Wyckoff letter* is a symbol that identifies each position of a space group. The letter a is assigned to the last position, and the assignment continues upward alphabetically to end with the general position. In this space group the general position is labeled i.

The two-dimensional projections of the three-dimensional space group are given under the heading "Symmetry of special projections." *Special projections* are along a crystallographic direction, such as [001], and are projected onto a plane perpendicular to that direction. In this case the projections along [001], [010], [100] are all $p2$.

Every *special projection* is always one of the 17 two-dimensional space groups, as identified in the *International Tables for Crystallography*. Furthermore, *any* projection is also one of the 17 two-dimensional space groups.

The maximal t-subgroups and the minimal t-supergroups were discussed in the construction of the two-dimensional space group trees earlier in the chapter.

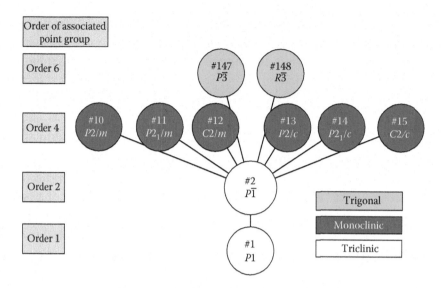

FIGURE 4.74   Part of the space group tree near $P\bar{1}$.

4.12.2 Space Group Tree Near $P\bar{1}$

The maximal t-subgroups and minimal t-supergroups allow construction of the space group tree, as in the two-dimensional case. $P\bar{1}$ has only one maximal t-subgroup, $P1$, No.1. Both are triclinic. Since $P\bar{1}$ has an inversion point, each of the minimal t-supergroups has an inversion point and hence is centrosymmetric. Space groups $P2/m$, No. 10; $P2_1/m$, No. 11; $C2/m$, No. 12; $P2/c$, No. 13; $P2_1/c$, No. 14; and $C2/c$, No. 15 are monoclinic. Space groups $P\bar{3}$, No. 147; and $R\bar{3}$, No. 148 are trigonal. See Figure 4.74, Exercises 4.49 through 4.51.

The complete space group tree in three dimensions contains 230 members.

4.12.3 HMB Crystal Data

The next step is to apply the space group information given in the *International Tables for Crystallography* to data on an actual crystal. HMB crystallizes in space group $P\bar{1}$. The crystallographic data for HMB are given in Table 4.13.

4.12.4 Unit Cell and the Asymmetric Unit for HMB

Figure 4.75 shows the HMB asymmetric unit, in red, superimposed on the unit cell.

The volume of the asymmetric unit, V_{AU}, is equal to the volume of the unit cell, V_{UC}, divided by the multiplicity of the general position.

For HMB the multiplicity of the general position is 2. The volume of the asymmetric unit, V_{AU}, is

$$V_{AU} = \tfrac{1}{2}V_{UC}.$$

TABLE 4.13   Crystallographic Data for HMB

Name of Compound	Hexamethylbenzene, HMB		
Chemical formula	$C_{12}H_{18}$		
Molecular weight	162.27 g/mol		
Crystal system	Triclinic		
Bravais lattice	aP		
Space group	$P\bar{1}$, No. 2		
a (Å)	9.010		
b (Å)	8.926		
c (Å)	5.344		
α (°)	44° 27′		
β (°)	116° 43′		
γ (°)	119° 34′		
Carbon A, fractional coordinates, $x\,y\,z$	0.0710	0.182	0
Carbon B, fractional coordinates, $x\,y\,z$	−0.109	0.073	0
Carbon C, fractional coordinates, $x\,y\,z$	−0.180	−0.109	0
Carbon D, fractional coordinates, $x\,y\,z$	0.145	0.371	0
Carbon E, fractional coordinates, $x\,y\,z$	−0.222	0.149	0
Carbon F, fractional coordinates, $x\,y\,z$	−0.367	−0.222	0
Volume (Å³)	251.84		
Density g/cm³	1.070		

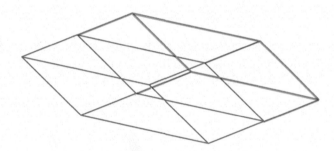

FIGURE 4.75   HMB asymmetric unit in red superimposed on the unit cell.

(See Exercise 4.52.)

4.12.5 Populated Cell for HMB

Table 4.13 lists the fractional coordinates for six carbon atoms *A* through *F*. The presence of the inversion point means that only these coordinates are necessary to describe the 12 carbon atoms of the HMB molecule. The hydrogen atoms are omitted. Carbon atoms *a* through *f* are generated by the inversion, which takes x,y,z into $\bar{x}, \bar{y}, \bar{z}$. For example, carbon atom *A* has fractional coordinates x,y,z equal to 0.0710, 0.182 0. Then carbon atom *a* has fractional coordinates $\bar{x}, \bar{y}, \bar{z}$ equal to −0.0710 −0.182 0. This molecule was examined in two dimensions in Chapter 2. Here every carbon atom has zero for the *z* coordinate, and the carbon atoms of the HMB molecule lie in a plane.

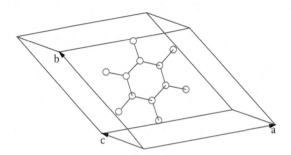

FIGURE 4.76   Unit cell for HMB showing the carbon atoms only.

Figure 4.76 shows the 12 HMB carbon atoms in the unit cell. Here 1/2, 1/2, 1/2 has been added to the fractional coordinates developed in the previous paragraph to bring the molecule into the center of the unit cell, as was done in Chapter 2. There HMB was considered as a two-dimensional case, and the molecule was drawn in the plane.

4.12.6 Special Projections for HMB

There are three special projections—along [001], [100], and [010]—all having two-dimensional space group p2. These views are gotten by rotating the three-dimensional unit cell in Figure 4.76. The views may be frozen in place.

Two-dimensional projections are easily done for three-dimensional figures when advanced computer languages like MATLAB are used.

Figure 4.77a shows the special projection along [001]. Note that this projection is different from the HMB molecule that was drawn in Chapter 2. That figure does not give a standard projection, but gives the molecule in the xy plane. In Figure 4.77b the symbol diagram for p2 is superimposed, showing that the HMB molecule obeys the symmetry operations.

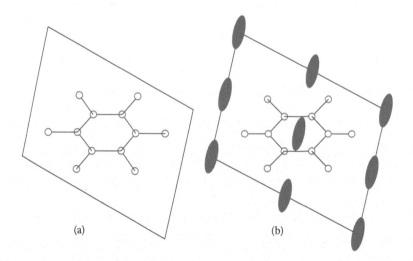

(a)   (b)

FIGURE 4.77   (a) HMB projected along [001], (b) symbol diagram p2 superimposed on HMB.

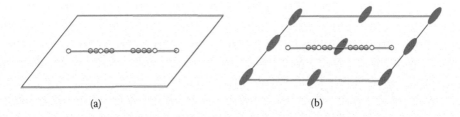

FIGURE 4.78 (a) HMB projected along [100], (b) symbol diagram *p*2 superimposed on HMB.

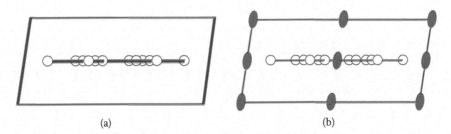

FIGURE 4.79 (a) HMB projected along [010], (b) symbol diagram *p*2 superimposed on HMB.

The next projection is along [100] and is given in Figure 4.78, showing another view of the HMB molecule. Since the molecule is planar, the edge-on view reduces the molecule to a line.

The next projection is along [010] and is presented in Figure 4.79, showing yet another edge-on view of the HMB molecule (see Exercise 4.53).

4.12.7 Interpretation of HMB Crystal Structure

Kathleen Lonsdale (Lonsdale, 1929b) solved the structure of HMB. Here is a list of questions that she posed in her paper. This list is quoted exactly to show some of the issues that were answered by various crystallographic studies:

1. Does the organic molecule exist as a separate entity in the crystalline state?

2. Are the carbon atoms in the benzene nucleus arranged in the form of a closed ring?

3. If so, is the ring hexagonal in shape?

4. What are the sizes of the atoms in the ring and the dimensions of the ring itself?

5. Is the ring planar, or are the atoms arranged on more than one plane?

6. What are the positions of the side-chain atoms or groups relative to the nucleus and to one another?

7. What is the arrangement of the valences in the 'aromatic' carbon atom?

8. How does the carbon atom in aromatic nuclei differ from the 'aliphatic' carbon atom?

9. Is the benzene nucleus always the same size and shape?

4.12.8 Discussion

1. Crystallographic techniques have provided examples of the existence of organic molecules as separate entities. Covalent bonds are within the molecule. Weaker Van der Waals bonds keep the molecules in place relative to each other. That fact is easy to be seen in HMB and the many thousands of organic structures, from the small molecules such as benzene and anthracene up to huge biochemical proteins.

2. The carbon atoms in benzene are in a closed ring, as seen in HMB. Benzene and its related compounds contain closed rings. Studies of HMB preceded studies of benzene because benzene has a more complicated crystal structure than HMB and because benzene is a liquid at room temperature.

3. In HMB the ring is *nearly* hexagonal in shape. This has also proved to be the case in many other aromatic organic compounds.

4. Figure 4.80 shows the HMB molecule. The centers of the atoms are determined by crystallographic analysis. In Chapter 2 the bond lengths were calculated using the **G** matrix. There are six bond lengths in the benzene ring: AB, BC, Ca, ab, bc, and cA. The molecule contains an inversion point, marked by the purple circle. This means that the number of crystallographically independent bonds is reduced from six to three because bond AB = ab; BC = bc; and Ca = cA. The three distinct bond lengths are AB = 1.4211 Å; BC = 1.4222 Å, and Ca = 1.4215 Å. The three bond lengths are not related by symmetry and so they are crystallographically distinct. However these bond lengths are not experimentally different. The average bond distance is 1.4216 Å. Since in the hard sphere model, the spheres are assumed to touch, the average value of the diameter of the carbon atom in the benzene nucleus is 1.4216 Å. The angle ABC is 120.01°.

5. All the fractional coordinates for z are zero for the entire benzene ring. This means the benzene ring is planar. This observation is seen in many examples of six-membered aromatic rings.

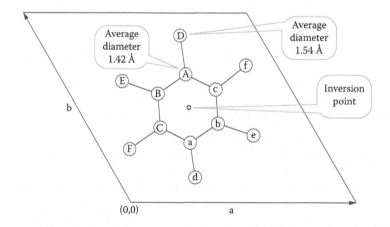

FIGURE 4.80   HMB, showing the diameters of the carbon atoms.

6. The side-chain carbon atoms—that is, the methyl carbons—are attached radially to their respective nuclear carbon atoms and lie in the plane of the ring.

7. The benzene ring contains an inversion point, eliminating the Kekulé *static* model with alternating double and single bonds. Three fixed double bonds are not possible because they cannot have an inversion point unless they are collinear.

8. There are six bond lengths between carbon atoms in the side chain and their adjacent carbon atoms in the benzene ring: AD, BE, CF, ad, be, and cf. Because the molecule contains an inversion point, the number of crystallographically independent bonds is reduced from six to three. Bond AD = ad = 1.4767 Å; BC = bc = 1.4759 Å; and Ca = cA = 1.4762 Å These values are crystallographically distinct, but not experimentally different. The average bond distance is 1.4764 Å.

The average interatomic distance between a side chain atom and a benzene nucleus atom is 1.4764 Å. This bond distance is equal to the average radius of the side-chain atom, $r_{SideChain}$, plus the average radius of the benzene nucleus atom, $r_{benzene}$.

$$r_{SideChain} = 1.4764 - r_{benzene} = 1.4764 - 0.5(1.4216) = 0.7656 \text{ Å}$$

The average diameter of the side chain carbon atom is 2×0.7656 Å = 1.5412 Å, which is the accepted value for a C—C single bond. The benzene ring carbons have a bond length of 1.42 Å, which lies between the double bond of 1.33 Å and the single bond of 1.54 Å, as expected. Carbon atoms of graphite also have a diameter of 1.42 Å.

9. In general, the size and shape of the benzene ring is preserved throughout many structures. For example, the benzene nucleus in HMB is nearly identical to the benzene ring in graphite.

4.13 AA AND SPACE GROUP NO. 150, *P*321

AA crystallizes in space group No. 150, *P*321.

4.13.1 Interpretation of the Space Group Symmetries from the *International Tables for Crystallography*

Figures 4.81 and 4.82 give the information for the space group No. 150, *P*321.

4.13.2 Space Group Tree Near *P*321

Information from Figure 4.82, showing the space group *P*321 from the *International Tables for Crystallography*, is used in Figure 4.83 to draw part of the space group tree near *P*321. The maximal *t*-subgroups are No. 143 (trigonal) *P*3 and No. 5, (monoclinic) *C*2. Neither of them contains an inversion point. Also No. 150 (trigonal) *P*321 does not contain an inversion point. The minimal *t*-supergroups are No. 164 (trigonal) $P\bar{3}m1$; No.165 (trigonal) $P\bar{3}c1$; No. 177 (hexagonal) *P*622; No. 182 (hexagonal) *P*6₃22; No. 189 (hexagonal) $P\bar{6}2m$; and No. 190 (hexagonal) $P\bar{6}2c$.

4.13.3 AA Crystal Data

The goal is to combine the space group information in Figure 4.82 and the data for anhydrous alum in Table 4.14 to construct the crystal structure.

*P*321 D_3^2 321   Trigonal

No. 150   *P*321   Patterson symmetry $P\bar{3}m1$

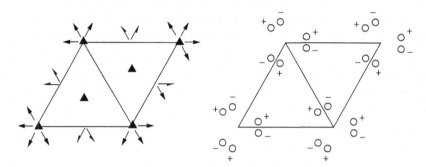

Origin at 321

Asymmetric unit $0 \le x \le \tfrac{2}{3};\ \ 0 \le y \le \tfrac{2}{3};\ \ 0 \le z \le \tfrac{1}{2};\ \ x \le (1+y)/2;\ \ y \le \min(1-x, (1+x)/2$

           Vertices $0,0,0 \ \ \tfrac{1}{2},0,0 \ \ \tfrac{2}{3},\tfrac{1}{3},0 \ \ \tfrac{1}{3},\tfrac{2}{3},0 \ \ 0,\tfrac{1}{2},0$

                           $0,0,\tfrac{1}{2} \ \ \tfrac{1}{2},0,\tfrac{1}{2} \ \ \tfrac{2}{3},\tfrac{1}{3},\tfrac{1}{2} \ \ \tfrac{1}{3},\tfrac{2}{3},\tfrac{1}{2} \ \ 0,\tfrac{1}{2},\tfrac{1}{2}$

Symmetry operations

   (1) 1   (2) 3^+ $0,0,z$ (3) 3^- $0,0,z$

   (4) 2 $x,x,0$ (5) 2 $x,0,0$ (6) 2 $0,y,0$

FIGURE 4.81 Space group No. 150, *P*321, first page. (From the *International Tables for Crystallography, A.*)

4.13.4 Unit Cell and the Asymmetric Unit for AA

The multiplicity of the general position in *P*321 is 6. Thus, the volume of the asymmetric unit is 1/6 that of the unit cell. Figure 4.84 shows the asymmetric unit, in red, superimposed on the unit cell for AA, in blue.

4.13.5 Calculation of the Number of Formula Units

Combine information from the density, formula weight, and volume to calculate the number of formula units. The formula for AA is $KAl(SO_4)_2$. See Chapter 2 for the density calculation. Solving the density equation for Z gives

$$ Z = \frac{\delta V N_A}{\sum A} = 1 $$

This means that there is one formula unit or, specifically, one potassium atom, one aluminum atom, two sulfur atoms, and eight oxygen atoms in the unit cell.

4.13.6 Wyckoff Positions

In Figure 4.82 look at the section called "Positions." The general position has a multiplicity of six. This means that if an atom is at a point in the general position, then a total of six atoms related by symmetry are in the unit cell. Since there is only one potassium atom in

Generators selected (1); $t(1,0,0)$; $t(0,1,0)$; $t(0,0,1)$; (2); (4)

Positions

Multiplicity, Wyckoff letter, Site symmetry	Coordinates		Reflection conditions

Coordinates (center), Reflection conditions (right):

General: no conditions
Special: no extra conditions

6 g 1   (1) x,y,z (2) $\overline{y}, x-y, z$ (3) $\overline{x}+y, \overline{x}, z$
             (4) $y, x, \overline{z}$ (5) $x-y, \overline{y}, \overline{z}$ (6) $\overline{x}, \overline{x}+y, \overline{z}$

3 f .2. $x, 0, \frac{1}{2}$ $0, x, \frac{1}{2}$ $\overline{x}, \overline{x}, \frac{1}{2}$

3 e .2. $x, 0, 0$ $0, x, 0$ $\overline{x}, \overline{x}, 0$

2 d 3.. $\frac{1}{3}, \frac{2}{3}, z$ $\frac{2}{3}, \frac{1}{3}, \overline{z}$

2 c 3.. $0.0, z$ $0, 0, \overline{z}$

1 b 32. $0, 0, \frac{1}{2}$

1 a 32. $0.0.0$

Symmetry of special projections

Along [001] $p3\,1\,m$

$\mathbf{a} = \mathbf{a'}$ $\mathbf{b'} = \mathbf{b}$

Origin at $0,0,z$

Along [100] $p2$

$\mathbf{a'} = \frac{1}{2}(\mathbf{a}+2\mathbf{b})$ $\mathbf{b'} = \mathbf{c}$

Origin at $x, 0, 0$

Along [210] $p\,1\,1\,m$

$\mathbf{a'} = \frac{1}{2}\mathbf{b}$ $\mathbf{b'} = \mathbf{c}$

Origin at $x, \frac{1}{2}x, 0$

Maximal non-isomorphic subgroups

I   [2]$P311(P3,143)$ 1;2;3

$\begin{cases} [3]P121(C2,5) & 1;4 \\ [3]P121(C2,5) & 1;5 \\ [3]P121(C2,5) & 1;6 \end{cases}$

IIa   none
IIb   [3] $P3_2\,2\,1$ ($\mathbf{c'} = 3\mathbf{c}$) (154); [3] $P3_1\,2\,1$ ($\mathbf{c'} = 3\mathbf{c}$) (152): [3] $H\,321$ ($\mathbf{a'} = 3\mathbf{a}.\mathbf{b'} = 3\mathbf{b}$) ($P3\,1\,2.\,149$)

Maximal isomorphic subgroups of lowest index

IIc   [2] $P321$ ($\mathbf{c'} = 2\mathbf{c}$) (150); [4] $P32\,1$ ($\mathbf{a'} = 2\mathbf{a}, \mathbf{b'} = 2\mathbf{b}$) (150)

Minimal non-isomorphic supergroups

I   [2] $P\overline{3}m\,1$ (164); [2]$P\overline{3}c1$ (165); [2] $P622$ (177); [2] $P6_322$ (182); [2] $P\overline{6}2m$ (189); [2]$P\overline{6}2c$(190)

II   [3]$H321$ ($P312,149$); [3] $R32$ (obverse)(155); [3] $R32$ (reverse) (155)

FIGURE 4.82 Space group No. 150, *P*321, second page. (From the *International Tables for Crystallography, A.*)

the unit cell, this atom does not occupy a point in the general position. It must occupy a point in a special position with a multiplicity of one. The same is true for the aluminum atom. There are two such special positions—one with Wyckoff letter a with coordinates $0,0,0$ and a second with Wyckoff letter b with coordinates $0,0,1/2$. Use Table 4.14 to assign Wyckoff positions a and b to potassium and aluminum, respectively.

There are two sulfur atoms in the unit cell. Since Wyckoff positions a and b have been used up and there are no more positions of multiplicity one, a position with multiplicity two is needed. In Figure 4.82, one sulfur atom has fractional coordinates 1/3, 2/3, 0.222, corresponding to a point with Wyckoff letter d with $z = 0.222$. A second sulfur atom has

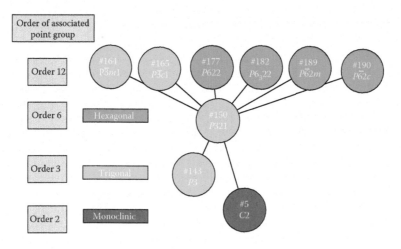

FIGURE 4.83 Part of the space group tree near $P321$.

fractional coordinates 2/3, 1/3, 0.778, corresponding to the other point with Wyckoff letter d with $\bar{z} = -0.222$. Note that -1 is added to $\bar{z}$ to keep it in the unit cell. The fractional coordinates for the sulfur atoms are found in Table 4.15. The reason why the fractional coordinates for only one sulfur atom are given in the crystallographic literature is that the *International Tables for Crystallography* contain the information for the coordinates of the other symmetry-related sulfur atom.

Now consider the eight oxygen atoms. In Table 4.14, the oxygen atoms are divided into Oxygen I and Oxygen II. The coordinates of Oxygen I have the form 1/3, 2/3, z, and the Wyckoff position d, with z equal to 0.016. Although both the sulfur atoms and the Oxygen I atoms are in Wyckoff position d, there is no contradiction because their z coordinates

TABLE 4.14 Crystallographic Data for Anhydrous Alum

Name of Compound	Anhydrous Alum
Chemical formula	$KAl(SO_4)_2$
Formula weight	258.1 g/mol
Crystal system	Trigonal
Bravais lattice	hP
Space group	$P321$, No. 150
a (Å)	4.709
c (Å)	7.984
Potassium fractional coordinates	0,0,0
Aluminum fractional coordinates	0,0,1/2
Oxygen II fractional coordinates	0.328, 0.344, 0.317
Sulfur fractional coordinates	1/3, 2/3, 0.222
Oxygen I fractional coordinates	1/3, 2/3, 0.016
Volume (Å³)	153.3
Density g/cm³	2.796

Source: From Vegard, L. and A. Maurstad. 1929. Die Kristallstruktur der wasserfreien Alaune R'R"$(SO_4)_2$, Z. *Kristallogr. Dtsch*, 69, 519–532. See the text for an explanation of the terms Oxygen I and Oxygen II.

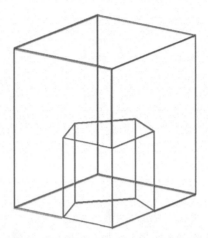

FIGURE 4.84 The unit cell for AA, with the asymmetric unit superimposed on it.

TABLE 4.15 Crystallographic Data for Sulfur and Oxygen in Wyckoff Position *d*

Wyckoff Letter	Element	1/3, 2/3, *z*	2/3, 1/3, $\bar{z}$
d	Sulfur	1/3, 2/3, 0.222	2/3, 1/3, 0.778
d	Oxygen I	1/3, 2/3, 0.016	2/3, 1/3, 0.984

differ (see Table 4.15). A second Oxygen I atom has fractional coordinates 2/3, 1/3, 0.984. Figure 4.85 shows the AA unit cell with the oxygens and sulfurs in the *d* special positions. The bond between each pair, consisting of an oxygen atom and a sulfur atom, is included.

The general position has a multiplicity of six. The Oxygen II atoms occupy points of the general position, Wyckoff *g*. The coordinates for one atom are given in Table 4.14. Then Table 4.16 shows how the symmetry operations generate all six atoms.

The eight oxygen atoms are made up of two atoms in Wyckoff position *d* and six atoms in Wyckoff position *g*. Oxygen atoms I and II are crystallographically distinct.

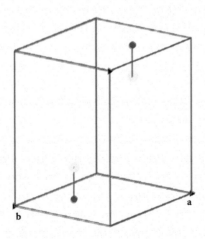

FIGURE 4.85 Unit cell of AA with the sulfur and oxygen atoms in Wyckoff special position *d*. Sulfur is yellow and oxygen is blue.

TABLE 4.16   Oxygen II Atoms, in the General Position

Label	Symmetry Operation	General Position Coordinates	Oxygen II Atoms, Fractional Coordinates
1	1, identity	x, y, z	0.328, 0.344, 0.317
2	$3^+ 0,0,z$	$\bar{y}, x - y, z$	0.656, 0.984, 0.317
3	$3^- 0,0,z$	$\bar{x} + y, \bar{x}, z$	0.016, 0.672, 0.317
4	$2\ x,x,0$	$y, x, \bar{z}$	0.344, 0.328, 0.683
5	$2\ x,0,0$	$x + \bar{y}, \bar{y}, \bar{z}$	0.984, 0.656, 0.683
6	$2\ 0,y,0$	$\bar{x}, \bar{x} + y, \bar{z}$	0.672, 0.016, 0.683

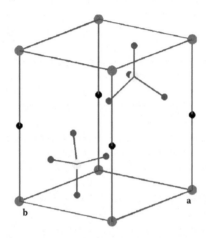

FIGURE 4.86   Unit cell of AA with sulfur in yellow, oxygen in blue, potassium in red, and aluminum in black.

Figure 4.86 shows the completely populated unit cell of anhydrous alum. A potassium atom is located at the origin. Due to the periodicity of the lattice, a potassium atom is located at each of the eight vertices. An aluminum atom is located at 0,0,1/2, along the edges of the unit cell. Due to the periodicity of the lattice, the aluminum atom is repeated four times. Two sulfate tetrahedra are also present (see Exercise 4.54).

4.13.7 Special Projections for AA

There are three different special projections—along [001] with the space group $p31m$, along [100] with $p2$, and along [210] with pm.

First consider the projection along [001]. This projection is gotten by rotating the structure in Figure 4.86 until the view is along [001] (see Figure 4.87a). This projection has the two-dimensional space group No. 15, $p31m$. Figure 4.87b shows the symbols for the symmetries superimposed on the projection.

Now rotate Figure 4.86 until the view is along [100]. This view is in Figure 4.88a. This projection has the two-dimensional space group No. 2, p2. Figure 4.88b shows the symbols for the symmetry superimposed on the projection. The view along [010] is identical to the view along [100] because the basis vectors **a** and **b** are related by a threefold rotation.

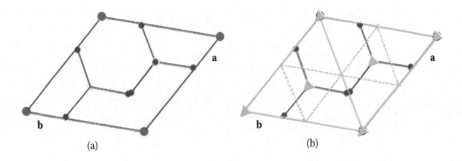

FIGURE 4.87 (a) Projection of AA along [001], (b) symbol diagram for *p*31*m* superimposed on the projection.

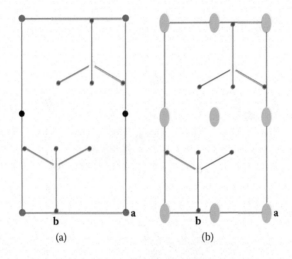

FIGURE 4.88 (a) Projection of AA along [100], (b) symbol diagram *p*2 superimposed on the projection.

The *International Tables for Crystallography* always give three distinct special projections for each three-dimensional space group.

In this case the third projection is along [210] and has the two-dimensional space group No. 3, *pm*. The projection and superimposed symbol diagram are shown in Figures 4.89a and b (see Exercise 4.55).

4.13.8 Polyhedrons in the AA Crystal Structure

In the unit cell there are two sulfate groups in AA, each consisting of a sulfur atom surrounded by four oxygen atoms in a tetrahedral configuration. Figure 4.90 shows the sulfates, SO_4^{-2}, represented by tetrahedrons (see Exercise 4.56).

Polyhedral structures are common in many crystals, particularly minerals. For example, the octahedral coordination is often seen in the 3*d* transition metal ions; and their role in the crystal structures may be represented in terms of octahedrons.

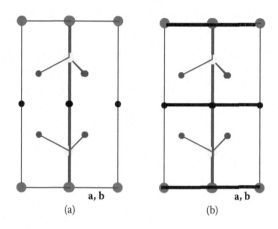

(a)   (b)

FIGURE 4.89   (a) Projection of AA along [210], (b) symmetry diagram *pm* superimposed on the projection.

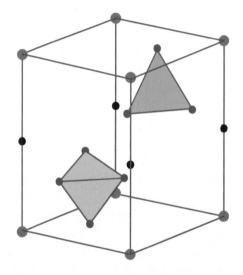

FIGURE 4.90   Unit cell of AA with sulfate groups shown as tetrahedrons. (Thanks to Katie Clark.)

A closer look at both Figures 4.86 and 4.91 shows that the two sulfate groups are related by a twofold rotation and thus are not crystallographically distinct. In the second figure consider the lower left tetrahedron with the three green oxygens. These are oxygen IIs in the Wyckoff general position *g*. These oxygens are related to each other by a threefold operation. The three S—O_{II} bonds are crystallographically identical and are equal in length to 1.6871 Å (see Table 4.17). (See Chapter 2 for the calculation of bond distances.) The large blue oxygen atom O_I in the Wyckoff position *d* is crystallographically distinct from the green oxygens. The S—O_I bond distance is 1.6447 Å. The difference between the two bond lengths is 0.042 Å, which is experimentally significant.

In a regular tetrahedron the four bond distances are equal. In AA the tetrahedron is distorted because the S—O_I bond distance is shortened by 0.042 Å. The blue O_I atom is

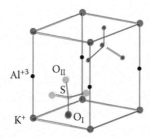

FIGURE 4.91 Distorted tetrahedron in the AA structure.

TABLE 4.17 Comparison between the Distorted Tetrahedron in AA and a Regular Tetrahedron

	Distorted Tetrahedron in AA	Regular Tetrahedron
Bond S–O_I	1.6447 Å	All bonds equal
Bond S–$O_{II,1}$	1.6871 Å	All bonds equal
Bond S–$O_{II,2}$	1.6871 Å	All bonds equal
Bond S–$O_{II,3}$	1.6871 Å	All bonds equal
Angle O_I–S–$O_{II,1}$	116.7158°	109.471°
Angle O_I–S–$O_{II,2}$	116.7158°	109.471°
Angle O_I–S–$O_{II,3}$	116.7158°	109.471°
Angle $O_{II,1}$–S–$O_{II,2}$	101.3521°	109.471°
Angle $O_{II,1}$–S–$O_{II,3}$	101.3521°	109.471°
Angle $O_{II,2}$–S–$O_{II,3}$	101.3521°	109.471°

closer to the S sulfur atom than are the green O_{II} atoms. The blue O_I repulses the three green O_{II} atoms. The O_I oxygen is associated with the K^+ ions while the O_{II} atoms are associated with the Al^{3+} ions. These two differently charged cations distort the sulfate tetrahedron.

The bond angles are also affected in the distortion. In the regular tetrahedron all six bond angles are equal to 109.471°. The O_I atom repulses the O_{II} atoms, causing the bond angles containing O_I to be expanded by 7.2° to 116.716°. The lengthening of the S–O_{II} bond permits the bond angles containing two oxygen atoms, both O_{II}, to collapse by 8.1° to 101.3521°.

The distortion of the sulfate tetrahedrons is calculated from the crystallographic data and is further justified by the environments of the +3 and +1 cations, Al^{3+} and K^+, respectively (see Exercise 4.57).

Start with the unit cell in Figure 4.86 and visualize an array of many unit cells in three dimensions. The crystal structure is described by sheets of $[Al(SO_4)_2]^-$ that are parallel to the (001) plane and are interconnected by K^+ ions. Each aluminum cation, Al^{3+}, is coordinated to six oxygens forming a trigonal prism encasing the Al^{3+}. Also each K^+ is coordinated to six oxygens.

4.13.9 Isotypic Crystal Structures

AA has several related isotypic crystal structures. *Isotypic crystals* (also called isomorphic crystals) are a series of crystals of different chemical compositions with the same structure. The variation is in the substitution of one atom for another atom or of one ion for another ion of the same charge. The crystals have approximately equal cell dimensions. Examples of isotypic crystals are $KAl(SO_4)_2$ and $RbFe(SeO_4)_2$. The K^+ in AA is replaced with Rb^+, the Al^{3+} is replaced with Fe^{3+}, and the S is replaced with Se. The tetrahedron is made up of oxygens in both cases (see Figure 4.92 and Table 4.18).

There are many other structures isotypic to $KAl(SO_4)_2$. Some examples are $KCr(SO_4)_2$, $RbAl(SO_4)_2$, $TlAl(SO_4)_2$, $CsFe(SeO_4)_2$, and $TlAl(SeO_4)_2$. These compounds are all classified as anhydrous alums. There are other compounds of the form $R_1^+ R_2^{3+} (R_3O_4)_2^{2-}$ that are anhydrous alums but are not isotypic with $KAl(SO_4)_2$ (see Exercises 4.58 through 4.62).

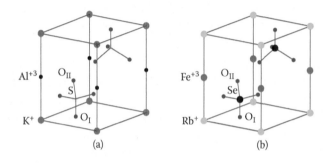

FIGURE 4.92   Structures of (a) $K^+ Al^{+3}(SO_4)_2$, (b) $Rb^+ Fe^{+3}(SeO_4)_2$.

TABLE 4.18   Comparison of $KAl(SO_4)_2$ and $RbFe(SeO_4)_2$

	$KAl(SO_4)_2$	$RbFe(SeO_4)_2$
Space group	$P321$	$P321$
Z	1	1
a, Å	4.709	5.005
c, Å	7.984	8.548
Volume, Å³	153.32	185.4
Wyckoff position a	K^+	Rb^+
Fractional coordinates	0,0,0	0,0,0
Wyckoff position b	Al^{3+}	Fe^{3+}
Fractional coordinates	0,0,1/2	0,0,1/2
Wyckoff position d	S, O_I	Se, O_I
Fractional coordinates S, Se	1/3,2/3,0.222	1/3, 2/3, 0.2968
Fractional coordinates O_I	1/3,2/3,0.016	1/3, 2/3, 0.1114
Wyckoff position g	O_{II}	O_{II}
Fractional coordinates O_{II}	0.016, 0.672, 0.317	0.0435, 0.745, 0.3614

Source: Giester, G. 1994. Crystal structure of anhydrous alum $RbFe(SeO_4)_2$
Monatshefte für Chemie/Chemical Monthly, **125**(11), 1223–1228.

4.14 CAFFEINE MONOHYDRATE AND TWO EFFECTIVE TOOLS FOR RELATING SYMMETRY AND STRUCTURE

This section presents two tools to relate the symmetry operations to the crystallographically related atoms and bonds. The first is the use of the CIF, the Crystallographic Information File, for a crystal; and the second is the Starter Program for this chapter. An alternative approach using CSD instead of the Starter Program is given in Chapter 8.

Consideration of caffeine monohydrate serves as an introduction to the crystal structures examined in Chapters 9 through 15. Caffeine monohydrate (Sutor, 1958) crystallizes in the space group number 14, $P2_1/a$. Both HMB and AA are too simple to illustrate the techniques needed. The caffeine structure is more complex than that of either HMB or AA. HMB, $C_{12}H_{18}$, has $Z = 1$ with 30 atoms in the unit cell and a volume of 251 Å^3. AA, $KAl(SO_4)_2$, also has $Z = 1$ with 12 atoms in the unit cell and a volume of 153 Å^3. By contrast, caffeine monohydrate, $C_8H_{10}N_4O_2 \cdot H_2O$, has $Z = 4$ with 108 atoms in the unit cell and a volume of 974 Å^3.

4.14.1 Using the Crystallographic Information File, CIF, for a Crystal

A critical step in the construction of a crystal structure is the extraction of a self-consistent set consisting of crystal data from the literature and symmetry information from the *International Tables for Crystallography*. This goal is accomplished with the crystallographic information file, CIF, which is combined with the Starter Program for this chapter. See Section 9.1.1 for an explanation of the CIF.

In addition to the *consistency* among the symmetry coordinates, the cell constants, and the fractional coordinates for a given crystal in the CIF, the electronic format is also helpful for accurate transferal of the data.

The CIF can be obtained from databases such as the Cambridge Structure Database (CSD), the Pearson Database of Inorganic Crystals and Minerals, and the American Mineralogist Crystal Structure Database.

The use of the CIF avoids some of the pitfalls that are found when the data from the *International Tables for Crystallography* is combined with the data from the literature, as seen in the following example. The space group for caffeine monohydrate is the most popular space group in the crystal world (see Chapter 8). The *standard symbol* for the space group #14 is $P2_1/c$. However, for this particular very popular space group *the International Tables for Crystallography* give two settings and several cell choices. The two settings are choice 1: *b*-axis unique and choice 2: *c*-axis unique. There are several cell choices. For choice 1 there are cell choices $P12_1/c1$, $P12_1/n1$, and $P12_1/a1$. For choice 2 there are cell choices $P112_1/a1$, $P112_1/n1$, and $P112_1/b$. The caffeine monohydrate structure was published in setting choice 1, with the *b*-axis unique, and the screw axis perpendicular to the *a*-glide plane. Each choice influences the cell constants, the fractional coordinates, and the symmetry operations. Fortunately, the CIF coordinates this information and prevents student confusion and instructor frustration.

4.14.1.1 Interpretation of the CIF for Caffeine Monohydrate

Table 4.19 gives the information for caffeine monohydrate from the CIF. The diffraction ambient temperature is the temperature of the crystal at which the crystal structure is

TABLE 4.19   Information for Caffeine Monohydrate from the CIF

Common Name, Chemical Formula, Systematic Chemical Name

Caffeine monohydrate; $C_8H_{10}N_4O_2 \cdot H_2O$; 1,3,7-Trimethyl-purine-2,6-dione monohydrate

Symmetry equivalent site ID and coordinates

1 x, y, z

2 $1/2-x, 1/2+y, -z$

3 $-x, -y, -z$

4 $1/2+x, 1/2-y, z$

Cell constants

$a = 14.8(1)$ Å $\alpha = 90°$

$b = 16.7(1)$ Å $\beta = 97.0(5)°$

$c = 3.97(3)$ Å $\gamma = 90°$

$Z = 4$; volume = 973.911 Å3, diffraction
ambient temperature = 295 K,
density = 1.447 g/cm3

Atom number, atomic site label, symbol, fractional coordinates					Atom number, atomic site label, symbol, fractional coordinates				
1 C1	C	0.24140	0.22250	−0.09980	13 O1	O	0.30630	0.24000	−0.23860
2 C2	C	0.10030	0.25330	0.12950	14 O2	O	0.13630	0.04040	0.16160
3 C3	C	0.08410	0.17590	0.19440	15 H1	H	−0.08700	0.26100	0.47400
4 C4	C	0.14630	0.11430	0.11550	16 H2	H	−0.01300	0.06200	0.59900
5 C5	C	−0.01990	0.25200	0.36380	17 H3	H	−0.06500	0.06300	0.27800
6 C6	C	0.28910	0.08320	−0.12100	18 H4	H	−0.10500	0.13700	0.51000
7 C7	C	0.19590	0.36380	−0.07910	19 H5	H	0.26300	0.36200	−0.14300
8 C8	C	−0.04640	0.10530	0.45840	20 H6	H	0.22800	0.39600	0.10500
9 N1	N	0.21960	0.14150	−0.02650	21 H7	H	0.14200	0.37700	−0.21700
10 N2	N	0.18010	0.27690	−0.01520	22 H8	H	0.34800	0.10000	−0.22800
11 N3	N	0.00200	0.17490	0.33760	23 H9	H	0.30000	0.03300	0.02200
12 N4	N	0.04030	0.30080	0.24400	24 H10	H	0.25700	0.06000	−0.32400
					25 O3	O	0.01840	0.47050	0.27050

experimentally measured. The number of molecules or formula units in the unit cell or Z, the density, and the volume of the unit cell are also given. Note that, in the fractional coordinate list, the final oxygen O3 is the oxygen atom in the water molecule. Its hydrogen atoms are not included. The crystal is precipitated from a water solution and a water molecule is incorporated into the crystal structure.

4.14.2 Overview of the Starter Program

Caffeine serves to illustrate the power of the Starter Program for this chapter. This MATLAB Starter Program has been used in several hundred student projects with very little fuss. It has also produced the diagrams used in Chapters 9 through 15. The MATLAB program is constructed not for efficiency but for transparency. Actually seeing an incomplete set of the symmetry operations and then adding the needed ones gives the student insights. Also by putting in atoms and bonds the student can see the symmetry operations creating symmetry-related atoms and bonds.

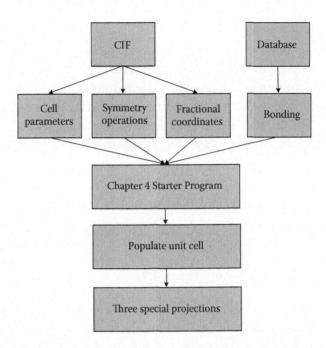

FIGURE 4.93 Flowchart for populating the unit cell and making the special projections.

Figure 4.93 is a flowchart for populating the unit cell. The CIF provides the cell parameters, the symmetry operations, and the fractional coordinates. They are inserted directly into the Starter Program. The CIF does not give a way to connect the atoms, that is, the bonding information. A bonding schematic for a crystal can be obtained by submitting the CIF to the International Union of Crystallography web site: http://checkcif.iucr.org/. A problem with this website is that sometimes the atoms are renamed. By contrast, the CSD provides a labeling diagram consistent with its CIF. Figure 4.94 is a diagram of caffeine monohydrate.

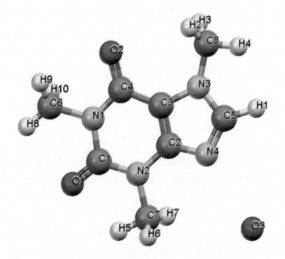

FIGURE 4.94 Labeled molecule of caffeine monohydrate. (CSD-CAFINE.)

4.14.3 Running the Starter Program

The Starter Program printed at the end of this chapter and available as an executable program on the web can be applied to a wide range of crystals. However, in order to help a student begin to understand the program, it is preloaded with information that is specific to caffeine monohydrate. Later, when the Starter Program is applied to another crystal, the information for that crystal is to be inserted in a manner immediately analogous to the insertion of the information for caffeine. The information for caffeine that is already loaded is

- The cell parameters—a, b, c, α, β, and γ—from the CIF (Table 4.19).

- Three of the four symmetry operations from the CIF (Table 4.19).

- The fractional coordinates from the first four carbon atoms from the CIF (Table 4.19).

- Two carbon–carbon bonds from Figure 4.94.

The following material is addressed to a student who is learning how to execute the Starter Program. Throughout these next sections the student is directed to a specific line of the code indicated by the comment symbol "%" followed by a short description of the topic. For example, the student can use the **find** command to find information about where to insert the cell parameters by using the expression "% CELL PARAMETERS."

4.14.3.1 Cell Parameters

The cell parameters have been preloaded. In the MATLAB code for the Starter Program

```
go to% CELL PARAMETERS
```

The cell parameters for caffeine are there. When another crystal is studied just substitute the appropriate a, b, c, alpha, beta, and gamma.

4.14.3.2 Symmetry Operations

Run the Starter Program of Chapter 4 to get Figure 4.95a, which shows the beginning of three molecules. There are four symmetry operations in the CIF (Table 4.19). Examine the code of the MATLAB Starter Program, and notice that only three symmetry operations are included.

```
Go to% SYMMETRY OPERATIONS
```

Add the fourth symmetry operation from Table 4.19, **run** the program, and see the beginnings of all four molecules appear. (See the associated web site for more explanations and demonstrations.) (see Exercise 4.63).

4.14.3.3 Atoms

Only the fractional coordinates of four atoms are preloaded. Those atoms are the first four carbon atoms from the CIF (Table 4.19). In the Starter Program code

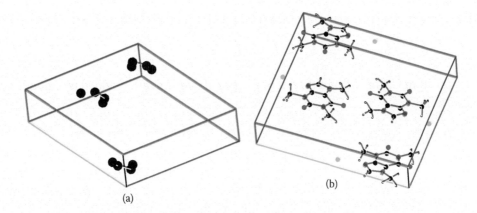

FIGURE 4.95 Three-dimensional unit cell. (a) From Starter Program, (b) completed populated cell of caffeine monohydrate.

```
go to% FRACTIONAL COORDINATES
```

Note that the atoms are commented out (%%%) and numbered.

```
% % % 1 C1 C 0.24140 0.22250 -0.09980
% % % 2 C2 C 0.10030 0.25330 0.12950
% % % 3 C3 C 0.08410 0.17590 0.19440
% % % 4 C4 C 0.14630 0.11430 0.11550
```

Each atom is described by fractional coordinates that give its position. Now add in the next atom, C5.

```
% % % 5 C5 C -0.01990 0.25200 0.36380
```

Only the first four atoms are accessible to the MATLAB code. These atoms appear in the code in a matrix labeled A

```
A = [0.24140 0.22250 -0.09980
0.10030 0.25330 0.12950
0.08410 0.17590 0.19440
0.14630 0.11430 0.11550];
```

The fractional coordinates of atom C5 are added and the code becomes:

```
A = [0.24140 0.22250 -0.09980
0.10030 0.25330 0.12950
0.08410 0.17590 0.19440
0.14630 0.11430 0.11550
-0.01990 0.25200 0.36380];
```

Before the program can be **run**, provision must be made to display or print the additional carbon atom. In the Starter Program
    go to% COLOR CODE AND SIZE ATOMS

The following line of code *calls* the *function* that colors the first four carbon atoms black

```
colorAtom(B,conversion,[1:4],'kO',8,'k')
```

Carbon atom C5, the fifth in the list of fractional coordinates, is added by extending the range as follows:

```
colorAtom(B,conversion,[1:5],'kO',8,'k')%
```

Run the code and note that there are now five atoms in each molecular entity. The four symmetry operations have added a fifth atom to each molecular entity. Each atom is represented by a colored circle which is color coded for identification. Black represents a carbon atom; red for oxygen; blue for nitrogen; and white for hydrogen. The oxygen atom belonging to the water molecule is green. The size of the atom can be adjusted as well. The white hydrogen atoms are much smaller than the other atoms.

Now put in the rest of the atoms, being careful to color code them appropriately. **Run** the code.

4.14.3.4 Bonds

The interatomic bonds must be inserted as indicated in Figure 4.94. The fractional coordinates are all numbered. In the Starter Program

```
go to% Create bonds
```

The following line of code traces the path from atom 2 to atom 3 to atom 4:

```
R = [2  3  4]
```

Atom C4 is connected to atom O2 as in Figure 4.94. Atom O2 is number 14 in the list of fractional coordinates in Table 4.19. The line of code becomes

```
R = [2  3  4  14]
```

Continue to trace a path through the caffeine molecule until Figure 4.95b is achieved. The path can be retraced without affecting the appearance of the diagram. Notice that as one bond is added in the asymmetric unit, the corresponding bond appears in each of the other molecules. This is controlled by the symmetry operations. The Starter Program draws the bonds first so that they lie underneath the circles representing the atoms. One hundred atoms are present including the four caffeine molecules and the oxygen atoms representing the four water molecules.

4.14.3.5 Projections

The next step is to understand symmetries operating in the unit cell that are not obvious in Figure 4.95b. The *International Tables for Crystallography* give the symmetries for the three special projections. Unfortunately, the two-dimensional space groups are for the special projections of $P2_1/c$, but not for $P2_1/a$. Consequently, the projections obtained from the Starter Programs do not match the projections from the *International Tables for Crystallography*. There is a *view* statement in the Starter Program which produces the relevant projection. The projection has to be carefully examined for its symmetry and then labeled correctly.

Figure 4.96a gives the two-dimensional projection along [001]. It was produced by activating a function, called *view*, in the Starter Program. First

```
go to% VIEW
```

Several choices are given for the argument of the *view* function. To get the projection along **c** put in

```
view(alongc)
```

(Be sure to remove the comments %%%%)

Figure 4.96a is the result of *running* the Starter Program. A twofold axis is clearly seen at the center of the unit cell, but the rest of the symmetries are not so obvious. The remaining symmetries can be seen by reproducing the unit cells and placing them side by side. There are several ways to do this. One way is to *import* the figure into *Paint* and then *select* the image (check *transparent*), *copy* it, and place the images side by side. Figure 4.96b shows a part of this enhanced image. Now all the symmetries may be identified.

Figure 4.96c shows the graphical symbols for the twofold axes, the red ovals, and for the glide planes, the purple dashed lines. They were added by using the tab *insert* and then *shape* from the menu in Microsoft Word. The asymmetric unit was also a *shape* which was *filled* and then the filled color was set at a *transparency* of 70%.

See Exercises 4.64 through 4.67.

DEFINITIONS

Asymmetric unit
Body-centered unit cell, *I*
C-faced centered unit cell
Cubic *F* lattice
Cubic *I* lattice
Cubic *P* lattice
Face-centered unit cell, *F*
General position
Glide (three dimensions)
Glide (two dimensions)
Hexagonal *P* lattice (three dimensions)
Hexagonal *P* lattice (two dimensions)
Hexagonal *R* lattice (Rhombohedral)
Isomorphic crystals

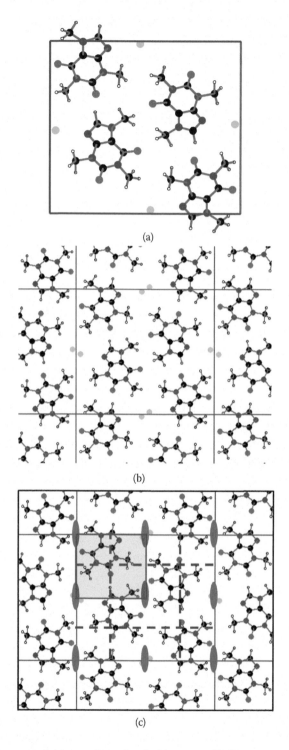

FIGURE 4.96   Caffeine monohydrate along [001]. (a) Unit cell, (b) augmented unit cell, (c) with symmetries and asymmetric unit.

Isotypic crystals
k-subgroups (*klassengleiche*)
k-supergroups (*klassengleiche*)
Lattice vector
Monoclinic *C* lattice
Monoclinic *P* lattice
Multiplicity
Nonsymmorphic space group
Oblique *P* lattice
Orthorhombic *C* lattice
Orthorhombic *F* lattice
Orthorhombic *I* lattice
Orthorhombic *P* lattice
Position
Primitive unit cell, *P*
Rectangular *C* lattice
Rectangular *P* lattice
Screw operation
Site symmetry
Special position
Square *P* lattice
Symmorphic space group
Tetragonal *I* lattice
Tetragonal *P* lattice
Translation
Triclinic *P* lattice
t-subgroups (*translationgleiche*)
t-supergroups (*translationgleiche*)
Wyckoff letter
Z, Z'

EXERCISES

4.1 How many lattice points are there in the unit cells associated with the lattices in Table 4.1? What is the area of each of these unit cells using the **G** matrix? For *oc* and *hp* compare the areas of the primitive and multiple cells.

🖳 4.2 Translate the butterfly (Figure 3.9 and Exercise 3.6) (or another computer-generated motif) five times.

4.3 Given space groups *p2*, *pg*, *p2mm*, *p4gm*, and *p31m*. For each give the number of lattice points in the unit cell, indicate if the space group is symmorphic or nonsymmorphic, and give the associated point group. Also indicate the crystal system. Make a table with the above space groups in the first column.

4.4 List the fractional coordinates of the red, blue, and lavender circles starting with the green circle at x,y in Figure 4.9.

▣ 4.5 Create a pattern showing space group *p1* symmetry. What is the asymmetric unit? You may use "clip art" or a motif generated by the computer. See the Starter Program for generating a lattice. (The program will have to be reduced to 2D.) Be sure the motif has point group 1 symmetry.

▣ 4.6 Create a pattern showing space group *p2* symmetry. What is the asymmetric unit? (Start with a motif with point group 2 and insert it into an oblique lattice. Note the generated twofold axes.)

4.7 Copy Figure 4.15 and on the figure indicate the fractional coordinates of all the circles starting with the green circle at x,y in the asymmetric unit.

▣ 4.8 Create a pattern showing space group *pm* symmetry. What is the asymmetric unit? (Start with a motif with point group *m* and insert it into a rectangular lattice. Note the generated mirror plane.)

4.9 Copy Figure 4.19 and on the figure indicate the fractional coordinates of all the circles starting with the red circle at x,y in the asymmetric unit.

▣ 4.10 Create a pattern showing space group *pg* symmetry. What is the asymmetric unit? Note the generated glide.

4.11 Copy Figure 4.23 and on the figure indicate the fractional coordinates of all the circles starting with the red circle at x,y in the asymmetric unit.

▣ 4.12 Create a pattern showing space group *cm* symmetry. What is the asymmetric unit? (Start with a motif with point group *m*, create a centered unit cell, and insert the motif into a rectangular lattice. Note the generated glide.)

4.13 Copy Figure 4.28a and on the figure indicate the fractional coordinates of all the circles starting with the red circle at x,y in the asymmetric unit.

▣ 4.14 Create a pattern showing space group *p2mm* symmetry. What is the asymmetric unit? (Start with a motif with point group *2mm* and insert the motif into a rectangular lattice. Note the generated symmetry operations.)

4.15 Copy Figure 4.32 and on the figure indicate the fractional coordinates of all the circles starting with the red circle at x,y in the asymmetric unit.

▣ 4.16 Create a pattern showing space group *p2mg* symmetry. What is the asymmetric unit? (Start with a motif with point group 2 or *m*, using a glide create a unit cell, and insert the motif into a rectangular lattice. Note the generated symmetry operations.)

4.17 Copy Figure 4.34b and on the figure indicate the fractional coordinates of all the circles starting with the green circle at x,y in the asymmetric unit.

4.18 Copy Figure 3.15 from Chapter 3. Find the unit cell, mark in the symmetry operations. Draw the general position diagram and symbol diagram superimposed on the unit cell. Check with the *International Tables for Crystallography* for space group *p*4.

4.19 Copy Figure 4.38. Find the unit cell, mark in the symmetry operations. Draw the general position diagram and symbol diagram superimposed on the unit cell. Check with the *International Tables for Crystallography* for space group *p*4*gm*.

🖥 4.20 Create a pattern showing space group *p*4*mm* symmetry. What is the asymmetric unit? (Start with a motif with point group 4*mm* and insert the motif into a square lattice. Note the generated symmetry operations.)

4.21 Given generators: 4^+ 0,0; and *m* 0,*y*, construct the symbol diagram and general position diagram. Is this space group symmorphic or nonsymmorphic? What is the Bravais lattice? Check your results with the *International Tables for X-ray Crystallography*.

4.22 Given generators: *b* 1/4,0; and *m* 0,*y*, construct the symbol diagram and general position diagram. Is this space group symmorphic or nonsymmorphic? What is the Bravais lattice? Check your results with the *International Tables for X-ray Crystallography*.

4.23 Given generators: *m* *x*,0; and *m* 0,*y*, construct the symbol diagram and general position diagram. Is this space group symmorphic or nonsymmorphic? What is the Bravais lattice? Check your results with the *International Tables for X-ray Crystallography*.

4.24 Given generators: 4^+ 0,0; and *b* 1/4,*y*, construct the symbol diagram and general position diagram. Is this space group symmorphic or nonsymmorphic? What is the Bravais lattice? Check your results with the *International Tables for X-ray Crystallography*.

4.25 Given generators: 2 0,0; and *b* 1/4,*y*, construct the symbol diagram and general position diagram. Is this space group symmorphic or nonsymmorphic? What is the Bravais lattice? Check your results with the *International Tables for X-ray Crystallography*.

4.26 Given generators: 2 0,0; *m* 0,*y* and 2 1/4,1/4, construct the symbol diagram and general position diagram. Is this space group symmorphic or nonsymmorphic? What is the Bravais lattice? Check your results with the *International Tables for X-ray Crystallography*.

4.27 Prove by geometry that the asymmetric unit for *p*3 (Figure 4.42) is one-third the area of the unit cell.

4.28 Create a pattern showing space group *p3* symmetry. What is the asymmetric unit? (Start with a motif with point group 3 and insert the motif into a hexagonal lattice. Note the generated symmetry operations.)

4.29 Copy Figure 4.42b and on the figure indicate the fractional coordinates of all the circles starting with the blue circle at *x*,*y* in the asymmetric unit.

4.30 Reproduce Figure 4.44, an example of *p3m1*, and Figure 4.45, an example of *p31m*. Include code.

4.31 Copy over Figure 4.48a. Find the unit cell, and mark in the symmetry operations. Draw the general position diagram and symbol diagram superimposed on the unit cell in a manner similar to Figure 4.36. Check with the *International Tables for Crystallography* for space group *p6*.

4.32 Copy over Figure 4.48b. Find the unit cell, and mark in the symmetry operations. Draw the general position diagram and symbol diagram superimposed on the unit cell in a manner similar to Figure 4.36. Check with the *International Tables for Crystallography* for space group *p6mm*.

4.33 Calculate the area of the asymmetric unit for all 17 space groups.

4.34 Demonstrate that the areas of the asymmetric units are the same for the primitive and centered cells for both *cm* and *c2mm*. In the primitive case relate the sides of the rhombus to the lattice parameters of the centered cell.

4.35 Construct the 2D space group tree in Figure 4.66 through the minimal t-supergroups and the maximal t-subgroups in the *International Tables for Crystallography*.

4.36 Construct a table of the minimal subgroups and maximal supergroups for space group *p3*.

4.37 Find the volumes of the unit cells of all 14 three-dimensional Bravais lattices using the conventional cell in each case.

4.38 Use the *International Tables for Crystallography* to find the symmorphic space groups for the trigonal R space groups. There are 5.

4.39 Use the *International Tables for Crystallography* to find the nonsymmorphic space groups for the trigonal R space groups. There are 2.

4.40 Given the Bravais lattices and point groups in Table 4.6 find the corresponding symmorphic space groups for the orthorhombic system as in Table 4.11.

4.41 Given the Bravais lattices and point groups in Table 4.6 and the *International Tables for Crystallography* find the corresponding nonsymmorphic space groups for the orthorhombic system as in Table 4.11.

4.42 Given the Bravais lattices and point groups in Table 4.6 find the corresponding symmorphic space groups for the tetragonal system as in Table 4.11.

4.43 Given the Bravais lattices and point groups in Table 4.6 and the *International Tables for Crystallography* find the corresponding nonsymmorphic space groups for the tetragonal system as in Table 4.11.

4.44 Given the Bravais lattices and point groups in Table 4.6 find the corresponding symmorphic space groups for the hexagonal system as in Table 4.11.

4.45 Given the Bravais lattices and point groups in Table 4.6 and the *International Tables for Crystallography* find the corresponding nonsymmorphic space groups for the hexagonal system as in Table 4.11.

4.46 Given the Bravais lattices and point groups in Table 4.6 find the corresponding symmorphic space groups for the cubic system as in Table 4.11.

4.47 Given the Bravais lattices and point groups in Table 4.6 and the *International Tables for Crystallography* find the corresponding nonsymmorphic space groups for the cubic system as in Table 4.11.

4.48 Given space groups $P3_1$, $P3m1$, $P6mm$, $I4_132$, and $P4_332$. For each give the number of lattice points in the unit cell, indicate if the space group is symmorphic or nonsymmorphic, and give the associated point group.

4.49 Using the maximal subgroups and minimal supergroup of space group No. 92, construct the part of the space group tree that relates space group No. 92 to its immediate subgroups and supergroup.

4.50 Using the maximal subgroups and minimal supergroup of space group No. 194, construct the part of the space group tree that relates space group No. 194 to its immediate subgroups and supergroup.

4.51 Using the maximal subgroups and minimal supergroup of space group No. 150, construct the part of the space group tree that relates space group No. 150 to its immediate subgroups and supergroup.

⌨ 4.52 Superimpose the asymmetric unit for HMB on the unit cell as in Figure 4.75. Use the Starter Program from Chapter 1.

⌨ 4.53 Read Section 4.14.3. Then adapt the program for HMB and reproduce Figures 4.76 through 4.79.

⌨ 4.54 Read Section 4.14.3. Then adapt the program for AA and reproduce Figures 4.85 and 4.86. For the CIF see the American Mineralogist Crystal Database under the mineral *Steklite*.

⌨ 4.55 Read Section 4.14.3. Then adapt the program for AA and reproduce Figure 4.87. For the CIF see the American Mineralogist Crystal Database under the mineral *Steklite*.

⌨ 4.56 Read Section 4.14.3. Then adapt the program for A A and reproduce Figure 4.90. Start with Exercise 4.54 and use the "fill" *function*.

⌨ 4.57 For RbFe(SeO$_4$)$_2$ calculate the distortion of the Se tetrahedron. Use the data in Table 4.18. Give the four bond lengths and six bond angles and compare with the regular tetrahedron as in Table 4.17.

4.58 Compare the ceramic materials NaCl, KCl, MgO, MnS, LiF, and FeO to check if they are isotypic by constructing a table similar to Table 4.18. Use the Internet.

4.59 Compare the ceramic materials ZnS, ZnTe, and SiC to check if they are isotypic by constructing a table similar to Table 4.18. Use the Internet.

4.60 Compare the ceramic materials CaF$_2$, UO$_2$, and Th O$_2$ to check if they are isotypic by constructing a table similar to Table 4.18. Use the Internet.

4.61 Compare the ceramic materials BaTiO$_3$, SrZrO$_3$, and SrSnO$_3$ to check if they are isotypic by constructing a table similar to Table 4.18. This is the perovskite series. Use the Internet.

4.62 Compare the ceramic materials MgAl$_2$O$_4$ and FeAl$_2$O$_4$ to check if they are isotypic by constructing a table similar to Table 4.18. These are spinels. Use the Internet.

⌨ 4.63 Use the Starter Program for Chapter 4 and create Figure 4.95b.

⌨ 4.64 Starting with Exercise 4.63 create the projection along [001] for caffeine monohydrate. Reproduce Figure 4.96a, b, and c. Include code.

⌨ 4.65 Starting with Exercise 4.63 create the projection along [100]. Reproduce figures analogous to Figure 4.96a, b, and c. Include code.

⌨ 4.66 Starting with Exercise 4.63 create the projection along [010]. Reproduce figures analogous to Figure 4.96a, b, and c. Include code.

⌨ 4.67 Use the Starter Program for Chapter 4 and create a populated cell of sucrose. See Chapter 10 for data.

MATLAB CODE: STARTER PROGRAM FOR CHAPTER 4: GRAPHIC OF POPULATED UNIT CELL AND PROJECTIONS

This Starter Program populates the unit cell with atoms. Four bonded atoms of the caffeine molecule are in the unit cell. The fourth molecule is inserted by adding a fourth symmetry operation. Then the rest of the atoms and the rest of the bonds are inserted. The unit cell can be rotated.

```
function PopulateCellCaffeine
clc
close all
clear
```

```
% % % How to use this program
% % % % % You will need the cif for caffeine. The identifier for
caffeine is
%%%%% CAFINE
%%%%% You will also need a diagram for caffeine which can be found
on the
%%%%% Cambridge Data Base
%%%%%%%%%%%%%%%%%%%%%%%%%%%%%%%%%%%%%%%%%%%%%%%%%%%%%%%%%%%%%%%%%%%%
% % % 1) Put in cell parameters from cif go to%1111111111
% % % 2) Put in fractional coordinates of asymmetric cell from cif
% % %     include numbered labels but comment this data,number
atoms in order this allows you to keep track of
% your atoms go to%222222222222222222222222222222222222222222
%%% 3)Copy fractional coordinates of asymmetric cell, but omit
labels go to%33333333333333333333333333333333
% % % 4)Put in symmetry operations from cif go to
% % 44444444444444444444444444444444444
% % % 5) connect atoms go to
% % 555555555555555555555555555555555555555555555555555555555555
% % % 6) group atoms according to type, all carbons, color code
circles
%%%%%%7) to get projections go to the view function at
7777777777777777777
% % % Script file: Population.m
%
% Purpose: This file creates the POPULATED 3D unit cell
% This file contains the crystallographic data
% Record of revisions:
%     Date  Programmer  Description of change
%     7/21  M. Julian   Original Code
% define variables:
% a = crystallographic a axis (Angstroms, Å)
% b = crystallographic b axis (Angstroms, Å)
% c = crystallographic c axis (Angstroms, Å)
% alpha angle between b and c (degrees)
% beta angle between a and c (degrees)
% gamma angle between a and b (degrees)
% c1 = c1 component of the cartesian conversion matrix
% c2 = c2 component of the cartesian conversion matrix
% c3 = c3 component of the cartesian conversion matrix
%11111111111111111111111111111111111111111111111111111111111111111
a = 14.8;
b = 16.7;
c = 3.97;
alpha = 90;
beta = 97;
gamma = 90;
```

```
%***********************************************************
%define c1, c2, c3 for cartesian conversion matrix
c1 = c*(cosd (beta));
c2 = ((c*((cosd (alpha))-((cosd (gamma)*cosd (beta)))))/(sind
(gamma)));
c3 = abs (sqrt ((c^2 - (c1)^2 - (c2)^2)));
%***********************************************************
%convert to Cartesian coordinates
conversion = [a b*cosd(gamma) c1
      0 b*sind(gamma) c2
      0 0    c3];
%plot cell outline
outline = [0 0 1 1 0 0 0 0 1 1 0 0 1 1 1 1
         0 0 0 0 0 1 1 0 0 1 1 1 1 1 1 0
         0 1 1 0 0 0 1 1 1 1 1 0 0 1 0 0];
plotline(outline, conversion, 'k', 2)%function to draw line
hold on
%plot red a axis
outline = [0 1
         0 0
         0 0];
plotline(outline, conversion, 'r', 2)%function to draw line
%plot green b axis
outline = [0 0
         0 1
         0 0];
plotline(outline, conversion, 'g', 2)%function to draw line
%plot black c axis
outline = [0 0
         0 0
         0 1];
plotline(outline, conversion, 'b', 2)%function to draw line
hold on
% format plot appearance
axis equal % preserves relative axis lengths
axis off    % turns off axes
% controlls width of window. increase if molecules cut off.
ww =.4;
axis([-ww*a a+ww*a -ww*b b+ww*b -ww*c c+ww*c])
%%7777777777777777777777777777777777777777777777777777777777777777
alonga = conversion*[1 0 0]';
alongb = conversion*[0 1 0]';
alongc = conversion*[0 0 1]';
along110 = conversion*[1 1 0]';
%insert other views as needed
%%%%view(alongb)
%***********************************************************
```

```
% part 1 enter fractional coordinates of all the atoms
%information from the cif
%Fractional coordinates with labels for atoms, first 4 entered-
you do
%2222222222222222222222222222222222222222222222222222222222222222
% % % #sequence AtomLabel atom (fractional coordinates) x y z
% % % 1 C1 C 0.24140 0.22250 −0.09980
% % % 2 C2 C 0.10030 0.25330 0.12950
% % % 3 C3 C 0.08410 0.17590 0.19440
% % % 4 C4 C 0.14630 0.11430 0.11550
% % % % select the fractional coordinates
%3333333333333333333333333333333333333333333333333333333333333333
A = [0.24140 0.22250 -0.09980
0.10030 0.25330 0.12950
0.08410 0.17590 0.19440
0.14630 0.11430 0.11550];
%Possible Problem:Look carefully at your data and notice that the
x and y fractional
%coordinates are mostly negative. Since one always can be added in
a perodic array
%add one to the first two columns. Here the third column is ok
%A = A +[ones(22,2) zeros(22,1)];
%%%%%%%%%%%%%%%%%%%%%%%%%%%%%%%%%%%%%%%%%%%%%%%%%%%%%%%%%%%%%%%%%%%%
%%%%
% Part II;
% data for Wyckoff letter for specific space group for atoms in
Part I.Here
% space group is #14 P21/a
[mask,w] = sym_equiv;
%w number of atoms in Wycoff a (multiplicity)
% go to the function GeneralPosition at the bottom of the program
% complete the table
%%%%%%%%%%%%%%%%%%%%%%%%%%%%%%%%%%%%%%%%%%%%%%%%%%%%%%%%%%%%%%%%%%%%
%%%
% Part III
% Create bonds by connectivity13
%5555555555555555555555555555555555555555555555555555555555555555
for i = 1:w
% call function which uses Wyckoff letter data
B = Wyckoff (A,w,i);
%use continuity to connect the atoms
R = [2 3 4];
r = length(R);
ii = 1:r;
outlineM = [B(R(ii),:)'];
        % Extract x,y,z fractional cooridnates
plotline(outlineM, conversion, 'r', 2)%function to plot bond
```

```
end
%%%%%%%%%%%%%%%%%%%%%%%%%%%%%%%%%%%%%%%%%%%%%%%%%%%%%%%%%%%%%%%%%%%%
% Part IV: identification, color code atoms of same atomic number—
all
% carbons black for example
for i = 1:w
    B = Wyckoff (A,w,i);
    colorAtom(B,conversion,[1:4],'kO',8,'k')%color carbon atoms
% % % colorAtom(B,conversion,[7:19],'KO',4,'W')%color hydrogen
atoms
% % % colorAtom(B,conversion,20,'bO',8,'b')%color nitrogen atom
% % % colorAtom(B,conversion,[21 22],'rO',8,'r')%color oxygen
atoms
end
%%%%%%%%%%%%%%%%%%%%%%%%%%%%%%%%%%%%%%%%%%%%%%%%%%%%%%%%%%%%%%%%
function B = Wyckoff (A,w,i)
    % atoms in Wyckoff a which depends on the particular space
      group
    m = length (A);% m is number of atoms in Wycoff a
    for n = 1:m %where n is number of atoms
            x = A(n,1);y = A(n,2);z = A(n,3);
            [mask,w] = sym_equiv;
            Wyk = GeneralPosition(x,y,z);
            B(n,:) = Wyk(i,:) + mask(i,:);
        end
    end
end
%%%%%%%%%%%%%%%%%%%%%%%%%%%%%%%%%%%%%%%%%%
function [mask,w] = sym_equiv
    x =.1;y =.1;z =.1;%testing parameter
Wyk = GeneralPosition(x,y,z);
[w,d] = size (Wyk);%w multiplicty, d dimension
mask = Wyk < 0;
end
%4444444444444444444444444444444444444444444444444444444444444444
function Wyk = GeneralPosition(x,y,z)
    Wyk = [x        y        z
          1/2-x    1/2+y    -z
          -x       -y       -z];
end
%%%%%%%%%%%%%%%%%%%%%%%%%%%%%%%%%%%%%%%%%%%%%%%%%%%%%%%%%%%%%%%%%%%%
function coloratoms = colorAtom(B,conversion,atomsNumbers,outlineco
lor, markersize,markerfacecolor)
    Rcolor = [atomsNumbers];%put in atom numbers of atoms to be colored
    rcolor = length(Rcolor);%count atoms
      ic = 1:rcolor;
          atomN = B(Rcolor(ic),:);
```

```matlab
        % convert to Cartesian Coordinates
        atomNC = (conversion*atomN');
        % pull out x,y,z for plotting
        xN = atomNC (1,:);
        yN = atomNC (2,:);
        zN = atomNC (3,:);
        % plot atoms
        plot3
(xN,yN,zN,outlinecolor,'MarkerSize',markersize, 'MarkerFaceColor',
markerfacecolor);
    end
    function PL = plotline(outline, conversion, Lcolor, linewidth)
cc = (conversion*outline);
% select points to plot from outline
x = cc (1,:);
y = cc (2,:);
z = cc (3,:);
plot3 (x,y,z,Lcolor,'LineWidth',linewidth);
    end
```

## CHAPTER 5

The Reciprocal Lattice

This is an x-ray diffraction pattern of a single quartz crystal recorded on an x-ray diffractometer. The white spots are the diffraction maxima. The distribution of the diffraction maxima is sparse. They are widely spaced compared to their size. They form a regular pattern described by the reciprocal lattice. Crystallographers begin with the reciprocal lattice and work back to the direct lattice. The reciprocal lattice is a property of the crystal, just like the direct lattice, the unit cell, the color, the chemical composition, or the index of refraction.

CONTENTS

CHAPTER OBJECTIVES

- Compare the direct lattice and the reciprocal lattice.

- Understand the relationship between $\mathbf{G}$ and $\mathbf{G}^*$.

- Use $\mathbf{G}^*$ to calculate V^*, $\mathbf{a}^*$, $\mathbf{b}^*$, and $\mathbf{c}^*$.

- Draw the reciprocal cell superimposed on the direct cell with normalized volumes.

- Relate the reciprocal lattice to the diffraction pattern.

- Relate the transformation matrix $\boldsymbol{P}$ of the direct lattice to the transformations of the reciprocal lattice.

- Show that the Bragg planes in direct space correspond to lattice points in reciprocal space.

- Using $\mathbf{G}^*$, calculate the d-spacings.

- Using $\mathbf{G}^*$, calculate the angle between crystal faces.

- Relate different sets of lattice parameters for a crystal.

5.1 INTRODUCTION

An understanding of the reciprocal lattice is essential to crystallography: Use of the reciprocal lattice unifies and simplifies crystallographic calculations.

The reciprocal lattice is a physical property of the crystal just like its other properties such as density, color, and electrical conductivity. The density is calculated from the molecular weight, the number of formula units, the volume of the unit cell, and Avogadro's number. Likewise, the reciprocal cell can be calculated from the direct cell, or alternatively the direct cell can be calculated from the reciprocal cell.

This chapter first defines the reciprocal lattice and then reviews the general properties of lattices. The relationships between the direct and reciprocal lattices are derived. Two invariant properties are demonstrated. A detailed example of the relationships is worked out for HMB. Then, the relationship between the reciprocal lattice and the diffraction pattern is explored. The planes of the direct lattice correspond to points of the reciprocal lattice. Mathematical applications of the reciprocal lattice give straightforward calculations of both the Bragg d-spacings and the interfacial angles of a crystal. Finally, the reciprocal lattice is used to show the relationship between two different sets of lattice parameters for HMB as observed in the PDF.

5.2 THE RECIPROCAL LATTICE

Associated with every lattice is another lattice called the reciprocal lattice. The lattice of the pattern discussed in Chapter 1 is called the *direct lattice* to distinguish it from the reciprocal lattice. If vectors $\mathbf{a}$, $\mathbf{b}$, $\mathbf{c}$ are basis vectors of the direct lattice, then $\mathbf{a}^*$, $\mathbf{b}^*$, $\mathbf{c}^*$ are basis vectors of the reciprocal lattice. The basis vectors of the *reciprocal lattice* are defined by the following matrix equation:

$$\begin{pmatrix} \mathbf{a}^* \cdot \mathbf{a} & \mathbf{a}^* \cdot \mathbf{b} & \mathbf{a}^* \cdot \mathbf{c} \\ \mathbf{b}^* \cdot \mathbf{a} & \mathbf{b}^* \cdot \mathbf{b} & \mathbf{b}^* \cdot \mathbf{c} \\ \mathbf{c}^* \cdot \mathbf{a} & \mathbf{c}^* \cdot \mathbf{b} & \mathbf{c}^* \cdot \mathbf{c} \end{pmatrix} = \begin{pmatrix} 1 & 0 & 0 \\ 0 & 1 & 0 \\ 0 & 0 & 1 \end{pmatrix} \tag{5.1}$$

The *dyad dot product* can be substituted in Equation 5.1.

$$\begin{pmatrix} \mathbf{a}^* \\ \mathbf{b}^* \\ \mathbf{c}^* \end{pmatrix} \cdot \begin{pmatrix} \mathbf{a} & \mathbf{b} & \mathbf{c} \end{pmatrix} = \begin{pmatrix} \mathbf{a}^* \cdot \mathbf{a} & \mathbf{a}^* \cdot \mathbf{b} & \mathbf{a}^* \cdot \mathbf{c} \\ \mathbf{b}^* \cdot \mathbf{a} & \mathbf{b}^* \cdot \mathbf{b} & \mathbf{b}^* \cdot \mathbf{c} \\ \mathbf{c}^* \cdot \mathbf{a} & \mathbf{c}^* \cdot \mathbf{b} & \mathbf{c}^* \cdot \mathbf{c} \end{pmatrix}$$

Thus, Equation 5.1 can be expressed in the condensed form

$$
\begin{pmatrix} \mathbf{a}^* \\ \mathbf{b}^* \\ \mathbf{c}^* \end{pmatrix} \cdot \begin{pmatrix} \mathbf{a} & \mathbf{b} & \mathbf{c} \end{pmatrix} = I
\tag{5.2}
$$

where I is the identity matrix.

From Equation 5.1

$$
\mathbf{a}^* \cdot \mathbf{a} = \mathbf{b}^* \cdot \mathbf{b} = \mathbf{c}^* \cdot \mathbf{c} = 1
$$

and

$$
\mathbf{a}^* \cdot \mathbf{b} = \mathbf{a}^* \cdot \mathbf{c} = \mathbf{b}^* \cdot \mathbf{a} = \mathbf{b}^* \cdot \mathbf{c} = \mathbf{c}^* \cdot \mathbf{a} = \mathbf{c}^* \cdot \mathbf{b} = 0
$$

Since the cosine of 90° is zero, then, for example, $\mathbf{a}^*$ is perpendicular to $\mathbf{b}$ and also $\mathbf{b}^*$ is perpendicular to $\mathbf{a}$. If the units of $\mathbf{a}$ are angstroms, then the units of $\mathbf{a}^*$ are angstroms^{-1}. Figure 5.1 illustrates the relationship between direct and reciprocal basis vectors in two dimensions. The angle between $\mathbf{a}$ and $\mathbf{b}$ is γ; the angle between $\mathbf{a}^*$ and $\mathbf{b}^*$ is γ^*. Since $\mathbf{a}$ and $\mathbf{b}$ are in angstroms and $\mathbf{a}^*$ and $\mathbf{b}^*$ are in inverse angstroms, the directions of the vectors are emphasized, not their relative lengths.

In the special case where $\gamma = 90°$, then $a^* = 1/a$, $b^* = 1/b$. In Figure 5.2, $\mathbf{a}$ and $\mathbf{a}^*$ are parallel, as are $\mathbf{b}$ and $\mathbf{b}^*$. Because of the reciprocal nature of the relationship, the long side in the direct cell becomes the short side in the reciprocal cell and vice versa. In this example, choose $a = 0.5$ Å and $b = 1.0$ Å, then $a^* = 2.0$ Å^{-1} and $b^* = 1.0$ Å^{-1}.

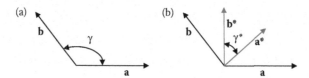

FIGURE 5.1   Two-dimensional nonorthogonal (a) basis vectors $\mathbf{a}$ and $\mathbf{b}$, (b) superimposed reciprocal basis vectors, $\mathbf{a}^*$ and $\mathbf{b}^*$, in red.

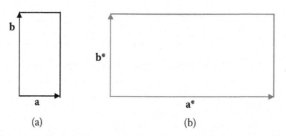

FIGURE 5.2   A two-dimensional unit cell in (a) direct lattice, (b) reciprocal lattice.

Given the basis vectors of a lattice, then the lattice vectors, the metric matrix, the volume of the unit cell, the magnitudes of the lattice vectors, and the angle between lattice vectors can be exhibited, as was done in Chapters 1 and 2. Table 5.1 presents the properties of the direct lattice, with vectors **a**, **b**, **c**, in *parallel* with the properties of the reciprocal lattice, with vectors **a***, **b***, **c***. Some new notation is introduced for the reciprocal lattice in anticipation of future applications. The reciprocal lattice vectors are labeled **H** and the corresponding integers are h, k, and l.

TABLE 5.1 Properties of a Direct Lattice in Parallel with Those of Its Reciprocal Lattice

Direct Lattice	Reciprocal Lattice
Direct lattice vector:	Reciprocal lattice vector:
$\mathbf{t}(uvw) = u\mathbf{a} + v\mathbf{b} + w\mathbf{c}$	$\mathbf{H}(hkl) = h\mathbf{a}^* + k\mathbf{b}^* + l\mathbf{c}^*$
where **a**, **b**, **c** are basis vectors for the direct lattice and u,v,w are integers between $-\infty$ and $+\infty$ (including zero).	where **a***, **b***, **c*** are basis vectors for the reciprocal lattice and h,k,l are integers between $-\infty$ and $+\infty$ (including zero).
G, the metric matrix, for the direct lattice:	**G***, the metric matrix, for the reciprocal lattice:
$$\mathbf{G} = \begin{pmatrix} \mathbf{a\cdot a} & \mathbf{a\cdot b} & \mathbf{a\cdot c} \\ \mathbf{b\cdot a} & \mathbf{b\cdot b} & \mathbf{b\cdot c} \\ \mathbf{c\cdot a} & \mathbf{c\cdot b} & \mathbf{c\cdot c} \end{pmatrix}$$	$$\mathbf{G}^* = \begin{pmatrix} \mathbf{a^*\cdot a^*} & \mathbf{a^*\cdot b^*} & \mathbf{a^*\cdot c^*} \\ \mathbf{b^*\cdot a^*} & \mathbf{b^*\cdot b^*} & \mathbf{b^*\cdot c^*} \\ \mathbf{c^*\cdot a^*} & \mathbf{c^*\cdot b^*} & \mathbf{c^*\cdot c^*} \end{pmatrix}$$
Volume, V, of a unit cell in direct space:	Volume, V^*, of a unit cell in reciprocal space:
$V = \sqrt{\det(\mathbf{G})}$	$V^* = \sqrt{\det(\mathbf{G}^*)}$
where $\det(\mathbf{G})$ is the determinant of **G**.	where $\det(\mathbf{G}^*)$ is the determinant of **G***.
The dot product of two vectors in direct space:	The dot product of two vectors in reciprocal space:
$$\mathbf{t}(u_1 v_1 w_1)\cdot \mathbf{t}(u_2 v_2 w_2) = (u_1 v_1 w_1)\,\mathbf{G}\begin{pmatrix} u_2 \\ v_2 \\ w_2 \end{pmatrix}$$	$$\mathbf{H}(h_1\,k_1\,l_1)\cdot \mathbf{H}(h_2\,k_2\,l_2) = (h_1\,k_1\,l_1)\,\mathbf{G}^*\begin{pmatrix} h_2 \\ k_2 \\ l_2 \end{pmatrix}$$
The magnitude squared of a vector in direct space:	The magnitude squared of a vector in reciprocal space:
$$t^2(u\,v\,w) = (u\,v\,w)\,\mathbf{G}\begin{pmatrix} u \\ v \\ w \end{pmatrix}$$	$$H^2(hkl) = (hkl)\,\mathbf{G}^*\begin{pmatrix} h \\ k \\ l \end{pmatrix}$$
The $\cos\theta$ of the angle between two vectors in direct space:	The $\cos\theta^*$ of the angle between two vectors in reciprocal space:
$$\cos\theta = \frac{(u_1 v_1 w_1)\,\mathbf{G}\begin{pmatrix} u_2 \\ v_2 \\ w_2 \end{pmatrix}}{t_1 t_2}$$	$$\cos\theta^* = \frac{(h_1\,k_1 l_1)\,\mathbf{G}^*\begin{pmatrix} h_2 \\ k_2 \\ l_2 \end{pmatrix}}{H_1 H_2}$$
where t_1 and t_2 are the magnitudes of the vectors $\mathbf{t}(u_1\,v_1\,w_1)$ and $\mathbf{t}(u_2\,v_2\,w_2)$, respectively.	where H_1 and H_2 are the magnitudes of the vectors $\mathbf{H}(h_1\,k_1\,l_1)$ and $\mathbf{H}(h_2\,k_2\,l_2)$ respectively.

5.3 RELATIONSHIPS BETWEEN DIRECT AND RECIPROCAL LATTICES

Table 5.1 reviews the general properties of lattices. The direct and reciprocal lattices are mathematically related by Equation 5.1. From this equation the relationships are now derived between the sets of basis vectors, the metric matrices, and the volumes in both direct and reciprocal lattices.

5.3.1 Relationships for Basis Vectors and Lattice Parameters

Since, by definition in Equation 5.1, $\mathbf{a}^*$ is perpendicular to both $\mathbf{b}$ and $\mathbf{c}$, $\mathbf{a}^*$ is parallel to the cross product of $\mathbf{b}$ and $\mathbf{c}$.

Thus,

$$\mathbf{a}^* = k(\mathbf{b} \times \mathbf{c})$$

where the scalar k is a proportionality constant.

Since

$$\mathbf{a}^* \cdot \mathbf{a} = 1$$

then, for a right-handed set of basis vectors,

$$\mathbf{a}^* \cdot \mathbf{a} = k\mathbf{a} \cdot (\mathbf{b} \times \mathbf{c}) = kV = 1$$

where V is the volume of the unit cell in direct space.

Then

$$k = 1/V$$

and therefore

$$\mathbf{a}^* = \frac{\mathbf{b} \times \mathbf{c}}{V} \tag{5.3}$$

In a similar way, the rest of the relationships in Table 5.2 can be derived.

TABLE 5.2   Some Derived Vector Relationships between the Direct and Reciprocal Lattices with Right-Handed Sets of Basis Vectors

$\mathbf{a} = \dfrac{\mathbf{b}^* \times \mathbf{c}^*}{V^*}$	$\mathbf{a}^* = \dfrac{\mathbf{b} \times \mathbf{c}}{V}$
$\mathbf{b} = \dfrac{\mathbf{c}^* \times \mathbf{a}^*}{V^*}$	$\mathbf{b}^* = \dfrac{\mathbf{c} \times \mathbf{a}}{V}$
$\mathbf{c} = \dfrac{\mathbf{a}^* \times \mathbf{b}^*}{V^*}$	$\mathbf{c}^* = \dfrac{\mathbf{a} \times \mathbf{b}}{V}$

Equation 5.3 in scalar form is

$$a^* = (b\,c\,\sin\alpha)/V \tag{5.4}$$

In a similar way, Table 5.3 is constructed.
Solve Equation 5.4 for $\sin\alpha$ using the equations in Table 5.3 and $V^* = 1/V$

$$\sin\alpha = \frac{a^* V}{bc} = a^*\, V \cdot \frac{V^*}{a^*\, c^* \sin\beta^*} \cdot \frac{V^*}{a^*\, b^* \sin\gamma^*}$$

$$= \frac{V^*}{a^* b^* c^* \sin\beta^* \sin\gamma^*}$$

Again, in a similar way Table 5.4 is constructed. Tables 5.3 and 5.4 have been popular in crystallography and are retained here to present one view of the relationships. See Exercises 5.1 and 5.2.

TABLE 5.3   Some Derived Scalar Relationships
between Direct and Reciprocal Lattices

$a = \dfrac{b^* c^* \sin\alpha^*}{V^*}$	$a^* = \dfrac{bc \sin\alpha}{V}$
$b = \dfrac{a^* c^* \sin\beta^*}{V^*}$	$b^* = \dfrac{ac \sin\beta}{V}$
$c = \dfrac{a^* b^* \sin\gamma^*}{V^*}$	$c^* = \dfrac{ab \sin\gamma}{V}$

TABLE 5.4   Some Derived Relationships between the Angles of
the Direct and the Reciprocal Lattices

$\sin\alpha = \dfrac{V^*}{a^* b^* c^* \sin\beta^* \sin\gamma^*}$	$\sin\alpha^* = \dfrac{V}{abc \sin\beta \sin\gamma}$
$\sin\beta = \dfrac{V^*}{a^* b^* c^* \sin\alpha^* \sin\gamma^*}$	$\sin\beta^* = \dfrac{V}{abc \sin\alpha \sin\gamma}$
$\sin\gamma = \dfrac{V^*}{a^* b^* c^* \sin\alpha^* \sin\beta^*}$	$\sin\gamma^* = \dfrac{V}{abc \sin\alpha \sin\beta}$
$\cos\alpha = \dfrac{\cos\beta^* \cos\gamma^* - \cos\alpha^*}{\sin\beta^* \sin\gamma^*}$	$\cos\alpha^* = \dfrac{\cos\beta \cos\gamma - \cos\alpha}{\sin\beta \sin\gamma}$
$\cos\beta = \dfrac{\cos\alpha^* \cos\gamma^* - \cos\beta^*}{\sin\alpha^* \sin\gamma^*}$	$\cos\beta^* = \dfrac{\cos\alpha \cos\gamma - \cos\beta}{\sin\alpha \sin\gamma}$
$\cos\gamma = \dfrac{\cos\alpha^* \cos\beta^* - \cos\gamma^*}{\sin\alpha^* \sin\beta^*}$	$\cos\gamma^* = \dfrac{\cos\alpha \cos\beta - \cos\gamma}{\sin\alpha \sin\beta}$

5.3.2 Relationship between the Metric Matrices **G** and **G***

The basis vectors **a***, **b***, **c*** of the reciprocal lattice are related to the basis vectors **a**, **b**, **c** of the direct lattice by Equation 5.1. Consider a change of basis between these.

With the first set of basis vectors, a vector **r** is given by

$$\mathbf{r} = x\mathbf{a} + y\mathbf{b} + z\mathbf{c} \tag{5.5}$$

and with the second set of basis vectors, the *same* vector **r** is given by

$$\mathbf{r} = x^*\mathbf{a}^* + y^*\mathbf{b}^* + z^*\mathbf{c}^* \tag{5.6}$$

From Equations 5.5 and 5.1

$$\mathbf{r} \cdot \mathbf{a}^* = x; \, \mathbf{r} \cdot \mathbf{b}^* = y; \, \mathbf{r} \cdot \mathbf{c}^* = z \tag{5.7}$$

From Equations 5.6 and 5.1

$$\mathbf{r} \cdot \mathbf{a} = x^*; \, \mathbf{r} \cdot \mathbf{b} = y^*; \, \mathbf{r} \cdot \mathbf{c} = z^* \tag{5.8}$$

From Equation 5.5

$$\begin{aligned}
\mathbf{r} \cdot \mathbf{a} &= x\mathbf{a} \cdot \mathbf{a} + y\mathbf{b} \cdot \mathbf{a} + z\mathbf{c} \cdot \mathbf{a} \\
\mathbf{r} \cdot \mathbf{b} &= x\mathbf{a} \cdot \mathbf{b} + y\mathbf{b} \cdot \mathbf{b} + z\mathbf{c} \cdot \mathbf{b} \\
\mathbf{r} \cdot \mathbf{c} &= x\mathbf{a} \cdot \mathbf{c} + y\mathbf{b} \cdot \mathbf{c} + z\mathbf{c} \cdot \mathbf{c}
\end{aligned} \tag{5.9}$$

Combining Equations 5.8 and 5.9 gives

$$\begin{pmatrix} \mathbf{r} \cdot \mathbf{a} \\ \mathbf{r} \cdot \mathbf{b} \\ \mathbf{r} \cdot \mathbf{c} \end{pmatrix} = \begin{pmatrix} x^* \\ y^* \\ z^* \end{pmatrix} = \begin{pmatrix} \mathbf{a} \cdot \mathbf{a} & \mathbf{a} \cdot \mathbf{b} & \mathbf{a} \cdot \mathbf{c} \\ \mathbf{b} \cdot \mathbf{a} & \mathbf{b} \cdot \mathbf{b} & \mathbf{b} \cdot \mathbf{c} \\ \mathbf{c} \cdot \mathbf{a} & \mathbf{c} \cdot \mathbf{b} & \mathbf{c} \cdot \mathbf{c} \end{pmatrix} \cdot \begin{pmatrix} x \\ y \\ z \end{pmatrix} \tag{5.10}$$

Substituting **G** for the metric matrix, Equation 5.10 becomes

$$\begin{pmatrix} x^* \\ y^* \\ z^* \end{pmatrix} = \mathbf{G} \begin{pmatrix} x \\ y \\ z \end{pmatrix} \tag{5.11}$$

Multiplying on each side by **G**$^{-1}$

$$\mathbf{G}^{-1} \begin{pmatrix} x^* \\ y^* \\ z^* \end{pmatrix} = \mathbf{G}^{-1} \cdot \mathbf{G} \cdot \begin{pmatrix} x \\ y \\ z \end{pmatrix} = \begin{pmatrix} x \\ y \\ z \end{pmatrix} \tag{5.12}$$

Rearranging Equation 5.12 gives

$$\begin{pmatrix} x \\ y \\ z \end{pmatrix} = \mathbf{G}^{-1} \begin{pmatrix} x^* \\ y^* \\ z^* \end{pmatrix} \tag{5.13}$$

A parallel computation is done starting with Equation 5.6. The first step is to take the dot product of $\mathbf{a}^*$ with $\mathbf{r}$ which is parallel to Equation 5.9

$$\mathbf{r} \cdot \mathbf{a}^* = x^*\mathbf{a}^* \cdot \mathbf{a}^* + y^*\mathbf{b}^* \cdot \mathbf{a}^* + z^*\mathbf{c}^* \cdot \mathbf{a}^* = x$$

$$\mathbf{r} \cdot \mathbf{b}^* = x^*\mathbf{a}^* \cdot \mathbf{b}^* + y^*\mathbf{b}^* \cdot \mathbf{b}^* + z^*\mathbf{c}^* \cdot \mathbf{b}^* = y$$

$$\mathbf{r} \cdot \mathbf{c}^* = x^*\mathbf{a}^* \cdot \mathbf{c}^* + y^*\mathbf{b}^* \cdot \mathbf{c}^* + z^*\mathbf{c}^* \cdot \mathbf{c}^* = z$$

Continue in a parallel way to get

$$\begin{pmatrix} x \\ y \\ z \end{pmatrix} = \mathbf{G}^* \begin{pmatrix} x^* \\ y^* \\ z^* \end{pmatrix} \tag{5.14}$$

Compare Equations 5.13 and 5.14

$$\mathbf{G}^* = \mathbf{G}^{-1} \tag{5.15}$$

Multiplying Equation 5.15 on both sides by $\mathbf{G}$ gives

$$\mathbf{G}\mathbf{G}^* = \mathbf{G}^*\mathbf{G} = \mathbf{I} \tag{5.16}$$

where $\mathbf{I}$ is the identity matrix.

Equation 5.16 is an invariant property of any lattice. An *invariant property* does not depend on the particular basis vectors chosen.

5.3.3 Relationship between the Volumes V and V^*

From Table 5.1

$$V = \sqrt{\det(\mathbf{G})} \text{ and } V^* = \sqrt{\det(\mathbf{G}^*)} \tag{5.17}$$

From Equation 5.16, since the determinant of a product equals the product of the determinants,

$$\det(\mathbf{G} \cdot \mathbf{G}^*) = \det(\mathbf{G}) \times \det(\mathbf{G}^*) = \det(\mathbf{I}) = 1 \tag{5.18}$$

Then from Equation 5.18

$$\det(\mathbf{G}) = \frac{1}{\det(\mathbf{G}^*)} \tag{5.19}$$

Combine Equations 5.17 and 5.19

$$V = \frac{1}{V^*}$$

or

$$VV^* = 1 \tag{5.20}$$

Equation 5.20 is an invariant property of a given lattice. See Exercises 5.3 and 5.4.

5.3.4 Relationship between Direct and Reciprocal Lattice Vectors Revisited

Again consider the relationship between $\mathbf{a}$, $\mathbf{b}$, $\mathbf{c}$, and $\mathbf{a}^*$, $\mathbf{b}^*$, $\mathbf{c}^*$, which was already given in Table 5.2 using cross products. There is another way to express that relationship using the metric matrices.

Introduce scalars q_1 through q_9 by

$$\begin{aligned}
\mathbf{a} &= q_1\mathbf{a}^* + q_2\mathbf{b}^* + q_3\mathbf{c}^* \\
\mathbf{b} &= q_4\mathbf{a}^* + q_5\mathbf{b}^* + q_6\mathbf{c}^* \\
\mathbf{c} &= q_7\mathbf{a}^* + q_8\mathbf{b}^* + q_9\mathbf{c}^*
\end{aligned} \tag{5.21}$$

Then

$$\begin{aligned}
\mathbf{a} \cdot \mathbf{a} &= q_1, \mathbf{b} \cdot \mathbf{b} = q_5, \mathbf{c} \cdot \mathbf{c} = q_9 \\
\mathbf{a} \cdot \mathbf{b} &= q_2 = q_4 \\
\mathbf{a} \cdot \mathbf{c} &= q_3 = q_7 \\
\mathbf{b} \cdot \mathbf{c} &= q_6 = q_8
\end{aligned} \tag{5.22}$$

Now substitute Equation 5.22 into Equation 5.21 and put into matrix form:

$$\begin{pmatrix} \mathbf{a} \\ \mathbf{b} \\ \mathbf{c} \end{pmatrix} = \mathbf{G} \begin{pmatrix} \mathbf{a}^* \\ \mathbf{b}^* \\ \mathbf{c}^* \end{pmatrix}$$

Since $\mathbf{G}^* = \mathbf{G}^{-1}$, then

$$\begin{pmatrix} \mathbf{a}^* \\ \mathbf{b}^* \\ \mathbf{c}^* \end{pmatrix} = \mathbf{G}^* \begin{pmatrix} \mathbf{a} \\ \mathbf{b} \\ \mathbf{c} \end{pmatrix} \tag{5.23}$$

TABLE 5.5 Summary of Transformations between Direct and Reciprocal Lattices

Objects Transformed	From Direct to Reciprocal Lattice	From Reciprocal to Direct Lattice
Coefficients	$$\begin{pmatrix} x^* \\ y^* \\ z^* \end{pmatrix} = \mathbf{G} \begin{pmatrix} x \\ y \\ z \end{pmatrix}$$	$$\begin{pmatrix} x \\ y \\ z \end{pmatrix} = \mathbf{G}^* \begin{pmatrix} x^* \\ y^* \\ z^* \end{pmatrix}$$
Metric matrix	$\mathbf{G}^* = \mathbf{G}^{-1}$	$\mathbf{G} = \mathbf{G}^{*-1}$
Volumes	$V^* = \dfrac{1}{V}$	$V = \dfrac{1}{V^*}$
Basis vectors	$$\begin{pmatrix} \mathbf{a}^* \\ \mathbf{b}^* \\ \mathbf{c}^* \end{pmatrix} = \mathbf{G}^* \begin{pmatrix} \mathbf{a} \\ \mathbf{b} \\ \mathbf{c} \end{pmatrix}$$	$$\begin{pmatrix} \mathbf{a} \\ \mathbf{b} \\ \mathbf{c} \end{pmatrix} = \mathbf{G} \begin{pmatrix} \mathbf{a}^* \\ \mathbf{b}^* \\ \mathbf{c}^* \end{pmatrix}$$

See Exercises 5.5 through 5.7. Although outside the scope of this book, the similarity between Equations 5.11, 5.14, and 5.23 has great significance.

5.3.5 Summary of Transformations between Direct and Reciprocal Lattices

The transformations between direct and reciprocal lattices are summarized in Table 5.5. Great care must be taken in applying these transformations. While they look simple, they are extremely powerful. Applications are given in the next section. Notice that the coefficients, x, y, z do not transform in the same way as the basis vectors, $\mathbf{a}$, $\mathbf{b}$, $\mathbf{c}$.

5.4 RECIPROCAL LATTICE CALCULATIONS FOR THREE CRYSTALS

Three examples of crystals are presented to show how to calculate reciprocal lattice parameters. HMB is triclinic and, as usual, presents the most general case while the calculations for the other crystals are special cases. The AA example continues. Finally, an orthorhombic crystal demonstrates what happens if α, β, and γ are all 90°.

🖥 5.4.1 Hexamethylbenzene

Here is an example with general significance discussed in detail, using HMB, to demonstrate calculations of $\mathbf{G}^*$, V^*, and the reciprocal lattice parameters a^*, b^*, c^*, α^*, β^*, and γ^*. The relationships of the basis vectors of the direct and reciprocal lattices are visualized by overlaying the reciprocal unit cell on the direct unit cell with the normalized volumes.

5.4.1.1 Calculation of $\mathbf{G}^*$ and V^*

Considering the direct lattice, start with the lattice parameters (Lonsdale 1929b) and calculate the metric matrix and the volume of the unit cell. This was done in Chapter 2. Now, with the aid of a computer, the metric matrix can be conveniently inverted. Since $\mathbf{G}^*$ is equal to $\mathbf{G}^{-1}$, the metric matrix can be used to calculate the reciprocal lattice parameters, bypassing the equations in Tables 5.3 and 5.4. The volume of the unit cell of the reciprocal lattice can be calculated either as the reciprocal of the direct volume or as the square root of the determinant of the metric matrix $\mathbf{G}^*$. Calculating the volume of the reciprocal unit cell both ways gives a good internal check of the calculations.

EXAMPLE 5.1

From the metric matrix **G** of HMB, calculate **G*** and the lattice parameters of the reciprocal lattice. Calculate the volume of the reciprocal cell in two ways: first, as the reciprocal of the volume of the direct cell, and second, as the square root of the determinant of the metric matrix **G***.

Solution

From Chapter 2, begin with the data for HMB in Table 5.6.

The crucial step is to take the inverse of **G** to get **G***. The result is in Table 5.7, along with the algebraic expression for **G*** from Table 5.1.

Now examine the individual entries in the **G*** matrix.
For example,

$$\mathbf{a}^* \cdot \mathbf{a}^* = (a^*)^2 = 0.0167 \text{ Å}^{-2} \quad \text{and} \quad a^* = 0.1293 \text{ Å}^{-1}$$

Similarly b^* and c^* can be calculated, obtaining b^* equal to 0.1664 Å^{-1}. Now consider the cross terms such as $\mathbf{a}^* \cdot \mathbf{b}^*$:

$$\mathbf{a}^* \cdot \mathbf{b}^* = a^* \ b^* \cos \gamma^* = 0.0059 = 0.1293 \times 0.1664 \times \cos \gamma^*$$

TABLE 5.6   HMB Data for the Direct Lattice

a (Å)	9.010		
b (Å)	8.926		
c (Å)	5.344		
α (degrees)	44 + 27/60		
β (degrees)	116 + 43/60		
γ (degrees)	119 + 34/60		
G (Å^2)	81.18	−39.68	−21.64
	−39.68	79.67	34.05
	−21.64	34.05	28.55
V (Å^3)	258.41		

TABLE 5.7   Metric Matrix, **G***, for HMB

G*(Å^2)$^{-1}$	0.0167	0.0059	0.0056	$\begin{pmatrix} \mathbf{a}^* \cdot \mathbf{a}^* & \mathbf{a}^* \cdot \mathbf{b}^* & \mathbf{a}^* \cdot \mathbf{c}^* \\ \mathbf{b}^* \cdot \mathbf{a}^* & \mathbf{b}^* \cdot \mathbf{b}^* & \mathbf{b}^* \cdot \mathbf{c}^* \\ \mathbf{c}^* \cdot \mathbf{a}^* & \mathbf{c}^* \cdot \mathbf{b}^* & \mathbf{c}^* \cdot \mathbf{c}^* \end{pmatrix}$
	0.0059	0.0277	−0.0285	
	0.0056	−0.0285	0.0733	

TABLE 5.8 Reciprocal Lattice Parameters for HMB	
a^* (Å^{-1})	0.1293
b^* (Å^{-1})	0.1664
c^* Å^{-1})	0.2707
α^* (degrees)	129.29
β^* (degrees)	80.80
γ^* (degrees)	73.99

Solving for γ^*

$$\gamma^* = 73.99°$$

Continuing in the same way, Table 5.8 can be completed.
More elegantly, Equation 5.23 can be used to complete Table 5.8.

$$\begin{pmatrix} \mathbf{a}^* \\ \mathbf{b}^* \\ \mathbf{c}^* \end{pmatrix} = \mathbf{G}^* \begin{pmatrix} \mathbf{a} \\ \mathbf{b} \\ \mathbf{c} \end{pmatrix}$$

Substituting the values for $\mathbf{G}^*$, the reciprocal basis vectors are

$$\mathbf{a}^* = 0.0167\,\mathbf{a} + 0.0059\,\mathbf{b} + 0.0056\,\mathbf{c}\ \text{Å}^{-1}$$

$$\mathbf{b}^* = 0.0059\,\mathbf{a} + 0.0277\,\mathbf{b} - 0.0285\,\mathbf{c}\ \text{Å}^{-1}$$

$$\mathbf{c}^* = 0.0056\,\mathbf{a} - 0.0285\,\mathbf{b} + 0.0733\,\mathbf{c}\ \text{Å}^{-1}$$

Take the dot products to get the magnitudes of the reciprocal basis vectors and show agreement with Table 5.8.

$$|\mathbf{a}^*| = 0.1293\ \text{Å}^{-1},\ |\mathbf{b}^*| = 0.1664\ \text{Å}^{-1},\ |\mathbf{c}^*| = 0.2707\text{Å}^{-1}$$

Finally, calculate the volume of the reciprocal cell:
First, from Table 5.6, $V = 258.4$ Å^3, then

$$V^* = 1/V = 0.0038\ (\text{Å}^{-3})$$

Second,

$$V^* = \sqrt{\det(\mathbf{G}^*)} = 0.0038\ (\text{Å}^{-3})$$

5.4.1.2 Overlaying the Reciprocal Unit Cell on the Direct Unit Cell

In order to plot the reciprocal unit cell superimposed over the direct unit cell, the sets of basis vectors for both the direct and reciprocal lattices are needed in Cartesian coordinates. Use the conversion matrix, **C**, derived in Chapter 1.

Namely,

$$\mathbf{C} = \begin{pmatrix} a & b\cos\gamma & c_1 \\ 0 & b\sin\gamma & c_2 \\ 0 & 0 & c_3 \end{pmatrix}$$

With

$$c_1 = c\cos\beta, \quad c_2 = \frac{c(\cos\alpha - \cos\gamma\cos\beta)}{\sin\gamma}, \quad \text{and} \quad c_3 = +\sqrt{c^2 - c_1^2 - c_2^2}$$

From Chapter 1 and Table 5.1

$$\mathbf{t} = x\,\mathbf{e}_1 + y\,\mathbf{e}_2 + z\,\mathbf{e}_3 = u\,\mathbf{a} + v\,\mathbf{b} + w\,\mathbf{c}$$

The transformation from crystallographic coordinates to Cartesian coordinates is

$$\begin{pmatrix} x \\ y \\ z \end{pmatrix} = \mathbf{C} \begin{pmatrix} u \\ v \\ w \end{pmatrix}$$

Since the column vector of the crystallographic coordinates of

$$\mathbf{a} \text{ is } \begin{pmatrix} 1 \\ 0 \\ 0 \end{pmatrix}, \quad \mathbf{b} \text{ is } \begin{pmatrix} 0 \\ 1 \\ 0 \end{pmatrix}, \quad \text{and} \quad \mathbf{c} \text{ is } \begin{pmatrix} 0 \\ 0 \\ 1 \end{pmatrix}$$

then the column vector of the Cartesian coordinates of

$$\mathbf{a} \text{ is } \mathbf{C} \begin{pmatrix} 1 \\ 0 \\ 0 \end{pmatrix}, \quad \mathbf{b} \text{ is } \mathbf{C} \begin{pmatrix} 0 \\ 1 \\ 0 \end{pmatrix}, \quad \text{and} \quad \mathbf{c} \text{ is } \mathbf{C} \begin{pmatrix} 0 \\ 0 \\ 1 \end{pmatrix}$$

Thus, for HMB

$$\mathbf{a} = 9.0100\mathbf{e}_1$$
$$\mathbf{b} = -4.4044\mathbf{e}_1 + 7.7637\mathbf{e}_2$$
$$\mathbf{c} = -2.4025\mathbf{e}_1 + 3.0230\mathbf{e}_2 + 3.6942\mathbf{e}_3$$

To get the corresponding basis vectors of the reciprocal lattice, use Equation 5.23

$$\begin{pmatrix} \mathbf{a}^* \\ \mathbf{b}^* \\ \mathbf{c}^* \end{pmatrix} = \mathbf{G}^* \begin{pmatrix} \mathbf{a} \\ \mathbf{b} \\ \mathbf{c} \end{pmatrix}$$

Note that $\mathbf{e}_1 = \mathbf{e}_1^*$; $\mathbf{e}_2 = \mathbf{e}_2^*$; $\mathbf{e}_3 = \mathbf{e}_3^*$.
Thus, the basis vectors for reciprocal lattice are

$$\mathbf{a}^* = 0.1110\mathbf{e}_1 + 0.0630\mathbf{e}_2 + 0.0207\mathbf{e}_3$$
$$\mathbf{b}^* = -0.0000\mathbf{e}_1 + 0.1288\mathbf{e}_2 + -0.1054\mathbf{e}_3$$
$$\mathbf{c}^* = +0.2707\mathbf{e}_3$$

The magnitudes of the basis vectors agree with Table 5.8.

$$|\mathbf{a}^*| = 0.1293 \text{ Å}^{-1}, \; |\mathbf{b}^*| = 0.1664 \text{ Å}^{-1}, \; |\mathbf{c}^*| = 0.2707 \text{ Å}^{-1}$$

Now that both sets of basis vectors are known, both unit cells can be plotted in the same diagram. The units of the basis vectors of the direct lattice are Å and the units of the basis vectors of the reciprocal lattice are Å^{-1}. These units can lead to some strange-looking diagrams. For example, in Figure 5.3a for HMB, $a = 9.010$ Å, $a^* = 0.1263$ Å^{-1}; and the reciprocal unit cell appears very small, in red, with respect to the direct unit cell, in blue. A scale equal to $V^{1/3}$ is used in Figure 5.3b. This scale is chosen because the volumes of both the direct and reciprocal unit cells are the same and are equal to 1. One way to implement this scale is to divide each of the direct lattice parameters, a, b, and c by $V^{1/3}$. When the reciprocal vectors are calculated from the scaled direct lattice parameters, they are in effect divided through by $V^{*1/3}$. The angles between $\mathbf{a}$ and $\mathbf{a}^*$ (green lines), between $\mathbf{b}$ and $\mathbf{b}^*$ (light blue lines), and between $\mathbf{c}$ and $\mathbf{c}^*$ (black lines) are highlighted. Note the relative shapes of the unit cells: The red reciprocal unit cell is elongated along c^* compared with the relatively squat direct unit cell in blue.

A Starter Program is provided to assist the student in overlaying the reciprocal unit cell on the direct unit cell (see Exercise 5.8).

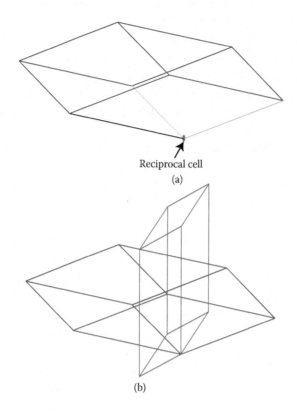

Reciprocal cell

(a)

(b)

FIGURE 5.3 HMB direct unit cell in blue and reciprocal cell in red (a) units Å and Å⁻¹, (b) volumes of both unit cells are equal.

🖳 5.4.2 Anhydrous Alum

Table 5.9 contains data for anhydrous alum. Notice in this case that, since $a = b$, then $a^* = b^*$. Also since $\alpha = \beta = 90°$, then $\alpha^* = \beta^* = 90°$. Consequently some of the off-diagonal entries in both the **G** and **G*** metric matrices are zero (see Exercise 5.9).

Figure 5.4 shows the reciprocal cell for anhydrous alum superimposed over the direct unit cell. The direct cell is blue and the reciprocal cell is red. Note that **c** and **c*** are

TABLE 5.9   Data for Anhydrous Alum for Both Direct and Reciprocal Lattices

Direct			Reciprocal			
a (Å)	4.709		a^* (Å⁻¹)	0.2452		
b (Å)	4.709		b^* (Å⁻¹)	0.2452		
c (Å)	7.940		c^* (Å⁻¹)	0.1259		
α (degrees)	90		α^* (degrees)	90		
β (degrees)	90		β^* (degrees)	90		
γ (degrees)	120		γ^* (degrees)	60		
G (Å²)	$\begin{pmatrix} 22.17 & -11.08 & 0 \\ -11.08 & 22.17 & 0 \\ 0 & 0 & 63.04 \end{pmatrix}$		**G*** (Å⁻²)	$\begin{pmatrix} 0.0601 & 0.0301 & 0 \\ 0.0301 & 0.0601 & 0 \\ 0 & 0 & 0.0159 \end{pmatrix}$		
V (Å³)	152.47		V^* (Å⁻³)	0.0066		

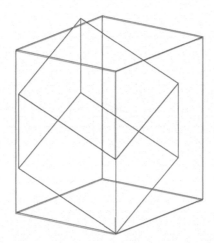

FIGURE 5.4   Reciprocal cell for AA, in red, superimposed over the direct cell, in blue.

parallel. Exercise 5.10 is to reproduce these cells. Both cells have been normalized to have their volumes equal to 1. Note the directions of **a**, **b**, and **c** relative to those of **a***, **b***, and **c***. The *shape* of the direct cell relative to the reciprocal cell is important too. In the direct cell the vector **c** is longer than the vector **a** while in the reciprocal cell just the opposite occurs. The vector **c*** is shorter than the vector **a***.

5.4.3 Crystal with α, β, and γ all 90°

An interesting simplification occurs if α, β, and γ are all 90°. This happens in the cubic, orthorhombic, and tetragonal crystal systems. In these systems **a** and **a***, **b** and **b***, **c** and **c*** are parallel to each other. The lattice parameters for $YBa_2Cu_3O_{7-x}$ are shown in Table 5.10. This high-temperature superconductor crystallizes in the orthorhombic system. The reciprocal unit cell superimposed on the direct cell is shown in Figure 5.5. Note that the longest vector in the direct cell **c** becomes the shortest vector in the reciprocal cell **c***. The volumes

TABLE 5.10   Data for Orthorhombic $YBa_2Cu_3O_{7-x}$ for Both Direct and Reciprocal Lattices

Direct				Reciprocal			
a (Å)	3.827			a^* (Å^{-1})	0.2613		
b (Å)	3.882			b^* (Å^{-1})	0.2576		
c (Å)	11.682			c^* (Å^{-1})	0.0856		
α (degrees)	90			α* (degrees)	90		
β (degrees)	90			β* (degrees)	90		
γ (degrees)	90			γ* (degrees)	90		
G (Å^2)	$\begin{pmatrix} 14.65 & 0 & 0 \\ 0 & 15.07 & 0 \\ 0 & 0 & 136.47 \end{pmatrix}$			G^*(Å^{-2})	$\begin{pmatrix} 0.0683 & 0 & 0 \\ 0 & 0.0664 & 0 \\ 0 & 0 & 0.0073 \end{pmatrix}$		
V (Å^3)	173.55			V^*(Å^{-3})	0.0058		

Source: David, W.I.F. et al. 1987. *Nature*, 327, 310–312.

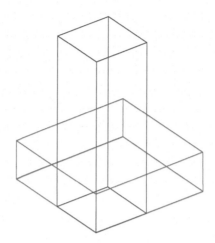

FIGURE 5.5 The reciprocal unit cell, in red, of the high-temperature superconductor $YBa_2Cu_3O_{7-x}$ superimposed on the direct cell, in blue.

are normalized to 1. Examples of the direct unit cell superimposed on the reciprocal cell for each of the seven crystal systems are found in Chapters 9 through 15 as follows with a, b, c, α, β, and γ included

- Triclinic system, DL-leucine: 14.12 Å, 5.19 Å, 5.39 Å, 111.1°, 86.4°, 97.0° in Section 9.4

- Monoclinic system, sucrose: 7.736(3) Å, 8.702(4) Å, 10.846 Å, 90°, 102.97°, 90° in Section 10.4

- Orthorhombic system, polyethylene: 7.400 Å, 4.930 Å, 2.534 Å, 90°, 90°, 90° in Section 11.4

- Tetragonal system, α-cristobalite: 4.9717 Å, 4.9717 Å, 6.9223 Å, 90°, 90°, 90° in Section 12.4

- Trigonal system, $H_{12}B_{12}{}^{2-}$, 3 K^+, Br^-: 6.871 Å, 6.871 Å, 6.871 Å, 86.38°, 86.38°, 86.38° (rhombohedral axes) in Section 13.4

- Hexagonal system, magnesium: 3.209 Å, 3.209 Å, 5.210 Å, 90°, 90°, 120° in Section 14.5

- Cubic system, acetylene: 6.105 Å, 6.105 Å, 6.105 Å, 90°, 90°, 90° in Section 15.4

See Exercises 5.11 through 5.14.

5.5 RELATIONSHIPS BETWEEN TRANSFORMATION MATRICES

In Chapter 2, the transformation matrix P for the direct lattice takes basis vectors $\mathbf{a}_1$, $\mathbf{b}_1$, $\mathbf{c}_1$ onto basis vectors $\mathbf{a}_2$, $\mathbf{b}_2$, $\mathbf{c}_2$

$$(\mathbf{a}_2, \mathbf{b}_2, \mathbf{c}_2) = (\mathbf{a}_1, \mathbf{b}_1, \mathbf{c}_1)P \tag{5.24}$$

Use the notation from the *International Tables for Crystallography*, and define $\mathbf{Q}$ as the transformation matrix for the reciprocal basis vectors

$$\begin{pmatrix} \mathbf{a}_2^* \\ \mathbf{b}_2^* \\ \mathbf{c}_2^* \end{pmatrix} = Q \begin{pmatrix} \mathbf{a}_1^* \\ \mathbf{b}_1^* \\ \mathbf{c}_1^* \end{pmatrix} \tag{5.25}$$

Substitute Equations 5.24 and 5.25 into Equation 5.2. Then use the definition of the dyad dot product for arbitrary vectors $\mathbf{v}_1$, $\mathbf{v}_2$, $\mathbf{v}_3$, $\mathbf{w}_1$, $\mathbf{w}_2$, and $\mathbf{w}_3$

$$\begin{pmatrix} \mathbf{v}_1 \\ \mathbf{v}_2 \\ \mathbf{v}_3 \end{pmatrix} \cdot \begin{pmatrix} \mathbf{w}_1 & \mathbf{w}_2 & \mathbf{w}_3 \end{pmatrix} = \begin{pmatrix} \mathbf{v}_1 \cdot \mathbf{w}_1 & \mathbf{v}_1 \cdot \mathbf{w}_2 & \mathbf{v}_1 \cdot \mathbf{w}_3 \\ \mathbf{v}_2 \cdot \mathbf{w}_1 & \mathbf{v}_2 \cdot \mathbf{w}_2 & \mathbf{v}_2 \cdot \mathbf{w}_3 \\ \mathbf{v}_3 \cdot \mathbf{w}_1 & \mathbf{v}_3 \cdot \mathbf{w}_2 & \mathbf{v}_3 \cdot \mathbf{w}_3 \end{pmatrix}$$

the ordinary distributive law of vector dot products, the standard definition of matrix multiplication by row and column vectors, and Equation 5.2 again to get

$$\begin{pmatrix} \mathbf{a}_2^* \\ \mathbf{b}_2^* \\ \mathbf{c}_2^* \end{pmatrix} \cdot \begin{pmatrix} \mathbf{a}_2, \mathbf{b}_2, \mathbf{c}_2 \end{pmatrix} = I = Q \begin{pmatrix} \mathbf{a}_1^* \\ \mathbf{b}_1^* \\ \mathbf{c}_1^* \end{pmatrix} \cdot \begin{pmatrix} \mathbf{a}_1, \mathbf{b}_1, \mathbf{c}_1 \end{pmatrix} P = Q\,P \tag{5.26}$$

Thus,

$$Q = P^{-1} \tag{5.27}$$

Therefore, Equation 5.25 can be written as

$$\begin{pmatrix} \mathbf{a}_2^* \\ \mathbf{b}_2^* \\ \mathbf{c}_2^* \end{pmatrix} = P^{-1} \begin{pmatrix} \mathbf{a}_1^* \\ \mathbf{b}_1^* \\ \mathbf{c}_1^* \end{pmatrix} \tag{5.28}$$

From Chapter 2 the transpose of a product of two matrices is the product of the transposes of the matrices in the reverse order: For matrices A and B, $(AB)^t = B^t A^t$. Also the transpose of a column vector is a row vector. Take the transpose of Equation 5.28. Thus,

$$\begin{pmatrix} \mathbf{a}_2^* & \mathbf{b}_2^* & \mathbf{c}_2^* \end{pmatrix} = \begin{pmatrix} \mathbf{a}_1^* & \mathbf{b}_1^* & \mathbf{c}_1^* \end{pmatrix} (P^{-1})^t \tag{5.29}$$

Go back to Equation 5.28, multiply both sides by P, and rearrange

$$\begin{pmatrix} \mathbf{a}_1^* \\ \mathbf{b}_1^* \\ \mathbf{c}_1^* \end{pmatrix} = P \begin{pmatrix} \mathbf{a}_2^* \\ \mathbf{b}_2^* \\ \mathbf{c}_2^* \end{pmatrix} \tag{5.30}$$

By analogy with the equation for $\mathbf{G}$ in Chapter 2, write $\mathbf{G}^*$ as a dyad dot product and substitute Equations 5.28 and 5.29 into the $\mathbf{G}^*$ matrix. Apply the same method used for Equation 5.26.

$$\begin{pmatrix} \mathbf{a}_2^* \\ \mathbf{b}_2^* \\ \mathbf{c}_2^* \end{pmatrix} \cdot (\mathbf{a}_2^* \ \ \mathbf{b}_2^* \ \ \mathbf{c}_2^*) = \mathbf{G}_2^* = P^{-1} \begin{pmatrix} \mathbf{a}_1^* \\ \mathbf{b}_1^* \\ \mathbf{c}_1^* \end{pmatrix} \cdot (\mathbf{a}_1^* \ \ \mathbf{b}_1^* \ \ \mathbf{c}_1^*) (P^{-1})^{\mathrm{t}} = P^{-1} \mathbf{G}_1^* (P^{-1})^{\mathrm{t}} \tag{5.31}$$

Multiplying Equation 5.31 by P on the left and P^{t} on the right gives

$$\mathbf{G}_1^* = P \mathbf{G}_2^* \, P^{\mathrm{t}} \tag{5.32}$$

where use is made of $(P^{\mathrm{t}})^{-1} = (P^{-1})^{\mathrm{t}}$.

Take the determinant of Equation 5.32, remembering that the determinant of a product is equal to the product of the determinants, and remembering that the determinant of a matrix is the same as the determinant of the transpose of that matrix

$$\det(\mathbf{G}_1^*) = \det(P \ \mathbf{G}_2^* \ P^{\mathrm{t}}) = \det(P)\det(\mathbf{G}_2^*)\det(P^{\mathrm{t}}) = \det{}^2(P)\det(\mathbf{G}_2^*) \tag{5.33}$$

Since $V^* = \sqrt{\det(\mathbf{G}^*)}$, then, using Equation 5.33

$$V_1^* = |\det(P)| \sqrt{\det(\mathbf{G}_2^*)} = |\det(P)| V_2^*$$

Table 5.11 summarizes the transformation relationships in terms of the transformation matrix P.

⌨ **EXAMPLE 5.2**

Given $\mathbf{a}_2 = 7\mathbf{a}_1 - 3\mathbf{b}_1$
   $\mathbf{b}_2 = -2\mathbf{a}_1 + \mathbf{b}_1$,
   $a_1 = 1\,\text{Å}$, $b_1 = 3\,\text{Å}$, $\gamma_1 = 120°$

Calculate $a_2, b_2, \gamma_2, P, Q, \mathbf{G}_1, \mathbf{G}_2, \mathbf{G}_1^*, \mathbf{G}_2^*, A_1, A_2, A_1^*, A_2^*$.

Solution

See Example 1.1 for obtaining the transformation matrix from the transformation equations. Here, $P = \begin{pmatrix} 7 & \overline{2} \\ \overline{3} & 1 \end{pmatrix}$. Note det($P$) = 1 so the transformation does not change the handedness of the system. Since det(P) is 1, if one cell is primitive the other cell is also primitive.

$$Q = P^{-1} = \begin{pmatrix} 1 & 2 \\ 3 & 7 \end{pmatrix}; G_1 = \begin{pmatrix} 1 & -1.5 \\ -1.5 & 9 \end{pmatrix} \mathring{A}^2$$

$$G_2 = P^t G_1 P = \begin{pmatrix} 7 & \overline{3} \\ \overline{2} & 1 \end{pmatrix} \begin{pmatrix} 1 & -1.5 \\ -1.5 & 9 \end{pmatrix} \begin{pmatrix} 7 & \overline{2} \\ \overline{3} & 1 \end{pmatrix} = \begin{pmatrix} 193 & -60.5 \\ -60.5 & 19 \end{pmatrix} \mathring{A}^2$$

$$G_2 = \begin{pmatrix} a_2^2 & a_2 b_2 \cos \gamma_2 \\ a_2 b_2 \cos \gamma_2 & b_2^2 \end{pmatrix} = \begin{pmatrix} 193 & -60.5 \\ -60.5 & 19 \end{pmatrix}; a_2 = 13.89 \, \mathring{A}, b_2 = 4.36 \, \mathring{A},$$

$$\gamma_2 = 177.54°$$

$$G_1^* = G_1^{-1} = \begin{pmatrix} 1.5 & 0.2222 \\ 0.2222 & 0.1481 \end{pmatrix} \mathring{A}^{-2}$$

$$G_2^* = G_2^{-1} = \begin{pmatrix} 2.81 & 8.96 \\ 8.96 & 28.59 \end{pmatrix} \mathring{A}^{-2} = P^{-1} G_1^* (P^{-1})^t$$

$$A_1 = \sqrt{\det(G_1)} = 2.591 \mathring{A}^2$$

$$A_2 = \sqrt{\det(G_2)} = 2.591 \mathring{A}^2$$

$$A_1^* = \sqrt{\det(G_1^*)} = 0.3849 \, \mathring{A}^{-2} = A_1^{-1}$$

$$A_2^* = \sqrt{\det(G_2^*)} = 0.3849 \, \mathring{A}^{-2} = A_2^{-1}$$

5.6 DIFFRACTION PATTERN AND THE RECIPROCAL LATTICE

The reciprocal lattice is used to interpret the x-ray diffraction pattern.

5.6.1 Motivation for Reciprocal Lattice

Figure 5.6 shows an experiment in which a crystal is placed in an x-ray beam. The crystal diffracts or bends the x-rays to form a pattern on a photographic plate or other suitable detector. The motivation for the reciprocal lattice is that the x-ray diffraction pattern is interpreted with it. Figure 5.7 is an annotated version of the x-ray diffraction image shown in the introduction to this chapter. The crystal is quartz and the $hk0$ plane is shown. The $\mathbf{a}^*$ and $\mathbf{b}^*$ directions are indicated. Each bright spot is interpreted by a point in reciprocal space. In particular, the points labeled are $\mathbf{a}^*$, $2\mathbf{b}^*$, and $3\mathbf{a}^* + 2\mathbf{b}^*$. In quartz $\gamma = 120°$ and so $\gamma^* = 60°$.

TABLE 5.11   Transformation Relationships in Terms of the Transformation Matrix, P, of the Direct Lattice

Direct Lattice	Reciprocal Lattice
$(\mathbf{a_2, b_2, c_2}) = (\mathbf{a_1, b_1, c_1})P$	$\begin{pmatrix} \mathbf{a_2^*} \\ \mathbf{b_2^*} \\ \mathbf{c_2^*} \end{pmatrix} = P^{-1} \begin{pmatrix} \mathbf{a_1^*} \\ \mathbf{b_1^*} \\ \mathbf{c_1^*} \end{pmatrix}$
$(\mathbf{a_1, b_1, c_1}) = (\mathbf{a_2, b_2, c_2})P^{-1}$	$\begin{pmatrix} \mathbf{a_1^*} \\ \mathbf{b_1^*} \\ \mathbf{c_1^*} \end{pmatrix} = P \begin{pmatrix} \mathbf{a_2^*} \\ \mathbf{b_2^*} \\ \mathbf{c_2^*} \end{pmatrix}$
$G_2 = P^t G_1 P$	$G_2^* = P^{-1} G_1^* (P^{-1})^t$
$G_1 = (P^{-1})^t G_2 P^{-1}$	$G_1^* = P G_2^* P^t$
$V_2 = \vert \det(P) \vert V_1$	$V_2^* = \vert \det(P^{-1}) \vert V_1^*$
$V_1 = \dfrac{1}{\vert \det(P) \vert} V_2$	$V_1^* = \dfrac{1}{\vert \det(P^{-1}) \vert} V_2^*$

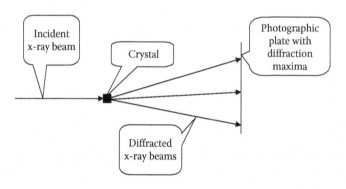

FIGURE 5.6   Diffraction pattern from a crystal.

5.6.2 Identification of the d-Spacings in Crystals

In Chapter 2 planes in the crystal were described by Miller indices. Three lattice points determine a crystallographic plane. The Miller indices of a plane are (hkl) and are calculated by taking the reciprocals of the fractional intercepts that the plane makes with the crystallographic axes. For *primitive* cells, the indices (hkl) are integers which do not contain a common factor. Every lattice can be described by a primitive unit cell.

The nature of crystallographic planes needs to be further explained. As indicated in the previous paragraph, three lattice points determine a crystallographic plane. It is natural to choose the origin on a lattice point. Because lattice points have identical neighborhoods, each crystallographic plane has a companion that is another crystallographic plane passing through the origin and parallel to the first crystallographic plane. Indeed, there is a whole set of crystallographic planes, one through every lattice point and parallel to the first crystallographic plane. Of all of these there will be two at equal distances on either side of

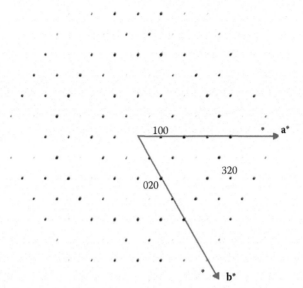

FIGURE 5.7 X-ray diffraction image of quartz showing the *hk*0 plane in reciprocal space. (Courtesy of Ross Angel.)

the origin and closest to the origin. The Miller indices of the crystallographic plane refer to these two closest planes. The distance, d_{hkl}, associated with the Miller indices, (hkl), is the distance from the origin to either one of these two closest planes. Indeed, the Miller indices describe the intercepts of one of these two closest planes with the basis vectors.

Consider the (11) plane in Figure 5.8a. This plane intersects the basis vector **a** at a times the reciprocal of 1 or **a** and the basis vector **b** at b times the reciprocal of 1 or **b**. A parallel plane through the origin is colored blue and also has the same Miller indices. The

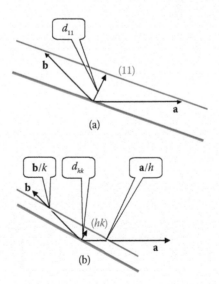

FIGURE 5.8 (a) d_{11}, (b) d_{hk}.

perpendicular distance, or d-spacing, between the parallel red and blue planes is d_{11}. In general, in Figure 5.8b the (hk) plane intersects the basis vector **a** at a times the reciprocal of h or **a**/h and the basis vector **b** at b times the reciprocal of k or **b**/k. The d-spacing for (hk) is d_{hk}.

In three dimensions the (hkl) plane intersects the basis vector **a** at a times the reciprocal of h or **a**/h, the basis vector **b** at b times the reciprocal of k or **b**/k, and the basis vector **c** at c times the reciprocal of l or **c**/l. The d-spacing for (hkl) is d_{hkl}.

5.6.3 Miller Indices, Laue Indices, and Bragg's Law

In Chapter 6 the physical properties of x-rays and their relationship to crystals will be explored.

Historically Bragg's law was developed by William Lawrence Bragg in terms of the optical law which states that the angle of incidence, θ, equals the angle of reflection. The use of the term "reflection" for a diffracted beam comes from the optical analog. In an optical mirror the angle of reflection is not restricted. By contrast, diffraction is an interference effect in which the angle of diffraction is related both to the wavelength of the incident beam and the periodicity of the crystal. In Figure 5.9 the incoming beam is diffracted from a set of planes with Miller indices (hk). The diffraction ONLY occurs when the following condition, called Bragg's law, is satisfied. Bragg's law is

$$n\lambda = 2d_{hk}\sin\theta_{hk}$$

where n is an integer, λ is the wavelength of the incoming beam, d_{hk} is the d-spacing, and θ_{hk} is the *Bragg angle*, in Figure 5.9, for the (hk) plane. The integer n is called the order of the diffracted beam.

When $n = 1$, for the first-order diffracted beam from the planes with Miller indices (hk), then

$$\sin\theta_{hk} = \frac{\lambda}{2d_{hk}}$$

When $n = 2$, for the second-order diffracted beam from the *same* planes with Miller indices (hk), then

$$\sin\theta_{hk} = \frac{2\lambda}{2d_{hk}}$$

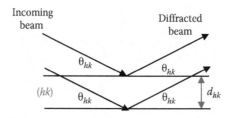

FIGURE 5.9 Bragg's law.

When $n = 3$, for the third-order diffracted beam from the *same* planes with Miller indices (hk), then

$$\sin\theta_{hk} = \frac{3\lambda}{2d_{hk}}$$

This notation is clumsy and is not used. Instead of listing the order of the diffracted beam of a given set of planes, the order is incorporated into the symbol for the d-spacing. This notation is called the Laue indices and is written without parentheses. The nth-order diffracted beam is written as $nh\ nk$, where nh is the product of n and h and nk is the product of n and k. For example, the second-order diffracted beam of the set of planes with Miller indices (11) is labeled 22. In another example, the third-order diffracted beam of the set of planes with Miller indices (21) is labeled 63. Use of the Laue indices causes the Bragg equation to be

$$\sin\theta_{63} = \frac{\lambda}{2d_{63}}$$

Or in general

$$d_{hk} = \frac{\lambda}{2\sin\theta_{hk}}$$

where now hk refers to the Laue indices.

The diffraction maxima are labeled by the *Laue indices*. Note the 020 diffraction maximum on the x-ray image of quartz in Figure 5.7.

5.6.4 Reciprocal Lattice Vectors and Sets of Crystallographic Planes

From the Bragg equation, which will be further developed in Chapter 6, the diffraction pattern is related to the d-spacings in the crystal. A vector in the reciprocal lattice has direction and magnitude. Each set of parallel crystallographic planes (hkl) is characterized by a vector in the reciprocal lattice whose direction is normal to the planes and whose magnitude is related to the interplanar d-spacing. The proof in two dimensions is given in two theorems. The first shows that the direction of the vector in reciprocal space is normal to the corresponding planes in direct space. The second theorem shows that the magnitude of the reciprocal vector is equal to the reciprocal of the d-spacing.

5.6.4.1 Direction of a Reciprocal Lattice Vector

Theorem 5.1

A vector in the reciprocal lattice $\mathbf{H}(h\ k) = h\mathbf{a}^* + k\mathbf{b}^*$ is normal to the family of planes (hk) of the direct lattice. In Chapter 2 the (hk) planes are defined.

Proof: In Figure 5.10a, the (hk) plane intersects the $\mathbf{a}$ axis at point $\mathbf{A}$ at $\mathbf{a}/h$ from the origin O, and the $\mathbf{b}$ axis at point B at $\mathbf{b}/k$ from the origin O. Vector $\mathbf{AB}$ is a vector that begins at $\mathbf{A}$ and ends at $\mathbf{B}$ and is in the (hk) plane.

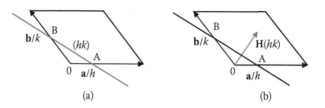

FIGURE 5.10   (a) (*hk*) plane in the direct lattice, in red, (b) **H**(*hk*) vector, in red.

Since

$$\mathbf{OA} + \mathbf{AB} = \mathbf{OB}$$

Then

$$\mathbf{AB} = \mathbf{OB} - \mathbf{OA} = \mathbf{b}/k - \mathbf{a}/h$$

From Table 5.1, the reciprocal vector **H**(*hk*) is

$$\mathbf{H}(hk) = h\mathbf{a}^* + k\mathbf{b}^*$$

Now form the dot product between **AB** and the reciprocal vector **H**.

$$\mathbf{AB} \cdot \mathbf{H}(hk) = (-\mathbf{a}/h + \mathbf{b}/k) \cdot (h\mathbf{a}^* + k\mathbf{b}^*) = 0$$

Since the dot product is zero, the reciprocal vector **H**(*h k*) is perpendicular (or normal) to the (*h k*) plane, as in Figure 5.10b.

5.6.4.2 Magnitude of a Vector in the Reciprocal Lattice

Theorem 5.2

If integers *h* and *k* have no common factor, then the length of the reciprocal vector is equal to the reciprocal of the *d*-spacing. Note that *h* and *k* cannot both be zero. In equation form

$$H(hk) = 1/d_{hk}$$

Proof: In Figure 5.11, d_{hk} is the distance between the origin O and the plane AB along the normal to AB. Since **H**(*hk*) is normal to the plane (*hk*), from Theorem 5.1, then **H**(*hk*)/*H*(*hk*) is a unit vector along the normal.

$$d_{hk} = \text{unit vector normal to plane} \cdot \mathbf{OA} = \frac{\mathbf{H}(hk)}{H(hk)} \cdot \frac{\mathbf{a}}{h} = \frac{h\mathbf{a}^* + k\mathbf{b}^*}{H(hk)} \cdot \frac{\mathbf{a}}{h} = \frac{1}{H(hk)}$$

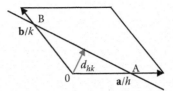

FIGURE 5.11 d_{hk}-spacing for (hk).

or

$$H(hk) = 1/d_{hk}.$$

Conclusion: For each set of crystallographic planes, (hk), there is a reciprocal lattice point (hk).

In three dimensions, the d-spacing is $d_{hkl} = 1/H(hkl)$.

5.7 THREE APPLICATIONS OF THE RECIPROCAL LATTICE

The next section gives three applications of the reciprocal lattice. First, it is used to calculate the d-spacings in a crystal. It is important to have a general procedure that can be systematized for computer calculations, as will be indicated in Chapter 7. A second application is the calculation of the angle between crystal faces. The Miller indices were originally designed to label crystal faces. Finally, the reciprocal lattice is used to solve the problem of relationship between two different basis vectors assigned to the same crystal, giving rise to two sets of indexing of the x-ray pattern, which means two different indexings of the reciprocal lattice.

🖳 5.7.1 Using the Reciprocal Lattice to Calculate d-Spacings

The tools are now in place to calculate the d-spacings in a crystal. Figure 5.12 shows a drawing of the reciprocal lattice. There are four reciprocal unit cells and the reciprocal vector **H** (21). This vector represents the (21) planes in direct space. In the reciprocal lattice this is

$$\mathbf{H}(21) = 2\mathbf{a}^* + \mathbf{b}^*$$

To find the d-spacing, calculate the magnitude of the vector and take the reciprocal.

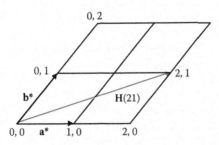

FIGURE 5.12   Reciprocal vector $\mathbf{H}(21) = 2\mathbf{a}^* + \mathbf{b}^*$, in red.

🖳 **EXAMPLE 5.3**

Find the d-spacing for the (21) planes for a two-dimensional oblique lattice if $a = 5.0$ Å, $b = 6.0$ Å, and $\gamma = 115°$.

Solution

First find **G**

$$\mathbf{G} = \begin{pmatrix} 25 & -12.68 \\ -12.68 & 36 \end{pmatrix}$$

then take the inverse of it to get **G***.

$$\mathbf{G}^* = \begin{pmatrix} .0487 & .0172 \\ .0172 & .0338 \end{pmatrix}$$

$$\frac{1}{d_{21}^2} = H^2(21) = (2\ 1)\mathbf{G}^*\begin{pmatrix} 2 \\ 1 \end{pmatrix} = (2\ 1)\begin{pmatrix} .0487 & .0172 \\ .0172 & .0338 \end{pmatrix}\begin{pmatrix} 2 \\ 1 \end{pmatrix} = 0.297\ \text{Å}^{-2}$$

$$d_{21} = 1/H(21) = 1.83\ \text{Å}$$

In general, for two dimensions

$$\frac{1}{d_{hk}^2} = (h\ k)\mathbf{G}^*\begin{pmatrix} h \\ k \end{pmatrix}$$

and for three dimensions

$$\frac{1}{d_{hkl}^2} = (h\ k\ l)\mathbf{G}^*\begin{pmatrix} h \\ k \\ l \end{pmatrix}$$

Using the metric matrix unifies and simplifies the calculations for all the crystal systems. Of course, for each crystal system the metric matrix gives the same results as the formulas that were designed for that system.

In the following example, the d-spacing is derived for the orthorhombic lattice.

EXAMPLE 5.4

Derive the d-spacing equation for an orthorhombic lattice in three dimensions.

Solution

$$\mathbf{G}^* = \begin{pmatrix} \dfrac{1}{a^2} & 0 & 0 \\ 0 & \dfrac{1}{b^2} & 0 \\ 0 & 0 & \dfrac{1}{c^2} \end{pmatrix}$$

and

$$\frac{1}{d_{hkl}^2} = \begin{pmatrix} h & k & l \end{pmatrix} \begin{pmatrix} \dfrac{1}{a^2} & 0 & 0 \\ 0 & \dfrac{1}{b^2} & 0 \\ 0 & 0 & \dfrac{1}{c^2} \end{pmatrix} \begin{pmatrix} h \\ k \\ l \end{pmatrix} = \begin{pmatrix} \dfrac{h}{a^2} & \dfrac{k}{b^2} & \dfrac{l}{c^2} \end{pmatrix} \begin{pmatrix} h \\ k \\ l \end{pmatrix} = \frac{h^2}{a^2} + \frac{k^2}{b^2} + \frac{l^2}{c^2}$$

See Exercises 5.15 through 5.20.

🖥 5.7.2 Calculating Angles between Crystal Faces

Figure 5.13 shows a diagram and a picture of a quartz crystal. While corresponding faces in two different crystals of the same substance may have different areas, the angle between

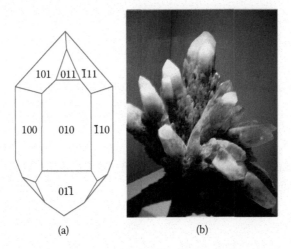

(a)   (b)

FIGURE 5.13   (a) Diagram of a quartz crystal, (b) quartz (amethyst) cluster. (Photograph by Elena Leshyn.)

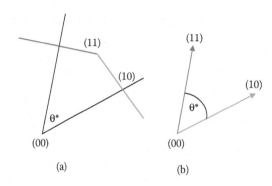

FIGURE 5.14 (a) Planes (11) and (10) in the direct lattice, (b) corresponding vectors in the reciprocal lattice.

any two corresponding faces is unique. This amazing experimental observation laid the foundations for crystallography. The faces of the crystals are crystallographic planes and are labeled by Miller indices. Mineralogists use the interfacial angles to identify crystals with well-developed faces.

As stated in Theorem 5.1, a crystallographic plane, represented by the Miller indices (hkl), is perpendicular to the reciprocal lattice vector $\mathbf{H}(hkl) = h\mathbf{a}^* + k\mathbf{b}^* + l\mathbf{c}^*$. The faces of the crystal are planes in direct space that translate into lattice points in reciprocal space. The use of the reciprocal lattice organizes and simplifies the calculations of the interfacial angles.

In Figure 5.14a faces (11) and (10) are indicated. The normals to the faces are drawn. The angle θ^*, the angle between the two normals, is the interfacial angle of the crystal. In Figure 5.14b the same information is given in reciprocal space. Note that the origin in direct space is indicated as 00 and the origin in reciprocal space as (00). Note that the problem now has been reduced to finding the angle between two vectors, $\mathbf{H}(h_1k_1) = h_1\mathbf{a}^* + k_1\mathbf{b}^*$ and $\mathbf{H}(h_2\,k_2) = h_2\mathbf{a}^* + k_2\mathbf{b}^*$. The cosine of the angle, $\cos\theta^*$, between two reciprocal lattice vectors is

$$\cos\theta^* = \frac{(h_1\ k_1)\mathbf{G}^*\begin{pmatrix} h_2 \\ k_2 \end{pmatrix}}{H_1 H_2}$$

where H_1 and H_2 are the magnitudes of the vectors $\mathbf{H}(h_1\ k_1)$ and $\mathbf{H}(h_2\ k_2)$. This formula is given in Table 5.1.

🖥 **EXAMPLE 5.5**

Find the interfacial angle, θ^*, between the (11) and (10) planes in a two-dimensional lattice where $a = 5$ Å, $b = 6$ Å, and $\gamma = 115°$.

Solution

The problem is done in reciprocal space. $\mathbf{G}^*$ has been calculated in Example 5.3. Here, $\mathbf{H}(11) = \mathbf{a}^* + \mathbf{b}^*$ and $\mathbf{H}(10) = \mathbf{a}^*$.

First, the magnitudes of $\mathbf{H}(11)$ and $\mathbf{H}(10)$ are calculated.

$$H(11) = \sqrt{(11)\begin{pmatrix} .0487 & .0172 \\ .0172 & .0338 \end{pmatrix}\begin{pmatrix} 1 \\ 1 \end{pmatrix}} = 0.3418\,\text{Å}^{-1}$$

Likewise, $H(10) = 0.2207$ Å^{-1}

$$\cos\theta^* = \frac{(11)\begin{pmatrix} .0487 & .0172 \\ .0172 & .0338 \end{pmatrix}\begin{pmatrix} 1 \\ 0 \end{pmatrix}}{H(11)H(10)}$$

Thus, $\theta^* = 29.2°$.

This technique is also easily extended to three dimensions.

EXAMPLE 5.6

Find the interfacial angle between $(h_1\,k_1\,l_1)$ and $(h_2\,k_2\,l_2)$ in the orthorhombic system.

Solution

Find the angle between the two reciprocal space vectors,

$$\mathbf{H}(h_1 k_1 l_1) = h_1\mathbf{a}^* + k_1\mathbf{b}^* + l_1\mathbf{c}^*$$

and

$$\mathbf{H}(h_2 k_2 l_2) = h_2\mathbf{a}^* + k_2\mathbf{b}^* + l_2\mathbf{c}^*$$

From Table 5.1, the cosine of the angle between two reciprocal lattice vectors $\cos\theta^*$ is

$$\cos\theta^* = \frac{(h_1 k_1 l_1)\,\mathbf{G}^* \begin{pmatrix} h_2 \\ k_2 \\ l_2 \end{pmatrix}}{H_1 H_2}$$

where H_1 and H_2 are the magnitudes of the reciprocal lattice vectors.

(continued)

EXAMPLE 5.6 (continued)

For an orthorhombic crystal

$$\mathbf{G}^* = \begin{pmatrix} \dfrac{1}{a^2} & 0 & 0 \\ 0 & \dfrac{1}{b^2} & 0 \\ 0 & 0 & \dfrac{1}{c^2} \end{pmatrix}$$

The magnitude of the vector $\mathbf{H}(h_1\, k_1\, l_1)$ is

$$H\left(h_1\, k_1\, l_1\right) = \sqrt{\begin{pmatrix} h_1 & k_1 & l_1 \end{pmatrix} \mathbf{G}^* \begin{pmatrix} h_1 \\ k_1 \\ l_1 \end{pmatrix}} = \sqrt{\dfrac{h_1^2}{a^2} + \dfrac{k_1^2}{b^2} + \dfrac{l_1^2}{c^2}}$$

And

$$H\left(h_2\, k_2\, l_2\right) = \sqrt{\dfrac{h_2^2}{a^2} + \dfrac{k_2^2}{b^2} + \dfrac{l_2^2}{c^2}}$$

Thus,

$$\cos\theta^* = \dfrac{\begin{pmatrix} h_1 & k_1 & l_1 \end{pmatrix} \begin{pmatrix} \dfrac{1}{a^2} & 0 & 0 \\ 0 & \dfrac{1}{b^2} & 0 \\ 0 & 0 & \dfrac{1}{c^2} \end{pmatrix} \begin{pmatrix} h_2 \\ k_2 \\ l_2 \end{pmatrix}}{\sqrt{(h_1^2/a^2) + (k_1^2/b^2) + (l_1^2/c^2)}\sqrt{(h_2^2/a^2) + (k_2^2/b^2) + (l_2^2/c^2)}}$$

Finally,

$$\theta^* = \arccos\left(\dfrac{(h_1 h_2/a^2) + (k_1 k_2/b^2) + (l_1 l_2/c^2)}{\sqrt{(h_1^2/a^2) + (k_1^2/b^2) + (l_1^2/c^2)}\sqrt{(h_2^2/a^2) + (k_2^2/b^2) + (l_2^2/c^2)}}\right)$$

This result has been worked out in full in order to demonstrate what is happening during the calculation. In practice a computational mathematical application would be used. See Exercises 5.21 and 5.22.

🖳 5.7.3 Relating Different Sets of Lattice Parameters for a Crystal

Each crystal has a diffraction pattern associated with it that forms a fingerprint of the crystal.

5.7.3.1 Powder Diffraction Files

Many x-ray diffraction patterns have been summarized in a library maintained by the Joint Committee on Powder Diffraction Standards (JCPDS). There are now over 250,000 entries. Figure 5.15 gives two PDFs for HMB. The two sets of lattice parameters a, b, c, α,

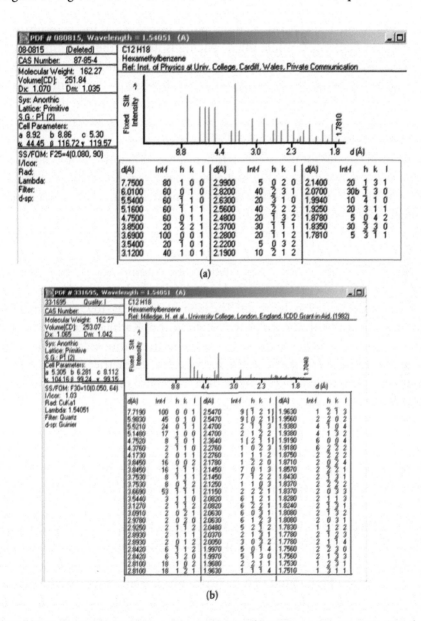

(a)

(b)

FIGURE 5.15   PDFs for HMB (a) PDF number 08-0815, (b) PDF number 33-1695. (Used with permission from the International Centre for Diffraction Data.)

β, and γ differ. In this section, the relation between the two sets is derived with the help of the reciprocal lattice.

Although HMB has only one diffraction pattern, there are different ways of labeling the diffraction maxima, relating to assignments of basis vectors, $\mathbf{a}^*$, $\mathbf{b}^*$, $\mathbf{c}^*$ for the reciprocal lattice. In a way similar to that treated in Chapter 1, the integers hkl assigned to a particular point in the reciprocal lattice depend on the basis vectors chosen to describe the lattice. In turn, through Equation 5.1, the assignment of $\mathbf{a}^*$, $\mathbf{b}^*$, $\mathbf{c}^*$ implies an assignment of $\mathbf{a}$, $\mathbf{b}$, $\mathbf{c}$ for the direct lattice, which, in turn, implies the lattice parameters a, b, c, α, β, and γ.

Each diffraction maximum of the crystal has a d-spacing that can be experimentally measured, as shown in the experimental setup in Figure 5.6 and in the resulting x-ray pattern for quartz in Figure 5.7. While the d-spacing is unique, the assignment of hkl to the diffraction maximum is not. As has been indicated in Section 5.6.3 and as will be shown in Chapter 6, the d-spacing can be calculated by the Bragg equation from the experimental data. It is through the d-spacing that the connection is made between the two sets of HMB data. The d-spacing is related to the reciprocal lattice vector because the magnitude of this reciprocal vector is equal to the reciprocal of the d-spacing.

There are valid reasons for the two assignments. As in Chapter 1, the assignment of basis vectors to a lattice is not unique. The first assignment was chosen in accordance with the external shape of the crystal, and the second is consistent with the rules suggested by the *International Tables for Crystallography*.

The data for the PDFs are entered in the order of their receipt by the PDF library; so the PDF file with the lower number precedes chronologically the file with the higher number. The reference for PDF 08-0815 is an undated private communication from the Institute of Physics at University College, Cardiff, Wales and will now be called the Cardiff data. This file uses the Lonsdale, 1929 assignment. The later file, PDF 33-1695, was reported in 1982 by H. Judith Milledge, a student and, later, a colleague of Kathleen Lonsdale. The Milledge file supersedes the older PDF 08-0815. The sets of lattice parameters and $h\,k\,l$ indices are different. This section will verify that they describe the same lattice. The intensities also are different and the newer file contains more data, but those differences represent improved experimental equipment.

5.7.3.2 Crystal Morphology

For the reason that HMB is triclinic with space group $P\bar{1}$, the only symmetry is the inversion. The choice of basis vectors for the lattice is not guided by the crystal symmetries, as happens with crystals with more symmetry. Kathleen Lonsdale (1929a) assigned her basis vectors by examining the external shape of the crystals. They precipitated in thin flat prisms (parallelepipeds) with three pairs of parallel faces (see Figure 5.16). The largest face was called (100). There was good cleavage parallel to one of the pairs of faces, and she assigned that face to be (001). This fortunate assignment facilitated her determination of the structure since she then found that the carbon atoms all lay in the (001) plane. That is to say, the plane containing the carbon atoms in every molecule is parallel to the plane with good cleavage. This structure was done before the standard guidelines were introduced in the *International Tables for Crystallography*. H. Judith Milledge followed these guidelines.

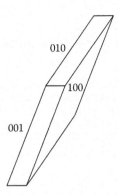

FIGURE 5.16   Crystal of HMB.

The guidelines were based on the *Niggli reduced cell*, which is unique for each lattice. The cell is primitive. Here, Milledge chose the negatively reduced Niggli cell, meaning that the angles α, β, and γ are all greater than 90°. Another condition of the guidelines is that a, b, and c are listed in the order of increasing length. (There is also a positively reduced Niggli cell where the angles α, β, and γ are all less than 90°.)

5.7.3.3 Comparing Indexings of the Reciprocal Lattice

Table 5.12 lists the lattice parameters for Kathleen Lonsdale's 1929 paper and the two PDF files. The data from the Lonsdale paper and PDF 08-0815 are nearly the same. Since the crystal is the same, the densities must be the same to within experimental error. The densities in Table 5.12 are 1.035, 1.070, and 1.065 g/cm³. Since the number of molecules to the unit cell, Z, is also the same, the volume for all three sets also must agree to within experimental error. The volumes calculated from the lattice parameters are 258.41, 254.84, and 253.07 Å³. The lattice parameters of Milledge a, b, c, α, β, and γ differ because Milledge chose the Niggli reduced cell.

In Table 5.13, three Miller indices hkl are matched between the Cardiff data and the Milledge data. The reason for three sets of Miller indices is that there are three unknown

TABLE 5.12   HMB Lattice Parameters, Comparing Lonsdale (1929a,b), PDF 08-0815, and PDF 33-1695

Lonsdale, 1929	PDF 08-0815 Cardiff	PDF 33-1695 Milledge 1982
$a_L = 9.010$ Å	$a_C = 8.92$ Å	$a_M = 5.305$ Å
$b_L = 8.926$ Å	$b_C = 8.86$ Å	$b_M = 6.281$ Å
$c_L = 5.344$ Å	$c_C = 5.30$ Å	$c_M = 8.112$ Å
$\alpha_L = 44° 27'$	$\alpha_C = 44.45°$	$\alpha_M = 104.16°$
$\beta_L = 116° 43'$	$\beta_C = 116.72°$	$\beta_M = 99.24°$
$\gamma_L = 119° 34'$	$\gamma_C = 119.57°$	$\gamma_M = 99.15°$
$V_L = 258.41$ Å³	$V_C = 254.84$ Å³	$V_M = 253.07$ Å³
$Z = 1$	$Z = 1$	$Z = 1$
Density$_L$ = 1.035 g/cm³	Density$_C$ = 1.070 g/cm³	Density$_M$ = 1.065 g/cm³

TABLE 5.13   Three Independent Matches between Cardiff and Milledge Data

d Å Cardiff	Intensity	hkl	d Å Milledge	Intensity	hkl
7.75	80	100	7.719	100	001
6.01	60	010	5.983	45	010
3.69	100	001	3.669	53	$\bar{1}\bar{1}1$

vectors; therefore, three sets of vector equations are needed. The idea is to get the Cardiff reciprocal basis vectors $\mathbf{a}^*_C$, $\mathbf{b}^*_C$, $\mathbf{c}^*_C$ in terms of the Milledge reciprocal basis vectors $\mathbf{a}^*_M$, $\mathbf{b}^*_M$, $\mathbf{c}^*_M$ and vice versa. The matching was done by this author using d-spacings and intensities, which are experimental data independent of the assignment of basis vectors. Once the basis vectors are assigned then the rest of the Miller indices are determined. For example, the (100) Cardiff corresponds to the (001) Milledge. The favored cleavage face, (001) Cardiff, becomes the ($\bar{1}\bar{1}1$) Milledge. Use the subscripts C and M to refer to the Cardiff and Milledge assignments so that

$$(001)_C = (\bar{1}\bar{1}1)_M \tag{5.34}$$

The Miller indices, hkl, represent points in reciprocal space and are expanded to $\mathbf{H}_{hkl} = h\,\mathbf{a}^* + k\,\mathbf{b}^* + l\,\mathbf{c}^*$, as in Table 5.1. Equation 5.34 becomes

$$\mathbf{c}^*_C = -\mathbf{a}^*_M - \mathbf{b}^*_M + \mathbf{c}^*_M \tag{5.35}$$

From the PDF files in Figure 5.15—considering that the smallest d-spacings are the most important for the determination of the structure and that in this case the d-spacings agree for the first two entries—use those entries

$$\mathbf{a}^*_C = \mathbf{c}^*_M \tag{5.36}$$

$$\mathbf{b}^*_C = \mathbf{b}^*_M \tag{5.37}$$

From Table 5.11 the equation using the transformation matrix, P^{-1}, between the two sets of reciprocal basis vectors is written as

$$\begin{pmatrix} \mathbf{a}^*_C \\ \mathbf{b}^*_C \\ \mathbf{c}^*_C \end{pmatrix} = P^{-1} \begin{pmatrix} \mathbf{a}^*_M \\ \mathbf{b}^*_M \\ \mathbf{c}^*_M \end{pmatrix} = \begin{pmatrix} 0 & 0 & 1 \\ 0 & 1 & 0 \\ \bar{1} & \bar{1} & 1 \end{pmatrix} \begin{pmatrix} \mathbf{a}^*_M \\ \mathbf{b}^*_M \\ \mathbf{c}^*_M \end{pmatrix}$$

The det(P^{-1}) = 1, therefore the transformation does not change the handedness.

🖳 *5.7.3.4 Transforming from the Reciprocal Lattice to the Direct Lattice*
The inverse of P^{-1} is P, thus

$$P = \begin{pmatrix} 1 & \bar{1} & \bar{1} \\ 0 & 1 & 0 \\ 1 & 0 & 0 \end{pmatrix}$$

The next goal is to relate the direct basis vectors for the Cardiff set to the direct basis vectors for the Milledge set. Now from Table 5.11

$$(\mathbf{a_C}, \mathbf{b_C}, \mathbf{c_C}) = (\mathbf{a_M}, \mathbf{b_M}, \mathbf{c_M})P = (\mathbf{a_M}, \mathbf{b_M}, \mathbf{c_M}) \begin{pmatrix} 1 & \bar{1} & \bar{1} \\ 0 & 1 & 0 \\ 1 & 0 & 0 \end{pmatrix}$$

or

$$(\mathbf{a_C}, \mathbf{b_C}, \mathbf{c_C}) = (\mathbf{a_M} + \mathbf{c_M}, -\mathbf{a_M} + \mathbf{b_M}, -\mathbf{a_M})$$

Table 5.14 gives the relationships between the Cardiff and Milledge direct lattice basis vectors (see Exercise 5.9).

An alternate method, which does not involve inverting a matrix, is to relate the direct basis vectors for the Cardiff set to the direct basis vectors for the Milledge set by using the relationships between the corresponding reciprocal basis vectors as given in Equations 5.35 through 5.37. Also needed is the relationship between direct basis vectors and reciprocal basis vectors given in Table 5.2. Because the volumes of the Milledge and Cardiff unit cells of the direct lattice are equal, the reciprocal volumes are equal; that is, $V_C^* = V_M^*$.

Thus,

$$\mathbf{a_C} = \frac{\mathbf{b_C^*} \times \mathbf{c_C^*}}{V_C^*} = \frac{\mathbf{b_M^*} \times (-\mathbf{a_M^*} - \mathbf{b_M^*} + \mathbf{c_M^*})}{V_M^*} = \frac{\mathbf{b_M^*} \times -\mathbf{a_M^*}}{V_M^*} + \frac{\mathbf{b_M^*} \times \mathbf{c_M^*}}{V_M^*} = \mathbf{c_M} + \mathbf{a_M}$$

Similarly,

$$\mathbf{b_C} = \frac{\mathbf{c_C^*} \times \mathbf{a_C^*}}{V_C^*} = \frac{(-\mathbf{a_M^*} - \mathbf{b_M^*} + \mathbf{c_M^*}) \times \mathbf{c_M^*}}{V_M^*} = \mathbf{b_M} - \mathbf{a_M}$$

TABLE 5.14   Relationships between Cardiff and Milledge Direct Lattice Basis Vectors

$\mathbf{a_C} = \mathbf{a_M} + \mathbf{c_M}$	$\mathbf{a_M} = -\mathbf{c_C}$
$\mathbf{b_C} = -\mathbf{a_M} + \mathbf{b_M}$	$\mathbf{b_M} = \mathbf{b_C} - \mathbf{c_C}$
$\mathbf{c_C} = -\mathbf{a_M}$	$\mathbf{c_M} = \mathbf{a_C} + \mathbf{c_C}$

and

$$c_C = \frac{a_C^* \times b_C^*}{V_C^*} = \frac{c_M^* \times b_M^*}{V_M^*} = -a_M$$

5.7.3.5 Comparison of PDF Files

The next goal is to show that the transformation P will generate the remaining hkl indices in the PDF file in Figure 5.15.

Equations 5.35 through 5.37 can be used to take the first 10 hkl indices for the Cardiff data and generate the corresponding hkl indices for the Milledge data.

A vector in the reciprocal lattice can be written in either the Cardiff basis or the Milledge basis.

Thus,

$$h_C a_C^* + k_C b_C^* + l_C c_C^* = h_M a_C^* + k_M b_C^* + l_M c_C^*$$

or

$$(h_C \; k_C \; l_C) \begin{pmatrix} a_C^* \\ b_C^* \\ c_C^* \end{pmatrix} = (h_M \; k_M \; l_M) \begin{pmatrix} a_M^* \\ b_M^* \\ c_M^* \end{pmatrix}$$

Substituting

$$\begin{pmatrix} a_C^* \\ b_C^* \\ c_C^* \end{pmatrix} = P^{-1} \begin{pmatrix} a_M^* \\ b_M^* \\ c_M^* \end{pmatrix}$$

$$(h_C \; k_C \; l_C) P^{-1} \begin{pmatrix} a_M^* \\ b_M^* \\ c_M^* \end{pmatrix} = (h_M \; k_M \; l_M) \begin{pmatrix} a_M^* \\ b_M^* \\ c_M^* \end{pmatrix}$$

Thus,

$$(h_C \; k_C \; l_C) P^{-1} = (h_M \; k_M \; l_M)$$

Take the transpose of both sides, remembering that the transpose of the product is the product of the transposes in the reverse order.

Thus,

$$\begin{pmatrix} h_M \\ k_M \\ l_M \end{pmatrix} = (P^{-1})^t \begin{pmatrix} h_C \\ k_C \\ l_C \end{pmatrix}$$

Note

$$(P^{-1})^t = \begin{pmatrix} 0 & 0 & \bar{1} \\ 0 & 1 & \bar{1} \\ 1 & 0 & 1 \end{pmatrix}$$

From Figure 5.15 put the first 11 $h_C\,k_C\,l_C$ in columns. They are

$$M_C = \begin{pmatrix} 1 & 0 & \bar{1} & \bar{1} & 0 & \bar{2} & 0 & \bar{1} & 1 & 0 & \bar{2} \\ 0 & 1 & 1 & 1 & 1 & 2 & 0 & 0 & 0 & 2 & 3 \\ 0 & 0 & 0 & 1 & 1 & 1 & 1 & 1 & 1 & 0 & 1 \end{pmatrix}$$

To get the hkl indices with the Milledge basis set, $\mathbf{M}_M$, then

$$M_M = P^{-1t}M_C = \begin{pmatrix} 0 & 0 & \bar{1} \\ 0 & 1 & \bar{1} \\ 1 & 0 & 0 \end{pmatrix}\begin{pmatrix} 1 & 0 & \bar{1} & \bar{1} & 0 & \bar{2} & 0 & \bar{1} & 1 & 0 & \bar{2} \\ 0 & 1 & 1 & 1 & 1 & 2 & 0 & 0 & 0 & 2 & 3 \\ 0 & 0 & 0 & 1 & 1 & 1 & 1 & 1 & 1 & 0 & 1 \end{pmatrix}$$

TABLE 5.15   Comparison of the d-Spacings, Intensities, and hkl between PDF 08-0815 and PDF 33-1695

PDF 08-0815 d (Å)	(Cardiff) Intensity	hkl	PDF 33-1695 d (Å)	(Milledge) Intensity	PDF 33-1695 hkl	Calculated from $(P^{-1})^t$ hkl
7.75	80	100	7.719	100	001	001
6.01	60	010	5.983	45	010	010
5.54	60	$\bar{1}$10	5.521	24	0$\bar{1}$1	$-(0\bar{1}1)$
5.16	60	$\bar{1}$10	5.148	17	100	$-(100)$
4.75	60	011	4.752	8	$\bar{1}$01	$\bar{1}$01
3.85	20	$\bar{2}$21	3.845	16	1$\bar{1}$1	$-(1\bar{1}1)$
		200			002	002
3.69	100	001	3.669	53	$\bar{1}\bar{1}$1	$\bar{1}\bar{1}$1
3.54	20	$\bar{1}$01	3.544	3	110	$-(110)$
3.12	40	101	3.127	2	$\bar{1}\bar{1}$2	$\bar{1}\bar{1}$2
2.99	5	020	2.978	2	020	020
2.82	40	$\bar{2}$31	2.810	18	1$\bar{2}$1	$-(1\bar{2}1)$
		3$\bar{1}\bar{1}$			102	102

Thus,

$$
\mathbf{M}_M = \begin{pmatrix}
0 & 0 & 0 & \bar{1} & \bar{1} & \bar{1} & \bar{1} & \bar{1} & \bar{1} & 0 & \bar{1} \\
0 & 1 & 1 & 0 & 0 & 1 & \bar{1} & \bar{1} & \bar{1} & 2 & 2 \\
1 & 0 & \bar{1} & 0 & 1 & \bar{1} & 1 & 0 & 2 & 0 & \bar{1}
\end{pmatrix}
$$

These results are now put into Table 5.15 remembering that, in general, for diffraction patterns $hkl = \bar{h}\bar{k}\bar{l}$. Compare this table with the PDF in Figure 5.15. See Exercises 5.23 through 5.25.

DEFINITIONS

Bragg's law
Direct lattice
Laue index, hkl
n, order of the diffracted beam
Niggli reduced cell
Powder Diffraction File, PDFs
Reciprocal lattice basis, $\mathbf{a}^*$, $\mathbf{b}^*$, $\mathbf{c}^*$

EXERCISES

💻 5.1 Calculate a^*, b^*, c^*, α^*, β^*, and γ^* for urea at 188 K using relationships from Tables 5.3 and 5.4. Data are available in Section 1.9.1.

💻 5.2 Calculate a^*, b^*, c^*, α^*, β^*, and γ^* for anthracene at 6.70 GPa using relationships from Tables 5.3 and 5.4. Data are available in Section 1.9.2.

💻 5.3 Calculate $\mathbf{G}$, $\mathbf{G}^*$, V, and V^* for urea at 188 K. Data are available in Section 1.9.1.

💻 5.4 Calculate $\mathbf{G}$, $\mathbf{G}^*$, V, and V^* for anthracene at 6.70 GPa. Data are available in Section 1.9.2.

💻 5.5 Calculate a^*, b^*, c^*, α^*, β^*, and γ^* for urea at 188 K using relationships in Equation 5.23 and compare with Exercise 5.1. Data are available in Section 1.9.1.

💻 5.6 Calculate a^*, b^*, c^*, α^*, β^*, and γ^* for anthracene at 6.70 GPa using relationships in Equation 5.23 and compare with Exercise 5.2. Data are available in Section 1.9.2.

💻 5.7 Calculate $\mathbf{G}^*$, V^*, a^*, b^*, c^*, α^*, β^*, and γ^* for HMB.

💻 5.8 Reproduce the reciprocal cell superimposed on the direct cell for HMB in Figure 5.3b. Normalize the volumes to 1.

💻 5.9 Calculate $\mathbf{G}^*$, V^*, a^*, b^*, c^*, α^*, β^*, and γ^* for anhydrous alum.

5.10 Reproduce the reciprocal cell superimposed on the direct cell for anhydrous alum in Figure 5.4. Normalize the volumes to 1.

5.11 Reproduce the reciprocal cell superimposed on the direct cell for the superconductor in Figure 5.5.

5.12 Construct the reciprocal cell superimposed on the direct cell for NaCl with the volume normalized to 1.

5.13 Construct the reciprocal cell superimposed on the direct cell for anthracene with the volume normalized to 1. See Exercise 5.1.

5.14 Construct the reciprocal cell superimposed on the direct cell for urea with the volume normalized to 1. See Exercise 5.2.

5.15 Find the d-spacing for 210, 420, and 630 in HMB.

5.16 Find the d-spacing for 210, 420, and 630 in AA.

5.17 Find the d-spacing for 210, 420, and 630 in urea. See Problem 5.1.

5.18 Find the d-spacing for 210, 420, and 630 in anthracene. See Problem 5.2.

5.19 Derive the d-spacing equation for a tetragonal lattice in three dimensions.

5.20 Derive the d-spacing equation for a hexagonal lattice in three dimensions.

5.21 Find the angle between the 101 and the 011 faces of quartz.

5.22 Look at Figure 5.13 and make a list of all the adjacent pairs of faces with their Miller indices. Find the interfacial angle for each pair.

5.23 Using the Milledge lattice constants for HMB, calculate the reciprocal lattice parameters, the reciprocal metric matrix, and the reciprocal volume. In a table compare these values with the corresponding Lonsdale values.

5.24 Using the Milledge lattice constants for HMB, construct the direct cell overlaid with the reciprocal cell and compare with Figure 5.3.

5.25 Use the Milledge lattice constants for HMB, the transformation matrix P, and Table 2.4 to transform the HMB coordinates for the carbon atoms. Populate the unit cell.

MATLAB CODE: STARTER PROGRAM FOR CHAPTER 5: GRAPHIC OF RECIPROCAL CELL SUPERIMPOSED ON DIRECT UNIT CELL

This program produces the reciprocal cell superimposed on the direct cell with the volumes normalized to 1. The example here is the triclinic crystal DL-leucine (Chapter 9). Also tables are produced giving a, b, c, α, β, γ, V, $\mathbf{G}$, a^*, b^*, c^*, α^*, β^*, γ^*, V^*, and $\mathbf{G}^*$.

```
%Chapter5StarterProgram
clc              %clears content of command window
close all   %
```

```
clear        %clears variables in workspace
%Script file: Chapter 5Starter program
%
% Purpose Plot reciprocal cell superimposed on direct cell
%
% Record of revisions:
% Date        Programmer          Description of change
% 1 12        M. Julian           Original Code
% 4 17        M. Julian           reciprocal lattice
%
% define variables:
% a = crystallographic a axis (Angstroms)
% b = crystallographic b axis (Angstroms)
% c = crystallographic c axis (Angstroms)
% alpha angle between b and c (degrees)
% beta angle between a and c (degrees)
% gamma angle between a and b (degrees)
% conversion converts from triclinc to cartesian
% outline outlines cell
% cc cartesian coordinates (direct basis) = conversion*outline
% ccstar cartesian coordinates (reciprocal
  basis) = conversion*outline
% astar, bstar, cstar reciprocal lattice vectors
% outline: outline of unit cell
% G metric matrix; Gstar is G*
% x, y, z, xx, yy, zz shorthand for plotting
%DLLeucine
a = 14.12
b = 5.19
c = 5.39
alpha = 111.1
beta = 86.4
gamma = 97.0
%calculate G matrix
G = [a^2             a*b*cosd(gamma) a*c*cosd(beta)
a*b*cosd(gamma) b^2     b*c*cosd(alpha)
a*c*cosd(beta) b*c*cosd(alpha) c^2]
V = sqrt(det(G));%calculate Volume
Gstar = inv(G);% calculate G*
Vstar = 1/V;% calculate V*
astar = sqrt(Gstar(1,1));% calculate a*
bstar = sqrt(Gstar(2,2));% calculate b*
cstar = sqrt(Gstar(3,3));% calculate c*
alphastar = acosd(Gstar(2,3)/(bstar*cstar));% calculate alpha*
betastar = acosd(Gstar(1,3)/(astar*cstar));% calculate beta*
gammastar = acosd(Gstar(1,2)/(astar*bstar));% calculate gamma*
%%%%%%%%%%%%%%%%%%%%%%%%%%%%%%%%%%%%%%%%%%%%
```

```
f = figure(1);
dat = {a,' a*',astar
b,' b*',bstar
c,' c*',cstar
alpha,' alpha*',alphastar
beta,' beta*',betastar
gamma,' gamma*',gammastar
V,' V*',Vstar};
cnames = {'Direct Lattice','','Reciprocal Lattice'};
rnames = {'a','b','c','alpha','beta','gamma','V'};
t = uitable('Parent',f,'Data',dat,'ColumnName',cnames,...
'RowName',rnames,'ColumnWidth',{75});
%G and G star etc
f = figure(2);

dat = [G
Gstar];
cnames = {'','',''};
rnames = {'G',' ',' ','Gstar',' ',' '};
t = uitable('Parent',f,'Data',dat,'ColumnName',cnames,...
'RowName',rnames,'ColumnWidth',{75});
%This is a scaling factor that normalizes volume.
%scaled by 1/cube root of the volume of crystal.
figure(3)
s = V^(1/3);
a = a/s;
b = b/s;
c = c/s;
%calculate normalized G matrix
G = [a^2              a*b*cosd(gamma) a*c*cosd(beta)
a*b*cosd(gamma) b^2       b*c*cosd(alpha)
a*c*cosd(beta) b*c*cosd(alpha) c^2];
V = sqrt(det(G));
%convert to Cartesian
c1 = c*cosd(beta);
c2 = c*(cosd(alpha)-cosd(gamma)*cosd(beta))/sind(gamma);
c3 = sqrt(c^2-c1^2 -c2^2);
conversion = [a b*cosd(gamma)    c1
            0 b*sind(gamma)    c2
            0 0                c3];
%%%%%conversion = CartesianC(a, b, c, alpha, beta, gamma)%function
outline = [0 1 1 0 0 0 1 1 1 1 1 1 0 0 0 0
            0 0 1 1 0 0 0 0 0 1 1 1 1 1 1 0
            0 0 0 0 0 1 1 0 1 1 0 1 1 0 1 1];
cc = (conversion*outline)';
x = cc(:,1)';
y = cc(:,2)';
```

```
z = cc(:,3)';
plot3(x,y,z,'LineWidth',2);
axis equal
axis off
%rescales axes by 70%
axis([-.7*a a+.7*a -.7*b b+.7*b -.7*c c+.7*c])
hold on
%********************************************************8
% reciprocal cell
%Calculate G* from G
Gstar = inv(G);
%Use Equation 5.23 in book to calculate vectors a*, b*, c*
astar = Gstar*[1 0 0]';
bstar = Gstar*[0 1 0]';
cstar = Gstar*[0 0 1]';
%create matrix of three reciprocal column vectors
star = [astar bstar cstar];
%now convert to cartesian and multiply by outline to create cell
ccc = (conversion*star*outline)';
xx = ccc(:,1)';
yy = ccc(:,2)';
zz = ccc(:,3)';
plot3(xx,yy,zz,'r','LineWidth',2);
```

## CHAPTER 6

Properties of X-Rays

Above is a picture of a modern x-ray tube. Most x-rays used in crystallography are produced by similar tubes. Synchrotrons also produce x-rays used in some protein studies. Nuclear bombs produce x-rays, which caused much damage to human tissue in Hiroshima. The sun produces x-rays, as do many other stars. X-ray astronomy is an important branch of star study. Certain radioactive materials decay to produce x-rays. For example, the low intensity x-rays from radioactive ^{55}Fe are used to test x-ray detectors.

CONTENTS

CHAPTER OBJECTIVES

- Discovery of x-rays

- Principle of superposition

- Constructive, destructive, and partial interference

- Young's double-slit experiment

- Properties of electromagnetic waves

- X-ray spectrum

- Bremsstrahlung and characteristic x-rays

- Absorption

- X-ray tubes and filters

- X-ray diffraction

- Bragg's law

- X-ray powder diffraction camera

- Synchrotron as source of x-rays

- Protein crystals and synchrotrons

6.1 INTRODUCTION

Chapters 1 through 5 have provided an introduction to the principles of symmetry in crystals. From the 1600s a connection between the characteristic interfacial angles in crystals and an underlying periodicity was hypothesized. The 230 space groups were worked out almost simultaneously in Russia, Germany, and England in the 1890s. What was needed was proof of the existence of the unit cell, including measurements of its size. The missing link was made by x-ray diffraction in 1912.

This chapter first discusses the discovery of x-rays in 1895. This discovery was made possible by technological advances that produced a vacuum of 1 millionth of an atmosphere, leading to the invention of the cathode-ray tube. Many scientists used the cathode-ray tube in their experiments. Unknown to them the tubes were giving off x-rays. The history of medicine, physics, metallurgy, chemistry, geology, and biology was forever changed by the discovery of x-rays. X-ray diffraction is the main technique for studying molecular structure and crystal structures.

6.2 THE DISCOVERY OF X-RAYS

The Crookes tube, a cathode-ray tube, is named after the English scientist, Sir William Crookes (1832–1919). This tube has a vacuum of 1 millionth of an atmosphere. Figure 6.1 shows a schematic Crookes tube. It consists of a *cathode*, or negatively charged electrode, and an *anode*, or positively charged electrode, both inserted into a glass tube which is then evacuated. The voltage needed is about 30,000 V. In Figure 6.2, when the switch is closed, a stream of electrons goes from the cathode to the anode. Because of the vacuum, the

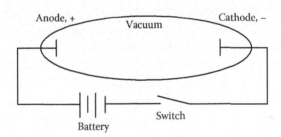

FIGURE 6.1   Crookes tube with switch open.

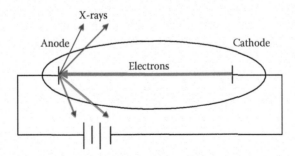

FIGURE 6.2   Crookes tube with switch closed.

electrons travel in a straight line to the anode. X-rays, which are invisible, are produced when the electrons slam into the highly positive anode. Many years passed between the production of the first cathode-ray tubes and the discovery of x-rays by Wilhelm Conrad Röntgen on November 8, 1895. On that day Röntgen had covered his cathode-ray tube with black opaque paper and had darkened the room. No visible light escaped from the shrouded cathode-ray tube. When Röntgen threw the switch, out of the corner of his eye he saw a light about a meter from the cathode-ray tube. This was startling! The light came from a fluorescent screen resting on a nearby table. When Röntgen turned off the switch, the light went away but returned again with the closing of the switch. Röntgen traced the source of the light back to the anode. X-rays produced at the anode caused the fluorescent screen to glow.

Röntgen named the phenomenon *x-rays*, where *x* stands for unknown. In a series of experiments done during 6 weeks, Röntgen proved that x-rays penetrated matter, were more transparent to wood than metal, and could be recorded on photographic film. He showed that x-rays traveled in a straight line, were not affected by a magnet, and could discharge both positively and negatively charged objects. Röntgen looked for the diffraction, or the bending, of x-rays but did not observe it. He took many pictures, including his own double-barrel shotgun. The outlines were visible because x-rays penetrate different materials differently. On December 22, 1895, Röntgen made the first x-ray photograph of a human hand, that of his wife, Bertha. On New Year's Day, 1896, Röntgen mailed copies of a report with the x-ray photograph of Bertha's hand, not only to the publisher but as preprints to his fellow scientists.

The Wiener Press in Germany picked up the story and put it in a Sunday edition in January 1896. This was a tremendous piece of public relations that immediately galvanized the medical world. Four months later Thomas Edison of New Jersey was selling x-ray apparatus. However, in 1904 his assistant Clarence Dally became the first person to die from x-ray exposure. Edison abandoned his x-ray work and even refused medical x-rays for the rest of his 84 years.

By January 1897, the Belgium government suggested that all hospitals be equipped with x-ray diagnostic tools. Within a year nine books and pamphlets and over 1000 articles were published. Between 1899 and 1902, x-rays were used to treat skin disease, tumors of body organs, and leukemia. One extraordinary aspect of the explosion of knowledge was the availability of cathode-ray tubes in many laboratories. Of course, these tubes had been giving off x-rays all along, even though no one had noticed them. Storing photographic film near a cathode-ray tube was considered bad practice! Arthur W. Goodspeed, at the University of Pennsylvania, looked back at his own data and found an x-ray photograph of two coins accidentally taken 5 years before Röntgen made his discovery. See Nitske (1971) and Julian (1997).

6.3 PROPERTIES OF WAVES

To understand the diffraction phenomena introduced in Chapter 5, a background in the properties of waves is needed. An example of a wave is the sound from striking middle C on a piano. First, the principle of superposition will be examined and then Young's double-slit experiment will be studied. Finally, the properties of electromagnetic waves will be sketched.

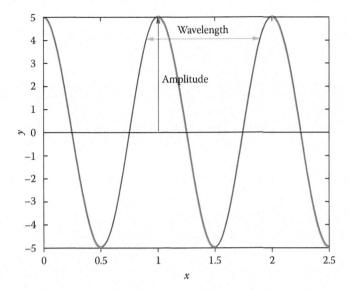

FIGURE 6.3   Plot of $y = 5 \cos (2\pi x)$ with wavelength $\lambda = 1$, in green, and amplitude $A = 5$, in red.

6.3.1 Principle of Superposition

The sound of middle C can be idealized as a sinusoidal wave. Mathematically, a sinusoidal wave can be represented by the periodic function

$$y(x,t) = A \cos \left(\frac{2\pi}{\lambda}(x - vt) + \varphi \right) \tag{6.1}$$

where y is the displacement, A is the amplitude, λ is the wavelength, x is the distance, v is the velocity, t is the time, and φ is the phase angle in radians. The *amplitude*, A, is the maximum displacement. In Figure 6.3, $A = 5$, $\lambda = 1$, $t = 0$, and $\varphi = 0$. The *wavelength* of a periodic function can be measured, for example, between two successive minima, as shown in the figure, or more generally between any two successive displacements, y, with identical neighborhoods. The slope, as well as the displacement, of the selected points must be the same for the two points to have identical neighborhoods.

The *principle of superposition* states that if two or more waves combine, the new wave is the algebraic sum of the original waves. There are two extreme behaviors called constructive interference and destructive interference. The cases in between are called partial interference.

6.3.1.1 Constructive Interference

Consider two waves in Figure 6.4, the first, $y = 5 \cos (2\pi x)$, with amplitude $A = 5$ (blue) and the second, $y = 6 \cos (2\pi x)$, with amplitude $A = 6$ (red). Both waves have wavelength $\lambda = 1$ and phase angle $\varphi = 0$. These waves are in phase. Two waves are in phase if $\varphi = n2\pi$, where n is an integer. When the red and blue waves are added together, the new wave (green) has $A = 11$ and $\lambda = 1$ and $\varphi = 0$. These two superimposed waves demonstrate constructive interference. *Constructive interference* is the phenomenon that occurs when superimposed

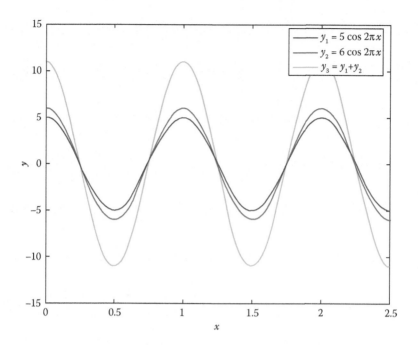

FIGURE 6.4   Plot of $y = 5 \cos (2\pi x)$, in blue, $y = 6 \cos (2\pi x)$, in red, and their sum, in green, showing constructive interference.

waves are in phase. The amplitude of the final wave is the largest possible. See Exercises 6.1 and 6.2.

6.3.1.2 Destructive Interference

The phase angle φ controls the placement of the wave along the x axis. Figure 6.5 compares two waves: The blue wave has the amplitude 5 and the phase angle $\varphi = 0$, and the red wave also has the amplitude 5 but a phase angle $\varphi = \pi$. The red wave is displaced by π radians, or in terms of degrees, is 180° out of phase. These two superimposed waves demonstrate *destructive interference* because the amplitude of the final wave is zero. See Exercises 6.3 and 6.4.

6.3.1.3 Partial Interference

The third case, partial interference, is demonstrated in Figure 6.6. The blue wave has amplitude $A = 5$ and phase angle $\varphi = 0$ and the red wave has amplitude $A = 6$ and phase angle $\varphi = 0.98\pi$. The red wave is displaced by almost π radians and is very close to being out of phase. These two superimposed waves demonstrate *partial interference* because the amplitude of the final wave (green) is neither maximized nor zero. The amplitude of the sum of the superimposed waves is less than the maximum possible value and not zero. See Exercises 6.5 and 6.6.

6.3.2 Young's Double-Slit Experiment

In 1802 Thomas Young, an English physicist and physician (1773–1829), discovered the principle of interference of light. (Young's modulus is also named after him.) As in

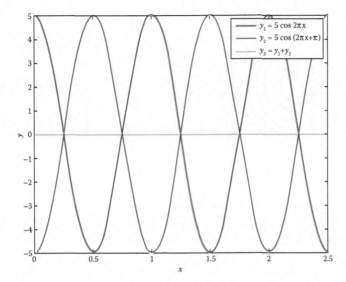

FIGURE 6.5   Plot of $y = 5 \cos (2\pi x)$, in blue, $y = 5 \cos (2\pi x + \pi)$, in red, and the sum of the blue and red waves in green showing destructive interference.

Figure 6.7, a source of light was first collimated by a single slit in the interference experiment. The light came out through the source slit (red waves). The distance between the concentric red waves is the wavelength of the source radiation, λ. When the light gets to the two slits, cylindrical waves spread out from the two slits (blue waves) forming two new sources of waves. These new waves have the same wavelength, λ, as the source radiation. These two waves (blue) superimpose. The green lines indicate constructive interference that causes bright spots on the screen. Between the bright spots there is destructive

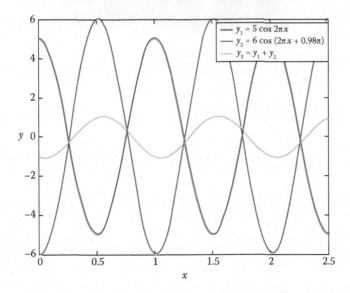

FIGURE 6.6   Plot of $y = 5 \cos (2\pi x)$, in blue, $y = 6 \cos (2\pi x + 0.98\pi)$, in red, and the sum of the red and blue waves, in green, showing partial interference.

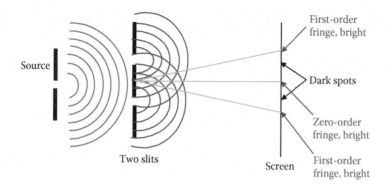

FIGURE 6.7   Young's experiment, the interference of light going through two slits.

interference forming dark spots. The central spot, or zero-order fringe, is always a bright spot. The first bright spot, on either side of the central spot, is called the first-order fringe (see Figure 6.8).

In Figure 6.9, let $r = CD$ be the distance between the centers of the slits and let θ be the angle OAB. A is the point midway between the slits. O is the central bright spot. B is another bright spot. Connect B to the center of each of the slits, forming lines BD and BC. Insert F such that line FB is equal in length to line DB. Use right triangles to show that θ = angle FDC. Then CF ≈ $r \sin \theta$. CF is the extra distance that the light travels. Since there

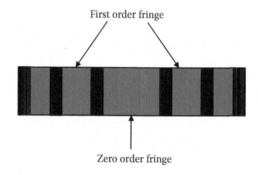

FIGURE 6.8   Interference fringes from two slits using red light.

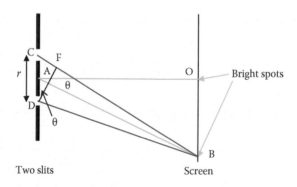

FIGURE 6.9   Calculations for Young's experiment.

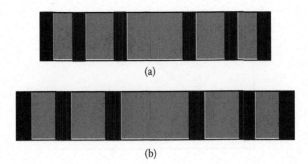

FIGURE 6.10 Interference fringes from two slits using red light, (a) larger r thus narrower fringe spacing, (b) smaller r thus larger fringe spacing.

is constructive interference at point B, the waves coming from the two slits must be in phase. In order for the waves to be in phase, the extra distance CF must be a multiple $m\lambda$ of the wavelength λ, where m is an integer.

Thus,

$$r \sin \theta = m\lambda \tag{6.2}$$

Because $\sin \theta \ll 1$, then $\sin \theta = m\lambda/r \ll 1$. Thus, $m\lambda \ll r$. Red light has $\lambda = 6500$ Å. At $m = 0$ there is no bending of the light and $\theta = 0°$. This is the zero-order fringe and is always bright. The first fringe has $m = 1$, and if $r = 13$ μm or 130,000 Å, then $\theta \approx 3°$. However, if r is 1 mm or 10^7 Å then θ is 0.003°, which is hard to observe. The result is that in order to get an interference pattern, the spacing between the slits must be of the same order as the wavelength of the radiation.

In 1895, Röntgen looked for the diffraction, or the bending, of x-rays but did not observe it. Over a decade and a half passed before the diffraction of x-rays was observed. In Equation 6.2, if λ and m are held constant, then as the r increases, $\sin \theta$ (and thus θ) decreases. In other words, the relationship between r and θ is reciprocal, as is demonstrated in Figure 6.10.

6.3.3 Properties of Electromagnetic Waves

X-rays, γ-rays, ultraviolet rays, visible light, and infrared rays are all examples of electromagnetic waves. *Electromagnetic waves* transmit energy through space. They have an electric component, **E**, and a perpendicular magnetic component, **H**. All electromagnetic waves travel at the speed of light, c, in a vacuum. Electromagnetic waves are transverse waves because the electric and magnetic vibrations are perpendicular to the direction the wave travels. For example, in Figure 6.11 the electromagnetic wave travels in the x direction. **E** is along the y axis and **H** is along the z axis.

For electromagnetic waves, Equation 6.1 becomes

$$\mathbf{E} = \mathbf{E}_0 \cos\left(\frac{2\pi}{\lambda}(x - vt) + \varphi\right) \tag{6.3}$$

where **E** is the electric field and $\mathbf{E}_0$ is the amplitude of the electric component.

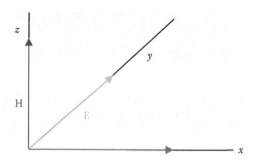

FIGURE 6.11   Electric and magnetic fields associated with a wave moving along x.

Polarization is a consequence of the nature of transverse waves. (Longitudinal waves cannot be polarized.) A *plane polarized wave* vibrates in a single direction; an *unpolarized wave* vibrates without favoring any direction in the plane perpendicular to the direction of motion. The magnetic component is always perpendicular to the electric wave and is omitted for clarity in Figure 6.12. In this figure, (a) represents a plane polarized electric wave and (b) represents an unpolarized electric wave. The direction of the wave is perpendicular to the page.

The frequency of an electromagnetic wave, v, equals c/λ. The units of frequency are sec^{-1} or, in an alternative notation, hertz, Hz. See Table 6.1 for values of the physical constants.

A *photon* is a quantum of electromagnetic radiation. Every photon has an energy, E, proportional to its frequency. The proportionality constant is Planck's constant, h.

Thus,

$$E = hv \tag{6.4}$$

(a)   (b)

FIGURE 6.12   Electric component of (a) a plane polarized wave, (b) an unpolarized wave.

TABLE 6.1   Physical Constants

Symbol	Name	Value	Units
c	Velocity of light in vacuum	2.998×10^8	m/sec
e	Charge of electron	1.602×10^{-19}	C
m	Mass of electron	9.11×10^{-31}	kg
h	Planck's constant	6.626×10^{-34}	J sec

TABLE 6.2   Comparison of Properties of X-Rays with Other Electromagnetic Radiations

Radiation	v, Frequency (Hz)	λ, Wavelength (m)	$E = hv$, Energy(J)
AM radio waves	10^6	300	7×10^{-28}
FM, TV	10^8	3	7×10^{-26}
Infrared	10^{13}	3×10^{-5}	7×10^{-21}
Visible	10^{15}	3×10^{-7}	7×10^{-19}
Ultraviolet	10^{17}	3×10^{-9}	7×10^{-17}
X-rays	10^{18}	3×10^{-10}	7×10^{-16}
γ-Rays	10^{22}	3×10^{-14}	7×10^{-12}

The energy units can be joules or electron volts. An *electron volt*, eV, is the energy acquired by an electron if its potential is increased by 1 V. Thus,

$$1 \text{ eV} = 1.602 \times 10^{-19} \text{ C (charge on the electron)} \times 1 \text{ V} = 1.602 \times 10^{-19} \text{ J.}$$

Since $v = c/\lambda$, then $E = hc/\lambda$ or $\lambda = hc/E$.

The frequency of x-rays used in crystallographic analysis is about 10^{18} Hz, or wavelength about 3 Å. Table 6.2 compares the frequency, wavelength, and energy of x-rays with other forms of electromagnetic radiation. As the frequency and energy increase, the wavelength decreases.

6.4 X-RAY SPECTRUM

Consider three variables that affect the x-ray spectrum. The first is the voltage across the tube and the second is the composition of the anode. The third is more subtle. Most crystallographic analyses use a single wavelength of x-rays. A wide range of x-rays is produced by the x-ray tube. X-rays can be selectively absorbed by filters. If the filter is properly chosen, much of the undesired x-rays are reduced.

6.4.1 Bremsstrahlung Radiation

High voltage causes electrons to leave the cathode and slam into the anode. X-rays are produced when electrons with sufficiently high kinetic energy decelerate by collision with the atoms of the anode. These x-rays are called *bremsstrahlung*, meaning braking radiation. Another name is *white or continuous radiation*, because many different wavelengths are produced.

Some electrons lose their energy in a single collision and thus transfer the maximum possible amount of energy to the photon. Thus,

$$KE = eV = hv_{max} = hc/\lambda_{min} \qquad (6.5)$$

where KE is kinetic energy in joules of an electron hitting the anode, e is the charge of an electron in coulombs, and V is the voltage across the x-ray tube in volts. The energy of the photon with the maximum amount of energy is h, Planck's constant, times the frequency,

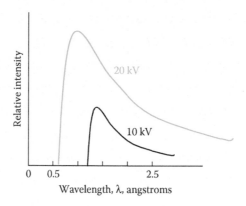

FIGURE 6.13   Bremsstrahlung radiation as a function of x-ray tube voltage.

V_{max}. Since the energy transferred is maximized, then the frequency is also maximized and the wavelength is minimized. Solving for λ_{min} from Equation 6.5 gives

$$\lambda_{min} = hc/eV = 12,398/V \tag{6.6}$$

where the voltage is in volts and the wavelength is in angstroms. A voltage of 10,000 V has $\lambda_{min} = 1.24$ Å, see Exercises 6.7 and 6.8.

Most of the electrons undergo multiple collisions, resulting in x-rays with energies less than the maximum energy. Hence, their wavelengths are greater than the minimum wavelength. Figure 6.13 shows the relative intensity of the bremsstrahlung radiation, which is zero up until the minimum wavelength, and quickly rises to a maximum, and then gradually tapers off. Intensity is energy per unit area per unit time.

The λ_{min} of the bremsstrahlung radiation is independent of the material of the target. However, the intensity of the radiation is directly proportional to the atomic number of the target. Thus, for the same voltage, Ni, with $Z = 28$, has a more intense bremsstrahlung radiation than Co, with $Z = 27$. However, for the same voltage both Ni and Co targets have the same λ_{min} (see Figure 6.14).

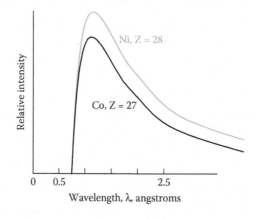

FIGURE 6.14   Bremsstrahlung radiations for Ni and Co anodes at the same accelerating voltage.

6.4.2 Characteristic Radiation

When the accelerating voltage exceeds a specific threshold, a *characteristic spectrum* of discrete lines appears superimposed on the bremsstrahlung. This characteristic x-ray spectrum depends on the atomic number, Z, of the material making up the anode. The energy of the bombarding electrons from the cathode causes an electron in an inner shell of an atom in the target to be ejected. A vacancy is formed. An outer-shell electron fills the vacancy, producing the characteristic radiation. The x-ray photon has an energy equal to the difference in energies of the two shells. Figure 6.15 shows atomic shells K, L, and M. If an electron is knocked out of the K shell, the vacancy can be filled from an electron in, for example, the L or M shell. The radiation is named after the shell where the vacancy occurs. The K lines represent radiation falling into the K shell. The subscripts indicate where the electron originates. K_α radiation comes from the next shell up; K_β comes from two shells away. A similar naming system applies to the L radiation. Figure 6.16 shows the K_α and K_β characteristic radiation of Ni superimposed over the bremsstrahlung radiation. The K_α line is always more intense than the K_β line because it is more likely that an electron falls down from the nearest shell.

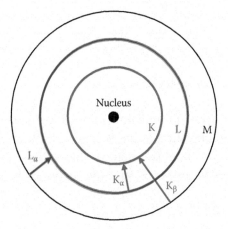

FIGURE 6.15   Atom showing the K, L, and M shells and x-ray emissions.

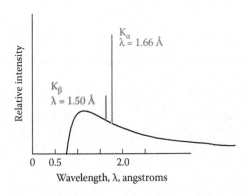

FIGURE 6.16   Characteristic x-ray spectrum superimposed over bremsstrahlung radiation for a nickel target.

TABLE 6.3   Characteristic Radiation for Several Elements

Atomic Number, Z	Element	K_α (Å)	K_β (Å)
24	Cr	2.29	2.08
25	Mn	2.10	1.91
26	Fe	1.94	1.76
27	Co	1.79	1.62
28	Ni	1.66	1.50
29	Cu	1.54	1.39

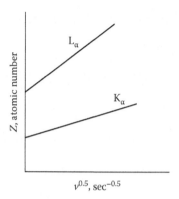

FIGURE 6.17   Moseley's relationship for K_α and L_α radiations.

Table 6.3 lists the characteristic radiations for six elements. If the anode is made of copper, Cu, then the characteristic K_α line is 1.54 Å. Each element has its unique set of characteristic radiations. The English scientist Henry Moseley (1887–1915) studied the relationship between atomic number, Z, and the corresponding characteristic radiation. Moseley found that the atomic number of an element, Z, is a linear function of the square root of the frequency of its characteristic radiation

$$Z = A\nu^{0.5} + B \tag{6.7}$$

where ν is the characteristic frequency associated with the characteristic wavelength; A is a constant depending on the specific transition between shells, that is, K_α or K_β; and B is another constant that shifts the origin. This equation had the startling effect of experimentally ordering the elements in the periodic table by *atomic number*. Until Moseley's work in 1913, the ordering of the periodic table was limited to measurements of *atomic weight*. The atomic number is unique to each element. Moseley's relationship sequencing the elements according to atomic number is shown in Figure 6.17. See Exercises 6.9 and 6.10.

6.4.3 Absorption

So far in this discussion fast electrons slam into matter producing x-rays. Now the discussion shifts to the interaction of x-rays with matter. When x-rays pass through matter, the x-rays are partially absorbed. The amount of *absorption* depends on three factors (a)

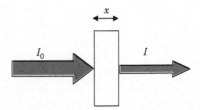

FIGURE 6.18   Absorption of x-rays by matter.

the type of x-rays, (b) the physical properties of the material, and (c) the thickness of the material. Figure 6.18 shows a beam of x-rays of intensity I_0 radiating a sample of nickel of thickness x. The transmitted beam I has been partially absorbed and partially transmitted.

For experiments using diffraction the characteristic radiation is utilized. In this case the absorption equation is

$$I = I_0 e^{-\mu x} \qquad (6.8)$$

where μ is the *linear absorption coefficient* with units cm^{-1} and is dependent on the absorber, its density, and the characteristic wavelength of the x-ray beam. Since μ is proportional to the density, ρ, then μ/ρ = constant, independent of the state of the matter; it is called the *mass absorption coefficient*. The mass absorption coefficient is the same for solid, liquid, or gaseous sodium as well as for sodium in sodium chloride or in elemental sodium. Equation 6.8 becomes

$$I = I_0\, e^{-(\mu/\rho)\, x\rho} \qquad (6.9)$$

Values for μ/ρ are listed in Table 6.4. The mass absorption coefficient, μ/ρ, for an element, depends on both its atomic number and the characteristic radiation.

Figure 6.19 shows the elements of Table 6.4 expanded to include atomic numbers from $Z = 20$ to $Z = 45$ for both Mo K_α and Cu K_α x-rays. There are two parts to each of the curves. Consider the Cu K_α x-rays. As the atomic number increases from $Z = 20$ (Ca) to $Z = 27$ (Co), the absorption coefficient increases until a maximum is reached. Here, the mechanism of absorption is the scattering of x-rays in all directions. For this part of the curve

$$\mu/\rho = k_1 \lambda^3 Z^3 \qquad (6.10)$$

where k_1 is a constant for this arm of the curve.

TABLE 6.4   Absorption Coefficients for Several Elements for Two Different Characteristic X-Rays

Atomic Number, Z	Element	μ/ρ cm²/g for Mo K_α	μ/ρ cm²/g for Cu K_α
24	Cr	29.25	255.3
25	Mn	31.86	272.6
26	Fe	37.74	304.4
27	Co	41.02	338.6
28	Ni	47.24	48.83
29	Cu	49.34	51.54

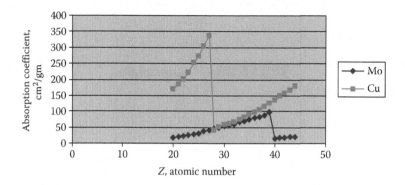

FIGURE 6.19   Absorption coefficient, μ/ρ, versus atomic number for Mo K_α and Cu K_α x-rays.

TABLE 6.5   Absorption Coefficients as a Function of Wavelength for Ni (Green) and Cu (Red)

Characteristic Radiation	Absorption Edge	Wavelength (Å)	Ni, $Z = 28$ (μ/ρ cm²/g)	Cu, $Z = 29$ (μ/ρ cm²/g)
Mo K_β		0.632	34.18	35.77
Mo K_α		0.711	47.24	49.34
	Cu K	1.38059		
Cu K_β		1.392	282.8	38.74
	Ni K	1.48807		
Cu K_α		1.542	48.83	51.54
Co K_β		1.621	56.05	59.22
Co K_α		1.790	73.75	78.11
Cr K_β		2.085	112.5	119.5
Cr K_α		2.291	145.7	155.2

There is a discontinuity where the absorption coefficient drops from 338.6 to 48.63 cm²/g between Co ($Z = 27$) and Ni ($Z = 28$). This is the absorption edge. The mechanism at the absorption edge is different. When sufficient energy is available, an electronic transition occurs. A photoelectron is emitted with a characteristic fluorescent radiation.

Alternatively, Equation 6.10 can be interpreted as holding the element constant and bathing it in different characteristic radiations. Table 6.5 gives the absorption coefficients as a function of wavelength for both nickel and copper filters. These data for nickel are plotted in Figure 6.20. The shorter wavelengths (high energies) have much greater penetration, or lower absorption, than the longer wavelengths.

6.4.4 Electrons and X-Rays

Note the parallel relationships between electrons and x-rays in Table 6.6. In the x-ray tube electrons (particles) with sufficient kinetic energy knock out electrons from the K shell; x-rays are emitted as the electrons fall back into the K shell. On the other hand, x-rays from the anode with sufficient energy, knock out photoelectrons at the absorption edge and

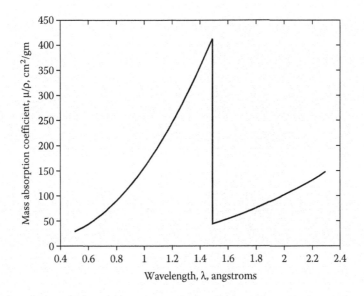

FIGURE 6.20   Absorption coefficient, μ/ρ, versus wavelength for Ni.

TABLE 6.6   Relationship between Electrons and X-Rays

		Origin	Action	Wavelength	Emission
Particles	Electrons	From the cathode of an x-ray tube	Knock out electrons from the K shell	$\lambda_{K\alpha}$ characteristic x-ray	X-rays
Waves	X-rays	From the anode of an x-ray tube	Knock out K photoelectrons	$\lambda_{K\alpha}$ at absorption edge	Fluorescent x-rays

fluorescent radiation is emitted. The characteristic x-ray wavelength, $\lambda_{K\alpha}$, needs exactly the same energy, $h\nu_{K\alpha}$, as fluorescent x-rays at the K_α absorption edge.

6.5 THE X-RAY TUBE

Figure 6.21 shows a modern x-ray tube. Much heat is generated in an x-ray tube. Over 99% of the energy in the x-ray tube becomes heat, leaving less than 1% to produce x-rays. Modern tubes are water-cooled to dissipate the heat. This is a difficult safety engineering problem because water is in proximity to high voltage.

The electrons boil off a filament cathode and smash into the anode. A contained water system cools the anode. The emitted x-rays travel through a beryllium window. Beryllium is used because of its low atomic number ($Z = 4$) and hence low mass absorption coefficient. There are four windows on the x-ray tube, allowing for experimental setups. Usually only one window is used.

6.5.1 Filters

The x-ray tube emits the spectrum shown in Figure 6.16, but usually only the K_α line is needed. Consider an x-ray tube with a Cu target, as in Figure 6.22. From Table 6.3,

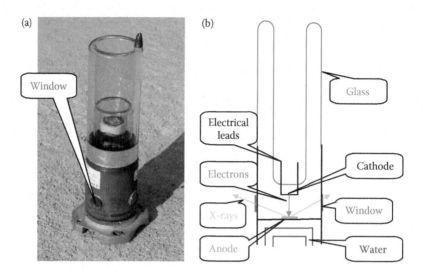

FIGURE 6.21   (a) X-ray tube, (b) schematic diagram. (Photograph by Maureen M. Julian.)

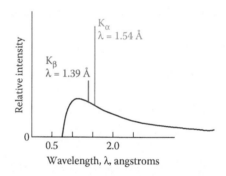

FIGURE 6.22   X-ray spectrum with a copper target.

the characteristic wavelength for Cu K_α is 1.54 Å and for Cu K_β is 1.39 Å. The K absorption edge for nickel, at 1.488 Å, is between the two characteristic wavelengths for copper. Hence, a nickel filter decreases the K_β much more drastically than the K_α. The filter does not change the characteristic wavelength of the copper but changes the relative intensity of the lines. Almost all x-ray tubes with copper anodes use a nickel filter (see Figure 6.23).

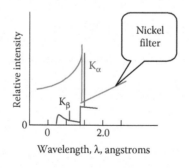

FIGURE 6.23   X-ray spectrum with copper target and nickel filter.

In x-ray diffraction work, the x-ray beam should be as close as possible to being monochromatic, that is, consisting of a single wavelength only.

6.6 X-RAY DIFFRACTION

In 1895, Röntgen knew that if x-rays were waves they should exhibit the diffraction effects demonstrated by Young's two-slit experiment. Röntgen was unable to exhibit diffraction for x-rays. Between 1895 and 1912 a vast improvement in x-ray apparatus—such as x-ray tubes, film, and high-voltage generators—was made, chiefly by medical and metallurgical users. These improvements were essential to the discovery of the diffraction effects, which are weak.

6.6.1 First X-Ray Diffraction Photograph

In 1912, Max Laue wondered if the wavelength of x-rays might be about the same as the distance between the atoms in a crystal. He asked W. Friedrich and P. Knipping to take an x-ray photograph of a crystal. Their apparatus is shown in Figure 6.24. The x-ray tube is on the left and the crystal is mounted on the right. A small telescope allows the crystal to be carefully aligned. The holder for the photographic plate is on the right.

A beautiful blue crystal of copper sulfate pentahydrate was chosen (see Figure 6.25). From the point of view of understanding the diffraction pattern, this was an extremely unfortunate choice. These crystals are triclinic and have a relatively large unit cell. Nevertheless, the result was astonishing. Figure 6.26 shows the first x-ray diffraction pattern. The large spot in the center of the photograph is the undeflected x-ray beam. The "smudges" away from the beam are the diffraction maxima. This was the effect that Roentgen looked for 17 years before in 1895.

Later Laue examined diffraction patterns from cubic crystals of sphalerite or zinc blende, ZnS, which crystallizes in space group No. 216, $F43m$. The crystal has both threefold and fourfold symmetry. The distribution of the diffraction maxima can be seen in the photograph in Figure 6.27. The dark central spot in both figures is the undeflected x-ray beam. In Figure 6.27a, the x-ray beam is along [111] and displays the threefold symmetry. The crystal is then turned and photographed along [100] in Figure 6.27b. The fourfold symmetry is seen.

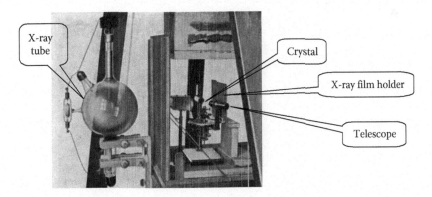

FIGURE 6.24 Friedrich and Knipping's x-ray apparatus. (From Ewald, P. P. 1962. *Fifty Years of X-ray Diffraction*, N. V. A. Oosthoek's Uitgeversmaatschappij, Utrecht, The Netherlands.)

FIGURE 6.25 $CuSO_4 \cdot 5H_2O$. (Crystals grown by Maureen M. Julian.)

Laue believed that the crystal acted as a three-dimensional grating. He assumed that the waves emitted by a radiated atom have a single frequency and that the phase of the emitted radiation depends on the position of the atom in the unit cell. In order for constructive interference to occur, the resulting x-rays had to be in phase. He developed three equations, one for each basis vector, which had to be satisfied simultaneously for a diffraction maximum to appear.

First, consider the basis vector **a** along the x axis in Figure 6.28. Vector S_0 represents the incoming x-rays and vector S represents the diffracted x-rays. Vectors S_0 and S have unit

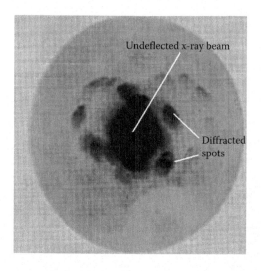

FIGURE 6.26   First x-ray diffraction photograph. (From Ewald, P. P. 1962. *Fifty Years of X-ray Diffraction*, N. V. A. Oosthoek's Uitgeversamaatschappij, Utrecht, The Netherlands.)

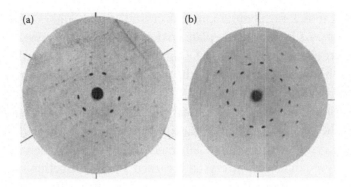

FIGURE 6.27 X-ray photographs of zinc blende (a) along threefold axis, (b) along fourfold axis. (From Ewald, P. P. 1962. *Fifty Years of X-ray Diffraction*, N. V. A. Oosthoek's Uitgeversamaatschappij, Utrecht, The Netherlands.)

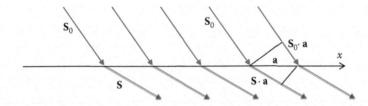

FIGURE 6.28 Diffraction along the x axis with repeat distance **a**.

magnitude. The incoming beam on the right travels an extra distance of $\mathbf{S_0} \cdot \mathbf{a}$ and the diffracted beam on the left travels an extra distance of $\mathbf{S} \cdot \mathbf{a}$. Note that the angle that $\mathbf{S_0}$ makes with **a** is different from the angle that **S** makes with **a**. In order for the diffracted beam, **S**, to be in phase with the incoming beam, $\mathbf{S_0}$, the difference in the paths must equal an integer times the wavelength. Thus,

$$(\mathbf{S} \cdot \mathbf{a}) - (\mathbf{S_0} \cdot \mathbf{a}) = (\mathbf{S} - \mathbf{S_0}) \cdot \mathbf{a} = n_a \lambda$$

This is the first Laue condition. Similar arguments can be made along the other two basis vectors **b** and **c**.

Thus,

$$(\mathbf{S} - \mathbf{S_0}) \cdot \mathbf{b} = n_b \lambda$$

$$(\mathbf{S} - \mathbf{S_0}) \cdot \mathbf{c} = n_c \lambda$$

Each diffracted beam can be identified by three integers n_a, n_b, and n_c. These integers are *hkl*, the Laue indices, from Chapter 5.

6.6.2 Bragg's Law

Laue was working in Germany. In England, William Lawrence Bragg was also interpreting the diffraction patterns. Many scientists thought that the x-rays "tunneled" through the crystal. Bragg took x-ray photographs of a crystal of mica and examined the shape of the diffracted spots. As he moved the plate farther away from the crystal the shape of the spots became more elongated. From this he concluded that the diffracted beam obeyed the optical rule that the angle of incidence of the x-rays equals the angle of reflection. The diffraction spots were called "reflections."

Another way of demonstrating the optical rule is to examine what happens to the diffraction spot as the crystal is rotated. If the x-ray beam tunnels through the "open channels" then, if the crystal is turned 2°, the diffraction spot turns by the same amount or, in this case, 2°. However, in fact, if the crystal is turned 2°, the diffraction spot turns by 4°, or double the amount.

This fact can be elegantly demonstrated with an optical analog. Shine a flashlight (x-ray beam) on a plane mirror (crystal); and, as the mirror is rotated by 45°, the reflection is rotated by 90°. This can be demonstrated in the classroom or the lecture hall (Julian, 1980).

Bragg's law goes way beyond the optical rule to specify the locations of the diffraction spots in terms of the arrangement of the atoms in the crystal. A heuristic version of why Bragg's law works is presented in Figure 6.29. The incident x-ray beam comes in at angle θ to a crystal plane and diffracts at angle θ. The beam acts as if it undergoes a mirror reflection. The angle of incidence equals the angle of reflection. The figure shows a rectangular lattice. The more general case is left as an exercise. The reflecting planes are crystallographic lattice planes.

The individual atoms within a unit cell do not determine the *location* of the diffraction maxima or diffracted spots, but rather the periodic nature of the unit cells—that is to say, the lattice—determines the location of the diffraction maxima.

From Chapter 5, the spacing between *hkl* planes is d_{hkl}. The angle θ is associated with a specific *hkl* and is indicated as θ_{hkl}. The extra distance that the lower beam in Figure 6.29 travels is $2d_{hkl}\sin\theta_{hkl}$. In order for a diffraction maximum (or "reflection") to occur the diffracted beam must undergo constructive interference. The extra distance traveled by the lower x-ray beam must be an integer multiple of the wavelength of the incident x-ray. The order of the reflection was discussed in Chapter 5.

Thus, Bragg's law

$$n\lambda = 2d_{hkl} \sin \theta_{hkl} \tag{6.11}$$

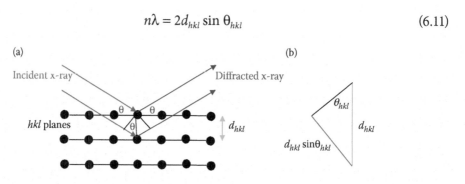

FIGURE 6.29 (a) Bragg's law, (b) enlargement of part of (a).

This remarkable formula unites the x-rays, λ, with the periodic nature of the crystal, d, and the experimental observation, θ. The angle θ is known as the *Bragg angle*.

6.6.3 X-Ray Powder Diffraction Camera

An x-ray powder diffraction camera is shown in Figure 6.30. The *diffraction angle* is the angle between the undeviated beam and the diffracted beam, 2θ, where θ is the Bragg angle.

The film is wrapped around the interior circumference of the circular camera. In this particular situation, the crystals are ground up to form a powder so that the crystals are randomly arranged in the x-ray beam. This avoids the careful alignment that is needed for single-crystal work. Many crystals at different angles contribute to the diffraction maxima, so that lines are recorded in Figure 6.31 instead of spots, as in the Laue photographs in Figure 6.27. These photographs are of aluminum, NaCl, and CaF_2. The placement of the lines is different and these differences form the fingerprint of the crystalline material. The powder diffraction files, from Chapter 5, are collected in this way.

The x-ray camera in Figure 6.30 has a radius of 58.81 mm. The diffraction angle, 2θ, can be measured directly from the film. The radiation used for the films in Figure 6.31 has a wavelength, λ, equal to 1.54 Å. Table 6.7 gives in column one 2θ, in radians, as measured on this film, in column two $\sin\theta$, and in column three $d = \lambda/(2\sin\theta)$ in Å for a ground-up aluminum sample. These are the experimental data that are collected directly from the crystals.

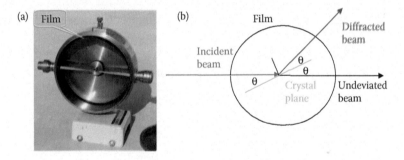

FIGURE 6.30 X-ray powder camera (a) photograph, (b) schematic. (Photograph by Maureen M. Julian.)

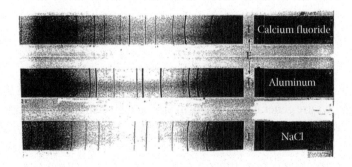

FIGURE 6.31 Powder diffraction photographs. (Photographs by Maureen M. Julian.)

TABLE 6.7   Comparison of Experimental Data with the PDF File for Aluminum

Data from Aluminum Sample Collected from Film			Data from PDF File 04-0787 (ICDD, 2000)	
2θ, in radians	$\sin\theta$	$d = \dfrac{\lambda}{2\sin\theta}(\text{Å})$	d from PDF (Å)	hkl
0.675	0.331	2.328	2.338	111
0.789	0.384	2.006	2.024	200
1.146	0.542	1.422	1.431	220
1.376	0.635	1.214	1.221	311
1.452	0.664	1.161	1.169	222
1.746	0.766	1.006	1.012	400
1.978	0.835	0.923	0.9289	331
2.054	0.856	0.901	0.9055	420
2.421	0.936	0.824	0.8266	422

Notice that none of the symmetry information, such as lattice parameters or space group, needs to be known in order to collect these data. The next step is to compare these collected data with the PDF file for aluminum. A comparison of the d-spacing from the experiment, in column 3, with the d-spacing in the PDF file, in column 4, confirms the identification of the aluminum used in the experiment. See Julian et al. (1986) for an interesting application to plant pathology. Also see Welberry (2004) for non-Bragg diffuse scattering.

As examples, for each of the seven crystals in Chapters 9 through 15, a powder diffraction pattern was created by the Starter Program for Chapter 6. These patterns are found as follows:

- DL-leucine, in the triclinic system, in Section 9.7.1

- Sucrose, in the monoclinic system, in Section 10.7.1

- Polyethylene, in the orthorhombic system, in Section 11.7.1

- α-Cristobalite, in the tetragonal system, in Section 12.7.1

- $H_{12}B_{12}^{2-}$, 3 K^+, Br^-, in the trigonal system, in Section 13.8.1

- Magnesium, in the hexagonal system, in Section 14.7.1

- Acetylene, in the cubic system, in Section 15.7.1

6.7 SYNCHROTRON X-RAYS

A *synchrotron* is an electron accelerator that uses a magnetic field to guide the electrons in a circular orbit. X-rays are generated by a synchrotron when electrons are accelerated at relativistic velocities. Figure 6.32 shows an aerial view of the Cornell University synchrotron, which has a circumference about one-half mile. X-rays generated by a synchrotron differ from x-rays generated by a conventional x-ray tube in several ways. Synchrotron x-rays are many thousands of times more intense. They have a continuous spectrum typically covering a range

FIGURE 6.32　Aerial view of synchrotron at Cornell Univeristy. (Image courtesy of CLASSE.)

from about 0.5 to 2.5 Å; the shape of the spectrum may be simple or complex, depending on the magnetic devices used to bend the electron beam. The radiation comes in rapid pulses. Synchrotrons are very expensive to operate and are found in only a few national and international laboratories. See Section 15.1.1 for a comparison of diffraction techniques using x-rays from conventional tubes, x-rays from synchrotrons, neutrons, and electrons.

Figure 6.33 shows an experimental station or "hutch" at Cornell. The x-ray beam from the synchrotron comes in the direction of the yellow arrow. The cryocooler cools the protein crystal with a stream of very cold nitrogen gas to reduce radiation damage. The sample

FIGURE 6.33　View of crystallographic experimental station. X-ray beam is indicated by yellow arrow and sample position by orange diamond. (Photograph by Doletha Marian Szebenyi.)

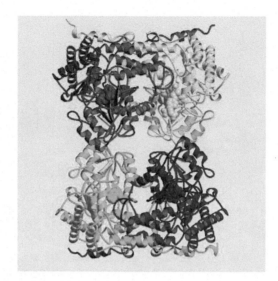

FIGURE 6.34 Serine hydroxymethyltransferase. There are four protein chains, shown in different colors. The small molecules bound to each chain, shown in space-filling form, are cofactors and ligands required for the enzyme to function.

mounting robot allows for remote operation and the option of a frequent change of sample. This particular setup is for macromolecular crystallography.

The intense synchrotron x-rays are particularly suited for protein work. Figure 6.34 shows an example of a protein, the enzyme serine hydroxymethyltransferase, whose structure was determined using data from the Cornell synchrotron source (Szebenyi et al., 2000). There are nearly 15,000 nonhydrogen atoms in the protein molecule. The crystal forms in the tetragonal space group #96, $P4_32_12$ with $a = b = 142.5$ Å and $c = 270.9$ Å. The volume of the unit cell is 5,502,507 Å^3.

For an interesting story on the search for the protein structures see Julian (2005).

DEFINITIONS

Absorption of x-rays
Amplitude
Anode
Bragg angle
Bremsstrahlung radiation
Cathode
Characteristic x-ray spectrum
Constructive interference
Continuous radiation
Destructive interference
Diffraction angle
Electromagnetic waves
Electron volt, eV
Linear absorption coefficient

Mass absorption coefficient
Partial interference
Photon
Plane polarized waves
Principle of superposition
Synchrotron
Unpolarized waves
Wavelength
White radiation

EXERCISES

6.1　Plot $y = 7 \cos (2\pi x + 0.75\pi)$, $y = -12 \cos (2\pi x + 0.75\pi)$, and their sum to show constructive interference.

6.2　Plot $y = 3.4 \cos (2\pi x - 0.75\pi)$, $y = -1.2 \cos (2\pi x - 0.75\pi)$, and their sum to show constructive interference.

6.3　Plot $y = 3.4 \cos (2\pi x - 0.75\pi)$, $y = 3.4 \cos (2\pi x + 0.25\pi)$, and their sum to show destructive interference.

6.4　Plot $y = -2.3 \cos (2\pi x + 0.63\pi)$, $y = -2.3 \cos (2\pi x + 1.63\pi)$, and their sum to show destructive interference.

6.5　Plot $y = -2.3 \cos (2\pi x + 0.63\pi)$, $y = -1.9 \cos (2\pi x + 1.58\pi)$, and their sum to show partial interference.

6.6　Plot $y = 3 \cos (2\pi x - 0.63\pi)$, $y = -1.9 \cos (2\pi x - 1.58\pi)$, and their sum to show partial interference.

6.7　In Equation 6.6, calculate the value of $\lambda_{min} \times V$.

6.8　What is the minimum wavelength of the radiation for a voltage of 30,000 V across an x-ray tube.

6.9　Plot the relationship between atomic number and the frequency of the K_α characteristic radiation for elements 15 through 35. Use the Internet. Use Equation 6.7.

6.10　Plot the relationship between atomic number and the frequency of the L_α characteristic radiation for elements 15 through 35. Use the Internet. Use Equation 6.7.

MATLAB CODE: STARTER PROGRAM FOR CHAPTER 6: GRAPHIC OF POWDER DIFFRACTION FILE

This program produced the powder diffraction patterns used in Chapters 9 through 15. The input consists of the cell constants of the crystal, in this example HMB, the *hkl* intensities, and the wavelength (lambda) of the x-rays. The program calculates the *d*-spacing using **G*** from Section 5.7.1. Then Bragg's law (Equation 6.11) is applied to calculate 2θ. Finally, the intensity is plotted versus 2θ.

```
clc                   %clears content of command window
close all        %
clear                   %clears variables in workspace
%Script file:   Chapter 6 Starter program
%
% Purpose  Plot powder diffraction pattern
%
% Record of revisions:
%  Date              Programmer            Description of change
%  10/30/            M. Julian             Original Code
%
%  define variables:
%  a  = crystallographic a axis   (Angstroms)
%  b  = crystallographic b axis   (Angstroms)
%  c  = crystallographic c axis   (Angstroms)
%  alpha    angle between b and c  (degrees)
%  beta     angle between a and c  (degrees)
%  gamma    angle between a and b  (degrees)
%  G  metric matrix, direct lattice
%   Gstar = G*  metric matrix, recipocal lattice
%  lambda   wavelength of x-rays
%  hkl  Miller indices
%  I intensity
%  n  number of reflections
%  dspacing     d-spacing  see section 5.7.1 of text
%  TwoT   2 theta diffraction angle  see section 6.6.3 of text
%  i
lambda = 1.542
%HMB
hklI=[ 1    0 0 80
       0    1 0 90
      -1    1 0 30
      -1   -1 1 20];
  [n, c]=size(hklI);
%hexamethylbenzene data
a=9.010;
b=8.926;
c=5.344;
alpha= 44 + 27/60;
beta= 116 + 43/60;
gamma=119 + 34/60;
%calculate G matrix
G=[a^2                a*b*cosd(gamma)   a*c*cosd(beta)
   a*b*cosd(gamma)    b^2               b*c*cosd(alpha)
   a*c*cosd(beta)     b*c*cosd(alpha)   c^2];
%%%%%%%%%%%%%%%%%%%%%%%%%%%%%%%%%%%%%%%%%%%%%%%%%%%%%%%%%%%%%%%%%%%
%%%%%%
```

```
%Calculate G* from
Gstar= inv(G);
for i = 1:n
h=hklI(i,1)
k=hklI(i,2)
l=hklI(i,3)
dspacing=([h k l]*Gstar*[h k l]')^(-.5)
sinT=lambda/(2*dspacing)
TwoT(i) = 2*asind(sinT)
I(i)=  hklI(i,4)
end
figure; bar(TwoT ,I,.01);%produces line
% ylim defines the bottom and top of the page
for i=1:n
str = num2str(hklI(i,1:3))
text(TwoT(i),I(i)+5,str,'FontSize',6)
%%%text(TwoT(i),I(i),str,'FontSize',6)
end
%adjust values of 2theta and intensity for graphing
xlim([5 30])
ylim([0 110])
xlabel('2\theta, degrees')
ylabel('Intensity')
title('Powder Diffraction Pattern of Hexamethylbenzene')
```

## CHAPTER 7

Electron Density Maps

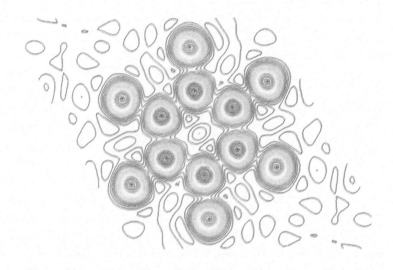

Above is an electron density map of hexamethylbenzene (HMB). The hydrogen atoms are omitted.

CONTENTS

CHAPTER OBJECTIVES

- Scattering by a single electron

- Scattering by a single atom and the atomic scattering factor

- Scattering by a crystal and the structure factor

- Structure factors and extinctions for cubic Bravais lattices

- Relationship of primitive, face-centered, and body-centered unit cells with their corresponding reciprocal unit cells

- Structure factors for unit cells with one or more different atoms

- Examples of crystals with real and complex structure factors

- Structure factors and extinctions for glides and screws

- Structure factors for centrosymmetric crystals in general

- Example of structure factors using HMB

- Generalized calculation of structure factors

- Friedel's law

- Fourier series and electron density maps in one dimension

- Electron density map for HMB

- Use the structure factors to compare the crystal model with the diffraction data

- Construct electron density maps from diffraction data

- Use R-value or the discrepancy index as a measure of the correctness of the structure

7.1 INTRODUCTION

Diffraction occurs when x-rays, neutrons, or electrons interact with matter. It is the result of both interference and scattering phenomena. This book emphasizes x-rays. In Chapter 6, the spatial distribution of the diffracted x-rays was discussed in terms of Bragg's law and the unit cell. The diffraction spots vary in intensity. The scattering and interference due to the individual atoms or ions located within the unit cell contribute to the variation in intensity.

This chapter discusses scattering by a single electron, by a single atom, and by a crystal. The scattering by a single atom is quantified by its atomic scattering factor, and the scattering by the unit cell of a crystal is quantified by its structure factor. The structure factors are calculated for the reciprocal lattice points. Sample calculations are made for several crystals. The structure factors are proportional to the coefficients in the Fourier series that are used to calculate an electron density map. Finally the electron density map is calculated for HMB.

7.2 SCATTERING BY AN ELECTRON

If an x-ray beam of frequency v falls on an electron, then that electron vibrates with the same frequency v. The electron is accelerating and decelerating as it undergoes oscillatory motion. From classical theory, a charged particle in accelerated motion is a source of electromagnetic radiation. Thus the electron in the path of an x-ray beam emits x-rays of the same frequency. *Scattering by an electron* is the process by which an electron, exposed to an x-ray beam, emits x-rays of the same frequency. The electron is much smaller than the wavelength of x-rays. The classical electron radius is 3×10^{-15} m while the wavelength of x-rays is about 10^{-10} m. When x-rays are scattered by an electron, the scattered radiation has the same intensity in all directions. See Figure 7.1. The intensity of this scattered radiation is only a miniscule fraction of the incident beam.

7.3 SCATTERING BY AN ATOM

An atom is a nucleus surrounded by electrons. When an x-ray beam falls on an atom, the electrons scatter a tiny fraction of the x-ray beam. Scattering by an atom is interpreted here

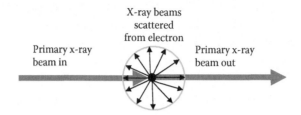

FIGURE 7.1   Electron in the path of an x-ray beam, emitting x-rays.

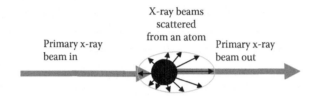

FIGURE 7.2   Atom in the path of an x-ray beam, emitting x-rays.

as simply the composite of the scattering by the individual electrons in the atom. However, the scattered radiation is no longer isotropic, because there is interference between the radiation scattered by the individual electrons. The atom is interpreted as being spherically symmetric. Consequently the scattered radiation is cylindrically symmetric about the direction of the x-ray beam and depends on the angle between the scattered radiation and the x-ray beam. As with the scattering from an electron, the intensity of this scattered x-ray is also only a miniscule fraction of that of the incident beam. See Figure 7.2.

There is an analogy between the scattered radiation from an atom and the first-order fringe in Young's two-slit experiment. See Chapter 6. In that experiment $\sin \theta = \lambda/r \ll 1$ where θ is the angle of deflection of the first fringe, λ is the wavelength of the incoming radiation, and r is the distance between the two slits.

For example, the sodium atom, Na, ($Z = 11$) has 11 electrons. In the forward direction all the scattered beams are in phase and constructive interference is observed. However, away from the forward direction, partial interference occurs; and the scattered beam is less than in the forward direction. As the angle increases away from the forward direction, the beams scattered by the individual electrons become more and more out of phase and the scattered radiation decreases. The diameter of the atom is of the same order of magnitude as the wavelength of x-rays. The atoms are bigger than 10^{-10} m while the wavelength of x-rays is about 10^{-10} m. The hydrogen atom has a diameter of about 0.5 Å or 0.5×10^{-10} m and a potassium atom has a diameter of about 4 Å or 4×10^{-10} m.

7.3.1 Atomic Scattering Factor

The *atomic scattering factor f* (also known as the *atomic form factor*) is the ratio of the amplitude of the wave scattered by the atom to the amplitude of the wave scattered by one electron. Its units are electrons. The atomic scattering factor is a measure of the efficiency of the atom in scattering x-rays. D. Cromer and J. Mann (1967) have fitted the atomic scattering factor to a 9-parameter function of ($\sin \theta/\lambda$):

$$f\left(\frac{\sin\theta}{\lambda}\right) = \sum_{i=1}^{4} a_i e^{-b_i\left(\frac{\sin\theta}{\lambda}\right)^2} + c \qquad (7.1)$$

where θ is the Bragg angle (or half the angle between the scattered radiation and the incident beam) and λ is the wavelength of the incident x-rays; and the nine Cromer–Mann coefficients are a_i and b_i, for $i = 1$ to 4, and c.

🖳 **EXAMPLE 7.1**

Calculate and plot the atomic scattering factor for carbon.

Solution

Figure 7.3 is a graph of the data from Table 7.1. Since the atomic number, Z, for carbon is 6, then $f(0) = 6$ electrons. See the starter program.
 See Exercise 🖳 7. 1.

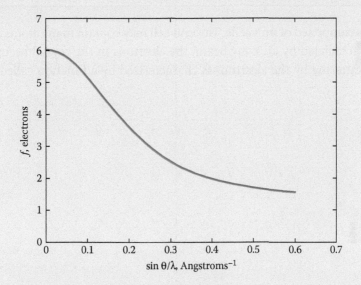

FIGURE 7.3   Atomic scattering factor curve for carbon.

TABLE 7.1   Cromer–Mann Coefficients for C, Z = 6

i	1	2	3	4
a_i	2.310	1.020	1.589	0.865
b_i	20.844	10.208	0.569	51.651
c	0.216	—	—	—

7.4 SCATTERING BY A CRYSTAL

There are two issues to be considered in the analysis of an x-ray diffraction pattern of a crystal. First, what are the spatial locations of the diffraction maxima and, second, how intense are they?

The locations of the diffraction maxima depend on the lattice of the crystal and not on the contents of the unit cell and are specified by Bragg's law, as described in Chapter 6.

Look at Figure 7.4, which is a diffraction pattern of a quartz crystal; and note that the diffraction spots vary in intensity.

The types and locations of atoms and ions in the unit cell contribute to the relative intensities of the reflections. The structure factor $F(hkl)$ which may be complex, has the information of both the type of atom and its position in the unit cell.

The square of the magnitude of the structure factor is proportional to the *intensity of the diffracted x-ray beam*. Thus the intensity of the hkl reflection, I_{hkl}, is proportional to $(F(hkl))^2$ if $F(hkl)$ is real or, more generally, to $F^*(hkl) F(hkl)$, where $F^*(hkl)$ is the complex conjugate of $F(hkl)$.

A crystal is comprised of unit cells. The unit cell may contain many atoms or ions. When the crystal is irradiated by an x-ray beam, the electrons in the crystal become sources of x-rays. The scattering by the electrons is characterized by a function called the *structure*

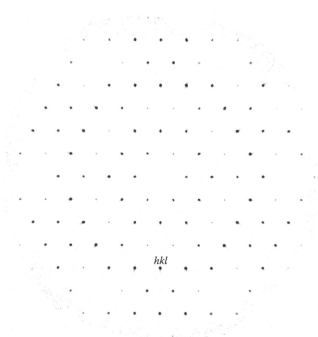

hkl

FIGURE 7.4   Diffraction pattern of quartz.

factor that depends on both the scattering power of an atom (the atomic scattering factor) and the position of the atom in the unit cell. In the forward direction, all the beams scattered by the electrons in all the atoms are in phase and constructive interference is observed. However, away from the forward direction, interference occurs, and the scattered beam is less than in the forward direction.

Because of the repetition of the unit cell, diffraction occurs only at very few directions in space, as in Figure 7.4. These directions *hkl* correspond to the reflections recorded on x-ray detectors. These reflections are also known as Bragg reflections because they obey Bragg's law, which may be written as $\lambda = 2d_{hkl} \sin \theta_{hkl}$. The *hkl* reflection is located at the Bragg angle θ_{hkl}. Bragg's law is defined in Chapter 6.

The *structure factor* for each atom or ion has two components for each *hkl*:

- A phase angle that contains information about the position of the atom in the unit cell, *xyz*.

- An amplitude that contains information about the scattering of the particular kind of atom, for example, carbon.

These components for each atom in the unit cell are combined to get the total scattering. Scattering by a crystal is interpreted here as the composite of the scattering by the individual atoms.

7.4.1 Phase Angle

To prepare for the calculation of the structure factor for each reflection, a phase angle, *in radians*, of the wave radiated from an atom for the *hkl* reflection is introduced that depends on the position of the atom or ion, but not on any of its other properties, such as its atomic number or ionization state.

Each atom in the unit cell is located by its fractional coordinates *x, y, z*. For atom *j* at position $\mathbf{r}_j$

$$\mathbf{r}_j = x_j\,\mathbf{a} + y_j\,\mathbf{b} + z_j\,\mathbf{c} \tag{7.2}$$

where **a**, **b**, and **c** are the basis vectors of the direct lattice.

The position of the *hkl* reflection in the reciprocal lattice is H(*hkl*) where

$$\mathbf{H}(hkl) = h\,\mathbf{a}^* + k\,\mathbf{b}^* + l\,\mathbf{c}^* \tag{7.3}$$

where **a***, **b***, and **c*** are the basis vectors of the reciprocal lattice.

The dot product H(*hkl*) · $\mathbf{r}_j$ is

$$\mathbf{H}(hkl) \cdot \mathbf{r}_j = hx_j + ky_j + lz_j \tag{7.4}$$

The *phase angle*, in *radians*, of the wave radiated from atom *j* is $2\pi\mathbf{H}(hkl) \cdot \mathbf{r}_j$.

EXAMPLE 7.2

If $hkl = 000$, calculate the phase angle.

Solution

If $hkl = 000$, this is the undeviated x-ray beam.

$$\mathbf{H}(000) \cdot \mathbf{r}_j = hx_j + ky_j + lz_j = 0.$$

The phase angle $2\pi\mathbf{H}(000) \cdot \mathbf{r}_j$ is 0 for all atoms.

EXAMPLE 7.3

If the atom j is at the origin, calculate the phase angle.

Solution

If the atom j is at the origin, then $x_j, y_j, z_j = 0, 0, 0$ and $\mathbf{r}_j = 0$.

$$\mathbf{H}(hkl) \cdot \mathbf{r}_j = h_{xj} + k_{yj} + lz_j = 0.$$

The phase angle $2\pi\mathbf{H}(hkl) \cdot \mathbf{r}_j$ is also 0 for atom j.

EXAMPLE 7.4

If an atom is at fractional coordinates 0.4, 0.6, 0.1, calculate the phase angle for the 212 reflection.

Solution

$$\mathbf{H}(212) \cdot \mathbf{r} = 2 \times 0.4 + 1 \times 0.6 + 2 \times 0.1 = 1.6$$

The phase angle $2\pi\mathbf{H}(hkl) \cdot \mathbf{r} = 3.2\pi$ radians.
See Exercises 7.2 through 7.4.

7.4.2 Amplitude

The amplitude contains information about the scattering of the particular kind of atom, such as a carbon atom, a potassium atom, or a potassium ion. From Section 7.3.1, when an isolated atom scatters, the atomic scattering factor is a continuous function of $(\sin \theta/\lambda)$. Figure 7.5 shows the atomic scattering curve for carbon, where λ is the wavelength and θ is any angle between zero and 90°. The variable $(\sin \theta/\lambda)$ is cleverly chosen to take advantage of the situation that the crystal scatters only at the Bragg angle, θ_{hlk}. For any given

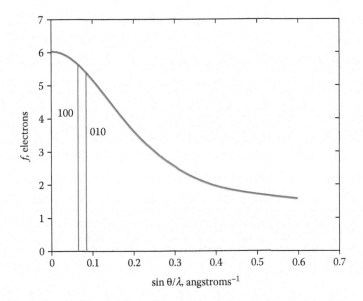

FIGURE 7.5   Scattering curve for carbon for reflections 100 and 010 of HMB.

crystal, the value of the scattering factor must be calculated specifically for each reflection. Rearranging Bragg's law, $\lambda = 2d_{hkl} \sin \theta_{hkl}$, gives

$$\frac{\sin \theta_{hkl}}{\lambda} = \frac{1}{2d_{hkl}}.$$

Thus the atomic scattering factor can be reparameterized

$$f\left(\frac{\sin \theta}{\lambda}\right) = f\left(\frac{1}{2d}\right)$$

For example, for HMB consider the data from PDF 08-0815 from Chapter 5 given in Table 7.2. The first column are the *hkl* values, the second column gives the *d*-spacing for each *hkl*. The third column gives $(1/2d_{hkl})$ for each value d_{hkl}. In the last column, the value for $(1/2d_{hkl})$ was applied to the Cromer–Mann equation, Equation 7.1, using the coefficients for carbon from Table 7.1. The values of the carbon atomic scattering factor for 100 and 010 are given in the last column and plotted in Figure 7.5.

TABLE 7.2   Atomic Scattering Factors for Carbon, f_C

hkl	d_{hkl} (Å)	$\dfrac{1}{2d_{hkl}}$ (Å^{-1})	Atomic Scattering Factor for Carbon (f_C)
100	7.75	0.0645	5.5947
010	6.01	0.0832	5.3538
$\bar{1}$10	5.64	0.0889	See Exercise 7.6
$\bar{1}$11	5.16	0.0969	See Exercise 7.7
011	4.75	0.1052	See Exercise 7.8

⌨ EXAMPLE 7.5

Calculate the value of the atomic scattering factor for a carbon atom in HMB where $hkl = 100$. Use $\lambda = 1.54$ Å.

Solution

Look at Table 7.2 (or calculate d_{hkl} using the metric matrix and the method described in Chapter 5) where $d_{100} = 7.75$ Å and $1/2d_{100} = 0.0645$ Å^{-1} = $\sin \theta_{100}/\lambda$.

Apply Equation 7.1 for carbon using the program developed in Example 7.1 and the coefficients in Table 7.1.

$$f(0.0645) = \sum_{i=1}^{4} a_i e^{-b_i (0.0645)^2} + c = 5.5947$$

See Exercise 7.5.

7.4.3 Structure Factor

In this section information from the phase angle and the atomic scattering factor for each atom in the unit cell is combined to get the total scattering. First consider that a wave can be represented, as in Figure 7.6, by the complex function $Ae^{\varphi i}$ where φ is the phase angle, in radians.

The wave scattered from atom j is represented by the complex function

$$f_j \left(\frac{1}{2d_{hkl}} \right) e^{2\pi i \mathbf{H}(hkl) \cdot \mathbf{r}_j}$$

where $f_j (1/2d_{hkl})$ is the atomic scattering factor. The atomic scattering factor is the amplitude of the wave associated with the atom.

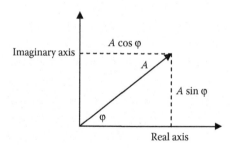

FIGURE 7.6 A wave vector $Ae^{\varphi i}$ in the complex plane.

To get the structure factor for the whole unit cell, all the contributions from the atoms are added. Thus for each Bragg reflection hkl, the *structure factor* $F(hkl)$ is

$$F(hkl) = \sum_{j=1}^{N} f_j \left(\frac{1}{2d_{hkl}} \right) e^{2\pi i \mathbf{H}(hkl) \cdot \mathbf{r}_j} \tag{7.5}$$

where N is the number of atoms in the unit cell and from Equation 7.4

$$\mathbf{H}(hkl) \cdot \mathbf{r}_j = hx_j + ky_j + lz_j.$$

7.5 SOME MATHEMATICAL IDENTITIES

There are several useful identities for working with complex waves.

The first identity is

$$e^{\varphi i} = \cos \varphi + i \sin \varphi \tag{7.6}$$

where φ is in radians and the cosine and sine must be computed accordingly. The mnemonic device is *cis*.

Some useful identities are

$$e^{\pi i} = e^{3\pi i} = e^{5\pi i} = e^{7\pi i} = -1 \tag{7.7}$$

$$e^{2\pi i} = e^{4\pi i} = e^{6\pi i} = e^{8\pi i} = 1 \tag{7.8}$$

$$e^{n\pi i} = e^{-n\pi i} = (-1)^n \tag{7.9}$$

where n is an integer.

$$e^{\pi i/2} = e^{5\pi i/2} = e^{9\pi i/2} = i \tag{7.10}$$

$$e^{3\pi i/2} = e^{7\pi i/2} = e^{11\pi i/2} = -i \tag{7.11}$$

7.6 STRUCTURE FACTORS FOR SOME CRYSTALS

Examples of structure factors are now calculated for several crystals. For convenience, all of them are cubic. The first is polonium, which crystallizes in a primitive cubic Bravais lattice with one atom in the unit cell. The second is chromium, which crystallizes in a body-centered cubic Bravais lattice with two atoms per unit cell. Third is copper, which crystallizes in a face-centered cubic Bravais lattice with four atoms per unit cell. Since the first three are all elements, there is only one type of atom in the unit cell. The fourth and fifth examples have two different atoms in the unit cell, and both crystallize in face-centered cubic Bravais lattices. However they differ in that NaCl has structure factors with real values while ZnS has structure factors with complex values. Table 7.3 contains Cromer–Mann coefficients for the atomic scattering factors of the elements that are needed here. The Web is a convenient source of other Cromer–Mann coefficients.

As examples, for each of the seven crystals in Chapters 9 through 15, atomic scattering curves for the elements, pointing out $\sin \theta / \lambda$ for the undeflected x-ray beam (000) and the most intense reflection, are found as follows

- DL-leucine, in the triclinic system, in Section 9.8. The elements are hydrogen, carbon, nitrogen, and oxygen. The most intense reflection is 100.

- Sucrose, in the monoclinic system, in Section 10.8. The elements are hydrogen, carbon, and oxygen. The most intense reflection is 112.

- Polyethylene, in the orthorhombic system, in Section 11.8. The elements are hydrogen and carbon. The most intense reflection is 110.

- α-Cristobalite, in the tetragonal system, in Section 12.8. The ions are O^{-2} and Si^{+4}. The most intense reflection is 101.

- $H_{12}B_{12}{}^{2-}$, 3 K^+, Br^-, in the trigonal system, in Section 13.9. The elements are hydrogen, boron, potassium, and bromine. The most intense reflection is 200.

- Magnesium, in the hexagonal system, in Section 14.8. The element is magnesium. The most intense reflection is 101.

- Acetylene, in the cubic system, in Section 15.8. The elements are hydrogen and carbon. The most intense reflection is 111.

See Exercises 7.9, through 7.14.

7.6.1 Simple Cubic, Polonium

Polonium, Po, was discovered in 1898 by Marie and Pierre Curie and was named in honor of Poland, the birthplace of Marie Curie. Po crystallizes in a primitive cubic Bravais lattice with one atom in the unit cell. See Figure 7.7. The lattice parameter for Po is $a = 3.359$ Å.

The single atom of Po has coordinates 0, 0, 0. The structure factor calculated from Equation 7.5 with $N = 1$ and an atom at position $x_1, y_1, z_1 = 0, 0, 0$ is

$$F(hkl) = \sum_{j=1}^{1} f_{Po}\left(\frac{1}{2d_{hkl}}\right) e^{2\pi i H(hkl)\cdot r_j} = f_{Po}\left(\frac{1}{2d_{hkl}}\right) e^{2\pi i (hx_1 + ky_1 + lz_1)} = f_{Po}\left(\frac{1}{2d_{hkl}}\right) \quad (7.12)$$

TABLE 7.3 Cromer–Mann Coefficients for the Atomic Scattering Factors of Some Elements

Element	Z	a_1	a_2	a_3	a_4	b_1	b_2	b_3	b_4	c
O^{2-}	8	4.758	3.637	0	0	7.831	30.05	0	0	1.594
Na^+	11	3.2565	3.9362	1.3998	1.0032	2.6671	6.1153	0.2001	14.039	0.404
Al^{3+}	13	4.17448	3.3876	1.20296	0.528137	1.93816	4.14553	0.228753	8.28524	0.706786
S	16	6.9053	5.2034	1.4379	1.5863	1.4679	22.2151	0.2536	56.172	0.8669
Cl^-	17	18.2915	7.2084	6.5337	2.3386	0.0066	1.1717	19.5424	60.4486	−16.378
K^+	19	7.9578	7.4917	6.359	1.1915	12.6331	0.7674	−0.002	31.9128	−4.9978
Cr	24	10.6406	7.3537	3.324	1.4922	6.1038	0.392	20.2626	98.7399	1.1832
Cu	29	13.338	7.1676	5.6158	1.6735	3.5828	0.247	11.3966	64.8126	1.191
Zn^{2+}	30	11.9719	7.3862	6.4668	1.394	2.9946	0.2031	7.0826	18.0995	0.7807
Po	84	34.6726	15.4733	13.1138	7.02588	0.700999	3.55078	9.55642	47.0045	13.677

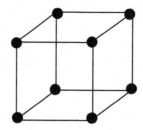

FIGURE 7.7   Primitive cubic unit cell of polonium.

The atomic scattering factor for Po, f_{Po}, is shown in Figure 7.8. The atomic scattering factor, $f_{Po}(hkl)$, is a function of $\sin \theta/\lambda$, as in Section 7.4.2. Rearranging Bragg's law, $\lambda = 2d_{hkl} \sin \theta_{hkl}$, gives

$$\frac{\sin \theta_{hkl}}{\lambda} = \frac{1}{2d_{hkl}}.$$

In this case the crystal is cubic. For any cubic crystal

$$d_{hkl} = \frac{a}{\sqrt{h^2 + k^2 + l^2}}.$$

See Section 5.7.1 for the calculation of the d-spacing. See Exercise 7.15.

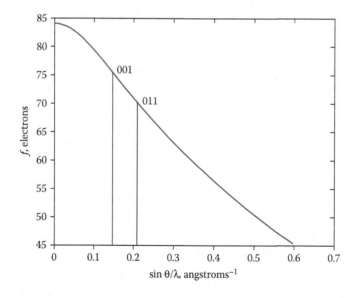

FIGURE 7.8   Atomic scattering factor curve for Po, $Z = 84$, with values for 001 and 011.

Thus for any cubic crystal

$$\frac{\sin \theta_{hkl}}{\lambda} = \frac{1}{2d_{hkl}} = \frac{\sqrt{h^2 + k^2 + l^2}}{2a}$$

and

$$f\left(\frac{\sin \theta_{hkl}}{\lambda}\right) = f\left(\frac{1}{2d_{hkl}}\right) = f\left(\frac{\sqrt{h^2 + k^2 + l^2}}{2a}\right) \qquad (7.13)$$

Then $\sin \theta_{hkl}/\lambda$ is a function only of the variables hkl. Table 7.4 shows the values of

$$\frac{\sqrt{h^2 + k^2 + l^2}}{2a}.$$

as a function of hkl. These are the only values needed for a cubic crystal, because of the symmetries of the cubic system. To get the last column, insert the value of

$$\frac{\sqrt{h^2 + k^2 + l^2}}{2a}.$$

into the atomic scattering factor expression, Equation 7.1, to calculate the structure factors. See Exercise 7.16.

TABLE 7.4   Structure Factor Calculations for Po, $a = 3.359$ Å

h	k	l	$\dfrac{\sqrt{h^2 + k^2 + l^2}}{2a}$	$F_{Po}(hkl) = f_{Po}$, Electrons
0	0	0	0	84
0	0	1	0.1489	75.20
0	1	1	0.2105	69.97
0	2	0	0.2977	63.28
0	2	1	0.3328	60.78
1	1	1	0.2578	66.24
2	1	1	0.3646	58.61
2	2	0	0.4210	54.95
2	2	1	0.4466	53.39
3	0	0	0.4466	53.39
3	0	1	0.4707	51.98
3	1	1	0.4937	50.69
3	2	0	0.5367	48.41
3	2	1	0.5570	47.39

Figure 7.8 shows the Po atomic scattering curve with appropriate values of *hkl* indicated. *F*(000) represents the forward direction, where all the waves are in phase. The atomic number of Po is 84; and, since there is only one atom to the unit cell, *F*(000) = 84 electrons. As the Bragg angle increases the scattering falls off. See Exercise 7.17.

7.6.2 Body-Centered Cubic, Chromium

The element chromium, Cr, has an atomic number *Z* equal to 24 and has a body-centered cubic Bravais lattice with *a* = 2.884 Å. Figure 7.9 shows a body-centered cell. There are two atoms in this unit cell, and they are located at fractional coordinates 0, 0, 0 and ½, ½, ½.

The structure factor is calculated from Equation 7.5 with *N* = 2

$$F(hkl) = \sum_{j=1}^{2} f_{Cr}\left(\frac{1}{2d_{hkl}}\right) e^{2\pi i \mathbf{H}(hkl)\cdot \mathbf{r}_j} = f_{Cr}\left(\frac{1}{2d_{hkl}}\right)(1 + e^{\pi i(h+k+l)}) \tag{7.14}$$

There are two cases: If *h* + *k* + *l* is even, then apply Equation 7.8; and if *h* + *k* + *l* is odd, then apply Equation 7.7.

For body-centered cubic
Case 1: *h* + *k* + *l* is even, such as 112, then *F*(*hkl*) = f_{Cr} (1/2d_{hkl}) (1 + 1) = 2f_{Cr}(1/2d_{hkl})
Case 2: *h* + *k* + *l* is odd, such as 102 then *F*(*hkl*) = f_{Cr} (1/2d_{hkl}) (1 − 1) = 0
In case 2, there is no reflection when *h* + *k* + *l* is odd. This is known as an extinction.

Look at the *International Tables for Crystallography*, space group No.141, *I*4₁*/amd*. The letter "*I*" means body-centered. Under "Positions," the third column gives the "Reflection Conditions" for this space group, which are for all *hkl*: *h* + *k* + *l* = 2*n*. This indicates the condition for body-centered cubic crystals. There are other reflection conditions, but they do not apply to all *hkl*. These conditions are discussed later. Check space groups No. 73, *Ibca* and No. 199, *I*2₁3 for their *hkl* reflection conditions.

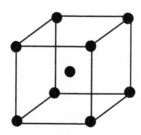

FIGURE 7.9 Body-centered cubic unit cell of chromium.

TABLE 7.5   Structure Factors for Cr, $a = 2.884$ Å

h	k	l	$\dfrac{\sqrt{h^2 + k^2 + l^2}}{2a}$	$F_{Cr}(hkl)$, Electrons
0	0	0	0	48
0	0	1	0.1734	0
0	1	1	0.2452	33.45
0	2	0	0.3467	27.19
0	2	1	0.3877	0
1	0	1	0.2452	33.45
1	1	1	0.3003	0
1	2	0	0.3877	0
1	2	1	0.4247	23.32
2	2	0	0.4904	20.70
2	2	1	0.5201	0
3	0	0	0.5201	0
3	0	1	0.5482	18.85
3	1	1	0.5750	0
3	2	0	0.6251	0
3	2	1	0.6487	16.46

As in the case of the polonium crystal, apply the Cromer–Mann coefficients for Cr to the atomic scattering factor formula, Equation 7.1. Then use Equation 7.14 to calculate the structure factors to produce Table 7.5. Then $F(000) = 48$ because there are two Cr atoms in the unit cell. The structure factors with $h + k + l$ odd are zero. See Exercises 7.18 and 7.19.

The extinguishing of the hkl reflections when the sum $h + k + l$ is odd has an extraordinary implication. Figure 7.10 shows the reciprocal lattice layers $hk0$, $hk1$, and $hk2$ omitting odd values of $h + k + l$, where the structure factors are equal to zero. Note that both 100 and $\overline{1}00$ have sum $h + k + l$ equal to an odd number and so are missing. In two dimensions, each layer considered separately has a face-centered lattice. Some specially selected lattice points are color-coded. Layers $hk0$ and $hk2$ are identical, as are all the layers with l even. On the other hand layers $hk1$ and $hk3$ are identical, as are all the layers with l odd.

Now assemble the three layers to form Figure 7.11. The color coding is only to clarify how the cell is assembled. This lattice is face-centered cubic.

A body-centered lattice in direct space transforms into a face-centered reciprocal lattice.

7.6.3 Face-Centered Cubic, Copper

The element copper, Cu, Figure 7.12, has atomic number Z equal to 29 and crystallizes in a face-centered cubic Bravais lattice with $a = 3.615$ Å. There are four atoms in the unit cell. The fractional coordinates of one atom, associated with the vertices of the unit cell, are 0, 0, 0. The fractional coordinates of the other three atoms, associated with the faces of the unit cell, are 0, ½, ½; ½, 0, ½; and ½, ½, 0.

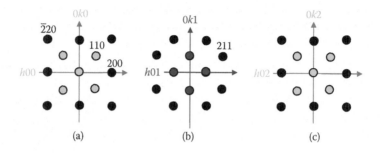

FIGURE 7.10 Reciprocal lattice layers (a) $hk0$, (b) $hk1$, (c) $hk2$ when $h + k + l$ is even. See Figure 7.11 for the color code used to select the reciprocal unit cell.

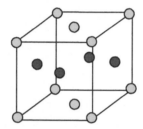

FIGURE 7.11 Reciprocal unit cell hkl, where green lattice points have l even and blue lattice points have l odd.

FIGURE 7.12 Copper ore from Michigan. (Photograph by Maureen M. Julian.)

The structure factor is calculated from Equation 7.5 with $N = 4$

$$F(hkl) = \sum_{j=1}^{4} f_{Cu}\left(\frac{1}{2d_{hkl}}\right) e^{2\pi i \mathbf{H}(hkl)\cdot \mathbf{r}_j} = f_{Cu}\left(\frac{1}{2d_{hkl}}\right)(1 + e^{\pi i(k+l)} + e^{\pi i(h+l)} + e^{\pi i(h+k)}) \quad (7.15)$$

There are two cases: Either hkl are unmixed or hkl are mixed.

For face-centered cubic

Case 1: *hkl* are unmixed. This means *hkl* are all even such as 224 or all odd such as 113. The number 0 is even. The sum of any two even numbers is even $(0 + 2 = 2)$, and the sum of any two odd numbers is also even $(1 + 3 = 4)$. In both cases apply Equation 7.8 to get

$$F(hkl) = f_{Cu}\left(\frac{1}{2d_{hkl}}\right)(1 + 1 + 1 + 1) = 4f_{Cu}\left(\frac{1}{2d_{hkl}}\right)$$

Case 2: *hkl* are mixed. This means that *hkl* are two odd numbers and an even number, such as (231), or two even numbers and an odd number, such as (221). Then

$$F(hkl) = 0$$

In case 2, there is no reflection when *hkl* is mixed. This is known as an extinction.

Look at the *International Tables for Crystallography*, space group No.43, *Fdd2*. The letter "*F*" means body-centered. Under "Positions," the third column gives the "Reflection Conditions" for this space group which is $h + k, h + l, k + l = 2n$. This is the condition for face-centered cubic crystals. There are other reflection conditions, but they do not apply to all *hkl*. These conditions are discussed later. Check space groups No. 225, *Fm3̄m* and No. 227, *Fd3̄m* for their *hkl* reflection conditions.

Apply the Cromer–Mann coefficients for Cu to the atomic scattering factor formula Equation 7.1. Then use Equation 7.15 to calculate the structure factors to produce Table 7.6. The structure factor $F(000) = 116$ electrons (4×29) because there are four Cu atoms in the unit cell. The structure factors with *hkl* mixed are zero. See Exercises 7.20 and 7.21.

Figure 7.13 shows the reciprocal lattice layers *hk*0, *hk*1, and *hk*2 with *hkl* mixed omitted. Layers *hk*0 and *hk*2 are identical, as are all the layers with *l* even. On the other hand, layers *hk*1 and *hk*3 are identical, as are all the layers with *l* odd.

Now assemble the three layers to form Figure 7.14. The color coding is only to clarify how the cell is assembled. This lattice is body-centered cubic.

A face-centered lattice in direct space transforms into a body-centered reciprocal lattice.

Consider the primitive cell in an earlier section. In this case there are no general restrictions on all *hkl*, thus the reciprocal cell is also primitive.

A primitive lattice in direct space transforms into a primitive reciprocal lattice.

Look at the *International Tables for Crystallography*, space group No.2, *P1̄*. The letter "*P*" means primitive. Under "Positions," the third column gives the "Reflection Conditions" and there are no conditions. This is characteristic of the primitive lattice. Check space groups No. 4, $P2_1$, and No. 53, *Pmna*, for their *hkl* reflection conditions.

TABLE 7.6 Structure Factor Calculations for Cu, $a = 3.615$ Å

h	k	l	$\dfrac{\sqrt{h^2 + k^2 + l^2}}{2a}$	$F_{Cu}(hkl)$, Electrons
0	0	0	0	116
0	0	1	0.1383	0
0	1	1	0.1956	0
0	2	0	0.2766	82.89
1	1	1	0.2396	88.31
1	2	0	0.3093	0
1	2	1	0.3388	0
2	2	0	0.3912	67.13
2	2	1	0.4149	0
3	0	0	0.4149	0
3	0	1	0.4374	0
3	1	1	0.4587	59.13
3	2	0	0.4987	0
3	2	1	0.5175	0

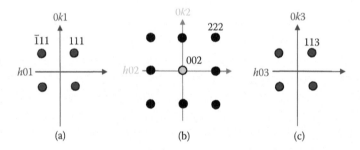

FIGURE 7.13 Reciprocal lattice layers (a) $hk1$, (b) $hk2$, (c) $hk3$ with hkl unmixed. See Figure 7.14 for color code used to select the reciprocal unit cell.

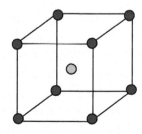

FIGURE 7.14 Reciprocal unit cell hkl, where green lattice points have l even and blue lattice points have l odd.

7.6.4 Crystal with Two Elements, NaCl

Sodium chloride, NaCl, is face-centered cubic. In Figure 7.15 the four chlorine ions are green and have fractional coordinates 0, 0, 0; 0, ½, ½; ½, 0, ½; and ½, ½, 0. There are four sodium ions, as expected to charge-balance the chlorines. The sodium ions are blue and occupy fractional coordinates 0, 0, ½; 0, ½, 0; ½, 0, 0; and ½, ½, ½. There is a total of eight ions. The sodium ion has a formal charge of +1 and the chlorine ion has a formal charge of −1.

The atomic scattering factors for atoms and ions of the same element are different because the atomic scattering factors depend on the number of electrons present.

The structure factor for NaCl is calculated from Equation 7.5 with $N = 8$. Apply Equation 7.9 to simplify the expression to get

$$
\begin{aligned}
F(hkl) &= \sum_{j=1}^{8} f_j \left(\frac{1}{2d_{hkl}} \right) e^{2\pi i \mathbf{H}(hkl)\cdot\mathbf{r}_j} = F_{Cl-}(hkl) + F_{Na+}(hkl) \\
&= f_{Cl-}(1 + e^{\pi i(k+l)} + e^{\pi i(h+l)} + e^{\pi i(h+k)}) + f_{Na+}(e^{\pi i(l)} + e^{\pi i(k)} + e^{\pi i(h)} + e^{\pi i(h+k+l)}) \\
&= (1 + e^{\pi i(k+l)} + e^{\pi i(h+l)} + e^{\pi i(h+k)})(f_{Cl-} + f_{Na+} e^{\pi i(h+k+l)})
\end{aligned}
\tag{7.16}
$$

The first factor on the last line of the equation, $(1 + e^{\pi i(k+l)} + e^{\pi i(h+l)} + e^{\pi i(h+k)})$, indicates that the lattice is face-centered cubic. The systematic extinctions for face-centered cubic Bravais lattices that were derived for the copper structure are universal and apply to all face-centered Bravais lattices. Thus if hkl is mixed then $F(hkl) = 0$. Compare with the calculations for the copper face-centered crystal in Table 7.6.

The chlorine ion has the same fractional coordinates as does the copper atom in the elemental copper crystal. Thus $F_{Cl-}(hkl) = f_{Cl-}(1 + e^{\pi i(k+l)} + e^{\pi i(h+l)} + e^{\pi i(h+k)})$

The sodium ions have $F_{Na+}(hkl) = f_{Na+}(e^{\pi i(l)} + e^{\pi i(k)} + e^{\pi i(h)} + e^{\pi i(h+k+l)})$.

Or alternatively the sodium ions are out of phase with the chlorine ions by $e^{\pi i(h+k+l)}$.

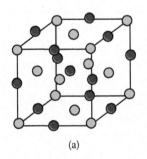

(a)   (b)

FIGURE 7.15 NaCl: (a) unit cell, Cl⁻ green, Na⁺ blue; (b) single crystal of NaCl from a mine in Michigan owned by Morton Salt and collected by Charles Deck and lent to the author by Carla Slebodnick. The crystal is about 7 inches across the bottom. (Photograph by Maureen M. Julian.)

TABLE 7.7   Structure Factor Calculations for NaCl, $a = 5.640$ Å

h	k	l	$1/2d$	$4F_{Cl^-}(hkl)$	$4F_{Na^+}(hkl)$	$F_{NaCl}(hkl)$, **Electrons**
0	0	0	0.0000	72	40	112
0	0	1	0.0887	0	0	0
0	1	0	0.0887	0	0	0
0	1	1	0.1254	0	0	0
0	2	0	0.1773	50.97	34.73	85.70
0	2	1	0.1982	0	0	0
1	0	0	0.0887	0	0	0
1	0	1	0.1254	0	0	0
1	1	0	0.1254	0	0	0
1	1	1	0.1536	54.42	−35.93	18.49
1	2	0	0.1982	0	0	0
1	2	1	0.2172	0	0	0
2	0	0	0.1773	50.97	34.73	85.70
2	0	1	0.1982	0	0	0
2	1	0	0.1982	0	0	0
2	1	1	0.2172	0	0	0
2	2	0	0.2507	42.27	30.54	72.81
2	2	1	0.2660	0	0	0
3	0	0	0.2660	0	0	0
3	0	1	0.2803	0	0	0
3	1	0	0.2803	0	0	0
3	1	1	0.2940	38.54	−27.93	10.61
3	2	0	0.3196	0	0	0
3	2	1	0.3317	0	0	0

Note: $4F_{Na^+}(hkl)$ has been multiplied by $e^{\pi i(h+k+l)}$ to give its sign.

In Table 7.7 the calculation is decomposed into $F_{NaCl}(hkl) = F_{Cl^-}(hkl) + F_{Na^+}(hkl)$. The sodium atoms are exactly in phase or exactly out of phase with the chlorine ions. The reason the waves that are out of phase do not cancel is that the atomic scattering factor for the chlorine ion is much greater than the atomic scattering factor for the sodium ion. The atomic number Z for chlorine is 17. The chlorine ion has a formal charge of plus one so that at hkl equal 000 then f_{Cl^-} is 18 electrons. Likewise the sodium ion has a formal charge of minus one and an atomic number Z of 11 so that at hkl equal 000 then f_{Na^+} is 10 electrons. See Figure 7.16.

The number of electrons in the unit cell is $4 \times 10 = 40$ from the sodium ions and $4 \times 18 = 72$ from the chlorine ions for a total of $F_{NaCl}(000) = 112$ electrons. See Exercise 7.22.

Figure 7.16 shows the atomic scattering factor curves for both the chlorine ion and the sodium ion. The $\sin \theta/\lambda$ values for reciprocal lattice points 111 and 020 are indicated. See Exercise 7.23.

7.6.5 Crystal with Two Elements, Sphalerite

In the examples so far, all the structure factors have been real-valued, as opposed to complex-valued. Structure factors may be complex, with both real and imaginary components,

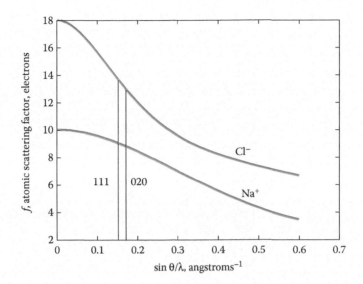

FIGURE 7.16 Atomic scattering factor curves for sodium ions and chlorine ions with values for 111 and 020.

as is seen in sphalerite. Sphalerite, or zinc blende, with the chemical formula ZnS, is the chief ore of zinc. The structure is face-centered cubic with zinc ions at 0, 0, 0; 0, ½, ½; ½, 0, ½; and ½, ½, 0 and sulfur ions at ¼, ¼, ¼; ¾, ¾, ¼; ¾, ¼, ¾; and ¼, ¾, ¾ and $a = 5.41$ Å. The zinc ion has a formal charge of +2 and the sulfur ion has a formal charge of −2.

The structure factor for sphalerite is calculated from Equation 7.5 with $N = 8$. Apply Equation 7.9 to simplify the expression to get

$$
\begin{aligned}
F(hkl) &= \sum_{j=1}^{8} f_j \left(\frac{1}{2d_{hkl}} \right) e^{2\pi i \mathbf{H}(hkl) \cdot \mathbf{r}_j} \\
&= f_{Zn^{++}} \left(1 + e^{\pi i(k+l)} + e^{\pi i(h+l)} + e^{\pi i(h+k)} \right) \\
&\quad + f_{S^{--}} \left(e^{(\pi i/2)(h+k+l)} + e^{(\pi i/2)(3h+3k+l)} + e^{(\pi i/2)(3h+k+3l)} + e^{(\pi i/2)(h+3k+3l)} \right) \\
&= \left(1 + e^{\pi i(k+l)} + e^{\pi i(h+l)} + e^{\pi i(h+k)} \right) \left(f_{Zn^{++}} + f_{S^{--}} e^{(\pi i/2)(h+k+l)} \right)
\end{aligned}
$$

The sulfur ions are out of phase with the zinc ions by $e^{\pi i/2(h+k+l)}$. From Equations 7.10 and 7.11 the phase difference is an imaginary number when $h + k + l$ is odd. Table 7.8 shows the structure factors for sphalerite. See Exercise 7.24.

7.6.6 Structure Factors for Glide Planes

Glide operations also have systematic extinctions. For example, space group No. 7, Pc, has a c glide in the plane of an xz mirror, that is $c\, x, 0, z$. The general position has points (1) x, y, z and (2) $x, \bar{y}, z + (1/2)$. Of course, the atoms must be identical to have the same atomic scattering factor.

TABLE 7.8 Structure Factors for Sphalerite

h	k	l	1/2d	$F_{Zn}(hkl)$	$F_S(hkl)$	$F_{ZnS}(hkl)$
0	0	0	0.0000	120	64	184
0	0	1	0.0924	0	0	0
0	1	0	0.0924	0	0	0
0	1	1	0.1307	0	0	0
0	2	0	0.1848	99.29	−46.11	53.18
0	2	1	0.2067	0	0	0
1	0	0	0.0924	0	0	0
1	0	1	0.1307	0	0	0
1	1	0	0.1307	0	0	0
1	1	1	0.1601	102.96	−49.07i	102.96 − 49.07i
1	2	0	0.2067	0	0	0
1	2	1	0.2264	0	0	0
2	0	0	0.1848	99.29	−46.11	53.18
2	0	1	0.2067	0	0	0
2	1	0	0.2067	0	0	0
2	1	1	0.2264	0	0	0
2	2	0	0.2614	88.32	38.80	127.12
2	2	1	0.2773	0	0	0
3	0	0	0.2773	0	0	0
3	0	1	0.2923	0	0	0
3	1	0	0.2923	0	0	0
3	1	1	0.3065	82.03	35.76i	82.03 + 35.76i
3	2	0	0.3332	0	0	0
3	2	1	0.3458	0	0	0

Two atoms related by a symmetry operation are necessarily identical because they are crystallographically equivalent.

The structure factor calculated from Equation 7.5 for these two points is

$$F(hkl) = \sum_{j=1}^{2} f\left(\frac{1}{2d_{hkl}}\right) e^{2\pi i(hx_j + ky_j + lz_j)} = f\left(\frac{1}{2d_{hkl}}\right) e^{2\pi i(hx + ky + lz)} (1 + e^{-4\pi iky} e^{\pi il})$$

If $k \neq 0$, (and, of course, $y \neq 0$) then $e^{-4\pi iky}$ will have some residual value preventing $1 + e^{-4\pi iky} e^{\pi il}$ ever to extinguish, even if l equal odd.

There are two cases for $k = 0$: Either l is even or l is odd.

Case 1: k is zero and l is even, such as $h04$. Then

$$F(h0l) = f\left(\frac{1}{2d_{h0l}}\right) e^{2\pi i(hx + lz)} (1 + 1) = 2f\left(\frac{1}{2d_{h0l}}\right) e^{2\pi i(hx + lz)}$$

Case 2: k is zero and l is odd, such as $h03$. Then

$$F(h0l) = f\left(\frac{1}{2d_{h0l}}\right)e^{2\pi i(hx+lz)}(1-1) = 0$$

In case 2, there is no reflection. This is known as an extinction.

In general, if a c glide is present, then the reflection conditions are $h0l: l = 2n$. Of course it also follows as a subset that if $h = 0$ then $00l: l = 2n$. This condition is included in the *International Table for Crystallography*. Check space group No. 73, *Ibca* for extinctions due to several glide planes.

7.6.7 Structure Factors for Screw Axes

Screw operations also have systematic extinctions. For example, space group No. 4, $P2_1$, has a screw operation The general position has points (1) x,y,z and (2) $\bar{x}, y + (1/2), \bar{z}$.

The structure factor calculated from Equation 7.5 for these two points is

$$F(hkl) = \sum_{j=1}^{2} f\left(\frac{1}{2d_{hkl}}\right)e^{2\pi i(hx_j+ky_j+lz_j)} = f\left(\frac{1}{2d_{hkl}}\right)e^{2\pi i(hx+ky+lz)}\left(1 + e^{-4\pi i(hx+lz)}e^{\pi ik}\right)$$

If $h \neq 0$ and $l \neq 0$, (and, of course, $x, z \neq 0$) then $e^{-4\pi i(hx+lz)}$ will have some residual value preventing $1 + e^{-4\pi i(hx+lz)}e^{\pi ik}$ for odd k ever to extinguish.

There are two cases when $h = l = 0$: Either k is even or k is odd.

Case 1: $h = l = 0$ and k is even, such as 040. Then

$$F(0k0) = \sum_{j=1}^{2} f\left(\frac{1}{2d_{0k0}}\right)e^{2\pi iky_j} = f\left(\frac{1}{2d_{0k0}}\right)e^{2\pi i(ky)}(1 + e^{\pi ik}) = 2f\left(\frac{1}{2d_{0k0}}\right)e^{2\pi iky}$$

Case 2: $h = l = 0$ and k is odd, such as 030. In this case

$$F(0k0) = f\left(\frac{1}{2d_{0k0}}\right)e^{2\pi i(ky)}(1-1) = 0$$

In case 2, there is no reflection. This is known as an extinction.

This condition is included in the *International Tables for Crystallography*. Check space group No. 19, $P2_12_12_1$ for extinctions due to screw axes.

7.6.8 Structure Factor at the Origin

There is a special case for $hkl = 000$, the undeviated x-ray beam. In general, from Section 7.4.3

$$F(hkl) = \sum_{j=1}^{N} f_j \left(\frac{\sin\theta_{hkl}}{\lambda} \right) e^{2\pi i(hx_j + ky_j + lz_j)}$$

where N is the number of atoms in the unit cell. Then, at $hkl = 000$,

$$F(000) = \sum_{j=1}^{N} f_j(0)$$

The atomic scattering factor at the origin, $f_j(0)$ is equal to the number of electons in the jth atom. Thus $F(000)$ is the total number of electons of all the atoms in the unit cell.

In the polonium crystal there is one atom of polonium in the unit cell, and polonium has atomic number 84; so $F(000)$ is 84 electrons. See Table 7.4. Likewise in the copper crystal there are four atoms in the unit cell, and copper has an atomic number of 29; so $F(000)$ is 116 electrons. As further examples, in Chapters 9 through 15, $F(000)$ was calculated for each of the seven crystals.

- In Section 9.9.1 for DL-leucine (triclinic system), $F(000)$ is 144 electrons.

- In Section 10.9.1 for sucrose (monoclinic system), $F(000)$ is 364 electrons.

- In Section 11.9.1for polyethylene (orthorhombic system), $F(000)$ is 64 electrons.

- In Section 12.9.1 for α-cristobalite (tetragonal system), $F(000)$ is 120 electrons.

- In Section 13.10.1 for $H_{12}B_{12}^{2-}$, 3 K^+, Br^- (trigonal system), $F(000)$ is 164 electrons.

- In Section 14.9.1 for magnesium (hexagonal system), $F(000)$ is 24 electrons.

- In Section 15.9.1 for acetylene (cubic system), $F(000)$ is 56 electrons.

7.7 STRUCTURE FACTORS FOR CENTROSYMMETRIC AND NONCENTROSYMMETRIC CRYSTALS

The presence of an inversion point in the unit cell means that the space group of the crystal contains the symmetry operation $\bar{1}$. These crystals are called centrosymmetric. From Chapter 3, the presence of an inversion point guarantees that these crystals are not piezoelectric. If the origin is placed at the inversion point, then the symmetry operation $\bar{1}$ takes fractional coordinates x, y, z into $\bar{x}, \bar{y}, \bar{z}$. Throughout the rest of this chapter the origin is placed at an inversion point for centrosymmetric crystals.

7.7.1 Partitioning the Space Groups

The space groups can be partitioned into two classes: those with inversion points and those without inversion points.

7.7.1.1 Centrosymmetric Space Groups

Centrosymmetric space groups contain an inversion point. Their general position can be partitioned in two equal parts related by the inversion operation. The *International Tables for Crystallography* list the inversion operation *last* in the list of generators to call the readers' attention to this extraordinarily important operation. For example, look at the cubic space group No. 227, $Fd\bar{3}m$, the space group of diamond. Use origin choice 2. The inversion operation is symmetry operation (25). Symmetry operations (1) through (24) are proper operations. Symmetry operations (25) through (48) are improper operations, each of which can be represented as a product of the inversion operation and one of the 24 proper symmetry operations. For example, multiply symmetry operation (2)—which is represented by the point with coordinates $\bar{x} + (3/4), \bar{y} + (1/4), z + (1/2)$, a twofold operation—by the inversion operation to get symmetry operation (26)—which is represented by the point with coordinates $x + (1/4), y + (3/4), \bar{z} + (1/2)$, a diamond glide.

7.7.1.2 Noncentrosymmetric Space Groups

The noncentrosymmetric space groups do not have an inversion point. These space groups can either have all proper operations or they can be mixed between proper and improper operations.

Because every space group contains the identity, a proper operation, no space group consists entirely of improper operations. An example of a space group with all proper operations is No. 92, $P4_12_12$. This space group has eight symmetry operations—the identity, 4^+, 4^-, and four twofold rotations.

On the other hand, space group No. 43, $Fdd2$, has no inversion point but has improper symmetry operations. This space group is face-centered and its general position has a multiplicity of 16, or 4 symmetry operations per lattice point. The proper operations are the identity and a twofold rotation. The two improper operations are both diamond, or d, glides. The determinants of the associated operations can also be checked as $+1$ for proper operations and -1 for improper operations. See Table 7.9. See Exercises 7.25 and 7.26.

7.7.2 Inversion Points and Structure Factors

Because the symmetry operation $\bar{1}$ takes every fractional coordinate x, y, z into $\bar{x}, \bar{y}, \bar{z}$, Equation 7.5, the structure factor equation, can be written as

$$F(hkl) = \sum_{j=1}^{N} f_j e^{2\pi i(hx_j + ky_j + lz_j)} = \sum_{j=1}^{N/2} f_j e^{2\pi i(hx_j + ky_j + lz_j)} + \sum_{j=1}^{N/2} f_j e^{-2\pi i(hx_j + ky_j + lz_j)}$$

TABLE 7.9 Space Groups, Inversion Points, Proper and Improper Operations

Space Group	Inversion Point	Proper Operations	Improper Operations
No. 227, $Fd\bar{3}m$	Yes	(1) to (24)	(25) to (48)
No. 43, $Fdd2$	No	(1), (2)	(3), (4)
No. 92, $P4_12_12$	No	(1) to (8)	none

Use the identity from Equation 7.6 to get

$$F(hkl) = \sum_{j=1}^{N/2} f_j \cos(2\pi(hx_j + ky_j + lz_j)) + i\sum_{i=1}^{N/2} f_j \sin(2\pi(hx_j + ky_j + lz_j))$$
$$+ \sum_{j=1}^{N/2} f_j \cos(2\pi(-hx_j - ky_j - lz_j)) + i\sum_{j=1}^{N/2} f_j \sin(2\pi(-hx_j - ky_j - lz_j))$$

Since $\cos\theta = \cos(-\theta)$ and $\sin\theta = -\sin(-\theta)$, then

$$F(hkl) = 2\sum_{j=1}^{N/2} f_j \cos(2\pi(hx_j + ky_j + lz_j)) \qquad (7.17)$$

If a crystal has an inversion point and the origin is placed on the inversion point, then its structure factors are all real.

NaCl crystallizes in cubic space group No. 225, $Fm\bar{3}m$, which has an inversion point. Note that in Table 7.7 the structure factors for NaCl are all real. On the other hand, sphalerite, zinc sulfide, crystallizes in space group No. 216, $F\bar{4}3m$ which does not contain an inversion point. Note that in Table 7.8 sphalerite has structure factors with real and imaginary parts.

7.7.3 Hexamethylbenzene (HMB): Example of a Centrosymmetric Crystal

The space group for HMB is $P\bar{1}$, which has two symmetry operations, the identity and the inversion. Hence HMB is centrosymmetric. There are six crystallographically independent carbon atoms in the asymmetric unit. The inversion point generates another six carbon atoms. Equation 7.17 applies. Since all the atoms are carbon atoms, the atomic scattering factor is taken outside the summation sign. For HMB

$$F(hkl) = 2f_C\left(\frac{1}{2d_{hkl}}\right)\sum_{j=1}^{6} \cos(2\pi(hx_j + ky_j + lz_j)) \qquad (7.18)$$

Part of the calculation of $F(100)$ is done in Table 7.10. Here

$$\sum_{j=1}^{6} \cos(2\pi(hx_j + ky_j + lz_j))$$

is equal to 2.219661. From Table 7.2 $f_{100} = 5.5947$.
Thus

$$F(100) = 2 \times 5.5947 \times 2.219661 = 24.84$$

See Exercises 7.27 through 7.30.

TABLE 7.10   Calculation of $\sum_{j=1}^{6} \cos(2\pi(hx_j + ky_j + lz_j))$ for HMB for $hkl = 100$

	x_i	y_i	z_i	$hx_i + ky_i + lz_i$	$\cos(2\pi(hx_i + ky_i + lz_i))$
C_1	0.071	0.182	0	0.071	0.9021
C_2	−0.109	0.073	0	−0.109	0.7745
C_3	−0.18	−0.109	0	−0.18	0.4257
C_4	0.145	0.371	0	0.145	0.6129
C_5	−0.222	0.149	0	−0.222	0.1750
C_6	−0.367	−0.222	0	−0.367	−0.6706
Sum over carbon atoms					2.2196

7.7.4 Generalized Calculation of Structure Factors

The purpose of this section is to combine previously derived expressions to automate the calculations. The steps are

1. In Equation 7.1 the atomic scattering factor, f, is calculated from the Cromer–Mann coefficients as a continuous function of $\sin\theta/\lambda$.

2. Bragg's law is applied to selected points hkl in the reciprocal lattice.

3. From Chapter 5, $d_{hkl} = 1/H_{hkl}$.

4. From Chapter 5, $(H_{hkl})^2 = (h\ k\ l)\mathbf{G}^*(h\ k\ l)^t$

5. Combine information from steps 1 to 4.

$$F(hkl) = \sum_{j=1}^{N} f_j e^{2\pi i(hx_j + ky_j + lz_j)}$$

Figure 7.17 is a flow chart that gives the steps necessary to calculate the structure factors for any crystal.

7.8 ELECTRON DENSITY MAPS

In the front of this chapter is an electron density map of hexamethylbenzene (HMB). One of the goals in crystallography is to produce a similar map for each crystal. An electron density map is calculated by using a Fourier series. The structure factors divided by the volume of the unit cell are the amplitudes in the Fourier series. The Fourier series relates the reciprocal lattice points to the electron density. The electron density map will be calculated for HMB.

7.8.1 Friedel's Law

X-ray diffraction patterns as recorded by x-ray detectors are centrosymmetric. For example, the *crystal structure* of sphalerite, ZnS, does not have an inversion point, as was seen ear-

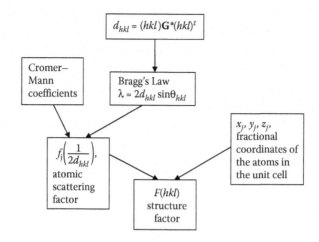

$d_{hkl} = (hkl)\mathbf{G}^*(hkl)^t$

Cromer–Mann coefficients

Bragg's Law $\lambda = 2d_{hkl} \sin\theta_{hkl}$

x_j, y_j, z_j, fractional coordinates of the atoms in the unit cell

$f_j\left(\dfrac{1}{2d_{hkl}}\right)$, atomic scattering factor

$F(hkl)$ structure factor

FIGURE 7.17   Flow chart for calculating the structure factors.

lier in this chapter in Section 7.6.5. However the *diffraction pattern* of ZnS does have an inversion point. Examine the x-ray pattern in Figure 7.18. This is an example of Friedel's law.

Friedel's law states that the intensity of the *hkl* reflection is equal to the intensity of the $\overline{hkl}$ reflection.

The atomic scattering factor will have the same value for both *hkl* and $\overline{hkl}$: Since *hkl* and $\overline{hkl}$ label planes with equal *d*-spacing on either side of the origin, they have the same Bragg angle, θ.

The structure factor Equation 7.5 can be broken down into real and imaginary parts.

$$F(hkl) = \sum_{j=1}^{N} f_j\left(\frac{1}{2d_{hkl}}\right) e^{2\pi i(hx_j + ky_j + lz_j)}$$

$$F(hkl) = \sum_{j=1}^{N} f_j\left(\frac{1}{2d_{hkl}}\right) \cos 2\pi(hx_j + ky_j + lz_j) + i\sum_{j=1}^{N} f_j\left(\frac{1}{2d_{hkl}}\right) \sin 2\pi(hx_j + ky_j + lz_j)$$

(7.19)

or

$$F(hkl) = A + iB \tag{7.20}$$

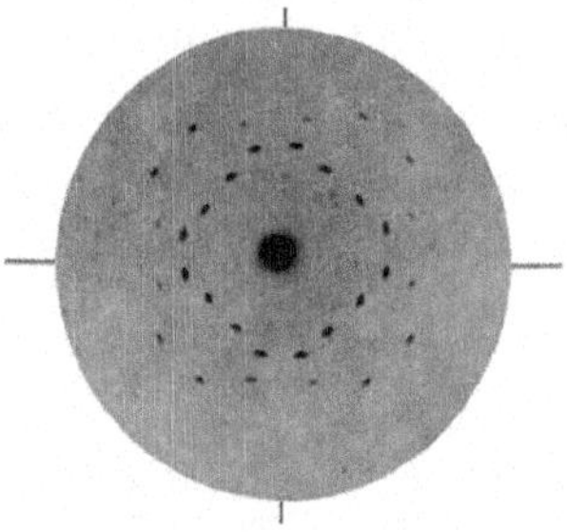

FIGURE 7.18   ZnS x-ray pattern down a fourfold axis. (Ewald, P. P. 1962 *Fifty Years of X-ray Diffraction*, N. V. A. Oosthoek's Uitgeversamaatschappij, Utrecht, The Netherlands.)

where

$$A = \sum_{j=1}^{N} f_j \left(\frac{1}{2d_{hkl}} \right) \cos 2\pi(hx_j + ky_j + lz_j)$$

and

$$B = \sum_{j=1}^{N} f_j \left(\frac{1}{2d_{hkl}} \right) \sin 2\pi(hx_j + ky_j + lz_j)$$

Consider the *hkl* related by an inversion point

$$F(\bar{h}\,\bar{k}\,\bar{l}) = \sum_{j=1}^{N} f_j \left(\frac{1}{2d_{hkl}} \right) \cos 2\pi(hx_j + ky_j + lz_j)$$
$$- i \sum_{j=1}^{N} f_j \left(\frac{1}{2d_{hkl}} \right) \sin 2\pi(hx_j + ky_j + lz_j)$$

(7.21)

$$F(\bar{h}\,\bar{k}\,\bar{l}) = A - iB \tag{7.22}$$

Note that the magnitude of a complex number is equal to the square root of the sum of the squares of the real part and the imaginary part. Thus for $F(hkl) = A + iB$, then

$$|F(hkl)| = \sqrt{A^2 + B^2}$$

Thus

$$|F(hkl)| = |F(\bar{h}\,\bar{k}\,\bar{l})|$$

or

$$|F(hkl)|^2 = |F(\bar{h}\,\bar{k}\,\bar{l})|^2$$

The intensities of the reflections are proportional the squares of the structure factors.

Thus the intensities, *I*, of diffraction maxima related by an inversion point are

$$I(hkl) = I(\bar{h}\,\bar{k}\,\bar{l})$$

The intensities are recorded by the x-ray detectors.

In the case of a centrosymmetric crystal the structure factors are real and Equations 7.20 and 7.22 give

$$F(hkl) = F(\bar{h}\,\bar{k}\,\bar{l})$$

7.8.2 Fourier Series

A periodic function can be expressed as an infinite sum of sine or cosine terms. This sum is called the *Fourier series* after the French mathematician J. B. Fourier (1768–1830). Since a crystal has periodicity, the electron density can be calculated by a complex Fourier series.

The *Fourier coefficients* are the structure factors divided by the volume of the unit cell.

7.8.2.1 One-Dimensional Fourier Series

A one-dimensional analog is considered here to prepare for the three-dimensional case. The volume becomes a length, the repeat distance a. The electron density $\rho(xa)$ in electrons per repeat unit length is given by the Fourier series

$$\rho(xa) = \sum_{all\,h} \frac{F(h)}{a} e^{-2\pi ihx} \tag{7.23}$$

where x is the fractional coordinate, a is the lattice parameter in angstroms, and xa is the linear coordinate in angstroms. The values of h are integers from minus infinity to plus infinity. $F(h)$ is the structure factor.

Or, using Equation 7.6,

$$\rho(xa) = \sum_{all\,h} \frac{F(h)}{a} \cos(2\pi hx) + i\sum_{all\,h} \frac{F(h)}{a} \sin(2\pi hx) \tag{7.24}$$

If $\rho(xa)$ has an inversion point, then the structure factors are real and Equation 7.24 can be written as

$$\rho(xa)\frac{F(0)}{a} + \sum_{h=1}^{\infty} \frac{F(h)}{a} \cos(2\pi hx) + \sum_{h=1}^{\infty} \frac{F(\bar{h})}{a} \cos(2\pi hx)$$

$$+ i\sum_{h=1}^{\infty} \frac{F(h)}{a} \sin(2\pi hx) - i\sum_{h=1}^{\infty} \frac{F(\bar{h})}{a} \sin(2\pi hx)$$

or

$$\rho(xa) = \frac{F(0)}{a} + 2\sum_{h=1}^{\infty} \frac{F(h)}{a} \cos(2\pi hx) \tag{7.25}$$

Note that $F(h) = F(\bar{h})$ by Friedel's law.

Consider a one-dimensional polonium crystal with a repeat distance of 3.359 Å. The atomic number of Po is 84 which means each atom has 84 electrons. The linear electron density of this crystal is a series of peaks with 84 electrons repeating every 3.359 Å. In general the integral of the one-dimensional electron density over the unit cell is equal to the number of electrons in the unit cell. For one atom in the unit cell, the number of electrons in the unit cell is equal to the atomic number of the element.

$$\int_{-\frac{a}{2}}^{\frac{a}{2}} \rho(xa)d(ax) = Z \tag{7.26}$$

▣ EXAMPLE 7.6

Model a one-dimensional row of polonium atoms with the function

$$\rho(xa) = A + B\cos 2\pi x.$$

Give the structure factors for this model and graph three unit cells. Assume $A/B = 3$.

Solution

This model has an inversion point because $\rho(xa) = \rho(-xa)$. Since the total number of electrons is 84 for a polonium atom, then from Equation 7.26

$$\int_{-\frac{a}{2}}^{\frac{a}{2}} (A + B\cos 2\pi x)d(xa) = a\int_{-\frac{1}{2}}^{\frac{1}{2}} (A + B\cos 2\pi x)dx = Aa = 84$$

The limits of integration are chosen in this way because the polonium atom is centered at the origin and the unit cell has a repeat distance of a. The polonium atom is modeled by a peak at the origin. The value of A/B must be ≥ 1 to keep $\rho(xa)$ positive. Since $A/B = 3$, then $A = 3B = 84/a$ and $B = 84/(3a)$.

Thus

$$\rho(xa) = \frac{84}{a} + \frac{84\cos(2\pi x)}{3a}$$

From Equation 7.25, the structure factors $F(0) = 84$ and $F(1) = 14$ and $F(h) = 0$ for positive integers $h > 1$. Figure 7.19 shows a plot of this model.

(continued)

EXAMPLE 7.6 (continued)

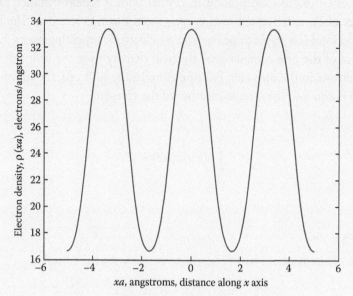

FIGURE 7.19   Three polonium atoms modeled with $\rho(xa) = A + B\cos 2\pi x$.

▪ **EXAMPLE 7.7**

Model a one-dimensional row of polonium atoms with the function

$$\rho(xa) = A + B\cos^4(\pi x).$$

Give the structure factors for this model and graph three unit cells. Assume $A/B = 0.05$.

Solution

This model also has an inversion point. Use the trigonometric identity

$$\cos^4(\pi x) = \frac{1}{8}(3 + 4\cos(2\pi x) + \cos(4\pi x))$$

Thus

$$\rho(xa) = A + \frac{1}{8}B(3 + 4\cos(2\pi x) + \cos(4\pi x))$$

Using Equation 7.26, $(A + 3/8\ B)\ a = 84$. The model requires $A \geq 0$ and $B \geq 0$. Thus

$$\rho(xa) = 25 + 29.42\cos(2\pi x) + 7.36\cos(4\pi x)$$

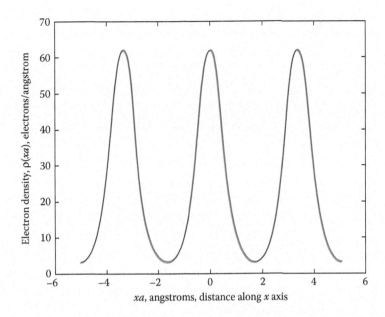

FIGURE 7.20 Three polonium atoms modeled with $\rho(xa) = A + B\cos^4(\pi x)$.

From Equation 7.25, the structure factors $F(0) = 84$, $F(1) = 49.48$, $F(2) = 12.35$ and $F(h) = 0$ for positive integers $h > 2$. Figure 7.20 shows a plot of this model.

See Exercises 7.31 and 7.32.

7.8.2.2 Three-Dimensional Fourier Series

The electron density, $\rho(xa,yb,zc)$, as a function of fractional coordinates xyz is

$$\rho(xa, yb, zc) = \frac{1}{V}\sum_{h=-\infty}^{h=\infty}\sum_{k=-\infty}^{k=\infty}\sum_{l=-\infty}^{l=\infty}F_{hkl}e^{-i2\pi(hx+ky+lz)} \qquad (7.27)$$

where V is the volume, F_{hkl} are the structure factors, and hkl are the indices. HMB has an inversion point. As in Equation 7.25, since $\rho(xa,yb,zc)$ has an inversion point then Equation 7.27 becomes

$$\rho(xa, yb, zc) = \frac{F(000)}{V} + \frac{2}{V}\sum_{h=-\infty}^{h=\infty}\sum_{k=-\infty}^{k=\infty}\sum_{l=1}^{l=\infty}F_{hkl}\cos(2\pi(hx + ky + lz))$$

Figure 7.21 shows HMB sliced through the xy plane. In this case $z = 0$ and so Equation 7.27 becomes

$$\rho(xa, yb, 0) = \frac{F(000)}{V} + \frac{2}{V}\sum_{h=-\infty}^{h=\infty}\sum_{k=-\infty}^{k=\infty}\sum_{l=1}^{l=\infty}F_{hkl}\cos(2\pi(hx + ky)) \qquad (7.28)$$

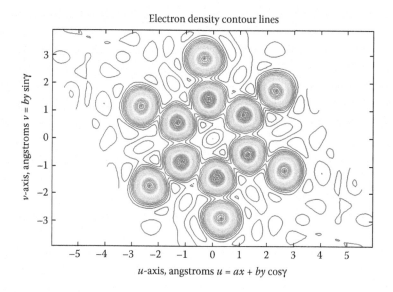

Electron density contour lines

v-axis, angstroms $v = by \sin\gamma$

u-axis, angstroms $u = ax + by \cos\gamma$

FIGURE 7.21   Electron density map of HMB.

Figure 7.17 shows a flow chart for systematically calculating structure factors. Of course, the calculation of the electron density cannot contain an infinite number of terms. For the calculation of Figure 7.21, *h* and *k* were chosen to go from −9 to 9, and *l* from 1 to 6. Another condition was that $d_{hkl} > 1$ Å. These conditions produced 620 reflections.

7.9 MAJOR USES OF STRUCTURE FACTORS

Major uses of the structure factors are the analysis of diffraction data and the construction of the electron density map from diffraction data. In general, these ideas lie outside the scope of this book. However, an overview seems appropriate.

7.9.1 Structure Factors for the Crystal Model

The structure factors for a crystal are used in the comparison of the crystal model with the diffraction data. The crystal model is a specification, made by the crystallographer, of the type and location of every atom in the unit cell. Each type of atom, such as a carbon atom, scatters differently; and the values are indicated by its atomic scattering factor curve. The science of crystallography has a number of methods for making a guess about what the model might be. Sometimes the crystal is part of an isotypic series (Sections 4.13.9 and 13.11), which gives a starting model. However, often with a crystal the model is unknown. Then the crystallographer has a number of ways to make a guess (Glusker, 2010).

Figure 7.22 is a flow chart that shows the crystal model or calculated values on the left side and the experimental or observed data on the right side.

To illustrate the structure factor calculations from the model, the contribution of just one atom to the structure factor of the most intense *hkl* reflection of the x-ray powder pattern is calculated for all seven crystals in Chapters 9 through 15

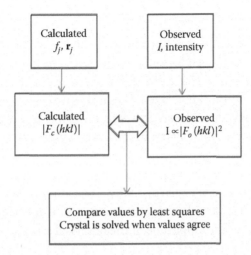

FIGURE 7.22   Flow chart for comparing structure factors between the crystal model and the diffraction data.

- In Section 9.9.2 for DL-leucine (triclinic system)

$$F_{C1}(100) = 5.87\ e^{-0.2162\pi i}\ \text{electrons}$$

- In Section 10.9.2 for sucrose (monoclinic system)

$$F_{C1}(112) = 4.48\ e^{1.124\pi i}\ \text{electrons}$$

- In Section 11.9.2 for polyethylene (orthorhombic system)

$$F_{C1}(110) = 4.77\ e^{-0.202\pi i}\ \text{electrons}$$

- In Section 12.9.2 for α-cristobalite (tetragonal system)

$$F_{O1^{-2}}(101) = 8.11\ e^{0.8358\pi i}\text{electrons}$$

- In Section 13.10.2 for $H_{12}B_{12}{}^{2-}\ 3K^+,\ Br^-$ (trigonal system)

$$F_{B1}(200) = 3.37\ e^{1.748\pi i}\ \text{electrons}$$

- In Section 14.9.2 for magnesium (hexagonal system)

$$F_{Mg1}(101) = 8.6742\ e^{1.1667\pi i}\ \text{electrons}$$

- In Section 15.9.2 for acetylene (cubic system)

$$F_{C1}(111) = 4.4413\ e^{0.3336\pi i}\ \text{electrons}$$

7.9.2 Observed Data and the Phase Factor

In Chapter 5 the reciprocal lattice is introduced to interpret the x-ray diffraction pattern. Figure 5.7 shows an x-ray pattern for quartz with the diffraction maxima labeled as *hkl*. The positions of 100, 020, and 320 are identified.

Each diffraction maximum can have a different relative intensity. The intensity is related to the square of the amplitude of the structure factor. Other issues enter such as the geometry of the diffraction apparatus, polarization factor, temperature factor, and multiplicity. See Suryanarayana (1998).

The important point is that the amplitude of the structure factor can be measured, but the phase factor is not accessible experimentally. Thus, on the right side of Figure 7.22, the top box indicates the experimental or observed value of the intensity. The square root of the intensity is proportional to the amplitude of the observed structure factor.

7.9.3 Comparison of Experimental and Model Structure Factors

The next step is to do a least squares analysis to compare the two sets of data. One data point is the pair of numbers for a specific *hkl*, $|F_c (hkl)|$, and $|F_o(hkl)|$. The amplitude or absolute value of the calculated structure factor is $|F_c (hkl)|$. The experimental or observed value, calculated from the diffraction pattern is $|F_o(hkl)|$. The least squares method uses a weighted sum over all the independent reflections, *hkl*, of $(|F_c (hkl)| - |F_o(hkl)|)^2$.

The *R-value* or *discrepancy index* is a measure of the correctness of the structure and is defined as

$$R = \frac{\Sigma \left| \left(|F_o| - |F_c| \right) \right|}{\Sigma |F_o|}$$

A smaller *R* indicates a better model. If the value of *R* is too large, then either the original model is modified or a new model is applied. A good value of *R* for small organic molecules is less than 0.06. Unfortunately, occasionally, a good *R* is calculated for an incorrect structure and a bad *R* is calculated for a good structure. There are other tests that can be used.

An illustration of a crystal structure is the α-cristobalite crystal in Chapter 14. A total of 146 crystallographically independent reflections were used in the least squares refinement for the two independent atoms. The α-cristobalite crystal had an *R* of 0.029.

7.9.4 Calculation of Electron Density Maps from Experimental Data

A major goal of a crystallographic study is to calculate the *electron density map*, which is an image of the scattering material. Equation 7.27 is

$$\rho(xa, yb, zc) = \frac{1}{V} \sum_{h=-\infty}^{h=\infty} \sum_{k=-\infty}^{k=\infty} \sum_{l=-\infty}^{l=\infty} F(hkl)e^{-i2\pi(hx+ky+lz)}$$

In general, $F_{hkl} = |F_{hkl}|e^{i\varphi}$, where $|F_{hkl}|$ can be experimentally measured but the phase angle, φ, cannot be measured. The phase angle must be known in order to calculate the electron density map. If the phase angle could be measured, then the electron density could be directly calculated and the structure could be deduced from the electron density pattern. Then the crystal model would not be needed.

A clever trick comes next. For each reflection, *hkl*, the structure factor $F_c(hkl)$ is calculated from contributions from all the atoms in the model

$$F_c(hkl) = \sum_{j=1}^{N} f_j e^{2\pi i(hx_j + ky_j + lz_j)} = \left| F_c(hkl) \right| e^{i\phi_c}$$

where ϕ_c is the calculated phase factor.

Now for the calculation of the electron density. The experimental structure factor is calculated by using the experimental or *observed* value of the amplitude with the *calculated* phase angle, ϕ_c, for the crystal model in question. Now the electron density equation, which includes the experimental data, becomes

$$\rho(xa, yb, zc) = \frac{1}{V} \sum_{h=-\infty}^{h=\infty} \sum_{k=-\infty}^{k=\infty} \sum_{l=-\infty}^{l=\infty} \left| F_o(hkl) \right| e^{i\phi_c} e^{-i2\pi(hx + ky + lz)}$$

This is the electron density map reported in the literature.

DEFINITIONS

Atomic form factor
Atomic scattering factor f
Electron density map
Fourier coefficient of an electron density map
Fourier series
Friedel's law
Intensity of a diffracted beam
Phase angle of the structure for atom j
R-value or discrepancy index
Scattering by an electron
Structure factor $F(hkl)$

EXERCISES

⌨ 7.1 Calculate and plot the atomic scattering factor curve for carbon using the Cromer–Mann coefficients in Table 7.1.

⌨ 7.2 If an atom is at fractional coordinates 0.1, 0.7, 0.3, calculate the phase angle for the 114 reflection.

⌨ 7.3 If an atom is at fractional coordinates 0.1, 0.7, 0.3, calculate the phase angle for the $11\bar{4}$ reflection.

7.4 If an atom is at fractional coordinates 1/2, 0, 1/2 calculate the phase angle for the $11\bar{4}$ reflection.

7.5 Calculate the value of the atomic scattering curve for a carbon atom in HMB where $hkl = 010$. Use Table 7.2.

7.6 Calculate the value of the atomic scattering curve for a carbon atom in HMB where $hkl = \bar{1}10$. Use Table 7.2.

7.7 Calculate the value of the atomic scattering curve for a carbon atom in HMB where $hkl = \bar{1}11$. Use Table 7.2.

7.8 Calculate the value of the atomic scattering curve for a carbon atom in HMB where $hkl = 011$. Use Table 7.2.

7.9 Calculate and plot the atomic scattering factor curve for Po using the Cromer–Mann coefficients in Table 7.3.

7.10 Calculate and plot the atomic scattering factor curve for Cr using the Cromer–Mann coefficients in Table 7.3.

7.11 Calculate and plot the atomic scattering factor curve for Cu using the Cromer–Mann coefficients in Table 7.3.

7.12 Calculate and plot the atomic scattering factor curves for both Na^+ and Cl^- superimposed on the same plot using the Cromer–Mann coefficients in Table 7.3.

7.13 Calculate and plot the atomic scattering factor curves for both Zn^{2+} and O^{2-} superimposed on the same plot using the Cromer–Mann coefficients in Table 7.3.

7.14 Calculate and plot the atomic scattering factor curves for K^+ and Cl^- superimposed on the same plot using the Cromer–Mann coefficients in Table 7.3.

7.15 Show that for cubic crystals, $d_{hkl} = a/\sqrt{h^2 + k^2 + l^2}$ Use the method developed in Chapter 5.

7.16 Calculate $F_{Po}(hkl)$ in Table 7.4 for the first 10 hkl for Po.

7.17 Superimpose the $F_{Po}(hkl)$ for the $hkls$ calculated in Exercise 7.16 on the graph plotted in Exercise 7.9.

7.18 Calculate $F_{Cr}(hkl)$, Table 7.5 for the first 10 hkl for Cr.

7.19 Superimpose the $F_{Cr}(hkl)$ for the $hkls$ calculated in Exercise 7.18 on the graph plotted in Exercise 7.10.

7.20 Calculate $F_{Cu}(hkl)$, in Table 7.6 for the first 10 hkl for Cu.

7.21 Superimpose the $F_{Cr}(hkl)$ for the $hkls$ calculated in Exercise 7.18 on the graph plotted in Exercise 7.11.

7.22 Calculate $F_{Cl-}(hkl)$, $F_{Na+}(hkl)$, and $F_{NaCl}(hkl)$ in Table 7.7 for the first 10 hkl.

🖥 7.23 Superimpose the $F_{NaCl}(hkl)$, for the *hkls* calculated in Exercise 7.22 on the graph plotted in Exercise 7.12.

🖥 7.24 Calculate $F_{Zn}(hkl)$, $F_S(hkl)$, and $F_{ZnS}(hkl)$ in Table 7.8 for the first 10 *hkl*.

🖥 7.25 Consider space groups Nos. 4, 12, 14, 15, and 18. Organize them in the same manner as Table 7.9.

🖥 7.26 Consider space groups Nos. 19, 53, 64, 164, and 205. Organize them in the same manner as Table 7.9.

🖥 7.27 Calculate $F(010)$ for HMB.

🖥 7.28 Calculate $F(\bar{1}10)$ for HMB.

🖥 7.29 Calculate $F(\bar{1}11)$ for HMB.

🖥 7.30 Calculate $F(011)$ for HMB.

🖥 7.31 Model a one-dimensional polonium row of atoms with the function $\rho(xa) = A + B\cos 2\pi x$. Give the structure factors for this model and graph three unit cells. Assume $A/B = 7$.

🖥 7.32 Model a one-dimensional polonium row of atoms with the function $\rho(xa) = A + B\cos^4(\pi x)$. Give the structure factors for this model and graph three unit cells. Assume $A/B = 5$.

MATLAB CODE: STARTER PROGRAM FOR CHAPTER 7: GRAPHIC OF ATOMIC SCATTERING CURVE

This program calculates the atomic scattering curves for multiple atoms from the Cromer–Mann coefficients found at http://www.ruppweb.org/xray/comp/scatfac.htm. This example is for bromine and potassium. Enter the atom with the highest Z first.

```
function ScatteringCurves
clc
clear
clear all
%Purpose: Graph atomic scattering factor
%Creates atomic scattering curves for multiple elements from
Cromer-Mann
%coefficients. This case is for Bromine and potassium.
%The elements can be changes and any number can be added.
%web page for Cromer Mann coefficients
%http://www.ruppweb.org/xray/comp/scatfac.htm
x = 0:.0001:.6;
%call Br data';
[a1,b1,c1] = Bromine
%%%%%call CromerMann function
[f1,x] = CromerMann(a1,b1,c1,x);
```

```
plot(x,f1,'r','LineWidth',2)
% create space along the y-axis, put element with highest Z first
ylim([0 f1(1)+3])
ylabel('f, electrons')
xlabel('sin\theta/\lambda, angstroms^-^1')
hold on
%call Potassium data';
[a2,b2,c2]=Potassium
%%%%%call CromerMann function
[f2,x]=CromerMann(a2,b2,c2,x);
plot(x,f2,'g','LineWidth',2)
legend('Bromine','Potassium')
%*****************************************************************
***********
function [a, b,c]=Bromine
%Cromer-Mann coefficients for Br
41.4328 2.9557
a=[17.1789 5.2358 5.6377 3.9851];
b=[2.1723 16.5796 0.2609 41.4328];
c=2.9557;
function [a, b,c]=Potassium
%Cromer-Mann coefficients for K
a=[8.2186 7.4398 1.0519 0.8659];
b=[12.7949 0.7748 213.187 41.6841];
c=1.4228;
function [f,x]=CromerMann(a,b,c,x)
% x=0:.001:.6;
f=0
for i=1:4
f=f+a(i)*exp(-b(i)*x.^2);
end
f=f+c;
```

## CHAPTER 8

Introduction to the Seven Crystals Exemplifying the Seven Crystal Systems

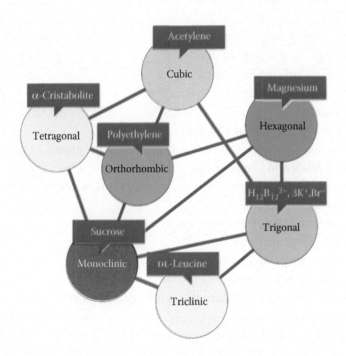

This chapter forms a bridge between the first seven chapters of this book, which give the foundations of crystallography, and the last seven chapters. Each of the last seven chapters is dedicated to a careful examination of one crystal from each of the seven crystal systems. The crystal for the triclinic system is DL-leucine, for monoclinic is sucrose, for orthorhombic is polyethylene, for tetragonal is alpha-cristabolite, for trigonal is $H_{12}B_{12}{}^{2-},3K^+,Br^-$, and for cubic is acetylene.

CONTENTS

CHAPTER OBJECTIVES

- Introduce Chapters 9 through 15, which give seven crystals that exemplify the seven crystal systems.

- Summarize the crystallographic data of each of the seven crystals.

- Explain the use of *parallel topics* and *special topics* in each chapter.

- Describe *parallel topics*, which are reoccurring topics, such as the asymmetric unit.

- Describe *special topics*, which are related to a given crystal, such as neutron diffraction.

- Explain the color-coding of the point group and space group diagrams.

- Give the criteria for choosing the seven crystals.

- Discuss Student Projects.

- Examine the distribution of organic, inorganic, and protein crystals by space group and crystal system.

8.1 INTRODUCTION

Chapters 1 through 7 introduce the foundations of crystallography. Chapters 1 through 4 present the concepts of crystal symmetry: specifically, lattices, unit cell calculations, point groups, and space groups. Chapter 5 discusses the reciprocal lattice, which is the link between the atomic world of crystal periodicity and the experimental world of diffraction patterns. Chapter 6 describes the properties of x-rays and Bragg's law. Finally, Chapter 7 introduces atomic scattering curves, calculates structure factors, and constructs electron-density maps.

This chapter introduces the seven crystals that serve as illustrations of the seven crystal systems, developed in Chapters 9 through 15. Section 8.2 gives the crystallographic data of the seven crystals. Section 8.3 shows how the chapters are constructed with parallel topics and special topics. Section 8.4 reviews the color coding of the general position diagrams and the symbol diagrams. Section 8.5 gives the criteria used for selecting the crystals. Section 8.6 gives suggestions for Student Projects. The chapter concludes with a comparison of organic, inorganic, and protein crystal structures by space group and crystal system.

8.2 CRYSTALLOGRAPHIC DATA FOR THE SEVEN CRYSTAL EXAMPLES

Seven crystals were chosen, one for each of the seven crystal systems. Table 8.1 presents the crystallographic data for the seven crystal examples. The entries in each of the columns are described below.

8.2.1 Column 1: Crystal System

The crystal systems are listed, starting with triclinic and ending with cubic.

8.2.2 Column 2: Space Group, Centrosymmetric Type, Symmorphic Type, Special Projections

Column 2 has the number of the space group as well as the short Hermann–Mauguin symbol. Giving the space group *number* as well as the symbol facilitates finding the space

TABLE 8.1   Crystallographic Data for the Seven Crystal Examples

Crystal System	Space Group [a,b]	Name, Formula, Volume, (Å^3) Z, Z'	Database Code Identifier[c]	Cell Parameters a, b, c Å	α, β, γ°
Triclinic	#2 $P\bar{1}$ C,S p2, p2, p2	DL-Leucine $C_6H_{13}NO_2$ V = 365 Z = 2, Z' = 1	DLLEUC[1]	14.12 5.19 5.39	111.1 86.4 97.0
Monoclinic	#4 $P2_1$ NC, NS pg, pg, p2	Sucrose $C_{12}H_{22}O_{11}$ V = 711 Z = 2, Z' = 1	SUCROS14[1]	7.736(3) 8.702(4) 10.846(5)	90 102.97(2) 90
Orthorhombic	#62 Pnam C, NS p2gg, c2mm, p2gm	Polyethylene $(C_2H_4)_n$ V = 92 Z = 2, Z' = 1/4	QILHUO01[1]	7.400 4.930 2.534	90 90 90
Tetragonal	#92 $P4_12_12$ NC, NS p4gm, p2gg, p2gm	α-Cristobalite SiO_2 V = 171 Z = 4, Z' = 1/2	0001629[2] entry number 1251919[3]	4.971 4.971 6.922	90 90 90
Trigonal (rhombohedral cell)	#166 $R\bar{3}m$ C, S p6mm, p2, p2mm	$H_{12}B_{12}^{2-}, 3K^+, Br^-$ V = 322 Z = 1, Z' = 1/12	GAZLEY[1]	6.871 6.871 6.871	86.38 86.38 86.38
Hexagonal	# 194 $P6_3/mmc$ C, NS p6mm, p2gm, p2mm	Magnesium Mg V = 46 Z = 2, Z' = 1/12	0011183[2]	3.209 3.209 5.210	90 90 120
Cubic	#205 $Pa\bar{3}$ C, NS p2gm, p6, p2gg	Acetylene C_2H_2 V = 227 Z = 4, Z' = 1/6	ACETYL03[1]	6.105 6.105 6.105	90 90 90

[a] Space groups restricted to the space groups in Brief Teaching Edition of the International Tables.

[b] C = centrosymmetric, NC = noncentrosymmetric, S = symorphic, NS = nonsymmorphic, the 2D symmetry of the special projections.

[c] [1]Structures found in the free Interactive Web CSD Teaching Database Demo. [2]Structures found in the free American Mineralogist Crystal Structure Database. [3]Pearson's Crystal Data, Demonstration version.

group in the *International Tables of Crystallography* and uniquely identifies those space groups that have more than one Hermann–Mauguin symbol.

A centrosymmetric space group is indicated by "C," and a noncentrosymmetric space group is indicated by "NC." As examples, DL-leucine and magnesium are centrosymmetric; and sucrose and α-cristobalite are noncentrosymmetric.

A symmorphic space group is indicated by "S," and a nonsymmorphic space group is indicated by "NS." As examples, DL-leucine and the boron compound are symmorphic; and sucrose and magnesium are nonsymmorphic.

The two-dimensional space groups of the three special projections are given. For example, the two-dimensional space groups of the three special projections for α-cristobalite are p4gm, p2gg, and p2gm.

8.2.3 Column 3: Name, Chemical Formula, Volume, Z, Z'

Column 3 has the name and chemical formula of each crystal. The volume of the unit cell is given in cubic angstroms. For example, the smallest volume of a unit cell is 46 Å^3 for magnesium.

The number of molecules (or formula units), Z, in the unit cell and the number of molecules (or formula units), Z', in the asymmetric unit are given. The concept is an experimental one. Once either Z or Z' is known, the other one can be calculated. For example, magnesium has $Z = 2$; so there are two magnesium atoms in the unit cell and $Z' = 1/12$.

8.2.4 Column 4: Database Code Identifier

Only free databases are used. The data for the five molecules—leucine, sucrose, polyethylene, $H_{12}B_{12}{}^{2-}$, $3K^+$, Br^- and acetylene—are found in the Teaching Subset of the CSD (Battle et al., 2010; Battle and Allen, 2012). The data for magnesium are found in the American Mineralogist Crystal Structure Database (Downs and Hall-Wallace, 2003). The data for α-cristobalite are found both in the demonstration version of the Pearson's Crystal Structure Database for Inorganic Compounds (http://www.crystalimpact.com/pcd/) and in the American Mineralogist Crystal Structure Database.

Column 4, the database code identifier, specifies the reference. The CSD uses an alphanumeric code such as SUCROS14 for sucrose. The American Mineralogy Association uses a numeric code such as 0011183 for magnesium. The Pearson's Crystal Structure Database for Inorganic Compounds uses a numeric code such as 1251919 for α-cristobalite.

8.2.5 Column 5: Cell Parameters

Column 5 has the cell parameters a, b, c in angstroms and α, β, γ in degrees. For example, sucrose has $a = 7.736(3)$, $b = 8.702(4)$, $c = 10.846(5)$ Å and $\alpha = 90$, $\beta = 102.97(2)$, $\gamma = 90°$. The number within parentheses refers to the accuracy of the data.

8.3 PRESENTATION OF CRYSTALS IN CHAPTERS 9 THROUGH 15

There are two parts to each chapter: one is the framework of parallel topics and the other consists of special topics that respond to crystallographic challenges. The parallel topics serve as a model for the Student Projects.

8.3.1 Parallel Topics

Chapters 9 through 15 are designed to be read independently or in parallel. When read in parallel, individual topics, such as the asymmetric unit, can be compared and contrasted for any of the seven crystals. All of the topics in the following list are treated in the following order for each of the seven crystals. The Exercises at the end of each chapter follow the order of this list. The section headings found in the seven chapters are in italics in this list.

- *Point group properties*
  - *Multiplication table*
    - Symmetry operations
    - Generators (matrices)
    - Identify if abelian or not
  - *Stereographic projections*
    - General position
      - List of proper operations
      - List of improper operations
    - Symbol
      - Identification of graphical symbols
- *Space group properties*
    - Bravais lattice
    - Symmetry operation requirements for lattice parameters
  - *Space group diagrams*
    - General position diagram—Color coded
    - Symbol diagram—Color coded
  - *Maximal subgroups and minimal supergroups*
  - *Asymmetric unit* superimposed on unit cell
- *Direct and reciprocal lattices*
    - Table of lattice parameters, metric matrix, and volume for both direct and reciprocal lattices
    - Diagram of reciprocal cell superimposed on the unit cell
- Crystal structure analysis
  - Schematic of the atoms and bonds needed to produce the populated cell
  - *Fractional coordinates*, multiplicity, and Wyckoff *letter*
  - *Crystal structure*
    - Diagram of populated unit cell
    - Table of general position coordinates and symmetry operations
    - *Symmetry of the three special projections*

– Diagram showing graphical symbols and asymmetric unit superimposed on each projection

- Diffraction analysis

  - *Reciprocal lattice and d-spacings*

    – Calculate sample *d*-spacing using $\mathbf{G}^*$ matrix

  - Diagram of *powder diffraction pattern*

    – Intensities from database as a function of *d*-spacings

  - *Atomic scattering curves*

    – Diagram containing Cromer–Mann scattering curve for each atomic element

  - *Structure factor*

    – *Calculation of the structure factor at 000*

      – Table of number of electrons in unit cell

    – *Calculation of the contribution of one atom to the structure factor* for the strongest reflection

8.3.2 Special Topics

Each crystal in Chapters 9 through 15 has its own challenges that give opportunities to introduce relevant topics not previously covered in Chapters 1 through 7. Indeed, often a crystal was chosen because of its special characteristics. For example, the element magnesium brings up the subject of close packing.

Table 8.2 lists the special topics that came up in the context of studying particular crystals.

TABLE 8.2 Special Topics

Crystal	Special Topics
DL-Leucine	CIF, chiral molecules, enantiomers, racemic mixtures
Sucrose	Proper point groups and space groups, chiral space groups, scanning electron microscope
Polyethylene	Low-temperature crystallography; short- and full Hermann–Mauguin symbols; relating the Hermann–Mauguin symbol to the cell choice, the asymmetric unit, and special projections
α-Cristobalite	Polygons, enantiomorphic space group pairs, experimental detection of enantiomers
$H_{12}B_{12}^{2-}, 3K^+, Br^-$	Rhombohedral and hexagonal axes, transforming crystallographic directions, space group diagrams for combined hexagonal and rhombohedral cells, boron icosahedron, space-filling model, $H_{12}B_{12}^{2-}, 3K^+, Br^-$: isotypic series
Magnesium	Close-packed structures, model of close-packed spheres, hexagonal close-packed structure (*hcp*), cubic close-packed structure (*ccp*), comparing hcp and ccp structures, interstitial spaces in a close-packed structure, sixfold axes, shifting the origin of the crystal, coordination number 12, site symmetry
Acetylene	Comparison of x-ray, neutron, and electron diffraction; neutron diffraction; neutron structure factors; symbols for threefold symmetry axes parallel to body diagonals

8.4 COLOR-CODING POINT GROUP AND SPACE GROUP DIAGRAMS

An important conceptual contribution of the second edition of this book is the introduction of color to the point group and space group diagrams. Color coding of the symbols according to the change in handedness has been introduced in Chapter 3 (see Section 3.7 and Table 3.15) for the point groups and in Chapter 4 (see Section 4.5) for the space groups. Color-coding clarifies the concepts of the diagrams. There were surprising results associated with the graphical symbols that needed two colors.

This section on color-coding is written for readers who wish to skip the earlier chapters.

The color-coding started out as a way to illuminate the information in the diagrams. A way was needed to test easily and quickly the students' understanding of the diagrams. Black and white figures were handed out to the class to color code as an exercise. As a result, the class became actively involved and comprehended these diagrams more fully.

In the first edition of this book many diagrams were colored to distinguish crystallographically distinct graphical symbols such as the four crystallographically distinct rotations in $p2$ (see Figures 4.14 and 4.15). Colors were also used to distinguish guide lines from crystallographically important symbols (see Table 3.23). Now a big step has been taken to insert more information into the diagrams and to increase student participation.

Figure 8.1 compares the black and white symbol diagram with the color-coded symbol diagram for the point group $6/mmm$. The color-coded symbol diagrams are shown in Figure 8.2 for the space group #92, $P4_12_12$, and for the space group #194, $P6_3/mmc$. The diagram with all red symbols describes the space group #92, $P4_12_12$. It contrasts dramatically with the purple and red diagram for the space group #194, $P6_3/mmc$.

8.4.1 General Position Diagrams

Figure 8.3a displays the point stereographic projection for point group $6/mmm$ in black and white as it appears in the *International Tables for Crystallography*, and Figure 8.3b shows the color-coded diagram. (See also de Graef, 1998.) Notice how much more information the colored diagram presents. The color red is used for the 12 proper operations. The color blue is used for the 12 improper operations. A proper operation does not change

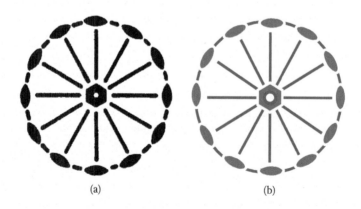

(a)   (b)

FIGURE 8.1   Symbol stereographic projection for point group $6/mmm$: (a) black and white; (b) color-coded.

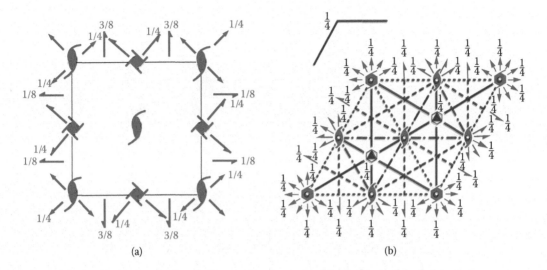

FIGURE 8.2 Symbol diagrams for space groups: (a) #92, $P4_12_12$; (b) #194, $P6_3/mmc$.

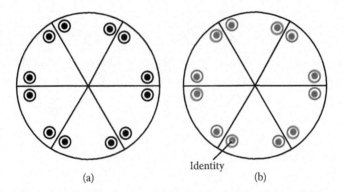

FIGURE 8.3 Point stereographic projections for point group $6/mmm$: (a) black and white; (b) color-coded.

handedness; and an improper operation changes handedness. The information that there are equal numbers of proper and improper operations becomes obvious. This contrasts with the proper point groups which are color-coded all red.

The color coding for the general position diagrams of the space groups is much easier because the notation for the space groups includes a comma to indicate a change of handedness. Figure 8.4 compares the general position diagram for the space group #194, $P6_3/mmc$, in black and white with the color-coded diagram. The color-coding emphasizes the relationships between the symmetry operations. For a space group such as #92, $P4_12_12$, in Figure 8.5, the red coloring emphasizes that there is no change of handedness in the operations.

8.4.2 Symbol Diagrams

The color coding of the symbol diagrams for the point groups and the space groups is more subtle. First, the colors of the symbols are discussed; then they are applied to the symbol diagrams.

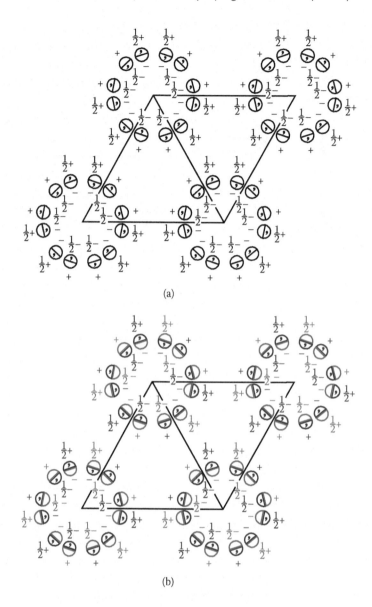

FIGURE 8.4 General position diagrams for space group #194, $P6_3/mmc$: (a) black and white; (b) color-coded.

A *graphical symbol* is a combination of the location of a geometrical object—a point, a line, a plane, or a line with a point—and its related set of symmetry operations. The color red indicates a proper operation, one that does not change handedness. The graphical symbols for the rotation axes are colored red because all their operations are proper. Similarly, the graphical symbols for the screw axes are colored red. The color blue indicates an improper operation, one that does change handedness. The graphical symbol for a mirror plane is colored purple because half its operations are proper (red), and half its operations are improper (blue). Similarly the graphical symbols for the rotoinversions and glides

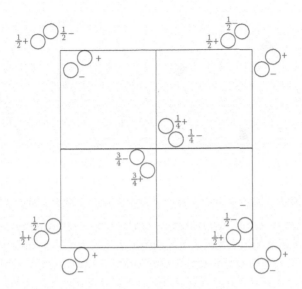

FIGURE 8.5   General position diagram for space group #92, $P4_12_12$.

are colored purple. For a discussion of the graphical symbols see deWolff et al. (1992), Flack et al. (2000), deWolff et al. (1989), and Aroyo (2013, personal communication).

My purpose in color coding is to emphasize the relationship between a general position diagram and its symbol diagram on an intuitive level.

First look at point group $\bar{1}$ (see Figure 8.6). The general position diagram for this point group has one red and one blue symmetry operation. The symbol diagram is interpreted as a purple circle composed of a combination of a red symmetry operation and a blue

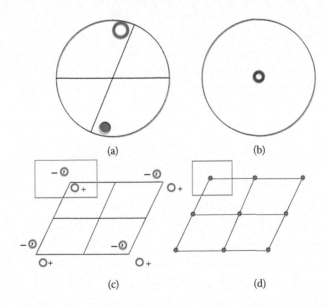

FIGURE 8.6   (a) Point group $\bar{1}$, general position diagram; (b) symbol diagram; (c) space group, #2, $P\bar{1}$, general position diagram; (d) symbol diagram.

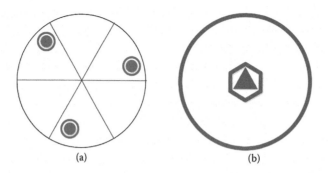

FIGURE 8.7   Point group, $\bar{6}$: (a) general position diagram; (b) symbol diagram.

symmetry operation. Now all inversions are colored purple wherever an inversion point appears in a symbol diagram either a point group diagram or a space group diagram. For example, consider the primitive symmorphic space group, #2, $P\bar{1}$. The relevant areas are outlined in red. Note that the *unit cell* in the general position diagram has one red and one blue symmetry operation.

Now look at the point group $\bar{6}$ in Figure 8.7. The general position diagram for this point group has three red and three blue symmetry operations. The symbol diagram is interpreted as a combination of a purple triangle inscribed in a purple hexagon and an accompanying mirror plane indicated by the large purple circle. These symbols are colored purple wherever they appear in a symbol diagram, either a point group diagram or a space group diagram. Now consider Figure 8.8 that shows the primitive symmorphic space group #174, $P\bar{6}$. The relevant areas are outlined in red. Note that the *unit cell* in the general position diagram has three red and three blue symmetry operations. Also note that the mirror plane, indicated in the upper left-hand corner of the symbol diagram, applies globally to the entire diagram.

In the case of the proper operations, Figure 8.5 shows the general position diagram for space group #92, $P4_12_12$, where all the symmetry operations are colored red; and Figure 8.2a shows the corresponding symbol diagram, where all the symbols are colored red.

In a similar manner the colors are assigned to each of the graphical symbols in Table 1.4.5 on page 9 of the *International Tables for Crystallography*, Volume A, except for the last six lines.

The last six lines in Table 1.4.5 on page 9 of the *International Tables for Crystallography*, Volume A, give a combination of a rotation or screw axis with an inversion point. For these

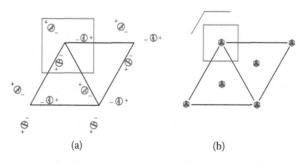

FIGURE 8.8   Space group, #174, $P\bar{6}$: (a) general position diagram; (b) symbol diagram.

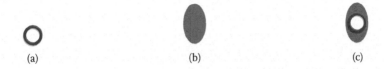

FIGURE 8.9   (a) Purple inversion point; (b) red twofold rotation axis; (c) combination.

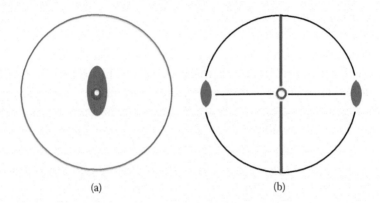

FIGURE 8.10   Point group 2/m: (a) unique axis c; (b) unique axis b.

symbols a purple inversion point is combined with a red rotation or red screw axis to give a two-colored symbol. For example, in Figure 8.9 a purple inversion point is combined with a red twofold rotation axis to give a two-color symbol.

The following example in Figure 8.10 is decisive in demonstrating that two colors must be used. Consider the point group 2/m and compare the diagram having the c-axis as the unique axis with the diagram having the b-axis as the unique axis. When the c-axis is the unique axis, the red twofold axis at the center of the diagram has the purple inversion point superimposed on it. The mirror plane is indicated by the large purple circle. However, when the b-axis is the unique axis, the twofold symbol separates into two red ovals connected by a black guide line. The mirror is represented by the vertical purple line. The inversion point is at the origin.

Tables 3.20 and 4.10 show the graphical symbols for three dimensions.

8.5 CRYSTAL SELECTION CRITERIA

The seven crystals in Chapters 9 through 15 are chosen to appeal to a wide readership. Their data must to be accessible to the readers of this book. That is, the students should be able to replicate the examples. The exercises have to be affordable and have to be small enough to be completed in a reasonable length of time. For these reasons requirements are placed on the space group, the database, the unit cell volume, and the size of the multiplication table.

A balanced choice of relatively ordinary crystals and some somewhat exotic ones is made to appeal to different disciplines. Some crystals are chosen because they introduce special topics. For example, acetylene is chosen because it was studied by neutron diffraction at low temperature. The magnesium structure allows a relatively sophisticated discussion of close packing.

8.5.1 Space Group Requirements

The space groups are restricted to the 24 space groups in the *Brief Teaching Edition of Volume A of the International Tables for Crystallography*. This book is more affordable than the larger volume from which it is extracted.

Unfortunately, many common crystals have to be excluded. For example, α-cristobalite with space group #92, $P4_12_12$, represents the mineral silicates instead of the more common α-quartz with space group #154, $P3_221$.

Furthermore, one task is to construct a multiplication table for the point group associated with the space group. The order of the point group determines the size of the multiplication table. See Chapter 3. The maximum order of the point groups in this book is 24. There is only one hexagonal space group, #194, $P6_3/mmm$. That space group has an associated point group of $6/mmm$ with order 24; so that became the maximum order in this book. In the cubic crystal system, acetylene, with order 24, was chosen over halite, NaCl, with order 48. The multiplication table of acetylene contains 576 entries. NaCl has a multiplication table with 2,304 entries! The Starter Program for Chapter 4 easily creates this large multiplication table, but it is clumsy to handle.

Two space groups were eliminated from the list in the *Brief Teaching Edition* because the associated point groups had order 48. The space groups are #225, $Fm\bar{3}m$, the space group of NaCl and #227, $Fd\bar{3}m$, the space group of diamond. After that elimination there were 22 space groups left.

8.5.2 Database Requirements

Only crystals in free databases are used, specifically, the Teaching Subset of the CSD, the American Mineralogist Crystal Structure Database, and the demonstration version of Pearson's Crystal Structure Database for Inorganic Compounds.

Five of the crystals presented here—leucine, sucrose, polyethylene, a boron compound, and acetylene—are found among the crystals in the teaching subset of the CSD. The mineral, magnesium, is found in the *American Mineralogist* Crystal Structure Database. The cristobalite data are found in the demonstration version of Pearson's Crystal Structure Database for Inorganic Compounds.

8.5.3 Unit Cell Volume Requirements

The volume of the unit cell was limited to restrain the number of atoms in the unit cell. All seven crystals have the volume of the unit cell less than 750 Å^3. Most students can handle volumes up to 2000 Å^3. The crystals suggested at the end of each of Chapters 7 through 15 have unit cell volumes less than 2000 Å^3.

8.5.4 Appealing to a Wide Base of Readers

A major effort was made to select crystals that can appeal to many readers, including biochemists, biologists, ceramicists, chemists, electronic engineers, geologists, materials engineers, materials scientists, metallurgists, mineralogists, physicists, and welders.

- DL-Leucine is an alpha amino acid that is a building block of proteins. In this case it occurs as a racemic mixture in the crystal.

- Sucrose is probably one of the best known and recognized crystals. Indeed, the SEM picture at the beginning of Chapter 10 was taken of ordinary grocery store sugar.

- Polyethylene is an example of a linear polymer.

- Cristobalite is a representative of framework minerals. Polyhedrons, in this case tetrahedrons, are used to organize and analyze the structures.

- The boron complex, containing icosahedrons, is an example of an organometallic compound of interest to those studying the interaction between metals and organic materials.

- The element magnesium is a prototype structure that is a close approximation to the ideal hexagonal close-packed structure.

- Acetylene is a highly flammable gas at ambient temperatures. Low-temperature techniques have been applied to determine its crystal structure. While most crystal structures are done using an x-ray probe, both polyethylene and acetylene were analyzed by neutron diffraction studies.

8.6 STUDENT PROJECTS

Another purpose of the next seven chapters is to prepare for Student Projects. The parallel topics in Section 8.3.1 serve as a model. These projects can be utilized to introduce a multitude of crystals to the class and involving everyone in the presentation adds sparkle. Each Student Project is made up of homework assignments accumulated through the course (see Table 8.3). All the assignments can be done with the aid of the Starter Programs in this book.

The author's most recent crystallography class had 42 students, each doing a different crystal. At the end of the semester each student gives a Power Point presentation at the *Festival of Crystals*. Examples are on the accompanying web site. The students were limited to the CSD, primarily to put the entire class on equal footing. CSD has a selection of organic crystals, polymers, organometallics, metal complexes, peptides up to 24 residues, and molecular

TABLE 8.3   Suggested Time Table for the *Festival of Crystals*

Chapter	Topic
Chapter 1	Select crystal; find *CIF*
Chapter 2	Plot unit cell
Chapter 3	Construct multiplication table; color stereographic projections
Chapter 4	Construct asymmetric unit, populated unit cell, projections, color space group diagrams, and make diagram of maximal subgroups and minimal supergroups
Chapter 5	Compile table comparing direct and reciprocal lattices; draw reciprocal cell superimposed on direct unit cell
Chapter 6	Plot powder diffraction file
Chapter 7	Plot atomic scattering curves; perform structure factor calculations

inorganics that provides a satisfactory variety for the students' interests. At the end of each of Chapters 9 through 15 there is a list of suggested crystals from the free teaching subset of the CSD. Dr. Peter A. Wood of the CSD has been very kind in taking suggestions for crystals from their main database and putting these crystals into the teaching subset.

If working with the author's Starter Programs is not practical, many of the exercises can be done with material directly obtained from the CSD.

8.6.1 Cambridge Structural Database

The Teaching Subset of the CSD is a wonderful resource. In my course, it is the source of the CIF, which is the starting point for a crystal study. Of course, the other databases also give CIFs. The CSD has excellent visual models of the crystals; so the students know where they are heading. Other features used in the course are the labeling of the atoms, a calculated powder diffraction file, and valuable information on the crystal structure, including literature references.

8.6.2 Using the CSD as a Substitute for the Starter Programs

Caffeine is used to illustrate how the CSD can be substituted for the use of the MATLAB® Starter Programs in many but not all of the assignments in this book. The crystal identifier for caffeine is CAFINE. When this identifier is put into the database, the information about caffeine is called up. The outline of the unit cell can be produced by deleting all the molecules.

First, consider the material from the Starter Program in Chapter 4 for populating the unit cell. When caffeine is crystallized from a water solution, it includes a molecule of water in addition to the caffeine molecule. Figure 8.11a shows the molecules taken directly from the CSD. Figure 8.11b shows the unit cell populated with four caffeine molecules and four water (oxygen) molecules. The three-dimensional structure can be rotated to get the projections. Figure 8.7c shows the projection along [001]. This projection can be imported into *Microsoft Windows Paint* (for example), and the unit can be repeated to complete the unit cell until all the two-dimensional symmetries are easily seen.

Alternatively, CSD provides a block of $3 \times 3 \times 3$ unit cells (27 in all) that can be rotated to get the enhanced projection shown in Figure 8.12. Now the exercise of putting in the

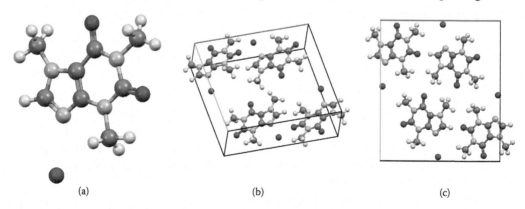

(a)   (b)   (c)

FIGURE 8.11   Caffeine monohydrate from the CSD-CAFINE: (a) molecules; (b) populated cell; (c) projection.

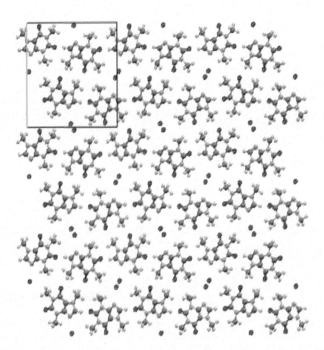

FIGURE 8.12   Caffeine monohydrate from the CSD for nine unit cells.

symmetries of the two-dimensional space groups and the asymmetric unit can be done in *Microsoft Windows Paint* (for example).

Other exercises can be made up with this resource. For example, Figure 8.13 shows the four caffeine molecules and the four water molecules, colored according to the symmetry operations. There are four symmetry operations. The identity is represented by the white

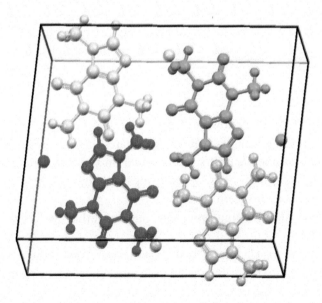

FIGURE 8.13   Caffeine monohydrate from CSD: unit cell with molecules colored by according to the symmetry operations.

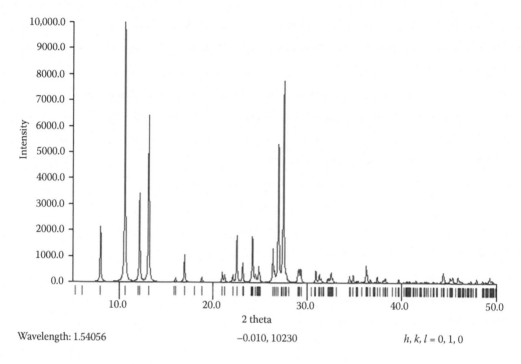

FIGURE 8.14　CSD powder diffraction pattern for caffeine monohydrate.

atoms. The green atoms are related to the white atoms by a twofold screw operation; the yellow atoms are related to the white atoms by an inversion; and the pink atoms are related to the white atoms by a glide.

Next consider the material from the Starter Program in Chapter 6 for making the powder diffraction pattern. In this book the intensities, taken as experimental data, are read from the CSD pattern, along with the *hkl* indexing. Alternatively, the diagram in Figure 8.14 can be imported into *Microsoft Windows Paint* (for example); and the reflections can be labeled by their *hkl* indices.

8.6.3 Approaches and Comments

The multiplication tables, also known as Cayley tables, are difficult to find.

The Starter Program for Chapter 4 greatly facilitates the exercises on the plotting of the three-dimensional asymmetric unit superimposed on the unit cell. However, the plot is easy enough to sketch for many of the space groups with lower symmetry. The two-dimensional space groups are no problem at all.

There are other crystal drawing programs, but the author prefers the transparency of her own programs.

The Starter Program for Chapter 5 facilitates the exercises on the plotting of the three-dimensional reciprocal cell superimposed on the unit cell. These are difficult except for the cubic case. Two-dimensional exercises could be substituted.

The atomic scattering curves from the Starter Program for Chapter 7 can be made with Microsoft Excel.

8.7 DISTRIBUTION OF CRYSTAL STRUCTURES AMONG SPACE GROUPS AND CRYSTAL SYSTEMS

There are exactly 230 ways to create three-dimensional periodic patterns. The 230 space groups were originally derived by the Russian geologist and crystallographer Evgraf Stepanovich Fedorov (1853–1919) working in parallel with the German mathematician Arthur Moritz Schoenflies (1853–1928). Fedorov and Schoenflies published their work separately in 1891. Three years later the English crystallographer William Barlow independently published his work. In other words, the mathematical foundation of space group theory preceded Max Laue's 1912 demonstration of diffraction in the blue copper sulfate pentahydrate crystals, Figure 6.25, by over 20 years. The mathematical tools for classifying crystals were ready for the experimentalists to use them (see Ewald, 1962, Section 3.3.b).

8.7.1 Distribution of Crystal Structures by Space Group

Crystal structures are found in all of the 230 space groups. However, the greatest numbers of crystal structures are concentrated in relatively few space groups. The distributions of three groups are displayed here. The groups are the organic crystals, the largest group; the inorganic crystals including minerals; and finally, the smallest group, the protein crystals.

8.7.1.1 Organic Crystals

Figure 8.15 shows the distribution of the 15 most populated of the space groups among 537,899 structures found in the CSD as of January 1, 2011 at http://www.ccdc.cam.ac.uk/products/csd/statistics/stats_sg11.pdf. The CSD includes organic and metal–organic compounds. The blue bars indicate that these space groups are found in the *International Tables for Crystallography, Brief Teaching Edition of Volume A*. The red bars mean that these space groups are not found there. The symmetry tends to increase from the left, beginning with the triclinic space group $P\bar{1}$ of HMB and ending in the orthorhombic space group #62, *Pnma,* illustrated by polyethylene. Note that 35% of the organic crystals form in the monoclinic space group #14, $P2_1/c$. Examples of organic compounds found in this space group from the student CSD are caffeine, aspirin, citric acid, cortisone (a steroid

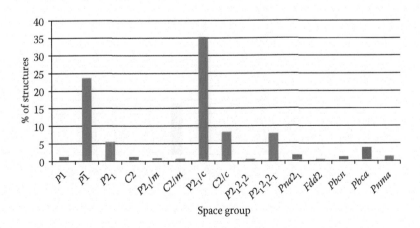

FIGURE 8.15   Distribution of organic crystals by space group. See text.

hormone), diazepam (valium), histidine (an essential amino acid), malic acid (which gives a tart taste to wine), n-butane, ethane, ethylene, boric acid, propane, naphthalene, toluene, DL-valine (an essential α-aminoacid), and vitamin A acid. The organics infrequently form in space groups that are more symmetric than orthorhombic *Pmna*. An exception is acetylene, the crystal chosen to represent the cubic crystal system. For a classic book with an overview of the organic crystals see Robertson (1953) and Mak et al. (1992). Sperling (2001) and Tadokoro (1990) introduce polymers.

8.7.1.2 Inorganic Crystals Including Minerals

Figure 8.16 shows the distribution of the most populated of the space groups among 15,574 inorganic crystals including minerals. Dr. Klaus Brandenburg kindly provided the data from Pearson's Crystal Structure Database for Inorganic Compounds. The blue bars indicate that these space groups are found in the *International Tables for Crystallography, Brief Teaching Edition of Volume A*. The red bar means that this space group is not found there. The symmetry tends to increase from left to right. An example of a low symmetry crystal is the beautiful blue copper sulfate pentahydrate in the triclinic space group #2, $P\bar{1}$. An example of a high symmetry crystal is diamond in the cubic space group #227, $Fd\bar{3}m$. See the American Mineralogist Crystal Database to look up the crystallographic data on a mineral. For classic books describing the crystallography of the minerals see McGraw (1973), Zoltai and Stout (1984), and Klein (1989).

8.7.1.3 Protein Crystals

Proteins and nucleic acids are important macromolecules essential for life (Rupp, 2010). Proteins have one or more polypeptides folded into a globular or fibrous form. Nucleic acids include DNA and RNA. The *Protein Data Bank (PDB)*, created in 1971 (http://www.rcsb.org/pd b/) collects three-dimensional structural data for large biological molecules. These structures are too big to be appropriate for the present book. Figure 8.17 shows the distribution of the most populated of the space groups from the protein database. The blue bars indicate that these space groups are found in the *International Tables for Crystallography, Brief Teaching Edition of Volume A*. The red bars mean that these space groups are not found there.

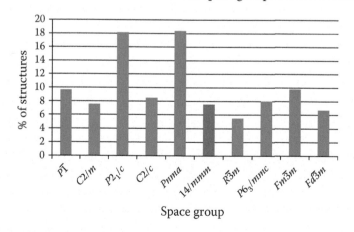

FIGURE 8.16   Distribution of inorganic crystals by space group. See text.

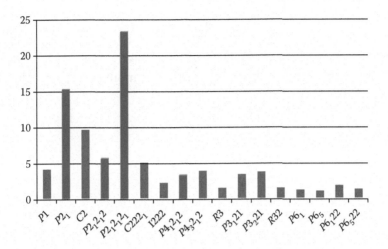

FIGURE 8.17   Distribution of protein crystals by space group. See text.

The biological crystal structures are important structures on the frontier of science. Several special considerations are needed. First, the production, purification, and crystallization of a protein are often particularly difficult. Second, the large biomolecules require not only special experimental apparatus, but also special software for the solution of the structures and special graphics for illustrating the structures. Because naturally occurring biological molecules are chiral, they have a handedness. Thus mirror planes, glide planes, and rotoinversion symmetry operations cannot occur. This restriction of the symmetry operations implies that only 11 of the 32 possible point groups can occur. As a result only 65 of the 230 space groups can occur.

8.7.2 Distribution of Crystal Structures by Crystal System

The data from Figures 8.15 through 8.17 are reinterpreted to produce Figure 8.18 that shows the distribution of organic, inorganic, and protein crystals among the seven

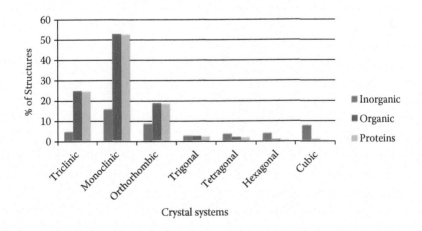

FIGURE 8.18   Distribution of all crystals by crystal system. See text.

crystal systems. The percent of crystals in a given crystal system is plotted. For all three groups, the monoclinic is the most populated, followed by the triclinic and orthorhombic. Next is the cubic system which is almost exclusively populated by the inorganics. In contrast, the inorganic structures are spread out more evenly over the entire seven systems.

## CHAPTER 9

Triclinic Crystal System

DL-Leucine

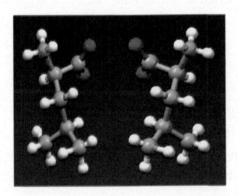

The picture shows two leucine molecules with different handedness. Chiral molecules have opposite handedness and cannot be superimposed on one another. Left and right hands are also chiral. Chiral molecules are of great interest in biochemistry. Only one of the leucine pair is found in nature. Our body often responds differently to molecules with different handedness. One form of carvone is found in caraway seeds and its chiral image is found in wintergreen. Humans can detect these differences because human smell receptors are sensitive to the handedness.

CONTENTS

CHAPTER OBJECTIVES

- Study the crystallographic properties of DL-leucine, space group #2, $P\bar{1}$, as a representative of the triclinic crystal system.

- Present the parallel crystallographic topics that recur in Chapters 9 through 15. A few examples are given below. A complete list is found in Section 8.3.1.

  - Multiplication table of the associated point group

  - Diagrams of the unit cell, the asymmetric unit, the populated unit cell, and the special projections

  - Diagram of the reciprocal cell superimposed on the direct unit cell

  - Calculation of the powder diffraction pattern

  - Calculation of the atomic scattering curves

  - Calculation of the structure factor at 000

  - Calculation of the contribution of one atom to the structure factor of the largest reflection

- Present the special crystallographic topics of this chapter, which are listed below. A complete list of the special topics found in Chapters 9 through 15 is in Section 8.3.2.

  - Crystallographic information files (CIF)

  - Chiral molecules and enantiomers

  - Racemic mixtures

9.1 INTRODUCTION

The example chosen for the triclinic crystal system is DL-leucine. Leucine is an alpha amino acid that is a building block of proteins and is of interest to chemists, biochemists, and biologists.

Leucine is a branched-chain alpha-amino acid necessary for the functioning of the liver and muscle tissue. Although the human body cannot make leucine, plants such as soy beans and peanuts can.

Leucine, $C_6H_{13}NO_2$, is a *zwitterion*, a neutral molecule with a positive and a negative electrical charge at different places within the molecule. See Figure 9.1. The nitrogen atom has charge +1 and one of the oxygen atoms has charge −1. Amino acids contain an ammonium group and a carboxylate group. The amine deprotonates the carboxylic acid in the following intramolecular reaction:

$$(NH_2)RCH(CO_2H) \rightleftharpoons (NH_3)^+RCH(CO_2^-)$$

where for leucine, R is a methyl group, CH_3.

Table 9.1 contains the crystal data for DL-leucine. These data are found in the CIF, in this case from the Cambridge Structural Database (CSD).

9.1.1 Special Topic: Crystallographic Information File, CIF

Lattice parameters and other crystallographic information are found in the *Crystallographic Information File*, CIF (see Section 11.3.1). The CIF has been used since 1990 by the International Union of Crystallography as a file structure for the archiving and distribution of crystallographic information (Brown and McMahon 2002). All crystal structures that are submitted to the crystallographic literature use this format (Julian, 2011). CIFs are found in the CSD and in the American Mineralogist Crystal Structure Database.

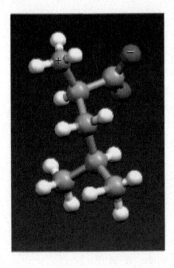

FIGURE 9.1 Leucine zwitterion with a +1 charge on the blue nitrogen atom and a −1 charge on the red oxygen atom.

TABLE 9.1   DL-Leucine Crystal Data

Compound	DL-Leucine
Formula	$C_6H_{13}NO_2$
Molecular weight	131.17 g/mol
Space group	$P\bar{1}$, #2
Bravais lattice	aP
Crystal system	Triclinic
Diffraction radiation probe	X-rays
Ambient temperature for data	Room temperature
Database and identifier	Cambridge Structural Database, DLLEUC
Reference	B. di Blasio, C. Pedone, A. Sirigu, *Acta Crystallographia B*: 1975, 31, 601
Molecular volume	182.844 g/mol
Z, Z'	$Z = 2, Z' = 1$
Density	1.191 g/cm^3 calculated

The *Crystallographic Information File, CIF*, is an electronic archiving file for crystallographic data. It has been widely adopted since 1990 (Brown et al., 2002). After a crystallographer completes a crystal study, the information is put into the standardized form of the CIF. The file is then submitted to the automated IUCr website (http://checkcif.iucr.org/) to be checked for internal consistency.

The CIF can be resubmitted as often as desired. Then the file is submitted with the paper on the structure to an appropriate journal such as *Acta Crystallographica* or the *American Mineralogist*. The format, although carefully controlled, is easy to be read. The CSD provides a CIF for every crystal in its database. Below part of the CIF for DL-leucine is given including two of the atoms (see Exercises 9.1 and 9.2).

```
_database_code_CSD DLLEUC
_chemical_formula_sum 'C6 H13 N1 O2'
_journal_volume 31
_journal_year 1975
_journal_page_first 601
_journal_name_full'Acta
Crystallogr.,Sect.B:Struct.Crystallogr.Cryst.Chem.'
loop_
_publ_author_name
"B.di Blasio"
"C.Pedone"
"A.Sirigu"
_cell_volume 365.688
_exptl_crystal_density_diffrn 1.285
_symmetry_space_group_name_H-M'P-1'
_symmetry_Int_Tables_number 2
loop_
```

```
_symmetry_equiv_pos_site_id
_symmetry_equiv_pos_as_xyz
1 x,y,z
2 -x,-y,-z
_cell_length_a 14.12
_cell_length_b 5.19
_cell_length_c 5.39
_cell_angle_alpha 111.1
_cell_angle_beta 86.4
_cell_angle_gamma 97.0
_cell_formula_units_Z 2
_atom_site_label
_atom_site_type_symbol
_atom_site_fract_x
_atom_site_fract_y
_atom_site_fract_z
C1 C -0.10810 -0.10890 0.44480
C2 C -0.13240 -0.12570 0.72030
```

9.1.2 Special Topic: Chiral Molecules, Enantiomers, and Racemic Mixtures

Figure 9.2 shows two seashells related by a mirror plane that form a pair of chiral images (Davis et al., 2005, p. 80). Left and right hands also form a chiral pair. The word "chiral" comes from the Greek meaning hand. A *chiral molecule* is a molecule whose handedness changes under an improper crystallographic operation to an image that cannot be super-imposed on the original. See Weng et al. (2003).

Two molecules consisting of a chiral molecule and its image under a change of hand-edness are called *enantiomers*. The two seashells are enantiomers. See Exercises 9.3 through 9.5.

If a carbon atom in a molecule has four different atoms or groups of atoms attached to it, then the molecule must be chiral. That special carbon atom is called the *chiral carbon atom*. Figure 9.3 shows the leucine molecule. The chiral carbon atom is labeled with "C." Attached to this carbon are four different groups: –H (inside the white circle),

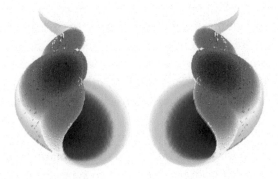

FIGURE 9.2   Chiral parametric seashells.

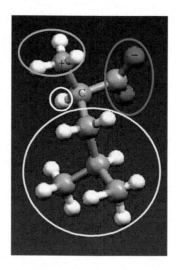

FIGURE 9.3   Chiral leucine molecule.

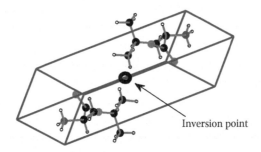

FIGURE 9.4   Chiral leucine molecules related by an inversion point in the crystal DL-leucine.

$-CH2-CH(CH3)_2$ (inside the yellow circle), $-COO^-$ (inside the red circle), and $-NH_3^+$ (inside the blue circle).

A right-handed image is labeled "D" as in the prefix "dextr-" or "dextro-," from the Latin meaning "right". A left-handed image is labeled "L" as in the prefix "lev-" or "levo-," from the Latin meaning "left". Only one of the leucine enantiomorphs, L-leucine, is found in nature.

Synthetic DL-leucine, containing both enantiomers, is a racemic mixture. A *racemic mixture* contains equal amounts of both enantiomers. In DL-leucine both enantiomers are in the same crystal and are related by an inversion point as shown in Figure 9.4. In this crystal Z is equal to two, which means there are two molecules in the unit cell. One is right-handed and is a D-leucine molecule, and the other is left-handed and is an L-leucine molecule. The crystal is called DL-leucine because it contains both enantiomers.

9.2   DL-LEUCINE: POINT GROUP PROPERTIES

The associated point group for DL-leucine is $\overline{1}$ in the Hermann–Mauguin notation and C_i in the Schönflies notation. This point group has two symmetry operations, 1 and $\overline{1}$; thus the order, or number of elements, is two.

TABLE 9.2   Multiplication Table for Point Group $\bar{1}$

	1	$\bar{1}$
1	1	$\bar{1}$
$\bar{1}$	$\bar{1}$	1

9.2.1 Multiplication Table

The generators shown in Appendix 3 are 1 and $\bar{1}$. However, because the point group is cyclic, it can be generated from a single operation, $\bar{1}$, as shown in Figure 3.58. The matrix for this generator is

$$\bar{1} = \begin{pmatrix} \bar{1} & 0 & 0 \\ 0 & \bar{1} & 0 \\ 0 & 0 & \bar{1} \end{pmatrix}$$

The Starter Program from Chapter 3 calculates the multiplication table in Table 9.2. This multiplication table is symmetric because the point group is abelian. It has an inversion, and thus its symmetry does not permit the piezoelectric effect. The relationships among the symmetry operations are seen in the multiplication table.

9.2.2 Stereographic Projections

The general position stereographic projection for the point group $\bar{1}$, Figure 9.5a, has two points corresponding to the two symmetry operations. See Section 3.9 for a general description of these projections. The point corresponding to the identity operation has a positive z-coordinate indicated by the dot, and the point corresponding to the inversion operation has a negative z-coordinate indicated by the open circle. The color red is used for the proper operation 1, and the color blue for the improper operation $\bar{1}$. A proper operation does not change handedness, and the determinant of the matrix is +1. An improper operation changes handedness and the determinant of the matrix is −1. Note that both dots and open circles are positioned interior to the circumference to allow any value for z.

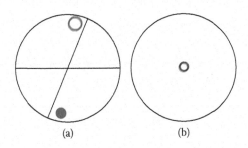

(a)   (b)

FIGURE 9.5   Point group $\bar{1}$ (a) point and (b) symbol stereographic projections.

Figure 9.5b is the symbol stereographic projection. The symbol for the point group $\bar{1}$ is colored purple. The purple symbol indicates that half the operations are proper, or red, and half the operations are improper, or blue.

Given the symbol diagram, the general position diagram can be created and *vice versa*.

9.3 DL-LEUCINE: SPACE GROUP PROPERTIES

DL-leucine crystallizes in the triclinic space group #2, $P\bar{1}$. The Bravais lattice is aP. The symmetry operations put no restrictions on the lattice parameters.

9.3.1 Space Group Diagrams

Figure 9.6 shows the effect of translation combined with the point group in Figure 9.5 to produce the symmorphic space group $P\bar{1}$. In the symbol diagram all the inversion points are indicated by small purple circles. There are eight crystallographically independent inversion points. If an inversion is placed on the origin, then the independent inversion points are at 000, 00½, 0½0, ½00, ½½0, ½0½, 0½½, and ½½½.

In the general position diagram the red circles indicate proper operations and the blue circles with commas indicate improper operations. The plus sign, (+), indicates that the z-coordinate is positive and the minus, (−), indicates that the z-coordinate is negative. Each of the four vertices is a lattice point and has an identical neighborhood. Now check out the operations from the symbol diagram. Given the symbol diagram, the general position diagram can be created and *vice versa*.

9.3.2 Maximal Subgroup and Minimal Supergroups

There are 230 space groups. Because the space groups contain translation operations, each space group has infinite order. Space groups can be arranged in trees similar to the trees of the point groups. However, there is no unique way of constructing the tree for the space groups. In a tree of type I, the centering type is kept and the order of the associated point group is changed. (Maximal subgroups and minimal supergroups are discussed in Section 4.8.) For example, the maximal type I subgroup of the space group #2, $P\bar{1}$, is space group #1, $P1$. Both these space groups have the same centering type, namely P; but for the subgroup the order of the associated point group is decreased. Specifically, the associated point group of the space group #2, $P\bar{1}$, is $\bar{1}$ and has order two while the associated point group of the space

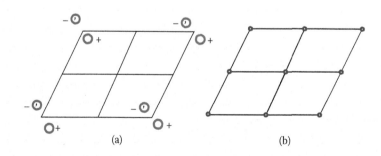

(a) (b)

FIGURE 9.6   Space Group $P\bar{1}$, #2, (a) general position diagram, (b) symbol diagram.

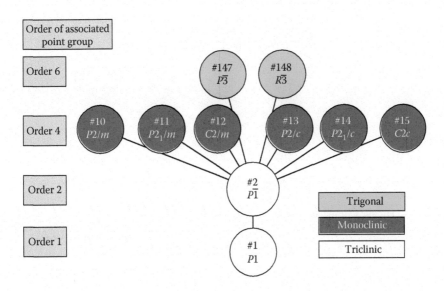

FIGURE 9.7 Space Group #2, $P\bar{1}$, maximal subgroup and minimal supergroups.

group #1, $P1$, is 1 and has order 1. The space group #2, $P\bar{1}$, has eight minimal supergroups, of which six have associated point groups of order 4 and two have associated point groups of order 6. The *International Tables for Crystallography* give the information necessary to construct Figure 9.7.

9.3.3 Asymmetric Unit

The asymmetric unit is the smallest part of the unit cell that, when operated on by the symmetry operations, produces the whole unit cell. The volume of the asymmetric unit is related to the volume of the unit cell by $V_{AU} = (V_{UC}/nh)$, where n is the number of lattice points in the unit cell and h is the number of operations in the associated point group. (Note that the product nh is equal to the multiplicity of the general position.) Because there are two symmetry operations in the associated point group of this primitive cell ($n = 1, h = 2$), the volume of the asymmetric unit is half the volume of the unit cell. Therefore, because the volume of the unit cell is 365.7 Å³, the volume of the asymmetric unit is 182.8 Å³. Figure 9.8 shows the unit cell drawn in blue for DL-leucine with the asymmetric unit in red superimposed on it. The asymmetric unit has

$$0 \leq x \leq 1/2, \quad 0 \leq y \leq 1, \quad \text{and} \quad 0 \leq z \leq 1$$

(Hahn, 2010, p. 90).

9.4 DL-LEUCINE: DIRECT AND RECIPROCAL LATTICES

In addition to the direct lattice, there is the reciprocal lattice, which is the lattice of the diffraction pattern. The diffraction pattern has the intensities of the Bragg reflections superimposed on the reciprocal lattice. Table 9.3 compares the direct and reciprocal lattices including the lattice parameters, the metric matrix, and the cell volume for both the direct

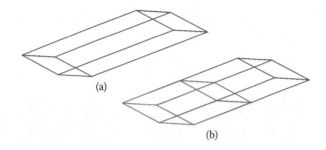

(a)

(b)

FIGURE 9.8   DL-Leucine (a) unit cell in blue (b) with the asymmetric unit in red.

TABLE 9.3   DL-Leucine: Comparison of Direct and Reciprocal Lattices

Direct Lattice	Reciprocal Lattice
$a = 14.12$ Å, $b = 5.19$ Å, $c = 5.39$ Å	$a^* = 0.0714$ Å^{-1}, $b^* = 0.2077$ Å^{-1}, $c^* = 0.1989$ Å^{-1}
$\alpha = 111.1°$, $\beta = 86.4°$, $\gamma = 97.0°$	$\alpha^* = 69.16°$, $\beta^* = 91.17°$, $\gamma^* = 83.88°$

$$\mathbf{G} = \begin{pmatrix} 199.37 & -8.93 & 4.78 \\ -8.93 & 26.9 & -10.07 \\ 4.78 & -10.07 & 29.05 \end{pmatrix} Å^2$$

$$\mathbf{G^*} = \begin{pmatrix} 0.0051 & 0.0016 & -0.0003 \\ 0.0016 & 0.0431 & 0.0147 \\ -0.0003 & 0.0147 & 0.0396 \end{pmatrix} Å^{-2}$$

$V = 365.7$ Å^3 $V^* = 0.0027$ Å^{-3}

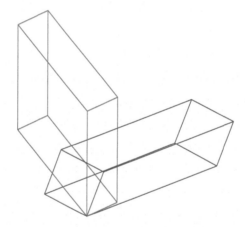

FIGURE 9.9   DL-Leucine direct cell (blue) and reciprocal cell (red) with volumes normalized to 1.

and reciprocal cells. Figure 9.9 shows the relationship between the direct unit cell (blue) and the reciprocal unit cell (red) with the volumes normalized to one.

9.5 DL-LEUCINE: FRACTIONAL COORDINATES AND OTHER DATA FOR THE CRYSTAL STRUCTURE

The crystal structure consists of the placement of the atoms in the unit cell and the bonds between the atoms. The fractional coordinates of the atoms in the asymmetric unit combined with the general position coordinates of the space group and the lattice parameters are sufficient to specify all the atoms in the unit cell. The bonds are discussed later in this section.

TABLE 9.4   DL-Leucine: Fractional Coordinates, Multiplicity, and Wyckoff Letter from the CIF

#	Atom	x Fractional Coordinate	y Fractional Coordinate	z Fractional Coordinate	Multiplicity Wyckoff Letter
1	C1	−0.1081	−0.1089	0.4448	2 i
2	C2	−0.1324	−0.1257	0.7203	2 i
3	C3	−0.2368	−0.2379	0.7416	2 i
4	C4	−0.3091	−0.0391	0.7518	2 i
5	C5	−0.4082	−0.2013	0.6827	2 i
6	C6	−0.3126	0.1902	1.0236	2 i
7	H1	−0.1157	0.0543	0.8464	2 i
8	H2	−0.4109	−0.3191	0.5273	2 i
9	H3	−0.3575	0.3163	1.0156	2 i
10	H4	−0.2543	0.2773	1.0730	2 i
11	H5	−0.3353	0.0933	1.1755	2 i
12	H6	−0.0749	−0.3153	0.9487	2 i
13	H7	−0.0086	−0.2696	0.7401	2 i
14	H8	−0.0899	−0.4936	0.6626	2 i
15	H9	−0.2503	−0.4201	0.5873	2 i
16	H10	−0.2459	−0.2818	0.9109	2 i
17	H11	−0.2865	0.0454	0.6123	2 i
18	H12	−0.4600	−0.0671	0.6996	2 i
19	H13	−0.4303	−0.2994	0.8727	2 i
20	N1	−0.0710	−0.3179	0.7663	2 i
21	O1	−0.1128	0.1154	0.4147	2 i
22	O2	−0.0871	−0.3291	0.2673	2 i

Z equals two or, in other words, there are two leucine molecules in the unit cell. Z' equals one; that is to say, there is one leucine molecule in the asymmetric unit. The fractional coordinates of the atoms for the asymmetric unit from the CIF are given in Table 9.4.

Table 9.5 gives the symmetry operations that are needed to populate the unit cell. There are 22 atoms in the molecule, all of which are in the asymmetric unit. These crystallographically independent atoms all have multiplicity 2 and have Wyckoff letter i, that is, are in the general position.

Now the bonds must be inserted. Figure 9.10 shows how the atoms in the asymmetric unit are connected using the labeling from Table 9.4. For emphasis, the red box outlines the asymmetric unit. In the Starter Program for populating the unit cell, this diagram is the key to putting in the intramolecular bonds.

TABLE 9.5   DL-Leucine General Position Coordinates and Symmetry Operations from CIF

#	General Position Coordinates	Symmetry Operations
1	x, y, z	1, identity
2	$\bar{x}, \bar{y}, \bar{z}$	$\bar{1}$, inversion operation

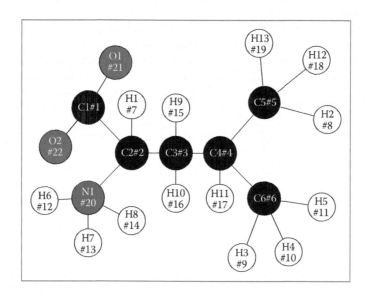

FIGURE 9.10   DL-Leucine bond schematic with atoms in the asymmetric unit outlined in red.

9.6 DL-LEUCINE: CRYSTAL STRUCTURE

Populate the unit cell by using the Starter Program from Chapter 4 to create the entire unit cell as shown in Figure 9.11a. This program combines the lattice parameters—*a, b, c,* α, β, and γ—(Table 9.3), the fractional coordinates (Table 9.4), the general position coordinates of the space group (Table 9.5), and the configuration of bonds (Figure 9.10). The lattice parameters, the fractional coordinates, and the general position coordinates are from the CIF.

The two symmetry operations, the identity and the inversion, are shown in Table 9.5. The fractional coordinates in Table 9.4, for all 22 atoms in the asymmetric unit, represent the identity with coordinates x, y, z in Table 9.5. Now the inversion operation is applied, producing the second molecule in the unit cell with coordinates $\bar{x}, \bar{y}, \bar{z}$. The unit cell has 44 atoms.

Notice how some of the atoms protrude through the walls of the unit cell. Also the molecules in a single cell appear to be sparsely packed. In Figure 9.11b there are three unit cells populated with molecules.

9.6.1 Symmetry of the Special Projections

The *International Tables for Crystallography* give the symmetry for three special projections for each space group in the standard orientation. These projections are orthogonal, which means that each projection is onto a plane normal to the direction of the projection. The basis vectors of the two-dimensional unit cell are labeled **a′** and **b′**, regardless of which two basis vectors of the three-dimensional unit cell (selected from **a, b, c**) are projected. For

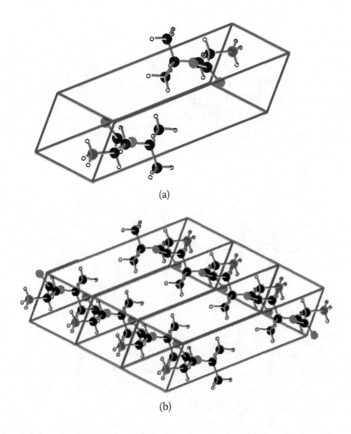

(a)

(b)

FIGURE 9.11   DL-leucine populated unit cell (a) one unit cell, (b) three unit cells.

the triclinic crystal system the special projections are along [001], [100], and [010]. In the Starter Program there is a "view" statement that selects the appropriate projection.

For the projection along [001], $\mathbf{a}_p$ is the projection of $\mathbf{a}$ onto the plane normal to [001] and likewise $\mathbf{b}_p$ is the projection of $\mathbf{b}$ onto the plane normal to [001]. For the projection along [100], $\mathbf{b}_p$ is the projection of $\mathbf{b}$ onto the plane normal to [100] and likewise $\mathbf{c}_p$ is the projection of $\mathbf{c}$ onto the plane normal to [100]. For the projection along [010], $\mathbf{a}_p$ is the projection of $\mathbf{a}$ onto the plane normal to [010], and likewise $\mathbf{c}_p$ is the projection of $\mathbf{c}$ onto the plane normal to [010]. This is the notation of the *International Tables for Crystallography*. Note especially that $\mathbf{a}_p$ for the projection along [001] is not equal to $\mathbf{a}_p$ for the projection along [010] and similarly for $\mathbf{b}_p$ and $\mathbf{c}_p$. Note also that in the triclinic system there is no shift in origin for any of the special projections. In $P\bar{1}$, the twofold rotation axes in special projections are on the vertices of the projections of the three-dimensional unit cell. See Figures 9.11 through 9.13. A careful examination of these projections can demonstrate that the atoms and bond lengths have been correctly implemented.

Quite generally, every two-dimensional projection has the symmetries of one of the 17 two-dimensional space groups.

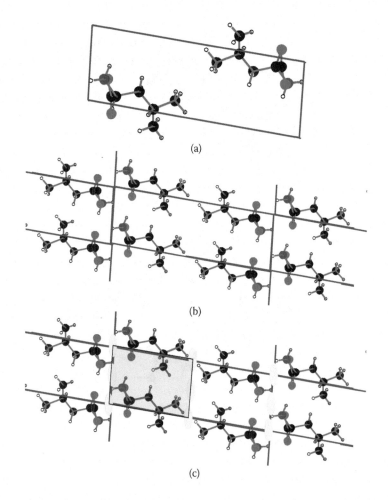

(a)

(b)

(c)

FIGURE 9.12 DL-Leucine projection along [001]: (a) unit cell, (b) enhanced unit cell, (c) with symmetries.

Along [001]

The two-dimensional space group along [001] is $p2$, where

$$\mathbf{a}' = \mathbf{a}_p \qquad \mathbf{b}' = \boldsymbol{b}_p$$

with the origin at 0, 0, z. Figure 9.12a is the direct output from the Starter Program. Here Z is equal to 2, and the two leucine molecules are clearly seen. The enhanced diagram in Figure 9.12b is obtained by repeatedly translating Figure 9.12a to form an array of adjacent figures in order to exhibit the symmetries of $p2$. This diagram shows clearly that the four vertices are lattice points and therefore have identical neighborhoods. Note that the opposite sides of the parallelogram are translations of one another and thus also have identical neighborhoods. Now examine the pattern for symmetries. The twofold symmetry operations are at each of the four vertices. Additionally three more independent twofolds can be located. Hence, as always when done correctly, all the information necessary to create the

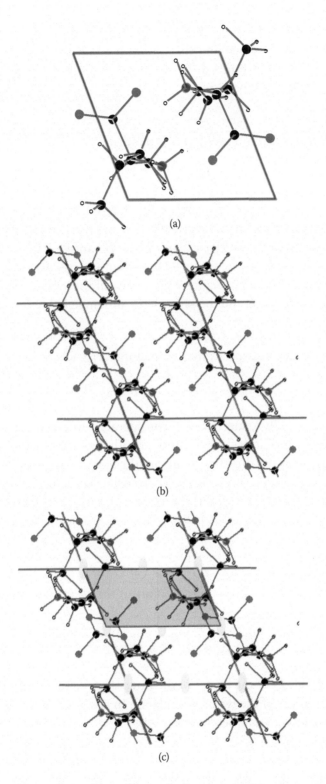

FIGURE 9.13 DL-Leucine projection along [100]: (a) unit cell, (b) enhanced unit cell, (c) with symmetries.

entire pattern is found in the unit cell. Figure 9.12c shows all the symmetry operations for the two-dimensional space group and the asymmetric unit in light blue (see Section 4.4.3). All the atoms of a leucine molecule can be accounted for in the asymmetric unit, although they do not all belong to the same molecule.

Along [100]
The two-dimensional space group along [100] is *p*2, where

$$\mathbf{a}' = \mathbf{b}_p \qquad \mathbf{b}' = \mathbf{c}_p$$

with the origin at *x*, 0, 0. Note the considerable difference in the appearance of the two molecules in this projection in contrast with their appearance along [001]. Figure 9.13a is the direct output from the Starter Program. The enhanced diagram in Figure 9.13b is obtained by repeatedly translating Figure 9.13a to form an array of adjacent figures in order to exhibit the symmetries of *p*2. Figure 9.13c shows all the symmetry operations for the two-dimensional space group and the asymmetric unit in light blue.

Along [010]
Finally, the two-dimensional space group along [010] is *p*2, where

$$\mathbf{a}' = \mathbf{c}_p \qquad \mathbf{b}' = \mathbf{a}_p$$

with the origin at 0, *y*, 0. Again note the considerable difference in the appearance of the two molecules in this projection in contrast with their appearance along either [001] or [100]. Figure 9.14a is the direct output from the Starter Program. The enhanced diagram in Figure 9.14b is gotten by repeatedly translating Figure 9.14a to form an array of adjacent figures in order to exhibit the symmetries of *p*2. Figure 9.14c shows all the symmetry operations for the two-dimensional space group and the asymmetric unit in light blue.

9.7 DL-LEUCINE: RECIPROCAL LATTICE AND *d*-SPACINGS

Planes in the direct lattice correspond to points of the reciprocal lattice. The reciprocal vector

$$\mathbf{H}(hkl) = h\mathbf{a}^* + k\mathbf{b}^* + l\mathbf{c}^*,$$

where *hkl* are the indices of the planes of the direct lattice. The *d*-spacing is the distance from the origin to either of the two closest planes. In general, the *d*-spacing can be calculated by the following equation (see Section 5.7.1):

$$\frac{1}{d_{hkl}^2} = H^2(hkl) = (h \quad k \quad l)\mathbf{G}^* \begin{pmatrix} h \\ k \\ l \end{pmatrix}$$

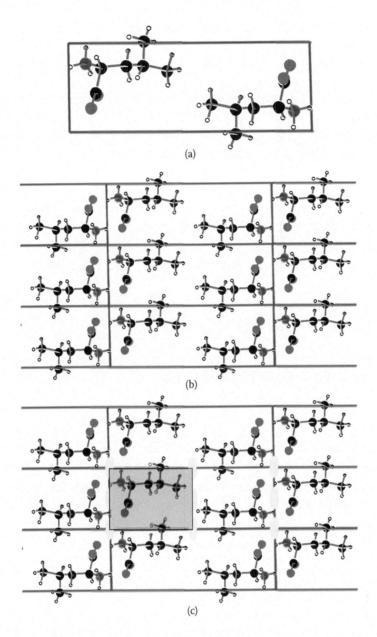

FIGURE 9.14 DL-Leucine Projection along [010]: (a) unit cell, (b) enhanced unit cell, (c) with symmetries.

The powder diffraction pattern in Figure 9.15 gives $hkl = 100$ as the reflection with the maximum intensity. The calculation of the d-spacing for this reflection is as follows:

$$\frac{1}{d_{100}^2} = H^2(100) = (1 \quad 0 \quad 0)\begin{pmatrix} 0.0051 & 0.0016 & -0.0003 \\ 0.0016 & 0.0431 & 0.0147 \\ -0.0003 & 0.0147 & 0.0396 \end{pmatrix}\begin{pmatrix} 1 \\ 0 \\ 0 \end{pmatrix} \tag{9.1}$$

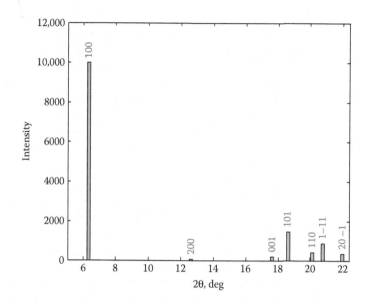

FIGURE 9.15   DL-Leucine powder diffraction pattern.

Now the d-spacing for the 100 can be calculated

$$d_{100} = 14.01 \text{ Å}. \tag{9.2}$$

9.7.1 Powder Diffraction Pattern

The diffraction angle is now calculated as a function of hkl. First, Bragg's law is applied to the d-spacing. Bragg's law is

$$\lambda = 2\, d_{hkl} \sin \theta_{hkl}$$

where λ is the wavelength of the incoming x-rays, d_{hkl} is the d-spacing of the hkl plane, and θ_{hkl} is the Bragg angle for the hkl diffraction maximum. The diffraction angle, $2\theta_{hkl}$, is twice the Bragg angle (see Section 6.6.3). Table 9.6 has hkl indices and intensities for the 10 most intense reflections calculated by the program *Mercury* of the CSD. The intensities are normalized so that the highest intensity is 10,000 arbitrary units. This information from Table 9.6 is the input for the Starter Program in Chapter 6. Figure 9.15 is the resulting powder diffraction pattern for DL-leucine.

9.8 DL-LEUCINE: ATOMIC SCATTERING CURVES

Figure 9.16 shows the atomic scattering curves from the Starter Program in Chapter 7 for the four elements found in DL-leucine. The atomic scattering curves are calculated from the Cromer–Mann coefficients. The atomic scattering factor (or form factor) is the ratio of the amplitude of the wave scattered by an atom to the amplitude of the wave scattered by

TABLE 9.6   DL-Leucine Intensity versus *hkl*

hkl	Intensity, Normalized to Highest Maxima
100	10,000
200	40
001	200
010	793
101	1467
10$\bar{1}$	491
1$\bar{1}$0	456
110	400
01$\bar{1}$	288
1$\bar{1}$1	867

one electron. Its units are electrons. The atomic scattering curve is a measure of the efficiency of an atom in scattering x-rays.

Bragg's law for (100) is

$$\sin \theta/\lambda = 1/2d = 1/(2 \times 14.012) = 0.0357 \text{ Å}^{-1}. \tag{9.3}$$

This reflection is labeled in Figure 9.16 in red. The values from the atomic scattering curves are

$$f_O(100) = 7.90, f_N(100) = 6.88, f_C(100) = 5.87, \text{ and } f_H(100) = 0.98 \text{ electrons.} \tag{9.4}$$

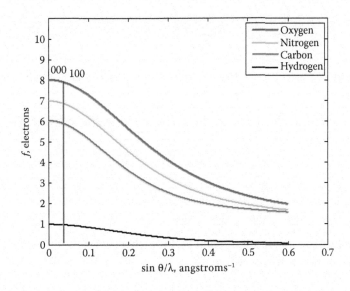

FIGURE 9.16   DL-Leucine atomic scattering curves with 000 and 100 diffraction maxima.

TABLE 9.7   DL-Leucine, $C_6H_{13}NO_2$, Calculation of Number of Electrons in Unit Cell

Atom	**C**	**H**	**N**	**O**	**Total/Molecule**
Atomic number	6	1	7	8	
Atoms/(molecule)	6	13	1	2	22
Total electrons/molecule	36	13	7	16	72
$Z = 2$					**Total/Unit Cell**
Atoms/unit cell	12	26	2	4	44
Electrons/unit cell	72	26	14	32	144

9.9 DL-LEUCINE: STRUCTURE FACTOR

The general formula for the structure factor, from Equation 7.5, is

$$F(hkl) = \sum_{j=1}^{N} f_j \left(\frac{\sin \theta_{hkl}}{\lambda} \right) e^{2\pi i(hx_j + ky_j + lz_j)}$$

where N is the number of atoms in the unit cell. In the case of DL-leucine there are 44 atoms in the unit cell (see Table 9.7). The running index j goes from 1 to 44. Table 9.4 includes the fractional coordinates x_j, y_j, z_j for the jth atom in the asymmetric unit.

The formula for the structure factor for DL-leucine can be broken down in the following way:

$$F(hkl) = F_O(hkl) + F_N(hkl) + F_C(hkl) + F_H(hkl) \tag{9.5}$$

The first term is a sum over 4 oxygen atoms, the second is over 2 nitrogen atoms, the third is over 12 carbon atoms, and the fourth is over 26 hydrogen atoms.

$$F(hkl) = \sum_{j}^{4} F_{Oj}(hkl) + \sum_{j}^{2} F_{Nj}(hkl) + \sum_{j}^{12} F_{Cj}(hkl) + \sum_{j}^{26} F_{Hj}(hkl) \tag{9.6}$$

Now put in the expressions for the individual atomic form factors and the phase factors

$$F(hkl) = \sum_{j}^{4} f_O(hkl)e^{2\pi i(hx_j + ky_j + lz_j)} + \sum_{j}^{2} f_N(hkl)e^{2\pi i(hx_j + ky_j + lz_j)}$$
$$+ \sum_{j}^{12} f_C(hkl)e^{2\pi i(hx_j + ky_j + lz_j)} + \sum_{j}^{26} f_H(hkl)e^{2\pi i(hx_j + ky_j + lz_j)} \tag{9.7}$$

9.9.1 Calculate the Structure Factor at 000

The special case of the undeviated beam occurs when $hkl = (0\ 0\ 0)$. See Example 7.2. Its position is labeled in red in Figure 9.16.

$$F(000) = 4f_O(000) + 2f_N(000) + 12f_C(000) + 26f_H(000)$$

Because the atomic scattering factor, f, is equal to the atomic number, $F(000)$ is the sum of all the electrons in the unit cell (Section 7.6.4). Table 9.7 shows how the total number of electrons is counted in the unit cell of DL-leucine to give

$$F(000) = 144 \text{ electrons} \qquad (9.8)$$

9.9.2 Calculate the Contribution of One Atom to a Structure Factor

The calculation in Equation 9.7 can be sampled by calculating only the contribution of one atom, C1, to the most intense hkl reflection found in the x-ray powder diagram, which is 100. Table 9.4 gives the fractional coordinates for C1 as

$$x = -0.1081, y = -0.1089, z = 0.4448$$

For the reflection 100

$$F_{C1}(100) = f_c(100)e^{2\pi i(-0.1081 + 0(-0.1089) + 0(0.4448))} = 5.87e^{-0.2162\pi i} \text{ electrons} \qquad (9.9)$$

Note that this is a complex number. The entire structure factor, including all 44 atoms, is real because the space group is centrosymmetric (see Section 7.7.2). The importance of the structure factors is that they become the coefficients of the Fourier expansion that is used to calculate the electron density (see Section 7.8.2 and Exercise 9.4).

DEFINITIONS

Chiral carbon atom
Chiral molecule
Crystallographic Information File, CIF
Enantiomers
Racemic mixture
Zwitterion

EXERCISES

9.1 Go to the CSD and get the CIF for DL-leucine. Now submit that file to the *International Union for Crystallography* for verification at http://checkcif.iucr.org/. The website responds with a final report.

9.2 Go to the CSD and get the CIF for a crystal of your choice. Now submit that file to the *International Union for Crystallography* for verification at the website responds with a final report. See Exercise 9.1 for the website.

9.3 Plot one seashell using the MATLAB® program below (Davis, p. 80)

$t = \text{linspace}(0,2*\text{pi},512); [u,v] = \text{meshgrid}(t); a = -0.2; b = 0.5; c = .1; n = 2;$
$x = (a*(1 - v/(2*\text{pi})).*(1 + \cos(u)) + C).*\cos(n*v);$
$y = (a*(1 - v/(2*\text{pi})).*(1 + \cos(u)) + C).*\sin(n*v);$

TABLE 9.8   Crystals with Space Group #1, $P\bar{1}$

Ref Code	Chemical Name	Formula	Volume (Å^3)
FUMAAC01	Fumaric acid	$C_4H_4O_4$	119.25
PAPVAD	Tetramethylethene	C_6H_{12}	149.9
HEXANE01	n-Hexane	C_6H_{14}	159.05
TEPHTH	Terephthalic acid	$C_8H_6O_4$	174.76
OCTANE01	n-Octane	C_8H_{18}	209.39
TARTAM	meso-Tartaric acid	$C_4H_6O_6$	294.54
DLILEU02	rac-2-Amino-3-methylpentanoic acid	$C_6H_{13}NO_2$	347.02
ZONYIK	4-Iodonitrobenzene	$C_6H_4INO_2$	354.63
CECMOM	1-Ethynyl-4-nitrobenzene	$C_8H_5NO_2$	360.1
HEPTAN02	n-Heptane	C_7H_{16}	368.05

$z = b*v/(2*\text{pi}) + a*(1 - v/(2*\text{pi})).*\sin(u)$;
surf(x,y,z,y)
shading interp

Then add its mirror image as in Figure 9.2.

9.4   Plot one seashell using the program from Exercise 9.3. Then add its image using the inversion point as the improper operation.

9.5   Plot one seashell using the parametric equations from Exericse 9.3. Then add its image using an improper operation other than mirror or inversion.

9.6   Suggested crystals from CSD with space group #2, $P\bar{1}$, are listed in ascending order of volume in Table 9.8. The website is http://webcsd.ccdc.cam.ac.uk/teaching_database_demo.php; a shortcut to an individual Reference code is: http://webcsd.ccdc.cam.ac.uk/display_csd_entry.php?identifier=FUMAAC01 for a crystal with Reference Code "FUMAAC01". Construct a paper or a PowerPoint presentation based on the parallel topics in this chapter. The parallel topics are discussed in Section 8.3.1. Examples are given at https://sites. google.com/a/vt.edu/foundations_of_crystallography/.

## CHAPTER 10

Monoclinic System

Sucrose

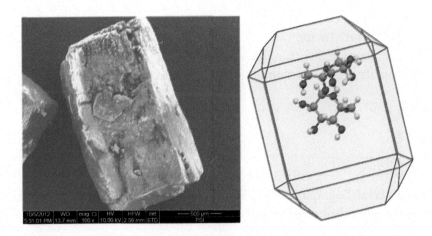

The crystal of sucrose on the left was taken directly from a packet of commercially obtained sugar. The image was created by a scanning electron microscope (SEM). The diagram on the right is an image from the Cambridge Structural Database (CSD) using the Bravais, Friedel, Donnay, and Harker (BFDH) crystal morphology model. The model is calculated from the unit cell parameters along with the point group symmetries. The equilibrium shape is obtained by minimizing the surface energies for the crystal faces at 0 K. A single molecule of sucrose is shown orientated within the crystal. Note that not all the faces calculated by the theoretical model necessarily grow in the actual sucrose crystal.

CONTENTS

CHAPTER OBJECTIVES

- Study the crystallographic properties of sucrose, space group #4, $P2_1$, as a representative of the monoclinic crystal system.

- Present the parallel crystallographic topics that recur in Chapters 9 through 15. A few examples are given below. A complete list is found in Section 8.3.1.

  - Multiplication table of the associated point group

  - Diagrams of the unit cell, the asymmetric unit, the populated unit cell, and the special projections

  - Diagram of the reciprocal cell superimposed on the direct unit cell

  - Calculation of the powder diffraction pattern

  - Calculation of the atomic scattering curves

  - Calculation of the structure factor at 000

  - Calculation of the contribution of one atom to the structure factor of the largest reflection

- Present the special crystallographic topics of this chapter, which are listed below. A complete list of the special topics found in Chapters 9 through 15 is in Section 8.3.2.

- Proper point groups

- Proper space groups

- Scanning electron microscope

10.1 INTRODUCTION

The example chosen for the monoclinic crystal system is sucrose, which is ordinary table sugar. There are two molecules in the unit cell ($Z = 2$) related by a screw symmetry operation. Both these molecules have the same handedness. Table 10.1 contains the crystal data for sucrose. These data are found in the CIF, in this case from the CSD.

10.1.1 Special Topic: Proper Point Groups

Many biologically active molecules such as sugars and proteins are chiral (see Section 9.1.2). If all the chiral molecules in a crystal have the same handedness, then the point groups associated with the crystal can only have proper operations. The proper operations for the crystallographic point groups are found in point groups 1, 2, 3, 4, and 6. A *proper point group* is a point group that has only proper operations. Proper point groups are also called *pure rotational point groups* (Burns and Glazer, 2013, p. 71). Biochemists who deal mainly with chiral molecules often use the term "*chiral point groups*" (Rupp, 2010, p. 227) for the proper point groups. The term "proper point group" is used in this book because molecules or formula units that are not chiral—that is, that do not have a handedness—can form in the proper point groups. An example is α-cristobalite with point group 222 (see Chapter 12).

TABLE 10.1   Sucrose Crystal Data

Compound	Sucrose
Formula	$C_{12}H_{22}O_{11}$
Molecular weight	342.3 g/mol
Space group	$P2_1$, #4
Bravais lattice	*mP*
Crystal system	Monoclinic
Ambient temperature for data	173 K
Radiation	X-ray $\lambda = 0.71073$
Database and identifier	Cambridge Structural Database, SUCROS14
Reference	D.M.M. Jaradat, S. Mebs, L. Checinska, P. Luger, *Carbohydrate Research* 2007, 342, 1480
Molecular volume	711.511 Å^3
Z, Z′	$Z = 2$, $Z′ = 1$
Density	1.598 g/cm
Color	Colorless
Habit	Prism
Size of crystal	$0.40 \times 0.35 \times 0.30$ mm^3
Absorption coefficient	0.143 mm^{-1}

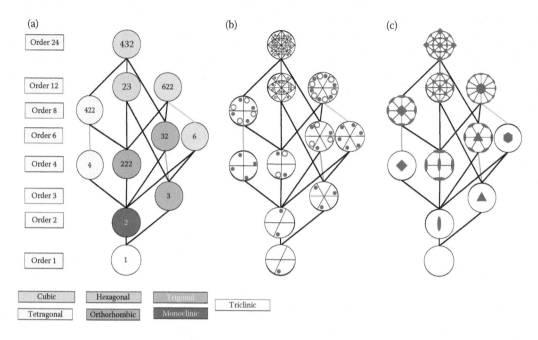

FIGURE 10.1 Proper point group trees: (a) name, (b) general position, and (c) symbol diagrams.

There are 11 proper crystallographic point groups in three dimensions. The Herman–Mauguin symbols for point groups make apparent which are the proper point groups. The symbols for the proper crystallographic point groups are 1, 2, 3, 4, 6, 222, 32, 422, 622, 23, and 432. On the other hand, their respective Schoenflies symbols are C_1, C_2, C_3, C_4, C_6, D_2, D_3, D_4, D_6, T, and O.

Figure 10.1 gives the point group trees, including the Herman–Mauguin symbols, the symbol diagram, and the general position diagram for all three-dimensional proper point groups. Note that the only color for the symbols and the operations is red. Red is the color of the proper operations and of the symbols that represent subgroups containing only proper operations.

10.1.2 Special Topic: Proper Space Groups

A *proper space group* is a space group that has only proper operations. There are 65 proper space groups. Again the biochemists who deal mainly with chiral molecules often use the term *"chiral space groups"* (Rupp, 2010, p. 227) for the proper space groups. The term "proper space group" is used in this book because molecules or formula units that are not chiral—that is, that do not have a handedness—can form in the proper space groups. In addition to the proper point group operations, the proper space group operations include the screw operations (see Section 4.11.4.2). There are 11 screw operations: 2_1, 3_1, 3_2, 4_1, 4_2, 4_3, 6_1, 6_2, 6_3, 6_4, and 6_5. These symbols are included in the Herman–Mauguin symbol for a space group. For example, the symbol for the space group of sucrose is $P2_1$ and indicates the presence of a screw axis.

EXAMPLE 10.1

Construct a table for the three-dimensional Bravais lattice hP that includes all the proper point groups, the corresponding symmorphic space groups, and the nonsymmorphic space groups. Use the web to find a complete list of space groups.

Solution

Bravais Lattice	Proper Point Group	Symmorphic Space Groups	Nonsymmorphic Space Groups
hP	3	$P3$	$P3_1, P3_2$
	32	$P321, P312$	$P3_121, P3_221, P3_112, P3_212$
	6	$P6$	$P6_1, P6_2, P6_3, P6_4, P6_5$
	622	$P622$	$P6_122, P6_222, P6_322, P6_422, P6_522$

Now look at the problem the other way. See what information can be obtained from the Herman–Mauguin symbol for a space group.

EXAMPLE 10.2

Out of the following list of space groups, show which space groups are proper or not proper (P, NP), which are symmorphic or nonsymmorphic (S, NS) (Section 4.11.5); and show the Bravais lattice, (BL), (Figure 6.68) and the associated point group (APG).

$P1$ $P2_1$ $C2/m$ $P2_1/c$ $C2/c$ $P2_12_12$ $Fdd2$ $Ibca$ $P4_12_12$ $P321$ $P6_3$ $I23$

Solution

	P1	$P2_1$	C2/m	$P2_1/c$	C2/c	$P2_12_12$	Fdd2	Ibca	$P4_12_12$	P321	$P6_3$	I23
P,NP	P	P	NP	NP	NP	P	NP	NP	P	P	P	P
S,NS	S	NS	S	NS	NS	NS	NS	NS	NS	S	NS	S
BL	aP	mP	mC	mP	mP	oP	oF	oI	tP	hP	hP	cI
APG	1	2	$2/m$	$2/m$	$2/m$	222	$mm2$	mmm	422	32	6	23

See Exercise 10.2.

10.1.3 Special Topic: Scanning Electron Microscope, SEM

In Figure 10.2 are two pictures of a crystal of sucrose. The scanning electron microscope, SEM, creates the image, not by optical lens, but by scanning the crystal with a focused beam of electrons. The electrons interact with the electrons in the crystal producing signals that can be interpreted in terms of the crystal's surface topography. The magnification of the right-hand photograph is much greater than the photograph on the left. These sugar

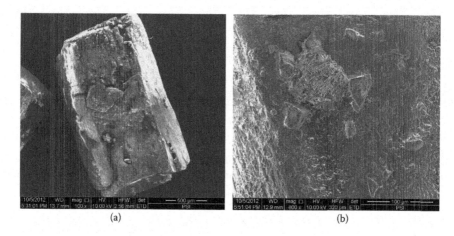

(a)   (b)

FIGURE 10.2 Scanning electron micrographs of sucrose. (a) Overall shape of crystal, (b) surface detail. (Courtesy of William Reynolds and Mitsuhiro Murayama, Materials Science and Engineering Department, Virginia Tech.)

crystals were packed in paper envelopes. A perfectly smooth surface would be defect free. However, under magnification all real crystals show defects. The faces on these crystals are rugged terrains.

10.2 SUCROSE: POINT GROUP PROPERTIES

The associated point group for sucrose is 2 in the Hermann–Mauguin notation and C_2 in the Schönflies notation. This point group has two symmetry operations, 1 and 2; thus the order, or number of elements, is two.

10.2.1 Multiplication Table

The generators shown in Appendix 3 are 1 and 2. However, the point group is cyclic and can be generated from one element, 2, as shown in Figure 3.58. In this chapter, the unique monoclinic axis is by choice the y-axis. The matrix for the generator with the twofold axis along the y-axis is

$$2 = \begin{pmatrix} \bar{1} & 0 & 0 \\ 0 & 1 & 0 \\ 0 & 0 & \bar{1} \end{pmatrix}$$

The multiplication table is symmetric because the point group is abelian. Its multiplication table is isomorphic to that of point group $\bar{1}$ shown in Table 9.2, even though the point groups are different. The relationships among the symmetry operations are seen in the multiplication table in Table 10.2.

TABLE 10.2   Multiplication Table for Point Group 2

	1	2
1	1	2
2	2	1

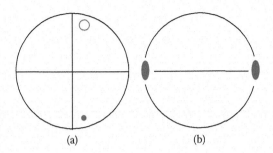

(a)   (b)

FIGURE 10.3   Point group 2 (a) point and (b) symbol stereographic projections with b-axis unique.

10.2.2 Stereographic Projections

In this chapter, the unique monoclinic axis is by choice the y-axis. The general position stereographic projection for the point group 2, Figure 10.3a, has two points corresponding to the two symmetry operations. The point corresponding to the identity operation has a positive z-coordinate indicated by the dot, and the point corresponding to the twofold operation has a negative z-coordinate indicated by the open circle. The color red is used for the proper operations 1 and 2. A proper operation does not change handedness, and the determinant of the matrix is +1. Note that both dots and open circles are positioned interior to the circumference to allow any value for z.

Figure 10.3b is the symbol stereographic projection. The symbol for group 2 is colored red because all the operations are proper. The symbol for the twofold along the y-axis consists of two red ovals on the circumference of the circle joined by a black guide line.

10.3 SUCROSE: SPACE GROUP PROPERTIES

Sucrose crystallizes in the monoclinic space group #4, $P2_1$. The Bravais lattice is mP. The symmetry operations, with the y-axis chosen to be the unique monoclinic axis, require that both the lattice parameters α and γ equal 90°.

10.3.1 Space Group Diagrams

Figure 10.4 shows diagrams for the nonsymmorphic space group $P2_1$ with the unique monoclinic axis, **b**, perpendicular to the plane of the page. All the operations are proper,

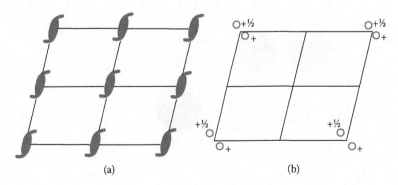

(a)   (b)

FIGURE 10.4   Space Group $P2_1$, #4, (a) symbol diagram, (b) general position diagram. The unique monoclinic axis, **b**, is perpendicular to the plane of the page.

hence the red coloring in both the symbol and general position diagrams. The 2_1 screw axes are along **b**, giving a translation of one-half cell length along **b** in both diagrams. There are four crystallographically independent 2_1 screw axes at $0y0$, $0y\frac{1}{2}$, $\frac{1}{2}y0$, and $\frac{1}{2}y\frac{1}{2}$.

10.3.2 Maximal Subgroup and Minimal Supergroups

There are 230 space groups. Because the space groups contain translation operations, each space group has infinite order. Space groups can be arranged in trees similar to the trees of the point groups. However, there is no unique way of constructing the tree for the space groups. In a tree of type I, the centering type is kept and the order of the associated point group is changed. (Maximal subgroups and minimal supergroups are discussed in Section 4.8.) For example, the maximal type I subgroup of space group #4, $P2_1$, is space group #1, $P1$. Both these space groups have the same centering type, namely P; but for the subgroup the order of the associated point group is decreased. Specifically, the associated point group of space group #4, $P2_1$, has order two while the associated point group of space group #1, $P1$, has order one. Space group #4, $P2_1$, has only 1 maximal subgroup and 16 minimal supergroups. The associated point groups of 13 of the supergroups have order 4, and 3 have order 6. The *International Tables for Crystallography* give the necessary information to construct Figure 10.5.

10.3.3 Asymmetric Unit

The asymmetric unit is the smallest part of the unit cell that, when operated on by the symmetry operations, produces the whole unit cell. The volume of the asymmetric unit is related to the volume of the unit cell by $V_{AU} = V_{UC}/nh$, where n is the number of lattice

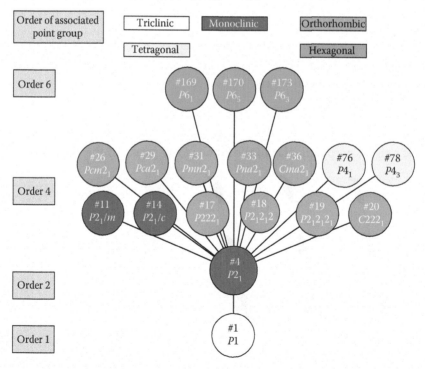

FIGURE 10.5 Space Group #4, $P2_1$, maximal subgroup and minimal supergroups.

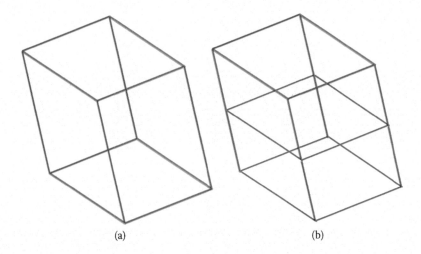

(a)   (b)

FIGURE 10.6   Sucrose (a) unit cell in blue with (b) the asymmetric unit in red.

points in the unit cell and h is the number of operations in the associated point group. (Note that that the product nh is equal to the multiplicity of the general position.) Because there are two symmetry operations in the associated point group and the unit cell is *primitive* ($n = 1$, $h = 2$), the volume of the asymmetric unit is half the volume of the unit cell. Therefore, because the volume of the unit cell is 711 Å³, the volume of the asymmetric unit is 355 Å³. Figure 10.6 shows the unit cell drawn in blue for sucrose with the asymmetric unit in red superimposed on it. The asymmetric unit has

$$0 \le x \le 1, \quad 0 \le y \le 1, \quad \text{and} \quad 0 \le z \le 1/2$$

(Hahn, 2010, p. 92).

10.4 SUCROSE: DIRECT AND RECIPROCAL LATTICES

In addition to the direct lattice, there is the reciprocal lattice, which is the lattice of the diffraction pattern. The diffraction pattern has the intensities of the Bragg reflections superimposed on the reciprocal lattice. Table 10.3 compares the direct and reciprocal lattices, including the lattice parameters, the metric matrix, and the cell volume for both the direct and reciprocal cells. Figure 10.7 shows the relationship between the direct unit cell (blue)

TABLE 10.3   Sucrose: Comparison of Direct and Reciprocal Lattices

Direct Lattice	Reciprocal Lattice
$a = 7.736$ Å; $b = 8.702$ Å; $c = 10.846$ Å	$a^* = 0.1326$ Å⁻¹; $b^* = 0.1149$ Å⁻¹; $c^* = 0.0946$ Å⁻¹
$\alpha = 90.0°$; $\beta = 102.97(2)°$; $\gamma = 90.0°$	$\alpha^* = 90°$; $\beta^* = 77.0°$; $\gamma^* = 90°$

$$\mathbf{G} = \begin{pmatrix} 59.9 & 0 & -18.8 \\ 0 & 75.7 & 0 \\ -18.8 & 0 & 117.6 \end{pmatrix} \text{Å}^2 \qquad \mathbf{G^*} = \begin{pmatrix} 0.0176 & 0 & 0.0028 \\ 0 & 0.0132 & 0 \\ 0.0028 & 0 & 0.0090 \end{pmatrix} \text{Å}^{-2}$$

$V = 711.511$ Å³ $V^* = 0.0014$ Å⁻³

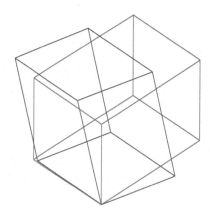

FIGURE 10.7 Sucrose direct cell (blue) and reciprocal cell (red) with volumes normalized to one.

and the reciprocal unit cell (red) with the volumes normalized to one. The Starter Program from Chapter 5 produces the reciprocal cell superimposed on the unit cell. The program produces a figure that can be rotated to observe that, for example, **a** is perpendicular to both **b*** and **c***. See Sections 5.2 and 5.3 for all the relevant relationships.

10.5 SUCROSE: FRACTIONAL COORDINATES AND OTHER DATA FOR THE CRYSTAL STRUCTURE

The crystal structure consists of the atoms in the unit cell and the bonds between the atoms.

The fractional coordinates of the atoms in the asymmetric unit combined with the general position coordinates of the space group and the lattice parameters are sufficient to specify all the atoms in the unit cell.

For sucrose Z equals two or, in other words, there are two sucrose molecules in the unit cell. Z' equals one; that is to say, there is one sucrose molecule in the asymmetric unit. The fractional coordinates of the atoms for the asymmetric unit from the CIF are given in Table 10.4.

Table 10.5 gives the symmetry operations that are needed to populate the unit cell. There are 45 atoms in the molecule, all of which are in the asymmetric unit. These crystallographically independent atoms all have multiplicity 2 and have Wyckoff letter a, that is, are in the general position.

Now the bonds must be inserted. Figure 10.8 shows how the atoms in the asymmetric unit are connected and labeled according to Table 10.4. For emphasis, the red box outlines the asymmetric unit. In the Starter Program from Chapter 4 for populating the unit cell, this diagram is the key to putting in the intramolecular bonds. The gray carbon atoms are outlined in black, the pink oxygen atoms are outlined in red, and the white hydrogen atoms are outlined in black.

10.6 SUCROSE: CRYSTAL STRUCTURE

Figure 10.9a shows a single molecule of sucrose. Populate the unit cell by using the Starter Program from Chapter 4 to create the entire unit cell as shown in Figure 10.9b. This program combines the lattice parameters—a, b, c, α, β, and γ—(Table 10.3), the fractional coordinates (Table 10.4), the general position coordinates of the space group (Table 10.5),

TABLE 10.4 Sucrose Fractional Coordinates, Multiplicity, and Wyckoff Letter for Atoms in the Asymmetric Unit

Atom	x Fractional Coordinate	y Fractional Coordinate	z Fractional Coordinate	Multiplicity Wyckoff Letter
O1	0.29812(5)	0.31150	0.57595(3)	2a
O2	−0.25150(5)	0.19379(5)	0.26982(3)	2a
O3	0.14483(6)	−0.18697(5)	0.15053(3)	2a
O4	0.59110(5)	0.53563(5)	0.52122(3)	2a
O5	0.21335(6)	0.08386(5)	−0.08416(3)	2a
O6	0.13098(5)	0.23078(5)	0.12144(3)	2a
O7	0.10786(5)	0.28166(5)	0.32894(3)	2a
O8	−0.20559(6)	−0.12131(5)	0.19131(3)	2a
O9	0.18042(5)	0.53604(5)	0.28797(3)	2a
O10	−0.12318(5)	0.38915(5)	0.47202(3)	2a
O11	0.45821(6)	0.39079(5)	0.17224(3)	2a
C1	0.20479(5)	0.07741(5)	0.13987(3)	2a
C2	0.05671(5)	−0.04325(5)	0.12443(3)	2a
C3	−0.06584(5)	−0.00917(5)	0.21424(3)	2a
C4	0.31622(5)	0.05766(5)	0.04086(3)	2a
C5	−0.13889(5)	0.15397(5)	0.18633(3)	2a
C6	0.44569(4)	0.46512(5)	0.43548(3)	2a
C7	0.12916(5)	0.43648(5)	0.37652(3)	2a
C8	−0.04735(5)	0.49653(5)	0.39877(3)	2a
C9	0.36911(5)	0.57042(5)	0.32265(3)	2a
C10	0.28545(4)	0.43952(5)	0.49425(2)	2a
C11	0.45065(5)	0.54879(5)	0.20854(3)	2a
C12	0.01367(5)	0.27141(5)	0.19967(3)	2a
H1	−0.02516	−0.03760	0.02724	2a
H2	0.36591	−0.05899	0.05270	2a
H3	−0.04335	0.38359	0.16468	2a
H4	−0.22103	0.15910	0.08902	2a
H5	0.27342	0.54471	0.54831	2a
H6	0.00661	−0.01548	0.31343	2a
H7	0.28970	0.06337	0.23555	2a
H8	0.48682	0.35593	0.40043	2a
H9	−0.34727	0.25995	0.22656	2a
H10	0.58897	0.58448	0.23177	2a
H11	0.33676	0.35387	0.15062	2a
H12	0.39193	0.69131	0.35235	2a
H13	0.43149	0.13342	0.05962	2a
H14	−0.13555	0.51681	0.30611	2a
H15	0.31995	0.22018	0.53087	2a
H16	−0.24297	−0.13773	0.26874	2a
H17	0.21027	0.18957	−0.10822	2a
H18	−0.02698	0.60685	0.44768	2a
H19	0.69935	0.48124	0.51556	2a
H20	−0.16093	0.30355	0.41454	2a
H21	0.06525	−0.26349	0.15298	2a
H22	0.37133	0.61416	0.12979	2a

TABLE 10.5 Sucrose General Position Coordinates and Symmetry Operations from CIF

#	General Position Coordinates	Symmetry Operations
1	x, y, z	1, identity
2	$\bar{x}, \frac{1}{2} + y, \bar{z}$	2_1, screw operation along the y-axis

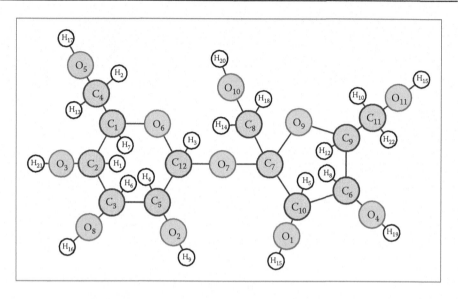

FIGURE 10.8 Sucrose bond schematic with atoms in asymmetric unit outlined with a red box.

and the configuration of bonds (Figure 10.8). The lattice parameters, the fractional coordinates, and the general position coordinates are from the CIF.

The two symmetry operations, the identity and the twofold screw, are shown in Table 10.5. The fractional coordinates in Table 10.4, for all 45 atoms in the asymmetric unit, represent the identity with coordinates x, y, z in Table 10.5. Now the twofold screw operation is applied, producing the second molecule in the unit cell with coordinates $\bar{x}$, $y + \frac{1}{2}$, $\bar{z}$. The unit cell has 90 atoms.

Notice how some of the atoms protrude through the walls of the unit cell. Also the molecules in a single cell appear to be sparsely packed. In Figure 10.9c there are three unit cells populated with molecules.

10.6.1 Symmetry of the Special Projections

The *International Tables for Crystallography* give the symmetry for three special projections for each space group in the standard orientation. These projections are orthogonal, which means that each projection is onto a plane normal to the direction of the projection. The basis vectors of the two-dimensional unit cells are labeled $\mathbf{a}'$ and $\mathbf{b}'$, regardless of which two basis vectors of the three-dimensional unit cell (selected from $\mathbf{a}$, $\mathbf{b}$, $\mathbf{c}$) are projected. For the monoclinic crystal system the special projections are along [001], [100], and [010]. The Starter Program has a "view" statement that selects the appropriate projections. In the monoclinic crystal system, with basis vector $\mathbf{b}$ for the unique axis, the basis vector $\mathbf{a}$ is replaced with the projected vector $\mathbf{a}_p$ when projecting along [001]; and the basis vector $\mathbf{c}$

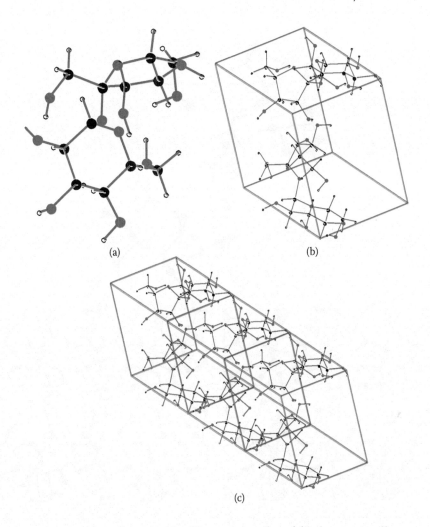

FIGURE 10.9 Sucrose: (a) single molecule, (b) one unit cell, and (c) three unit cells.

is replaced with the projected vector $\mathbf{c}_p$ when projecting along [100]. Note also that in the monoclinic system there is no shift in origin for any of the special projections. A careful examination of these projections can demonstrate that the atoms and bond lengths have been correctly implemented.

Quite generally, every two-dimensional projection has the symmetries of one of the 17 two-dimensional space groups.

Along [001]

The two-dimensional space group along [001] is *pg*, where

$$\mathbf{a}' = \mathbf{a}_p \quad \mathbf{b}' = \mathbf{b}$$

with the origin at 0, 0, z. Figure 10.10a is the direct output from the Starter Program. Here Z is equal to two, and the two sucrose molecules are discernible with care. The enhanced diagram in Figure 10.10b is obtained by repeatedly translating Figure 10.10a to form an array

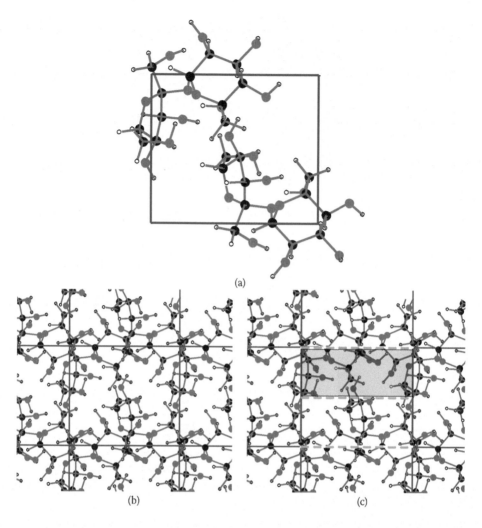

FIGURE 10.10 Sucrose projection along [001]: (a) unit cell, (b) enhanced unit cell, and (c) with symmetries.

of adjacent figures in order to exhibit the symmetries of *pg* (see Section 4.4.5). This diagram shows clearly that the four vertices are lattice points and therefore have identical neighborhoods. Note that the opposite sides of the rectangle are translations of one another and thus also have identical neighborhoods. Now examine the pattern for symmetries. The glides are visible. Hence, as always when done correctly, all the information necessary to create the entire pattern is found in the unit cell. Figure 10.10c shows all the symmetry operations for the two-dimensional space group and the asymmetric unit in light blue. With careful attention, all the atoms of a sucrose molecule can be accounted for in the asymmetric unit, although they do not all belong to the same molecule.

Along [100]

The two-dimensional space group along [100] is *pg* where

$$\mathbf{a}' = \mathbf{b} \quad \mathbf{b}' = \mathbf{c}_p$$

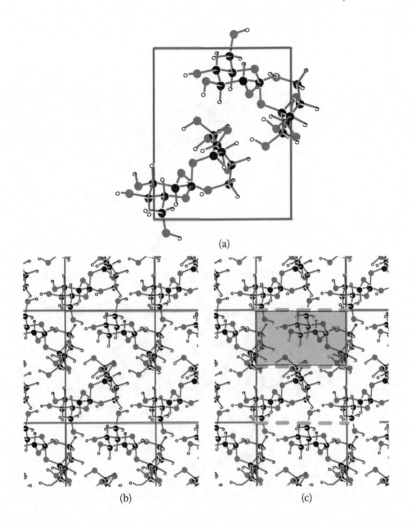

FIGURE 10.11   Sucrose projection along [100]: (a) unit cell, (b) enhanced unit cell, (c) with symmetries.

with the origin at x, 0, 0. Note the difference in the appearance of the two molecules in this projection in contrast with their appearance along [001]. Figure 10.11a is the direct output from the Starter Program. The enhanced diagram in Figure 10.11b is obtained by repeatedly translating Figure 10.11a to form an array of adjacent figures in order to exhibit the symmetries of *pg*. Figure 10.11c shows all the symmetry operations for the two-dimensional space group and the asymmetric unit in light blue.

Along [010]
Finally, the two-dimensional space group along [010] is *p2*, where

$$\mathbf{a'} = \mathbf{c} \quad \mathbf{b'} = \mathbf{a}$$

with the origin at 0, y, 0. Again note the difference in the appearance of the two molecules in this projection in contrast with their appearance along either [001] or [100]. Figure 10.12a is the direct output from the Starter Program. The enhanced diagram in Figure 10.12b is

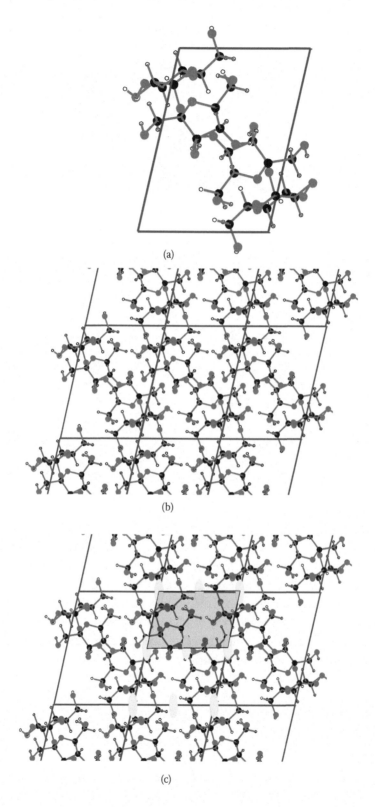

(a)

(b)

(c)

FIGURE 10.12   Sucrose projection along [010]: (a) unit cell, (b) enhanced unit cell, (c) with symmetries.

obtained by repeatedly translating Figure 10.12a to form an array of adjacent figures in order to exhibit the symmetries of *p2* (see Section 4.4.3). Figure 10.12b clearly shows the twofold rotation axes. Figure 10.12c shows all the symmetry operations for the two-dimensional space group and the asymmetric unit in light blue.

10.7 SUCROSE: RECIPROCAL LATTICE AND *d*-SPACINGS

Planes in the direct lattice correspond to points of the reciprocal lattice. The reciprocal vector

$$\mathbf{H}(hkl) = h\mathbf{a}^* + k\mathbf{b}^* + l\mathbf{c}^*$$

where *hkl* are the indices of the planes of the direct lattice. The *d*-spacing is the distance from the origin to either of the two closest planes. In general, the *d*-spacing can be calculated by the following equation (see Section 5.7.1.):

$$\frac{1}{d_{hkl}^2} = H^2(hkl) = (h \quad k \quad l)\mathbf{G}^* \begin{pmatrix} h \\ k \\ l \end{pmatrix}$$

The powder diffraction pattern in Figure 10.13 gives *hkl* = 112 as the reflection with the maximum intensity. The calculation for the *d*-spacing for this reflection is as follows:

$$\frac{1}{d_{112}^2} = H^2(112) = (1 \quad 1 \quad 2) \begin{pmatrix} 0.0176 & 0 & 0.0028 \\ 0 & 0.0132 & 0 \\ 0.0028 & 0 & 0.0090 \end{pmatrix} \begin{pmatrix} 1 \\ 1 \\ 2 \end{pmatrix} \qquad (10.1)$$

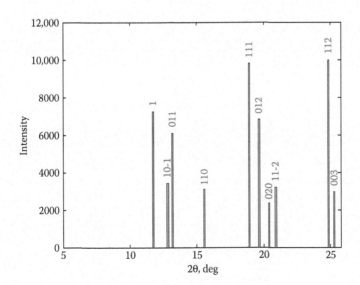

FIGURE 10.13 Sucrose powder diffraction pattern.

TABLE 10.6   Sucrose Diffraction Peak Intensity versus *hkl*

hkl	Intensity, Normalized to Highest Maximum
100	7241
$10\bar{1}$	3423
011	6105
110	3105
111	9797
012	6832
020	2378
$11\bar{2}$	3196
112	10,000
003	3809

Now the *d*-spacing for the 112 can be calculated

$$d_{112} = 3.583 \text{ Å} \tag{10.2}$$

10.7.1 Powder Diffraction Pattern

The diffraction angle is now calculated as a function of *hkl*. First, Bragg's law is applied to the *d*-spacing. Bragg's law is

$$\lambda = 2d_{hkl} \sin \theta_{hkl}$$

where λ is the wavelength of the incoming x-rays, d_{hkl} is the *d*-spacing of the *hkl* plane, and θ_{hkl} is the Bragg angle for the *hkl* diffraction maximum. The diffraction angle, $2\theta_{hkl}$, is twice the Bragg angle (see Section 6.6.3). Table 10.6 has *hkl* indices and intensities for the 10 most intense reflections. The reflections are not actual experimental data but were calculated by the program *Mercury* of the CSD. The intensities are normalized so that the highest intensity is 10,000 arbitrary units. This information is the input for the Starter Program in Chapter 6. Figure 10.13 is the resulting powder diffraction pattern for sucrose.

10.8 SUCROSE: ATOMIC SCATTERING CURVES

Figure 10.14 shows the atomic scattering curves from the Starter Program in Chapter 7 for the three elements found in sucrose. The atomic scattering curves are calculated from the Cromer–Mann coefficients. The atomic scattering factor (or form factor) is the ratio of the amplitude of the wave scattered by an atom to the amplitude of the wave scattered by one electron. Its units are electrons. The atomic scattering curve is a measure of the efficiency of an atom in scattering x-rays.

Bragg's law for (112) is

$$\sin \theta/\lambda = 1/(2d) = 1/(2 \times 3.58) = 0.140 \text{ Å}^{-1} \tag{10.3}$$

This reflection is labeled in Figure 10.14 in red. The values from the atomic scattering curves are

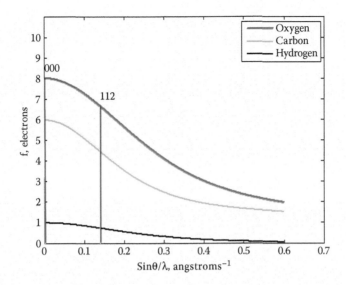

FIGURE 10.14   Sucrose atomic scattering curves with 000 and 112 diffraction maxima.

$$f_O(112) = 6.64, f_C(112) = 4.48, \text{ and } f_H(112) = 0.744 \text{ electrons} \qquad (10.4)$$

10.9 SUCROSE: STRUCTURE FACTOR

The general formula for the structure factor, Equation 7.5, is

$$F\left(hkl\right) = \sum_{j=1}^{N} f_j \left(\frac{\sin \theta_{hkl}}{\lambda} \right) e^{2\pi i \left(hx_j + ky_j + lz_j \right)}$$

where N is the number of atoms in the unit cell. In the case of sucrose there are 90 atoms in the unit cell (see Table 10.7). The running index j goes from 1 to 90. Table 10.4 includes the fractional coordinates x_j, y_j, z_j for the jth atom in the asymmetric unit. The formula for the structure factor for sucrose can be broken down in the following way:

$$F(hkl) = F_O(hkl) + F_C(hkl) + F_H(hkl) \qquad (10.5)$$

TABLE 10.7   Sucrose, $C_{12}H_{22}O_{11}$, Calculation of Number of Electrons in Unit Cell

Atom	C	H	O	Total/Molecule
Atomic number	6	1	8	
Atoms/molecule	12	22	11	45
Total electrons/molecule	72	22	88	182
$Z = 2$				Total/unit cell
Atoms/unit cell	24	44	22	90
Electrons/unit cell	144	44	176	364

The first term is a sum over 24 carbon atoms, the second is over 44 hydrogen atoms, and the third is over 22 oxygen atoms

$$F(hkl) = \sum_{j}^{24} F_{Cj}(hkl) + \sum_{j}^{44} F_{Hj}(hkl) + \sum_{j}^{22} F_{Oj}(hkl) \tag{10.6}$$

Now put in the expressions for the individual atomic form factors and the phase factors

$$F(hkl) = \sum_{j}^{24} f_C(hkl)e^{2\pi i(hx_j + ky_j + lz_j)}$$

$$+ \sum_{j}^{44} f_H(hkl)e^{2\pi i(hx_j + ky_j + lz_j)} + \sum_{j}^{22} f_O(hkl)e^{2\pi i(hx_j + ky_j + lz_j)} \tag{10.7}$$

10.9.1 Calculate the Structure Factor at 000

The special case of the undeviated beam occurs when $hkl = 000$. Its position is labeled in red in Figure 10.14.

$$F(000) = 24f_C(000) + 44f_H(000) + 22f_O(000)$$

The atomic scattering factor, f, at the origin is equal to the number of electrons; therefore $F(000)$ is the sum of all the electrons in the unit cell (Section 7.6.4). Table 10.7 shows how the total number of electrons is counted in the unit cell of sucrose to give

$$F(000) = 364 \text{ electrons} \tag{10.8}$$

10.9.2 Calculate the Contribution of One Atom to the Structure Factor

The calculation in Equation 10.7 can be sampled by calculating only the contribution of one atom, C1, to the most intense hkl reflection found in the x-ray powder diagram, which is 112. Table 10.4 gives the fractional coordinates for C1 as

$$x = 0.2048, \quad y = 0.0774, \quad z = 0.1399$$

For the reflection 112

$$F_{C1}(112) = f_C(112)e^{2\pi i(0.2048 + 0.0774 + 2\times(0.1399))} = 4.48e^{1.124\pi i} \text{electrons} \tag{10.9}$$

Note that this is a complex number. The entire structure factor, including all 90 atoms, may be real or complex because this space group is noncentrosymmetric (see Section 7.7.2). The importance of the structure factors is that they become the coefficients of the Fourier expansion that is used to calculate the electron density (see Section 7.8.2).

DEFINITIONS

Chiral point group
Chiral space group
Proper point group
Proper space group

EXERCISES

10.1 Construct a table for the three-dimensional Bravais lattice mC which includes all the proper point groups, the corresponding symmorphic space groups, and the nonsymmorphic space groups. Use the web to find a complete list of space groups.

10.2 Out of the following list of space groups, show which space groups are proper and improper (P, IP), which are symmorphic or nonsymmorphic (S, NS) (Section 4.11.5); and show the Bravais lattice, (BL), (Figure 6.68) and the associated point group (APG).

| $P\bar{1}$ | $P2$ | $C2/m$ | $P2_1/c$ | $C2/c$ | $P2_12_12$ | $Fdd2$ | $Ibca$ | $P4_12_12$ | $R3$ | $P6_3$ | $I23$ |

10.3 Research the uses and operation of the scanning electron microscope, SEM.

10.4 Suggested monoclinic crystals from CSD with space group #4, $P2_1$ are listed in ascending order of volume in Table 10.8. The website is http://webcsd.ccdc.cam. ac.uk/teaching_database_demo.php; a shortcut to an individual Reference code is: http://webcsd.ccdc.cam.ac.uk/display_csd_entry.php?identifier=GLYCIN for a crystal with Reference Code "GLYCIN". Construct a paper or a PowerPoint presentation based on the parallel topics in this chapter. The parallel topics are discussed in Section 8.3.1. Examples are given at https://sites.google.com/a/vt.edu/foundations_of_crystallography/.

10.5 Suggested monoclinic crystals from CSD with space group #14, $P2_1/c$ are listed in ascending order of volume in Table 10.9. The website is http://webcsd.ccdc.cam.

TABLE 10.8 Suggested Crystals from CSD with Space Group # 4, $P2_1$

Ref. Code	Chemical Name	Formula	Volume Å³
GLYCIN	Glycine	$C_2H_5NO_2$	157.36
TARTAC	Tartaric acid	$C_4H_6O_6$	283.73
WURTOS	3-Iodo-1-nitrobenzene	$C_6H_4INO_2$	362.1
ZEXQAU	1,3,5-Tricyanobenzene	$C_9H_3N_3$	408.83
NESCAQ	1,4-Dihydroxy-4-phenylbutan-2-one	$C_{10}H_{12}O_3$	453.83
ADRENL	(-)-Adrenaline	$C_9H_{13}NO_3$	457.25
LUPNIN	Lupinine	$C_{10}H_{19}NO$	511.98
TRYPTC	L-Tryptophan hydrochloride	$C_{11}H_{13}N_2O_2{}^{1+},Cl^{1-}$	572.43
ADENOS10	Adenosine	$C_{10}H_{13}N_5O_4$	578.84
JETLID	Sodium bis(5-oxoproline)-platinum(ii) dihydrate	$C_{10}H_{11}N_2O_6Pt^{1-},Na^{1+},2(H_2O)$	724.21

TABLE 10.9   Suggested Crystals from CSD with Space Group #14, $P2_1/c$

Ref. Code	Chemical Name	Formula	Volume Å^3
ETDIAM12	1,2-Diaminoethane	$C_2H_8N_2$	181.45
DUCKOB04	n-Butane	C_4H_{10}	239.11
BNZQUI03	p-Benzoquinone	$C_6H_4O_2$	262.86
CADVEI	(E)-3-Hexenedinitrile	$C_6H_6N_2$	297.45
DCLBEN03	p-Dichlorobenzene	$C_6H_4Cl_2$	305.38
FACETC10	Monofluoroacetic acid	$C_2H_3FO_2$	327.98
DERWAY	trans,trans-1,3,5,7-Octatetraene	C_8H_{10}	348.03
TAQYUG	p-Dibromobenzene	$C_6H_4Br_2$	352.82
XOMHUC	1,7-Octadiene	C_8H_{14}	395.47
ANTQUO08	Anthraquinone	$C_{14}H_8O_2$	480.21

ac.uk/teaching_database_demo.php; a shortcut to an individual Reference code is: http://webcsd.ccdc.cam.ac.uk/display_csd_entry.php?identifier=GLYCIN for a crystal with Reference Code "GLYCIN". Construct a paper or a PowerPoint presentation based on the parallel topics in this chapter. The parallel topics are discussed in Section 8.3.1. Examples are given at https://sites.google.com/a/vt.edu/foundations_of_crystallography/.

## CHAPTER 11

Orthorhombic Crystal System

Polyethylene

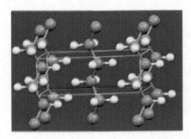

The picture above shows how polymer chains of polyethylene are accommodated in the crystal structure. Polyethylene is a common plastic. The material analyzed in this chapter is a high-density organic polymer. Toys, milk containers, water pipes, and garbage cans are all examples of this highly useful and economically important material. The first analysis of the crystal structure of polyethylene was done by C. W. Bunn in 1939. Much work has been done on this classic structure over the past three-quarters of a century. The work reported here is a remarkable study done with neutron diffraction at a variety of temperatures, beginning at 10 K.

CONTENTS

CHAPTER OBJECTIVES

- Study the crystallographic properties of polyethylene, space group #62, *Pnam*, as a representative of the orthorhombic crystal system.

- Present the parallel crystallographic topics that recur in Chapters 9 through 15. A few examples are given below. A complete list is found in Section 8.3.1.

  - Multiplication table of the associated point group

  - Diagrams of the unit cell, the asymmetric unit, the populated unit cell, and the special projections

  - Diagram of the reciprocal cell superimposed on the direct unit cell

  - Calculation of the powder diffraction pattern

  - Calculation of the atomic scattering curves

  - Calculation of the structure factor at 000

- Calculation of the contribution of one atom to the structure factor of the largest reflection

- Present the special crystallographic topics of this chapter, which are listed below. A complete list of the special topics found in Chapters 9 through 15 is in Section 8.3.2.

  - Low-temperature experiments

  - Short and full Hermann–Mauguin symbols

  - Relating the Hermann–Mauguin symbol to the cell choice

  - Comparison of symbol diagrams *Pnma* and *Pnam*

  - Comparison of general position coordinates between *Pnma* and *Pnam*

  - General position diagram for *Pnma* only

  - Advantages of using CIF

  - Relating the Hermann–Mauguin symbol to the asymmetric unit

  - Special projections and the Hermann–Mauguin symbols

11.1 INTRODUCTION

The example chosen for the orthorhombic crystal system is a thermoplastic polymer, polyethylene. The structure given here is for the deuterated form, which means that deuterium isotopes are substituted for the ordinary hydrogen atoms. There are two long chains $(C_2D_4)_n$ running through the unit cell. As is customary in crystallography, the chains are analyzed as if they were infinitely long, in the same manner in which crystals are analyzed as if they were infinitely large. A *mer* is the repeat unit in a polymer, which in this case is C_2D_4. The unit cell contains two mers ($Z = 2$); and the asymmetric unit contains a quarter mer, consisting of a half carbon atom and two half deuterium atoms ($Z' = 1/4$). The diffraction radiation probe was neutron diffraction. Neutron diffraction and the use of deuterium with neutron diffraction are discussed in Section 15.1. The study was undertaken to examine the translational and rotational displacements of the molecular chain. The study concluded that the displacements were mainly translational. Neutron radiation was used to elucidate the role of the deuterium atoms with respect to the carbon atoms (Takahashi, 1998).

Table 11.1 contains the crystal data taken at room temperature for polyethylene. These data are found in the Crystallographic Information File (CIF), in this case from the Cambridge Structural Database (CSD).

11.1.1 Special Topic: Low-Temperature Experiments

Cryogenics is the study of materials at temperatures below 123 K. The National Institute of Standards and Technology (2012) has chosen this temperature because the normal boiling points of the common gases such as helium, hydrogen, neon, nitrogen, oxygen, and air are all below this value. Liquid nitrogen boils at 77 K at 1 atm of pressure. Therefore, liquid nitrogen can be used to study crystals at temperatures down to 77 K. Temperatures below

TABLE 11.1   Polyethylene: Crystal Data

Compound	Polyethylene
Formula	$(C_2D_4)_n$
Molecular weight	28.10
Space group	#62, *Pnam*
Bravais lattice	*oP*
Crystal system	Orthorhombic
Diffraction radiation probe	Neutron radiation; at 300 K
Ambient temperature for data	Room temperature (283–303) K
Data base and identifier	CSD QILHUO05
Molecular volume	46.523 Å³
Z, Z′	$Z = 2$; $Z' = 1/4$
Density	1.143 g/cm³
Author	Y. Takahashi
Publication	*Macromolecules* (1998), 31, 3868
Polymorph	High-density organic polymer

TABLE 11.2   Temperature Variation of Unit Cell Dimensions *a* and *b*

Temperature(K)	10	100	200	300
a, Å	7.120	7.162	7.258	7.417
ln $(a$, Å)	1.962	1.968	1.982	2.003
b, Å	4.842	4.863	4.898	4.939
ln $(b$, Å)	1.577	1.581	1.588	1.597

77 K are more difficult to be attained. The boiling point of helium-4 at 1 atm pressure is 4.2 K. To cool crystals to temperatures below 77 K, the crystals are first cooled with liquid nitrogen and then further cooled with liquid helium. Liquid nitrogen is also used in crystallography as a coolant for the x-ray detectors used to record the x-ray diffraction data. See Exercise 11.1.

The crystal studied in this chapter was a polyethylene film that was made by a hot-press process and stretched in boiling water (Takahashi, 1998). This study was performed at 10, 100, 200, and 300 K (see Table 11.2). Figure 11.1 shows that the dependence of the natural logarithm of the lattice parameter *a* on the temperature is greater than the dependence of the natural logarithm of *b* over the same range (see Section 1.9.1). For both graphs the axis for the logarithm of the lattice parameter has increments of 0.01 and a range of 0.05. The linear thermal expansion is plainly anisotropic; that is, the lattice parameter *a* expands more rapidly than *b* over the same temperature range. Neither of these curves is linear.

11.2 POLYETHYLENE: POINT GROUP PROPERTIES

The associated point group for polyethylene is *mmm* in the Hermann–Mauguin notation and D_{2h} in the Schönflies notation. This point group has eight symmetry operations, 1, 2_z, 2_y, 2_x, $\bar{1}$, m_{xy}, m_{xz}, and m_{yz}; thus the order, or the number of elements, is 8.

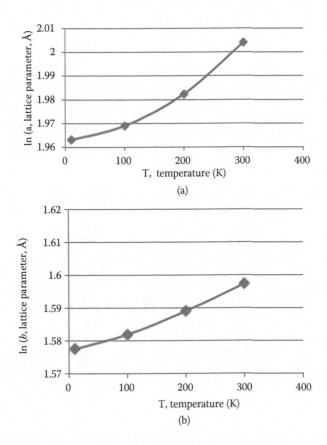

FIGURE 11.1 Temperature data for polyethylene: (a) expansion along a; (b) expansion along b.

11.2.1 Multiplication Table

The generators from Appendix 3 are 1, 2_z, 2_y, and $\bar{1}$. However, the point group mmm can be generated from just three operations, as shown in Figure 3.58. There are many choices for these generators. Possibilities are 2_z, 2_y, and $\bar{1}$. The matrices for these generators are

$$2_z = \begin{pmatrix} \bar{1} & 0 & 0 \\ 0 & \bar{1} & 0 \\ 0 & 0 & 1 \end{pmatrix}, \quad 2_y = \begin{pmatrix} \bar{1} & 0 & 0 \\ 0 & 1 & 0 \\ 0 & 0 & \bar{1} \end{pmatrix}, \quad \text{and} \quad \bar{1} = \begin{pmatrix} \bar{1} & 0 & 0 \\ 0 & \bar{1} & 0 \\ 0 & 0 & \bar{1} \end{pmatrix}.$$

This is the smallest crystallographic point group with three generators. The Starter Program from Chapter 3 calculates the multiplication table in Table 11.3. This multiplication table is symmetric because the point group is abelian. It has an inversion, and thus its symmetry does not permit the piezoelectric effect. The relationships among the symmetry operations are seen in the multiplication table.

11.2.2 Stereographic Projections

The general position stereographic projection for the point group mmm, Figure 11.2a, has eight points corresponding to the eight symmetry operations. See Section 3.9 for a general

TABLE 11.3   Multiplication Table for Point Group *mmm*

	1	2_z	2_y	2_x	$\bar{1}$	m_{xy}	m_{xz}	m_{yz}
1	1	2_z	2_y	2_x	$\bar{1}$	m_{xy}	m_{xz}	m_{yz}
2_z	2_z	1	2_x	2_y	m_{xy}	$\bar{1}$	m_{yz}	m_{xz}
2_y	2_y	2_x	1	2_z	m_{xz}	m_{yz}	$\bar{1}$	m_{xy}
2_x	2_x	2_y	2_z	1	m_{yz}	m_{xz}	m_{xy}	$\bar{1}$
$\bar{1}$	$\bar{1}$	m_{xy}	m_{xz}	m_{yz}	1	2_z	2_y	2_x
m_{xy}	m_{xy}	$\bar{1}$	m_{yz}	m_{xz}	2_z	1	2_x	2_y
m_{xz}	m_{xz}	m_{yz}	$\bar{1}$	m_{xy}	2_y	2_x	1	2_z
m_{yz}	m_{yz}	m_{xz}	m_{xy}	$\bar{1}$	2_x	2_y	2_z	1

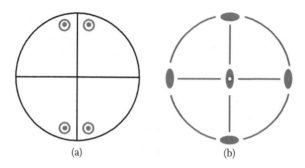

FIGURE 11.2   Point group *mmm*: (a) point; (b) symbol stereographic projections.

description of these projections. The point corresponding to the identity operation has a positive z-coordinate indicated by the dot in the lower right-hand quadrant. The twofold along the z-axis, 2_z, takes the identity dot into a dot in the upper left-hand quadrant. The horizontal twofold along the y-axis, 2_y, produces the open circles in the upper right and lower left quadrants. Open circles indicate negative z-coordinates. The color red is used for all these proper operations. On the contrary, the inversion operation is an improper operation and produces the blue dots and blue open circles. A proper operation does not change handedness, and the determinant of the matrix is +1; and an improper operation changes handedness and the determinant of the matrix is –1. Half the operations are proper, red, and half the operations are improper, blue. Note that both dots and open circles are positioned interior to the circumference to allow any value for z.

Figure 11.2b is the symbol stereographic projection. There are three twofold axes represented. Each twofold group is of order 2 with two operations, a twofold and the identity. The directions of the twofolds are indicated by the subscripts: 2_z is indicated by the oval in the center of the circle; 2_y is indicated by the two ovals along the y-axis; and finally 2_x is indicated by the two ovals along the x-axis. All the symmetry operations in these groups have positive determinants; therefore, their symbols are colored red. There are four more groups, three mirror groups and one inversion point. These groups have equal numbers of proper and improper operations; so they are colored purple. The three purple mirrors are m_{xy}, m_{xz}, and m_{yz}. The symbol for the mirror in the xy-plane, m_{xy}, is the purple circumference. The symbol for the mirror in the xz-plane, m_{xz}, is the vertical purple line. The symbol for the mirror in the yz-plane, m_{yz}, is the horizontal purple line. The symbol for the inversion point is the small purple open circle superimposed on the red oval at the center.

11.3 POLYETHYLENE: SPACE GROUP PROPERTIES

Polyethylene crystallizes in the orthorhombic space group #62. The CIF uses the setting *Pnam*. The Bravais lattice is *oP*. The symmetry operations impose the restrictions

$$\alpha = \beta = \gamma = \text{exactly } 90°$$

11.3.1 Space Group Diagrams

The presentation of the space group diagrams for the space group #62 in the *International Tables for Crystallography* favors the setting indicated by the symbol "*Pnma*" (Hahn, 2010, pp. 122–123). The CIF from the Cambridge Structural Database uses the setting "*Pnam*." The Hermann–Mauguin symbol favored in the *International Tables for Crystallography* differs from the symbol used in the CIF. This difference needs the explanation given in the following special topic. Figure 11.3 shows the symbol diagrams for *Pnma* and *Pnam*, and Figure 11.4 shows the general position diagram for *Pnma*.

11.3.1.1 Special Topic: Short and Full Hermann–Mauguin Symbols

Certain space groups have both a short form and a full form of the Hermann–Mauguin symbols. These space groups belong to one of eight geometric classes (Section 8.3.1). These classes have associated point groups m, $2/m$, mmm, $4/mmm$, $\bar{3}m$, $6/mmm$, $m3$, and $m\bar{3}m$. The *full form of the Hermann–Mauguin symbol* has both the symmetry axes and the symmetry planes of each symmetry direction listed. The information is given for the x-axis, y-axis, and z-axis in that order. The *short form of the Hermann–Mauguin symbol* compresses the symbols for the symmetry axes as much as possible. For example, for space group #62, *Pnma* is the short form and $P2_1/n\,2_1/m\,2_1/a$ is the full form. The full form gives

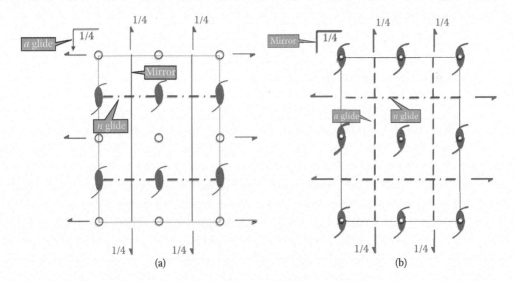

FIGURE 11.3   Space group #62 symbol diagrams: (a) *Pnma*; (b) *Pnam*.

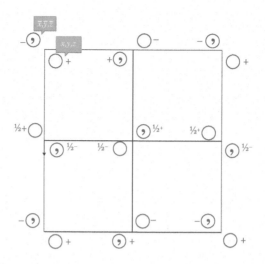

FIGURE 11.4   Space group #62 general position diagram for *Pnma*.

more information about the axes. In this case, there are twofold screws indicated to be parallel to each of the three axes. The full form also gives the symmetry planes normal to the appropriate axis. For example, the *n* glide plane is normal to the *x*-axis.

11.3.1.2 Special Topic: Relating the Hermann–Mauguin Symbol to the Cell Choice

There are six descriptions of the space group #62 called "settings" corresponding to the six permutations of the basis vectors **a**, **b**, and **c**. These are *Pnma*, *Pnam*, *Pcmn*, *Pmnb*, *Pbnm*, and *Pmcn*. In the *International Tables for Crystallography* the symbol diagrams for these settings are presented in a condensed manner. Note that each of the six symbols contains, in addition to the centering type, *P*; one diagonal glide plane, *n*; one mirror plane, *m*; and one axial glide plane, *a*, *b*, or *c*. See Table 4.8 for a description of glide operations.

Some crystallographers like to list the lattice parameters in descending order of length. In the crystal for this chapter

$$a = 7.417 \text{ Å}, \quad b = 4.939 \text{ Å}, \quad \text{and} \quad c = 2.540 \text{ Å}$$

In this case, the appropriate setting is *Pnam*.

11.3.1.3 Special Topic: Comparison of Symbol Diagrams Pnma and Pnam

Figure 11.3 shows the symbol diagrams for *Pnma* and *Pnam*. Both diagrams have an inversion point at the origin, indicated by a small purple circle. The "*P*" in the Hermann–Mauguin symbol means that the unit cell is chosen to be primitive. In the orthorhombic crystal system, the next three positions in the Hermann–Mauguin symbol refer to the symmetries associated with the *x*-axis, *y*-axis, and *z*-axis, respectively.

In the symbol "*Pnma*," the "*n*" in the first position indicates a diagonal glide in the *yz*-plane that is represented in the diagram by a dash-dot line drawn normal to the *x*-axis at *x* = 1/4. The "*m*" in the second position indicates a mirror in the *xz*-plane that is represented by a solid line drawn normal to the *y*-axis at *y* = 1/4. The "*a*" in the third position indicates an *a*-glide in the *xy*-plane normal to the *z*-axis that is represented in the upper left-hand corner by a bent arrow pointing downward; the glide is at $z = 1/4$. These three symbols are purple because the corresponding groups contain both improper and proper operations.

On the other hand, in the symbol "*Pnam*," the "*n*" in the first position again indicates a *n*-glide in the *yz*-plane that is represented in the diagram by a dash-dot line drawn normal to the *x*-axis; the *n*-glide is at $x = 1/4$. The "*a*" in the second position indicates an *a*-glide in the *xz*-plane that is represented by a dashed line drawn normal to the *y*-axis; the *a*-glide is at $y = 1/4$. The "*m*" in the third position indicates a mirror in the *xy*-plane normal to the *z*-axis that is represented in the upper left-hand corner by two short perpendicular lines; the mirror is at $z = 1/4$.

11.3.1.4 Special Topic: Comparison of General Position Coordinates between Pnma and Pnam

Table 11.4 shows a comparison between the general position coordinates of *Pnma* and *Pnam*. Notice that the *x*-coordinates are the same in both orientations, but the *y*- and *z*-coordinates are interchanged. This permutation of the axes that happens in the space group #62 also arises in other popular space groups such as #12, #14, #15, #43, #53, #64, and #73 (Hahn, 2010). The Schoenflies symbol for the space group #62 is D_{2h}^{16} and is independent of the setting.

11.3.1.5 Special Topic: General Position Diagram for Pnma Only

In Figure 11.4 the general position diagram is given for *Pnma* only, because that is the only general position diagram given in the *International Tables for Crystallography* for the space group #62. The red circles indicate proper operations, and the blue circles with commas indicate improper operations. The plus sign indicates that the *z*-coordinate is positive, and the minus sign indicates that the *z*-coordinate is negative. Within the unit cell there are eight circles indicating the eight operations, or the multiplicity of the general position of this space group. The red circle representing the coordinates, *xyz*, of the identity operation is pointed out in the diagram. The blue circle with a comma representing the coordinates, $\bar{x}, \bar{y}, \bar{z}$, of the inversion operation is also pointed out in the diagram. Each of the four vertices is a lattice point and has an identical neighborhood. Now check out other operations

TABLE 11.4   Comparison of General Position Coordinates between *Pmna* and *Pnam*

Space Group #62 *Pnma* International Tables for Crystallography (Hahn 2010, pp. 122–3)				Space Group #62 *Pnam* CIF Cambridge Structural Database QILHUO04			
Number				Number			
1	x	y	z	1	x	y	z
3	$-x$	$1/2 + y$	$-z$	2	$-x$	$-y$	$1/2 + z$
4	$1/2 + x$	$1/2 - y$	$1/2 - z$	3	$1/2 + x$	$1/2 - y$	$1/2 - z$
2	$1/2 - x$	$-y$	$1/2 + z$	4	$1/2 - x$	$1/2 + y$	$-z$
5	$-x$	$-y$	$-z$	5	$-x$	$-y$	$-z$
7	x	$1/2 - y$	z	6	x	y	$-1/2 - z$
8	$1/2 - x$	$1/2 + y$	$1/2 + z$	7	$-1/2 - x$	$-1/2 + y$	$-1/2 + z$
6	$1/2 + x$	y	$1/2 - z$	8	$-1/2 + x$	$-1/2 - y$	z

from the symbol diagram. Given the symbol diagram, the general position diagram can be created and vice versa. See Exercises 11.2 through 11.4.

11.3.1.6 Special Topic: Advantages of Using the CIF

Fortunately this problem about the interchange of axes can be alleviated by using the CIF.

The CIF requires that the lattice parameters—a, b, c, α, β, γ—the fractional coordinates, and the general position coordinates all are consistent with one choice of the orientation of the axes. This is a powerful reason to use the crystal data from the CIF. By contrast, the papers directly from the crystallographic literature include the lattice parameters and the fractional coordinates, but the reader is expected to go to the *International Tables for Crystallography* to retrieve the general position coordinates.

11.3.2 Maximal Subgroups for *Pnma*

The *International Tables for Crystallography* give the maximal subgroups of *Pnma*, but not those of *Pnam*. This section gives the maximal subgroups for *Pnma* only in Figure 11.5.

There are 230 space groups. Because the space groups contain translation operations, each space group has infinite order. Space groups can be arranged in trees similar to the trees of the point groups. However, there is no unique way of constructing the tree for the space groups. In a tree of Type I, the centering type is kept and the order of the associated point group is changed. (Maximal subgroups and minimal supergroups are discussed in Section 4.8.)

For example, a maximal Type I subgroup of the space group #62, *Pnma*, is space group #14, $P2_1/c$. Both these space groups have the same centering type, namely *P*; but for the subgroup the order of the associated point group is decreased. Specifically, the associated point group of *Pnma* is *mmm* and has order eight, while the associated point group of $P2_1/c$ is $2/m$ and has order four. *Pnma* has six maximal subgroups of which the associated point groups all have order four. *Pnma* has no supergroups; it is at the top of its branch of the space group tree. The *International Tables for Crystallography* give the information necessary to construct Figure 11.5.

11.3.3 Asymmetric Unit

The *International Tables for Crystallography* give the asymmetric unit of *Pnma* but not, of *Pnam*.

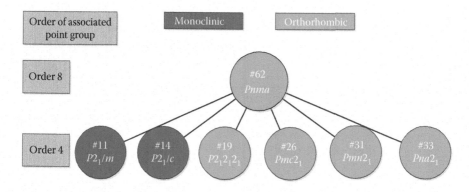

FIGURE 11.5 Space group #62, *Pnma*, maximal subgroups.

The asymmetric unit is the smallest part of the unit cell that, when operated on by the symmetry operations, produces the whole unit cell. The volume of the asymmetric unit is related to the volume of the unit cell by $V_{AU} = V_{UC}/nh$, where n is the number of lattice points in the unit cell and h is the number of operations in the associated point group. (Note that the product nh is equal to the multiplicity of the general position.) Because there are eight symmetry operations in the associated point group of this primitive cell ($n = 1$, $h = 8$), the volume of the asymmetric unit is one-eighth of the volume of the unit cell. Therefore, because the volume of the unit cell is 92 Å^3, the volume of the asymmetric unit is 11.3 Å^3.

11.3.3.1 Special Topic: Relating the Hermann–Mauguin Symbol to the Asymmetric Unit

The labeling of the unit cell for *Pnam* can be obtained from the labeling of the unit cell for *Pnma* by interchanging **b** and **c**. The relationship among the axes is

$$\mathbf{a} = \mathbf{a'}; \quad \mathbf{b} = -\mathbf{c'}; \quad \mathbf{c} = \mathbf{b'}$$

where the unprimed letters refer to *Pnma* and the primed letters refer to *Pnam*. The negative sign is chosen so that both basis sets have the same handedness (see Section 1.5). The determinant of the **P** matrix of the above transformation is +1. Figure 11.6b shows the unit cell for *Pnam* in blue where

$$a' = 7.417 \text{ Å}, \quad b' = 4.939 \text{ Å}, \quad \text{and} \quad c' = 2.540 \text{ Å from the CIF}$$

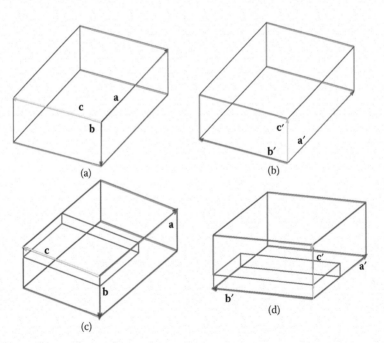

FIGURE 11.6 Polyethylene: (a) unit cell in blue *Pnma*; (b) unit cell *Pnam*; (c) with the asymmetric unit in red *Pnma*; (d) with asymmetric unit *Pnam*.

while Figure 11.6a shows the unit cell for *Pnma* in blue where

$$a = 7.417 \text{ Å}, \quad b = 2.540 \text{ Å}, \quad \text{and} \quad c = 4.939 \text{ Å}$$

Note that there is a change of origin. Both **a** and **a′** are red, both **b** and **b′** are blue, and both **c** and **c′** are green.

The asymmetric unit given by the *International Tables for Crystallography* for *Pnma* (unprimed coordinates) is

$$0 \leq x \leq 1/2, \quad 0 \leq y \leq 1/4, \quad \text{and} \quad 0 \leq z \leq 1$$

as shown in Figure 11.6c in red superimposed on the blue unit cell.

The asymmetric unit for *Pnam* (primed coordinates) is

$$0 \leq x' \leq 1/2, \quad 0 \leq y' \leq 1, \quad \text{and} \quad 0 \leq z' \leq 1/4$$

as shown in Figure 11.6d in red superimposed on the blue unit cell.

11.4 POLYETHYLENE: DIRECT AND RECIPROCAL LATTICES

In addition to the direct lattice, there is the reciprocal lattice, which is the lattice of the diffraction pattern. The diffraction pattern has the intensities of the Bragg reflections superimposed on the reciprocal lattice. Table 11.5 compares the direct and reciprocal lattices including the lattice parameters, the metric matrix, and the cell volume for both the direct and reciprocal cells. The data are from the CIF for this crystal and therefore use the setting *Pnam*. Figure 11.7 shows the relationship between the direct unit cell (blue) and the reciprocal unit cell (red) with the volumes normalized to one.

11.5 POLYETHYLENE: FRACTIONAL COORDINATES AND OTHER DATA FOR THE CRYSTAL STRUCTURE

The crystal structure consists of the placement of the atoms in the unit cell and the bonds between the atoms. The fractional coordinates of the atoms in the asymmetric unit combined with the general position coordinates of the space group and the lattice parameters are sufficient to specify all the atoms in the unit cell.

TABLE 11.5 Polyethylene, *Pnam*: Comparison of Direct and Reciprocal Lattices

Direct Lattice	Reciprocal Lattice
$a = 7.417$ Å; $b = 4.939$ Å; $c = 2.540$ Å	$a^* = 0.1348$ Å^{-1}; $b^* = 0.2025$ Å^{-1}; $c^* = 0.3937$ Å^{-1}
$\alpha = 90°$; $\beta = 90°$; $\gamma = 90°$	$\alpha^* = 90°$; $\beta^* = 90°$; $\gamma^* = 90°$
$\mathbf{G} = \begin{pmatrix} 55.01 & 0 & 0 \\ 0 & 24.39 & 0 \\ 0 & 0 & 6.45 \end{pmatrix}$ Å^2	$\mathbf{G}^* = \begin{pmatrix} 0.0182 & 0 & 0 \\ 0 & 0.0410 & 0 \\ 0 & 0 & 0.1550 \end{pmatrix}$ Å^{-2}
$V = 93.047$ Å^3	$V^* = 0.0107$ Å^{-3}

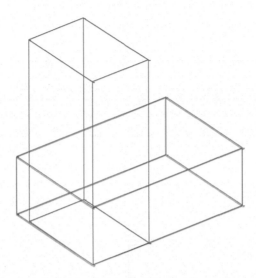

FIGURE 11.7   Polyethylene direct cell in blue and reciprocal cell in red with volumes normalized to 1.

Z equals two or, in other words, there are two mers in the unit cell. Z' equals 1/4; that is to say, there is one-fourth of a mer in the asymmetric unit. The fractional coordinates of the atoms from the CIF are given in Table 11.6.

Crystallographically independent atoms are atoms not related by symmetries. Examples of crystallographically independent atoms are the atoms in the asymmetric unit. The asymmetric unit has one half of the carbon atom C1, one half of the deuterium atom D1, and one half of a second deuterium atom D2. These atoms all have multiplicity 4 and Wyckoff letter c, that is, are in a special position. When the symmetry operations are applied to these atoms, all the atoms in the unit cell are produced. Table 11.7 gives the symmetry operations that are needed to populate the unit cell.

Now the bonds must be inserted. Figure 11.8 shows how the atoms are connected using the labeling from Table 11.6. In addition to the three crystallographically independent atoms—C1, D1, and D2—five more atoms are given to facilitate drawing the bonds. The red box outlines the asymmetric unit. In the Starter Program for populating the unit cell, this diagram is the key to putting in the bonds.

TABLE 11.6   Polyethylene: Fractional Coordinates

Atom	x Fractional Coordinate	y Fractional Coordinate	z Fractional Coordinate	Multiplicity, Wyckoff Letter
C1	0.040	0.061	0.25	4c
D1	0.185	0.023	0.25	4c
D2	0.015	0.278	0.25	4c
C1A	−0.040	−0.061	−0.25	
C1A*	−0.040	−0.061	0.75	
C1*	0.040	0.061	−0.75	
D1A	−0.185	−0.023	−0.25	
D2A	−0.015	−0.278	−0.25	

TABLE 11.7   Polyethylene General Position Coordinates and Symmetry
Operations from CIF

#	General Position Coordinates	Symmetry Operations
1	x, y, z	1, identity
2	$-x, -y, 1/2 + z$	Twofold screw
3	$1/2 + x, 1/2 - y, 1/2 - z$	Twofold screw
4	$1/2 - x, 1/2 + y, -z$	Twofold screw
2	$\bar{x}, \bar{y}, \bar{z}$	$\bar{1}$, inversion
6	$x, y, -1/2 - z$	Mirror normal to [001]
7	$-1/2 - x, -1/2 + y, -1/2 + z$	n-Glide normal to [100]
8	$-1/2 + x, -1/2 - y, z$	a-Glide normal to [010]

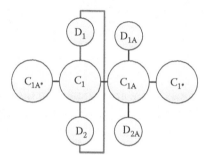

FIGURE 11.8   Polyethylene bond schematic with atoms in asymmetric unit outlined in red.

11.6 POLYETHYLENE: CRYSTAL STRUCTURE

Populate the unit cell by using the Starter Program from Chapter 4 to create the entire
unit cell as shown in Figure 11.9a. This program combines the lattice parameters—a, b,
c, α, β, and γ—(Table 11.5), the fractional coordinates (Table 11.6), the general position
coordinates of the space group (Table 11.7), and the configuration of bonds (Figure 11.8).
The lattice parameters, the fractional coordinates, and the general position coordinates are
from the CIF.

The multiplicity of the general Wyckoff position is eight. That is, there are eight symmetry operations—the identity, three twofold screws, the inversion, two glides, and a mirror.
There are two mers, $Z = 2$, in the unit cell. The carbon and deuterium atoms are in a special
position and have multiplicity four.

Beginning with the general position coordinates the Starter Program automatically
includes all the atoms. No special procedure is required for atoms in special positions.

Note that the mers in the corner of Figure 11.9a are related by a translation. Figure 11.9b
shows two unit cells and indicates how the polymer is extended along the z-axis, which is
vertical in the diagrams.

11.6.1 Symmetry of the Special Projections

The *International Tables for Crystallography* give the symmetry for three special projections for each space group in the standard orientation, which is *Pnma*. These projections

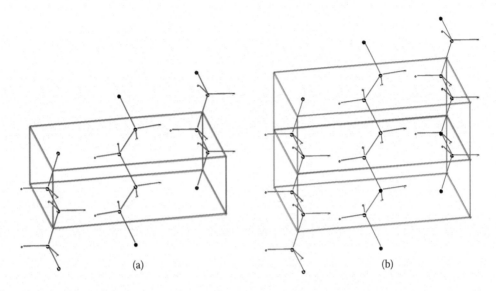

FIGURE 11.9   Polyethylene populated unit cell: (a) one unit cell; (b) two unit cells.

are orthogonal, which means that each projection is onto a plane normal to the direction of the projection. The basis vectors of the two-dimensional unit cell are labeled **a′** and **b′**, regardless of which two basis vectors of the three-dimensional unit cell (selected from **a**, **b**, **c**) are projected. For the orthorhombic crystal system the special projections are along [001], [100], and [010]. In the Starter Program there is a "view" statement that selects the appropriate projections. A careful examination of these projections can demonstrate that the atoms and bond lengths have been correctly implemented.

Quite generally, every two-dimensional projection has the symmetries of one of the 17 two-dimensional space groups.

11.6.1.1 Special Topic: Special Projections and the Hermann–Mauguin Symbols

Care must be taken in examining the symmetries of the special positions because the *CIF* is consistent with *Pnam* and the *International Tables for Crystallography* are consistent with *Pnma*. The labeling of the unit cell for *Pnam* can be obtained from the labeling of the unit cell for *Pnma* by interchanging **b** and **c**. See Section 11.3.3.1. Thus, the projection along [100] for both settings has the symmetries of *c2mm*. On the contrary, the projection along [001] for *Pnma* is *p2gm*, while the projection along [010] for *Pnam* is *p2mg*. The projection along [010] for *Pnma* is the same as the projection along [001] for *Pnam*, namely *p2gg*.

Along [001]

The two-dimensional space group along [001] for *Pnam* is *p2gg*, where

$$\mathbf{a′} = \mathbf{a} \qquad \mathbf{b′} = \mathbf{b}$$

with the origin at $0,0,z$. Figure 11.10a contains the basis vectors. Figure 11.10b is the direct output from the Starter Program. It shows a projection looking down the length of the

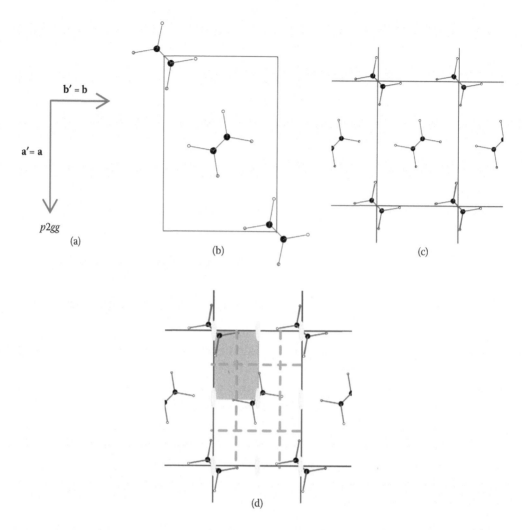

FIGURE 11.10 Polyethylene projections along [001]: (a) basis vectors; (b) from Starter Program; (c) enhanced unit cell; (d) with symmetries and asymmetric unit.

polymer. The mers on the corners are related by a translation. Z is equal to two. The mer at the center of the unit cell is related by a glide to the corner mer. The enhanced diagram in Figure 11.10c is obtained by repeatedly translating Figure 11.10b to form an array of adjacent figures in order to exhibit the symmetries of $p2gg$. Figure 11.10c shows that the four corner mers are related by translations. In the projection there is a twofold between each of the two carbon atoms in a mer. Figure 11.10d shows all the symmetries for the two-dimensional space group and the asymmetric unit in light blue.

Along [100]
The two-dimensional space group along [100] is $c2mm$, where

$$a' = b \qquad b' = c$$

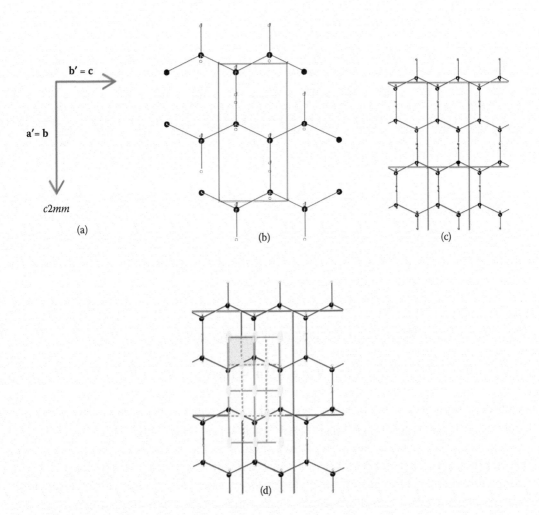

b' = c

a' = b

c2mm

(a)

(b)

(c)

(d)

FIGURE 11.11   Polyethylene projections along [100]: (a) basis vectors; (b) from Starter Program; (c) enhanced unit cell; (d) with symmetries.

with the origin at x,1/4,1/4. Note the considerable difference in the appearance of the two polymers in this projection contrasting with their appearance along [001]. The overlapping of the polymers in this projection makes it difficult to distinguish the distinct polymer chains.

Figure 11.11a contains the basis vectors. Figure 11.11b is the direct output from the Starter Program. The enhanced diagram in Figure 11.11c is gotten by repeatedly translating Figure 11.11b to form an array of adjacent figures in order to exhibit the symmetries of *c2mm*. In *c2mm* in the *International Tables for Crystallography*, the axis of the twofold through the origin is at the intersection of two perpendicular mirror planes (Hahn, 2010, p. 78). In Figure 11.11c the twofold at the projected origin is at the intersection of two perpendicular glides. Move 1/4 in x' and 1/4 in y'; there is the twofold with the two perpendicular mirrors. All of the information necessary to create the entire pattern is

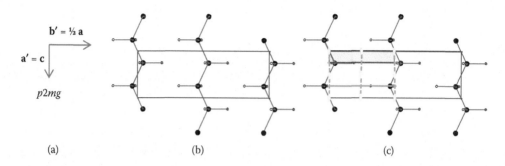

b′ = ½ a

a′ = c

p2mg

(a)   (b)   (c)

FIGURE 11.12   Polyethylene projections along [010]: (a) basis vectors; (b) from Starter Program; (c) with symmetries.

found in the unit cell. Figure 11.11d shows all the symmetries for the two-dimensional space group and the asymmetric unit in light blue.

Along [010]

The two-dimensional space group along [010] is *p2mg*, where

$$\mathbf{a}' = \mathbf{c} \qquad \mathbf{b}' = 1/2\mathbf{a}$$

with the origin at 0, *y*, 0. Note the considerable difference in the appearance of the two polymers in this projection contrasting with their appearance along [001].

Figure 11.12a contains the basis vectors. Figure 11.12b is the direct output from the Starter Program. Examine this diagram to match it with the symmetries for *p2mg*. The mirror planes are normal to **a′**, and the glide planes are normal to **b′**. Figure 11.12c shows all the symmetries for the two-dimensional space group and the asymmetric unit in light blue.

11.7 POLYETHYLENE: RECIPROCAL LATTICE AND *d*-SPACINGS

Planes in the direct lattice correspond to points of the reciprocal lattice. The reciprocal vector

$$\mathbf{H}(hkl) = h\mathbf{a}^* + k\mathbf{b}^* + l\mathbf{c}^*$$

where *hkl* are the indices of the planes of the direct lattice. The *d*-spacing is the distance from the origin to either of the two closest planes. In general, the *d*-spacing can be calculated by the following equation (see Section 5.7.1)

$$\frac{1}{d_{hkl}^2} = H^2(hkl) = \begin{pmatrix} h & k & l \end{pmatrix} \mathbf{G}^* \begin{pmatrix} h \\ k \\ l \end{pmatrix}$$

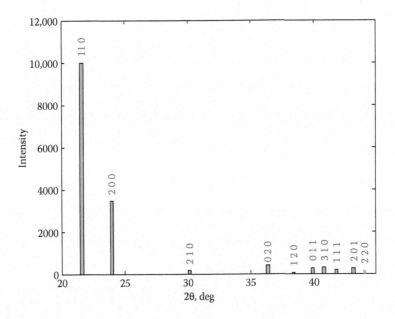

FIGURE 11.13   Polyethylene powder diffraction pattern.

The powder diffraction pattern in Figure 11.13 gives $hkl = 110$ as the reflection with the maximum intensity. The calculation of the *d*-spacing for this reflection is as follows:

$$\frac{1}{d_{110}^2} = H^2(110) = \begin{pmatrix} 1 & 1 & 0 \end{pmatrix} \begin{pmatrix} 0.0182 & 0 & 0 \\ 0 & 0.0410 & 0 \\ 0 & 0 & 0.1550 \end{pmatrix} \begin{pmatrix} 1 \\ 1 \\ 0 \end{pmatrix} \tag{11.1}$$

Now the *d*-spacing for the 110 can be calculated

$$d_{110} = 4.111 \text{ Å}. \tag{11.2}$$

11.7.1 Powder Diffraction Pattern

The diffraction angle is now calculated as a function of *hkl*. First, Bragg's law is applied to the *d*-spacing. Bragg's law is

$$\lambda = 2\, d_{hkl} \sin \theta_{hkl}$$

where λ is the wavelength of the incoming x-rays, d_{hkl} is the *d*-spacing of the *hkl* plane, and θ_{hkl} is the Bragg angle for the *hkl* diffraction maximum. The diffraction angle, $2\theta_{hkl}$, is twice the Bragg angle (see Section 6.6.3). Table 11.8 has *hkl* indices and intensities for the 10 most intense reflections for *Pnam* calculated from the CIF by the program *Mercury*

TABLE 11.8   Polyethylene *hkl* and
Powder Diffraction Intensities

hkl	Intensity
110	10,000
200	3,460
210	167
020	418
120	46
011	278
310	287
111	191
201	249
220	143

of the Cambridge Structural Database. The intensities are normalized so that the highest intensity is 10,000 arbitrary units. The information from Table 11.8 is the input for the Starter Program in Chapter 6. Figure 11.13 is the resulting powder diffraction pattern for polyethylene.

11.8 POLYETHYLENE: ATOMIC SCATTERING CURVES

The diffraction radiation probe for the study used in this chapter was neutron radiation. However, from this point the calculations proceed as if x-rays had been used. This is so that the calculations can be compared with the calculations for the other crystal systems. Figure 11.14 shows the atomic scattering curves from the Starter Program in Chapter 7 for

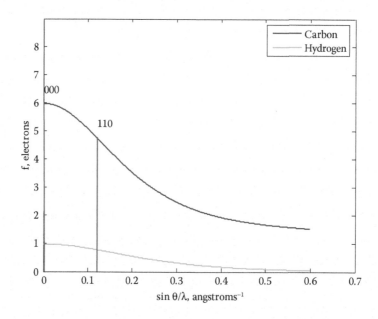

FIGURE 11.14   Polyethylene atomic scattering curves with *hkl* equal to 000 and 110.

the two elements found in polyethylene. The atomic scattering curves are calculated from the Cromer–Mann coefficients. The atomic scattering factor (or form factor) is the ratio of the amplitude of the wave scattered by an atom to the amplitude of the wave scattered by one electron. Its units are electrons. The atomic scattering curve is a measure of the efficiency of an atom in scattering x-rays.

Bragg's law for (110) is

$$\sin \theta / \lambda = 1/(2d) = 1/(2 \times 4.111) = 0.1216 \text{ Å}^{-1}. \quad (11.3)$$

$$f_C(110) = 4.77 \quad \text{and} \quad f_H(110) = 0.79 \text{ electrons.} \quad (11.4)$$

11.9 POLYETHYLENE: STRUCTURE FACTOR

The general formula for the structure factor, Equation 7.5, is

$$F\left(hkl\right) = \sum_{j=1}^{N} f_j \left(\frac{\sin \theta_{hkl}}{\lambda} \right) e^{2\pi i(hx_j + ky_j + lz_j)}$$

where N is the number of atoms in the unit cell. In the case of polyethylene there are 24 atoms in the unit cell (see Table 11.9). The running index j goes from 1 to 24. Table 11.6 includes the fractional coordinates x_j, y_j, z_j for the jth atom in the asymmetric unit. The formula for the structure factor for polyethylene can be broken down in the following way:

$$F(hkl) = F_C(hkl) + F_D(hkl) \quad (11.5)$$

The first term is a sum over 8 carbon atoms, and the second term is a sum over 16 deuterium atoms

$$F(hkl) = \sum_{j}^{8C} F_{Cj}(hkl) + \sum_{j}^{16D} F_{Dj}(hkl) \quad (11.6)$$

Note that the atomic scattering factor for deuterium is indistinguishable from the atomic scattering factor for hydrogen because the number of electrons is the same, namely

TABLE 11.9   Polyethylene $(C_4D_8)_n$, Calculation of Number of Electrons in Unit Cell

Atom	C	D	Total/Mer
Atomic number	6	1	
Atoms/mer	4	8	12
Total electrons/mer	24	8	32
$Z = 2$			Total/Unit Cell
Atoms/unit cell	8	16	24
Electrons/unit cell	48	16	64

one electron. Now put in the expressions for the individual atomic form factors and the phase factors

$$F(hkl) = \sum_{j}^{8C} f_C(hkl)e^{2\pi i(hx_j+ky_j+lz_j)} + \sum_{j}^{16D} f_D(hkl)e^{2\pi i(hx_j+ky_j+lz_j)}$$ (11.7)

11.9.1 Calculate the Structure Factor at 000

The special case of the undeviated beam occurs when $hkl = (0\ 0\ 0)$. Its position is labeled in red in Figure 11.14.

$$F(000) = 8f_C(000) + 16f_D(000)$$

Because the atomic scattering factor, f, at the origin is equal to the atomic number, $F(000)$ is the sum of all the electrons in the unit cell (Section 7.6.4). Table 11.9 shows how the total number of electrons is counted in the unit cell of polyethylene to give

$$F(000) = 64 \text{ electrons.}$$ (11.8)

11.9.2 Calculate the Contribution of One Atom to the Structure Factor

The calculation in Equation 11.7 can be sampled by calculating only the contribution of one atom, C1, to the most intense hkl reflection found in the x-ray powder diagram, which is 110. Table 11.6 gives the fractional coordinates for C1 as

$$x = 0.040, \quad y = 0.061, \quad z = 1/4$$

For the reflection 110

$$F_{C1}(110) = f_c(110)e^{2\pi i(0.040+0.061+0(1/4))} = 4.77e^{0.202\pi i} \text{ electrons.}$$ (11.9)

Note that this is a complex number. The entire structure factor, including all 24 atoms, is real because the space group is centrosymmetric (see Section 7.7.2). The importance of the structure factors is that they become the coefficients of the Fourier expansion that is used to calculate the electron density (see Section 7.8.2). See Exercises 11.5 and 11.6.

DEFINITIONS

Cryogenics
Full form of the Hermann–Mauguin symbol
Mer
Short form of the Hermann–Mauguin symbol

EXERCISES

11.1 Write a report describing in detail the low-temperate apparatus used to achieve 200, 100, and 10 K.

11.2 The third symbol diagram for space group #62, *Pnma* is for *Pcmn*. Analyze the symbols in a way similar to the text.

11.3 For space group #62 draw the general position diagram for *Pnam*.

11.4 Space group #18 is $P2_12_12$. Analyze the three possibilities as done in the text.

11.5 Figure 11.14 shows the atomic scattering curves for polyethylene for x-rays. Draw the corresponding curves for neutron radiations. See Section 15.1.3.

11.6 Equation 11.9 shows the contribution of C1 to the 110 reflection using x-rays. Calculate the contribution of C1 to the 110 reflection using neutrons. See Section 15.1.3.

11.7 Suggested crystals from CSD with space group #62, *Pnma*, are listed in ascending order of volume in Table 11.10. Construct a paper or a PowerPoint presentation based on the parallel topics in this chapter. The website is http://webcsd.ccdc.cam.ac.uk/teaching_database_demo.php; a shortcut to an individual Reference code is: http://webcsd.ccdc.cam.ac.uk/display_csd_entry.php?identifier=MBRMET10 for a crystal with Reference Code "MBRMET10". The parallel topics are discussed in Section 8.3.1. Examples are given at https://sites.google.com/a/vt.edu/foundations_of_crystallography/.

11.8 Suggested crystals from CSD with space group #19, $P2_12_12_1$, are listed in ascending order of volume in Table 11.11. Construct a paper or a PowerPoint presentation based on the parallel topics in this chapter. The website is http://webcsd.ccdc.cam.ac.uk/teaching_database_demo.php; a shortcut to an individual Reference code

TABLE 11.10 Crystals with Space Group #62, *Pnma*

Ref. code	Chemical Name	Formula	Volume (Å³)
MBRMET10	Methyl bromide	CH_3Br	262.82
MIMETH10	Methyl iodide	CH_3I	324.95
BUPQAE01	4-Ethynylaniline	C_8H_7N	656.15
BARBAD01	Barbituric acid dihydrate	$C_4H_4N_2O_3, 2(H_2O)$	711.47
FURROZ	Tricarbonyl-(η^4-cyclobutenyl)-iron(0)	$C_7H_4FeO_3$	725.8
TAQBUI	((Trimethylammonio)methyl)tetrafluorosilicate	$C_4H_{11}F_4NSi$	794.11
KOVSOD	Hexacarbonyl-tungsten	C_6O_6W	877.15
FUBYIK	Hexacarbonyl-molybdenum	C_6MoO_6	890.13
FLUREN01	Fluorene	$C_{13}H_{10}$	916.56

TABLE 11.11   Crystals with Space Group #19, $P2_12_12_1$

Ref. code	Chemical Name	Formula	Volume (Å^3)
NTROMA05	Nitromethane	CH_3NO_2	263.44
ALUCAL05	D-Alanine	$C_3H_7NO_2$	421.47
YILLAG	L-(+)-Lactic acid	$C_3H_6O_3$	432.07
LSERIN01	L-(−)-Serine	$C_3H_7NO_3$	451.59
LCYSTN22	L-Cysteine	$C_3H_7NO_2S$	526.44
PROLIN	L-Proline	$C_5H_9NO_2$	541.74
LTHREO01	L-Threonine	$C_4H_9NO_3$	545.49
ABINOS	beta-L-Arabinose	$C_5H_{10}O_5$	615.86
BERTOH01	4-Aminobenzonitrile	$C_7H_6N_2$	615.96
LGLUAC01	L-Glutamic acid	$C_5H_9NO_4$	618.05

is: http://webcsd.ccdc.cam.ac.uk/display_csd_entry.php?identifier=MBRMET10 for a crystal with Reference Code "MBRMET10". The parallel topics are discussed in Section 8.3.1. Examples are given at https://sites.google.com/a/vt.edu/foundations_of_crystallography/.

## CHAPTER 12

Tetragonal System

α-Cristobalite

Single octahedral crystal of α-cristobalite was found in 1941 by a young soldier, Bennett Frank Buie in the Ellora Caves located in the northeast corner of Hyderabad State, India. The crystals were in a cool cave associated with the zeolite, mordenite. Van Valkenburg did the laboratory analysis while Buie was on active duty with the U.S. Army. Buie gave a specimen to Harvard University. Years later Carl Francis of the Harvard University Mineralogical Museum allowed Robert T. Downs to shake out the box the sample was stored in. That was where he got the single crystal of α-cristobalite that was used in the study for this chapter. (Van Valkenburg and Buie, 1945. Photograph courtesy of the RRUFF project.)

CONTENTS

CHAPTER OBJECTIVES

- Study the crystallographic properties of α-cristobalite, space group #92, $P4_12_12$, as a representative of the tetragonal crystal system.

- Present the parallel crystallographic topics that recur in Chapters 9 through 15. A few examples are given below. A complete list is found in Section 8.3.1.

  - Multiplication table of the associated point group

  - Diagrams of the unit cell, the asymmetric unit, the populated unit cell, and the special projections

  - Diagram of the reciprocal cell superimposed on the direct unit cell

  - Calculation of the powder diffraction pattern

  - Calculation of the atomic scattering curves

  - Calculation of the structure factor at 000

  - Calculation of the contribution of one atom to the structure factor of the largest reflection

- Present the special crystallographic topics of this chapter, which are listed below. A complete list of the special topics found in Chapters 9 through 15 is in Section 8.3.2.

- Organization of crystal structures with polyhedrons

- Enantiomorphic space group pairs

- Experimental detection of enantiomers

12.1 INTRODUCTION

The example chosen for the tetragonal crystal system is the silicate mineral α-cristobalite, SiO_2. It is a high-temperature form of silica. Other polymorphs of silica are quartz, coesite, tridymite, and stishovite. The space group for β-(high)-cristobalite is the cubic space group #227, $Fd\overline{3}m$, the same space group as diamond. Each of these polymorphs can be described in terms of SiO_4^{4-} tetrahedrons. The space group for α-cristobalite is #92, $P4_12_12$, which is one member of an enantiomorphic space group pair. This crystal was chosen to represent the Pearson's Crystal Data Base (PCD) of inorganic materials including minerals. This data base complements the Cambridge Structural Database. PCD has a free demonstration version that includes α-cristobalite.

Table 12.1 contains the crystal data for α-cristobalite.

The special topics considered in this chapter are polyhedrons, enantiomorphic space group pairs, and the experimental detection of enantiomers.

12.1.1 Special Topic: Polyhedrons

Polyhedrons are introduced in Section 4.13.8 to describe the SO_4^{2-} moiety in anhydrous alum. The populated unit cell and all the special projections of α-cristobalite are shown in

TABLE 12.1 α-Cristobalite Crystal Data

Compound	α-Cristobalite
Formula	SiO_2
Molecular weight	60.1 g mol^{-1}
Space group	$P4_12_12$, #92
Bravais lattice	tP
Crystal system	Tetrahedral
Diffraction radiation probe	X-rays Mo Kα; λ = 0.07093 nm
Ambient temperature for data	Room temperature
Data base and identifier	Pearson's Crystal Structure Database—demonstration version; 1251919
	American Mineralogist Data Base; 0001629
Reference	Downs R.T. and Palmer D.C. The pressure behavior of α-cristobalite.
	American Mineralogist (1994) 79, 9–14
Molecular volume	171.10 Å^3
Sample locality	India, Maharashtra State, Ellora caves
Geological description	Crystals formed in vesicles of Deccan basalt, perched on fibers of mordenite in association with paramorphs of quartz
Stability	Metastable phase, may be maintained up to ~773 K and up to 1.26–1.60 GPa
Experimental: Pressure range	P = 0.0001–1.60 GPa, a = 4.972–4.834, c = 6.922–6.642 Å
Z, Z'	Z = 4, Z' = 4/8 = 1/2
Density	2.33 g/cm^3

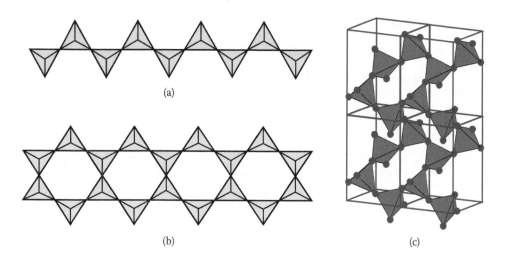

(a)

(b)   (c)

FIGURE 12.1   Silicate tetrahedrons: (a) single chain; (b) double chain; (c) linked framework.

this chapter with both the ball-and-stick model and the polyhedral model. In Chapter 13 boron forms an icosahedron and in Chapter 14 the figure made by the 12 nearest neighbors in close-packed structures is a cuboctahedron.

Tetrahedrons are found in many silicates. Octahedrons also occur. Silicates are the most abundant mineral, making up over 90% of the earth's crust (C. Sorrell, 1973). The tetrahedrons can occur separately, for example, in garnets. In such cases metal atoms balance the charge. An example of a garnet is $Fe_3Al_2(SiO_4)_3$. Single-chain silicates have tetrahedrons that link to form a chain by sharing two oxygen ions (see Figure 12.1a). Pyroxenes, such as $MgSiO_3$, are an example. Double-chain silicates have half the tetrahedrons sharing two oxygen ions and the other half sharing three oxygen ions (see Figure 12.1b). Amphiboles are examples. In the framework silicas, SiO_2, such as quartz and cristobalite, the tetrahedrons share all four oxygen ions with adjacent tetrahedrons. They link in a three-dimensional framework structure. Figure 12.1c shows the tilted tetrahedrons in α-cristobalite.

12.1.2 Special Topic: Enantiomorphic Pairs of Space Groups

The space group #92, $P4_12_12$, contains 4_1 screw axes. Figure 12.2 shows the fourfold axes 4, 4_1, 4_2, 4_3, and $\bar{4}$. Note that 4_1 and 4_3 are mirror images of one another. So are 3_1 and 3_2. In Figure 14.7, 6_1 and 6_5 as well as 6_2 and 6_4 also form pairs of mirror images. Screws 3_1, 4_1, 6_1, and 6_2 form right-handed spirals or helixes; and screws 3_2, 4_3, 6_4, and 6_5 form left-handed spirals or helixes. Space groups #144, $P3_1$, and #145, $P3_2$, are mirror images of each other and are together called an *enantiomorphic or chiral pair of space groups*. There are 11 such pairs that are given in Table 12.2.

Because of the intimate relationship between the enantiomorphic pairs of space groups, some classifications do not consider them distinct space groups. There are 219 (230-11) *affine space group types*. The *International Tables for Crystallography* use all 230 space groups, and they are formally called the *crystallographic space group types*. All 230 space groups are indexed with right-handed bases.

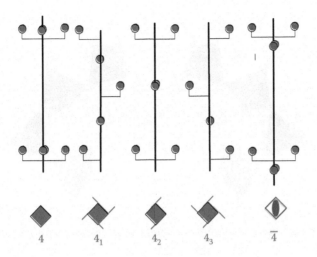

4 4_1 4_2 4_3 $\bar{4}$

FIGURE 12.2 Fourfold axes.

TABLE 12.2 Enantiomorphic Space Group Pairs

Crystal System/ Handedness	Right-Handed	Left-Handed
Tetragonal	#76 $P4_1$	#78 $P4_3$
Tetragonal	#91 $P4_12\,2$	#95 $P4_32\,2$
Tetragonal	#92 $P4_12_12$	#96 $P4_32_12$
Trigonal	#144 $P3_1$	#145 $P3_2$
Trigonal	#151 $P3_112$	#153 $P3_212$
Trigonal	#152 $P3_121$	#154 $P3_221$
Hexagonal	#169 $P6_1$	#170 $P6_5$
Hexagonal	#171 $P6_2$	#172 $P6_4$
Hexagonal	#178 $P6_122$	#179 $P6_522$
Hexagonal	#180 $P6_222$	#181 $P6_422$
Cubic	#210 $P4_132$	#212 $P4_332$

The crystal structure of each member of the enantiomorphic pair of space groups is chiral even though the formula units may not be. Such is the case with the tetrahedrons in α-cristobalite which is found in both space groups #92, $P4_12_12$, and its mirror #96, $P4_32_12$ (Deer et al., 1992) (see Figure 12.3). In Section 9.1.2 molecular enantiomers are discussed.

12.1.3 Special Topic: Experimental Detection of Enantiomers

Crystals of α-quartz form in the trigonal space groups #152, $P3_121$, and #154, $P3_221$. Crystals of the two space groups can be distinguished because they rotate polarized light along the threefold axis in opposite directions. The unfortunate assignment is such that right-handed α-quartz rotates polarized light in the left-handed direction and vice versa (Lang, 1965). Above 550°C α-quartz undergoes a transformation to β-quartz, which is in the hexagonal crystal system. There are two enantiomorphic forms of β-quartz, #180,

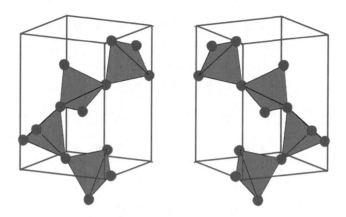

FIGURE 12.3 α-Cristobalite space groups: (a) #92, $P4_12_12$; (b) its mirror #96, $P4_32_12$.

$P6_222$, and #181, $P6_422$; and they rotate polarized light along the sixfold axis in opposite directions (Glazer, 2013, 207–208).

The instrument that measures the angle of rotation of polarized light after it has passed through an optically active material is a *polarimeter*. An *enantiomer*, that is, one of the members of an enantiomorphic pair, is named by the direction in which it rotates the plane of polarized light. If the light is rotated clockwise, the enantiomer is labeled by (+). On the other hand, if it rotates the light counterclockwise, it is labeled by (−). Polarimetry has been used since long before the structure of molecules was experimentally determined by diffraction methods. Therefore, the assignment of D- and L-, determined by the structure, is independent of the direction in which the material rotates polarized light, that is the assignment of (+) or (−). For example, the CSD gives the crystal structure of L-(+)-lactic acid, $C_3H_6O_3$, (YILLAG). This crystal rotates plane-polarized light in the clockwise direction, and its structure is left handed as defined in crystallography (Rupp, 28). Another use of a polarimeter is measuring the concentration of an optically active compound like sugar.

12.2 α-CRISTOBALITE: POINT GROUP PROPERTIES

The associated point group of α-cristobalite is 422 in the Hermann–Mauguin notation and D_4 in the Schönflies notation. This point group has eight symmetry operations: 1, 2_z, 4^+, 4^-, and four twofolds perpendicular to [001]; thus the order, or number of elements, is eight.

12.2.1 Multiplication Table

The generators from Appendix 3 are 1, 2_z, 4^+, and 2_y. However, the point group can be generated from just two symmetry operations (see Figure 3.58). There are several choices for these two generators. Possible examples are 4^+ and 2_y. The matrices for these generators are

$$4^+ = \begin{pmatrix} 0 & \bar{1} & 0 \\ 1 & 0 & 0 \\ 0 & 0 & 1 \end{pmatrix} \text{ and } 2_y = \begin{pmatrix} \bar{1} & 0 & 0 \\ 0 & 1 & 0 \\ 0 & 0 & \bar{1} \end{pmatrix}$$

TABLE 12.3   Multiplication Table for Point Group 422

	1	**2_z**	**4^+**	**4^-**	**2_y**	**2_x**	**2_{xx}**	**$2_{x\bar{x}}$**
1	1	2_z	4^+	4^-	2_y	2_x	2_{xx}	$2_{x\bar{x}}$
2_z	2_z	1	4^-	4^+	2_x	2_y	$2_{x\bar{x}}$	2_{xx}
4^+	4^+	4^-	2_z	1	$2_{x\bar{x}}$	2_{xx}	2_y	2_x
4^-	4^-	4^+	1	2_z	2_{xx}	$2_{x\bar{x}}$	2_x	2_y
2_y	2_y	2_x	2_{xx}	$2_{x\bar{x}}$	1	2_z	4^+	4^-
2_x	2_x	2_y	$2_{x\bar{x}}$	2_{xx}	2_z	1	4^-	4^+
2_{xx}	2_{xx}	$2_{x\bar{x}}$	2_x	2_y	4^-	4^+	1	2_z
$2_{x\bar{x}}$	$2_{x\bar{x}}$	2_{xx}	2_y	2_x	4^+	4^-	2_z	1

The Starter Program from Chapter 3 calculates the multiplication table in Table 12.3. Because

$$2_y \times 4^- = 2_{x\bar{x}} \quad \text{and} \quad 4^- \times 2_y = 2_{xx},$$

the matrix of the point group is not symmetric, and the point group is not abelian. Because the point group does not have an inversion operation, the symmetries of the space group permit the piezoelectric effect. Compare the multiplication table for *mmm* in Table 11.4 with the multiplication table for 422. Both point groups have order 8, but they are not isomorphic. Note that *mmm* has the identity operator, 1, down the diagonal while 422 does not. The relationships among the symmetry operations are seen in the multiplication table.

12.2.2 Stereographic Projections

The general position stereographic projection for the point group 422, Figure 12.4a, has eight points corresponding to eight symmetry operations. See Section 3.9 for a general description of these projections. The point corresponding to the identity operation has a positive *z*-coordinate indicated by the dot in the lower right-hand quadrant. The fourfold along the *z*-axis, generates the three other dots in the diagram. The open circle in the lower right-hand quadrant corresponds to the twofold rotation 2_{xx}. Open circles indicate

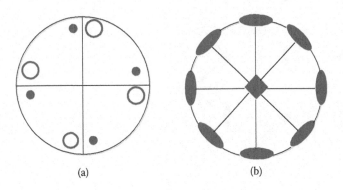

(a)   (b)

FIGURE 12.4   Point group 422: (a) point; (b) symbol stereographic projections.

negative z-coordinates. The color red is used for the proper operations. All the operations in point group 422 are proper. A proper operation does not change handedness, and the determinant of the matrix is +1. Note that dots and open circles are positioned interior to the circumference to allow any value for z.

Figure 12.4b is the symbol stereographic projection. The red square in the center represents the fourfold axis with the cyclic point group 4. In addition, there are four two-fold axes perpendicular to the z-axis, each with the cyclic point group 2. Each twofold axis is represented by two red ovals connected by a black guide line. All the operations are proper; thus all the symbols are colored red.

Given the symbol diagram, the general position diagram can be created and vice versa.

12.3 α-CRISTOBALITE: SPACE GROUP PROPERTIES

α-Cristobalite crystallizes in the tetragonal space group #92, $P4_12_12$. The Bravais lattice is tP. The symmetry operations impose the restrictions

$$a = b \quad \alpha = \beta = \gamma = \text{exactly } 90°$$

12.3.1 Space Group Diagrams

Figure 12.5a shows the symbol diagram for the nonsymmorphic space group #92, $P4_12_12$. All the operations are proper; hence all the symbols are red. An explanation of the Hermann–Mauguin symbol follows. The capital letter refers to the centering type. The next three parts of the Hermann–Mauguin symbol describe the symmetries in their relationship to the crystallographic directions in the space group. In the tetragonal system, the first place describes a fourfold axis parallel to [001], the second place describes symmetries related to [100] and [010], and the third place describes symmetries related to [110] and [1$\overline{1}$0].

For space group #92, $P4_12_12$, the "P" means that the cell is primitive. The "4_1" refers to the 4_1 screw axes parallel to [001], which are shown at the midpoint of each of the sides of the unit cell (see Figure 12.2). There are also twofold screw axes parallel to [001] at the vertices

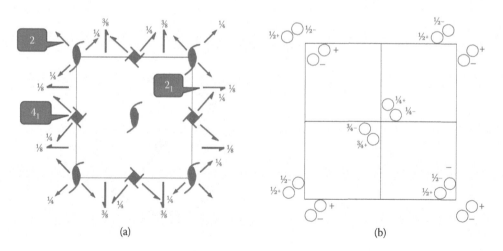

(a) (b)

FIGURE 12.5 Space group #92, $P4_12_12$: (a) symbol diagram; (b) general position diagram.

and the center of the unit cell that are generated by the 4_1 screw axes and the translation operations. The "2_1" refers to the twofold screw axes parallel to [100] and [010], indicated by the half-arrows. The "2" refers to the twofold axes along [110] and [1$\overline{1}$0] indicated by the arrows.

Figure 12.5b shows the general position diagram for the space group. The eight red circles within the unit cell represent the eight proper operations. The plus sign shows that the z-coordinate is positive, and the minus sign shows that the z-coordinate is negative. The red circle in the upper left-hand quadrant with the plus sign represents the coordinates, xyz, of the identity operation. Each of the four vertices of the unit cell is a lattice point and has an identical neighborhood. Now check out other operations from the symbol diagram. Given the symbol diagram, the general position diagram can be created and vice versa.

12.3.2 Maximal Subgroups and Minimal Supergroups

There are 230 space groups. Because the space groups contain translation operations, each space group has infinite order. Space groups can be arranged in trees similar to the trees of the point groups. However, there is no unique way of constructing the tree for the space groups. In a tree of Type I, the centering type is kept and the order of the associated point group is changed. (Maximal subgroups and minimal supergroups are discussed in Section 4.8.) For example, a maximal Type I subgroup of the space group #92, $P4_12_12$, is the space group #76, $P4_1$. Both these space groups have the same centering type, namely P; but for the subgroup the order of the associated point group is decreased. Specifically, the associated point group of the space group #92, $P4_12_12$, is 422 and has order eight, while the associated point group of the space group #76, $P4_1$, is 4 and has order four. The space group #92, $P4_12_12$, has three maximal subgroups all of which have associated point groups of order four. The space group has one minimal supergroup, which is cubic, and has an associated point group of order 24. The *International Tables for Crystallography* give the information necessary to construct Figure 12.6.

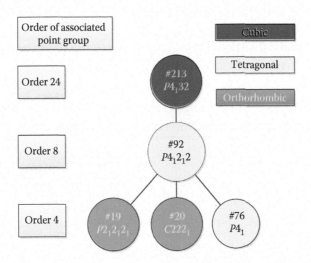

FIGURE 12.6   Space group #92, $P4_12_12$, maximal subgroups and minimal supergroup.

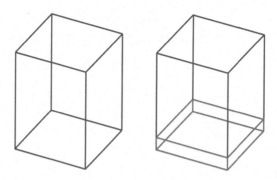

FIGURE 12.7 α-Cristobalite: (a) unit cell in blue; (b) with the asymmetric unit in red.

12.3.3 Asymmetric Unit

The asymmetric unit is the smallest part of the unit cell that, when operated on by the symmetry operations, produces the whole unit cell. The volume of the asymmetric unit is related to the volume of the unit cell by $V_{AU} = (V_{UC}/nh)$, where n is the number of lattice points in the unit cell and h is the number of operations in the associated point group. (Note that the product nh is equal to the multiplicity of the general position.) Because there are eight symmetry operations in the associated point group of this primitive cell ($n = 1$, $h = 8$), the volume of the asymmetric unit is one eighth the volume of the unit cell. Therefore, because the volume of the unit cell is 171.1 Å³, the volume of the asymmetric unit is 21.4 Å³. Figure 12.7 shows the unit cell drawn in blue for α-cristobalite with the asymmetric unit in red superimposed on it. The asymmetric unit has

$$0 \leq x \leq 1, \quad 0 \leq y \leq 1, \quad \text{and} \quad 0 \leq z \leq 1/8 \text{ (Hahn, 2010, p. 128).}$$

12.4 α-CRISTOBALITE: DIRECT AND RECIPROCAL LATTICES

In addition to the direct lattice, there is the reciprocal lattice, which is the lattice of the diffraction pattern. The diffraction pattern has the intensities of the Bragg reflections superimposed on the reciprocal lattice. Table 12.4 compares the direct and reciprocal lattices, including the lattice parameters, the metric matrix, and the cell volume for both the direct and reciprocal cells.

TABLE 12.4 α-Cristobalite: Comparison of Direct and Reciprocal Lattices

Direct Lattice	Reciprocal Lattice
$a = 4.9717$ Å, $b = 4.9717$ Å, $c = 6.9223$ Å	$a^* = 0.2011$ Å⁻¹, $b^* = 0.2011$ Å⁻¹, $c^* = 0.1445$ Å⁻¹
$\alpha = 90°$, $\beta = 90°$, $\gamma = 90°$	$\alpha^* = 90°$, $\beta^* = 90°$, $\gamma^* = 90°$
$\mathbf{G} = \begin{pmatrix} 24.712 & 0 & 0 \\ 0 & 24.712 & 0 \\ 0 & 0 & 47.918 \end{pmatrix}$ Å²	$\mathbf{G^*} = \begin{pmatrix} 0.0405 & 0 & 0 \\ 0 & 0.0405 & 0 \\ 0 & 0 & 0.0209 \end{pmatrix}$ Å⁻²
$V = 171.104$ Å³	$V^* = 0.0058$ Å⁻³

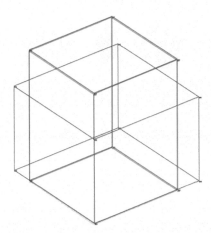

FIGURE 12.8   α-Cristobalite direct unit cell (red) and the reciprocal cell (blue) with volumes normalized to 1.

Figure 12.8 shows the relationship between the direct unit cell (blue) and the reciprocal unit cell (red) with the volumes normalized to one.

12.5 α-CRISTOBALITE: FRACTIONAL COORDINATES AND OTHER DATA FOR THE CRYSTAL STRUCTURE

The crystal structure consists of the placement of the atoms in the unit cell and the bonds between the atoms. The fractional coordinates of the atoms in the asymmetric unit combined with the general position coordinates of the space group and the lattice parameters are sufficient to specify all the atoms in the unit cell.

Z equals 4; that is, there are four formula units, SiO_2, in the unit cell. Thus Z′, the number of formula units in the asymmetric unit, is 1/2. The asymmetric unit has one-half of a silicon atom and one oxygen atom. The silicon atom has multiplicity 4 and Wyckoff letter *a*; and the oxygen atom has multiplicity 8 and Wyckoff letter *b* (see Table 12.5).

Crystallographically independent atoms are atoms not related by symmetries. Examples of crystallographically independent atoms are the atoms in the asymmetric unit. In the case of α-cristobalite, the two atoms—Si and O—are (at least partially) in the asymmetric unit. The red box in Figure 12.9 outlines the asymmetric unit. When the symmetry operations are applied to these atoms, all the atoms in the unit cell are produced. Table 12.6 gives the symmetry operations needed to populate the unit cell.

TABLE 12.5   α-Cristobalite Fractional Coordinates, Multiplicity, and Wyckoff Letter for Atoms in Asymmetric Unit

Atom	*x* Fractional Coordinate	*y* Fractional Coordinate	*z* Fractional Coordinate	Multiplicity Wyckoff Letter
Si	0.30028(9)	0.30028(9)	0	4a
O	0.2392(2)	0.1044(2)	0.1787(1)	8b
O1*	−0.2392 + .5	0.1044 + .5	−0.1787 + .25	

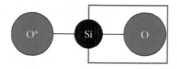

FIGURE 12.9 α-Cristobalite bond schematic with atoms in asymmetric unit outlined in red.

TABLE 12.6   α-Cristobalite General Position Coordinates and Symmetry Operations from *CIF*

#	General Position Coordinates	Symmetry Operations
1	x, y, z	1, identity
2	$1/2 - x, 1/2 + y, 1/4 - z$	Twofold screw perpendicular to **c**-axis
3	$-x, -y, 1/2 + z$	Twofold screw along **c**-axis
4	$1/2 - y, 1/2 + x, 1/4 + z$	4_1^+
5	$-y, -x, 1/2 - z$	Twofold perpendicular to **c**-axis
6	$1/2 + x, 1/2 - y, -1/4 - z$	Twofold screw perpendicular to **c**-axis
7	$1/2 + y, 1/2 - x, -1/4 + z$	4_1^-
8	$y, x, -z$	Twofold perpendicular to **c**-axis

Next, the bonds need to be inserted. The bonds also obey the symmetry operations. There are two types of bonds, the bonds that are contained entirely within the asymmetric unit and the bonds where one atom is in the asymmetric unit and the other atom is outside the asymmetric unit. Oxygen atom O1* is outside the asymmetric unit and is related by symmetry to oxygen O. Atoms Si, O, and O1* are incorporated in the schematic diagram (Figure 12.9). The bond Si–O is within the asymmetric unit and the bond Si–O1* is partially in the asymmetric unit. In the Starter Program for populating the unit cell, this diagram is the key to putting in the intramolecular bonds.

12.6 α-CRISTOBALITE: CRYSTAL STRUCTURE

Populate the unit cell by using the Starter Program from Chapter 4 to create the entire unit cell as shown in Figure 12.10a. This program combines the lattice parameters—a, b, c, α, β, and γ—(Table 12.4), the fractional coordinates (Table 12.5), the general position coordinates of the

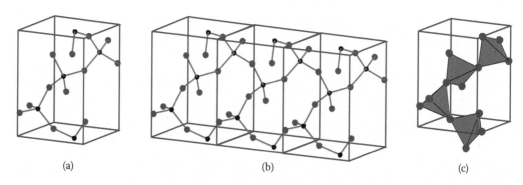

(a)   (b)   (c)

FIGURE 12.10   α-Cristobalite populated unit cell: (a) one unit cell; (b) three unit cells; (c) one unit cell with tetrahedrons.

space group (Table 12.6), and the configuration of bonds (Figure 12.9). The lattice parameters, the fractional coordinates, and the general position coordinates are from the CIF.

Beginning with the general position coordinates the Starter Program automatically includes all the atoms. No special procedure is required for atoms in special positions.

Z equals 4, that is, there are four formula units in the unit cell. The multiplicity of the general Wyckoff position is eight. That is, there are eight symmetry operations described in Section 12.3.1. $Z' = 4/8 = 1/2$. Notice how some of the atoms partially protrude through the walls of the unit cell. Also the atoms in a single cell appear to be sparsely packed. In Figure 12.10b there are three unit cells packed with molecules. Figure 12.10c shows each tetrahedron with a hidden silicon atom in the center attached to four oxygen atoms. The tetrahedrons are linked through the oxygen atoms in this framework structure.

12.6.1 Symmetry of the Special Projections

The *International Tables for Crystallography* give the symmetry for three special projections for each space group in the standard orientation. These projections are orthogonal, which means that each projection is onto a plane normal to the direction of the projection. The basis vectors of the two-dimensional unit cell are labeled **a′** and **b′**, regardless of which two basis vectors of the three-dimensional unit cell (selected from **a**, **b**, **c**) are projected. For the tetragonal crystal system the special projections are along [001], [100], and [110]. In the Starter Program there is a "view" statement that selects the appropriate projection. Here, the [010] is not useful because the same information is found along the [100] projection. Note also that there can be a shift in origin in the projections. For #92, $P4_12_12$, there is a shift of origin along both [010] and along [100]. A careful examination of these projections can demonstrate that the atoms and bond lengths have been correctly implemented.

Quite generally, every two-dimensional projection has the symmetries of one of the 17 two-dimensional space groups.

Along [001]

The two-dimensional space group along [001] is *p4gm*, where

$$a' = a \qquad b' = b$$

with the origin at $0,1/2,z$. Figure 12.11a is the direct output from the Starter Program. Z equals four. All four silicon atoms, shown in black, and all eight oxygen atoms, shown in blue, are discernible here in the unit cell. On an average there are two blue oxygen atoms to a silicon atom. The oxygen atoms are at the vertices of the linked tetrahedrons. Figure 12.11b shows that the structure is a framework structure made up of these linked tetrahedrons.

The enhanced diagrams in Figure 12.12 are gotten by repeatedly translating the diagrams in Figure 12.11 to form an array of adjacent figures in order to exhibit the symmetries of *p4gm*. In *p4gm* in the *International Tables for Crystallography*, the axis of the fourfold is through the origin (Hahn, 2010, 81). In Figure 12.12, the origin does not have fourfold symmetry, but now move 1/2 in y' and there is the fourfold in both figures. The

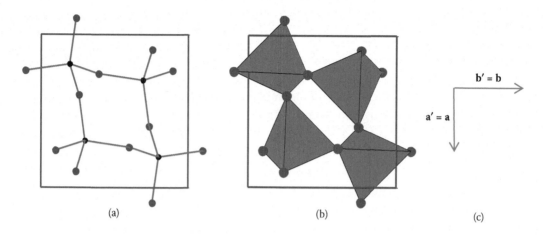

FIGURE 12.11   α-Cristobalite along [001]: (a) unit cell; (b) with tetrahedrons; (c) basis vectors.

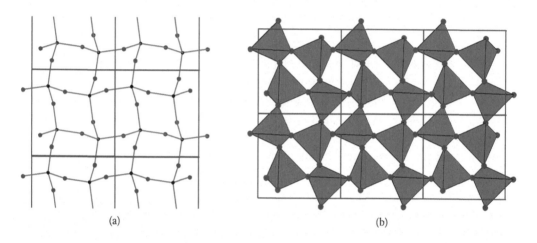

FIGURE 12.12   α-Cristobalite along [001]: (a) expanded unit cell; (b) with tetrahedrons.

extended figures also show that the opposite sides of the unit cell are identical translations of one another. All the information necessary to create the entire pattern is found in the unit cell. Figure 12.13a shows all the symmetries for the two-dimensional space group and the asymmetric unit in light blue. Figure 12.13b shows the corresponding figure with the tetrahedrons drawn in.

Along [100]

The two-dimensional space group along [100] is *p2gg*, where

$$\mathbf{a'} = \mathbf{b} \qquad \mathbf{b'} = \mathbf{c}$$

with the origin at x, 1/4, 3/8. Figure 12.14a is the direct output from the Starter Program, and Figure 12.14b shows the tetrahedrons. The enhanced diagrams in Figure 12.15 are gotten by repeatedly translating the diagrams in Figure 12.14 to form an array of adjacent figures in order to exhibit the symmetries of *p2gg*. In *p2gg* in the *International Tables*

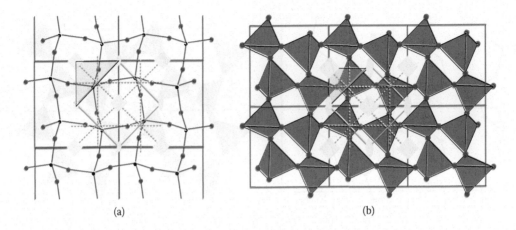

FIGURE 12.13   α-Cristobalite along [001]: (a) with symmetries; (b) with tetrahedrons.

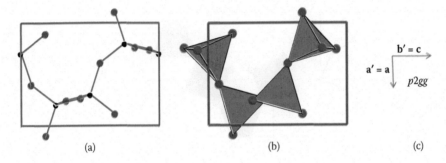

FIGURE 12.14   α-Cristobalite along [100]: (a) unit cell; (b) with tetrahedrons; (c) basis vectors.

for Crystallography, the axis of the twofold is through the origin (Hahn, 2010, p. 77). In Figure 12.15 the origin does not have twofold symmetry, but now move 1/4 in x' and 3/8 in y' and there is the twofold in both figures. Finally, the diagrams in Figure 12.16 show all the symmetries for the two-dimensional space group each with their asymmetric unit in light blue.

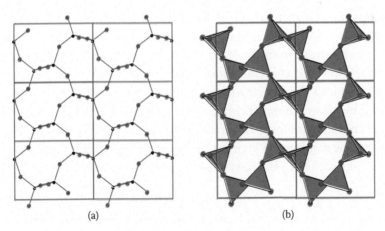

FIGURE 12.15   α-Cristobalite along [100]: (a) expanded unit cell; (b) with tetrahedrons.

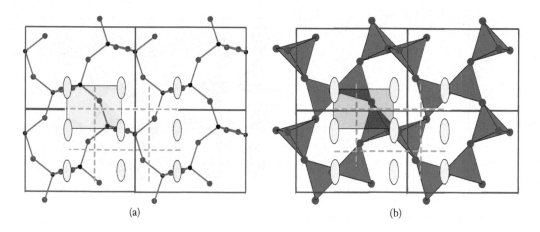

FIGURE 12.16   α-Cristobalite along [100]: (a) with symmetries; (b) with tetrahedrons.

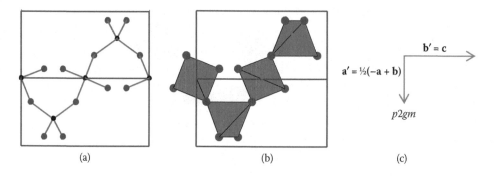

FIGURE 12.17   α-Cristobalite along [110]: (a) unit cell; (b) tetrahedrons; (c) basis vectors.

Along [110]

The two-dimensional space group along [110] is *p2gm*, where

$$\mathbf{a}' = 1/2(-\mathbf{a} + \mathbf{b}) \qquad \mathbf{b}' = \mathbf{c}$$

with the origin at *x*, *x*, 0, where *x* is an arbitrary parameter that could be chosen to be zero. Again note the subtle differences in the appearance of the four formula units in this projection contrasted with their appearance along [100] and [010]. The diagrams in Figure 12.17 are the direct output from the Starter Program. The direct output *partially* portrays two unit cells. However, it contains enough information to portray the cells completely. Here, Figure 12.18 uses the information to complete a unit cell. The completion is made by overlapping, by integral numbers of **a'**, one partially portrayed unit cell on another. The overlapping is continued to form an array of adjacent figures that exhibit the symmetries of *p2gm*. Finally, the diagrams in Figure 12.19 show all the symmetries for the two-dimensional space group, each with their asymmetric unit in light blue.

12.7 α-CRISTOBALITE: RECIPROCAL LATTICE AND *d*-SPACINGS

Planes in the direct lattice correspond to points of the reciprocal lattice. The reciprocal vector

$$\mathbf{H}(hkl) = h\mathbf{a}^* + k\mathbf{b}^* + l\mathbf{c}^*$$

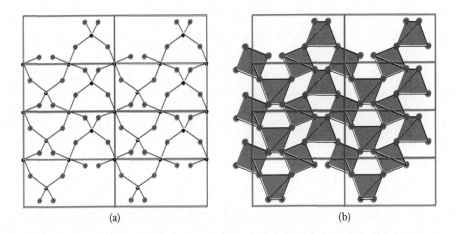

FIGURE 12.18 α-Cristobalite along [110]: (a) expanded unit cell; (b) tetrahedrons.

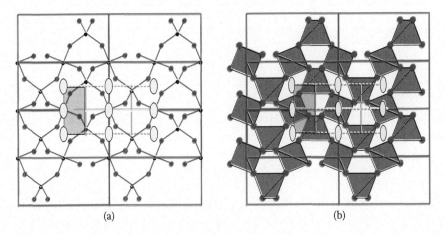

FIGURE 12.19 α-Cristobalite along [110]: (a) with symmetries; (b) with tetrahedrons.

where *hkl* are the indices of the planes of the direct lattice. The *d*-spacing is the distance from the origin to either of the two closest planes. In general, the *d*-spacing can be calculated by the following equation (see Section 5.7.1):

$$\frac{1}{d_{hkl}^2} = H^2(hkl) = (h \quad k \quad l)\mathbf{G}^*\begin{pmatrix} h \\ k \\ l \end{pmatrix}$$

The powder diffraction data in Table 12.7 give *hkl* = 101 as the reflection with the maximum intensity. The calculation of the *d*-spacing for this reflection follows:

$$\frac{1}{d_{101}^2} = H^2(101) = (1 \quad 0 \quad 1)\begin{pmatrix} 0.0405 & 0 & 0 \\ 0 & 0.0405 & 0 \\ 0 & 0 & 0.0209 \end{pmatrix}\begin{pmatrix} 1 \\ 0 \\ 1 \end{pmatrix} \qquad (12.1)$$

TABLE 12.7   α-Cristobalite Intensity
versus *hkl* from Pearson's Data Base

hkl	Intensity, Normalized to Highest Maximum
101	1000
110	7.1
111	92.6
102	117.2
200	151.6
201	0.8
211	29.9
202	28.8
113	60.1
212	62.2

Now, the *d*-spacing for the 101 can be calculated as

$$d_{101} = 4.0381 \text{ Å} \tag{12.2}$$

12.7.1 Powder Diffraction Pattern

The diffraction angle is now calculated as a function of *hkl*. First, Bragg's law is applied to the *d* spacing. Bragg's law is

$$\lambda = 2 \, d_{hkl} \sin \theta_{hkl}$$

where λ is the wavelength of the incoming x-rays, d_{hkl} is the *d*-spacing of the *hkl* plane, and θ_{hkl} is the Bragg angle for the *hkl* diffraction maximum. The diffraction angle, $2\theta_{hkl}$, is twice the Bragg angle (see Section 6.6.3). Table 12.7 has *hkl* indices and intensities for the 10 most intense reflections from Pearson's Data Base. The intensities are normalized so that the highest intensity is 1000 arbitrary units. This information from Table 12.7 is the input for the Starter Program in Chapter 6. Figure 12.20 is the resulting powder diffraction pattern for α-cristobalite.

12.8 α-CRISTOBALITE: ATOMIC SCATTERING CURVES

Figure 12.21 shows the atomic scattering curves from the Starter Program in Chapter 7 for the two elements found in α-cristobalite. Note that the Si^{4+} ion has 10 electrons and that the O^{2-} ion also has 10 electrons. The atomic scattering curves are calculated from the Cromer–Mann coefficients. The atomic scattering factor (or form factor) is the ratio of the amplitude of the wave scattered by an atom to the amplitude of the wave scattered by one electron. Its units are electrons. The atomic scattering curve is a measure of the efficiency of an atom in scattering x-rays.

Bragg's law for (101) is

$$\sin \theta / \lambda = 1/(2d) = 1/(2 \times 4.0381) = 0.1238 \text{ Å}^{-1} \tag{12.3}$$

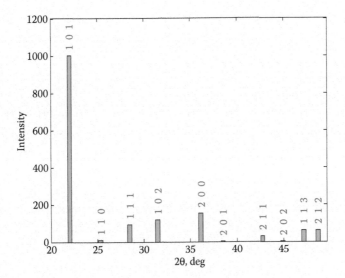

FIGURE 12.20 α-Cristobalite powder diffraction pattern.

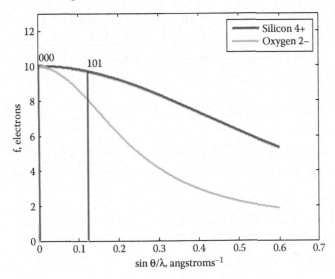

FIGURE 12.21 α-Cristobalite atomic scattering curves with 000 and 101 diffraction maxima.

This reflection is labeled in Figure 12.21 in red. The values from the atomic scattering curves are

$$f_{Si^{+4}}(101) = 9.68 \quad \text{and} \quad f_{O^{-2}}(101) = 8.11 \, \text{electrons} \tag{12.4}$$

12.9 α-CRISTOBALITE: STRUCTURE FACTOR

The general formula for a structure factor, from Equation 7.5, is

$$F(hkl) = \sum_{j=1}^{N} f_j \left(\frac{\sin\theta_{hkl}}{\lambda} \right) e^{2\pi i(hx_j + ky_j + lz_j)}$$

TABLE 12.8   α-Cristobalite, SiO_2, Calculation of Number of Electrons in Unit Cell

Ion	Si^{4+}	O^{2-}	Total/Ion
Number of electrons per ion	10	10	
Atoms/(formula unit)	1	2	3
Total electrons/formula unit	10	20	30
$Z = 4$			Total/Unit Cell
Ions/unit cell	4	8	12
Electrons/unit cell	40	80	120

where N is the number of atoms in the unit cell. In the case of α-cristobalite there are 12 ions in the unit cell (see Table 12.8). The running index j goes from 1 to 12. Table 12.5 includes the fractional coordinates x_j, y_j, z_j for the jth ion in the asymmetric unit. The formula for the structure factor for α-cristobalite can be broken down in the following way:

$$F(hkl) = F_{Si^{4+}}(hkl) + F_{O^{2-}}(hkl) \qquad (12.5)$$

The first term is a sum over 4 silicon ions, and the second term is over 8 oxygen ions

$$F(hkl) = \sum_{j}^{4} F_{Si_j^{4+}}(hkl) + \sum_{j}^{8} F_{O_j^{2-}}(hkl) \qquad (12.6)$$

Now put in the expressions for the individual atomic form factors and the phase factors

$$F(hkl) = \sum_{j}^{4} f_{Si^{4}}(hkl)e^{2\pi i(hx_j + ky_j + lz_j)} + \sum_{j}^{8} f_{O^{-2}}(hkl)e^{2\pi i(hx_j + ky_j + lz_j)} \qquad (12.7)$$

12.9.1 Calculate the Structure Factor at 000

The special case of the undeviated beam occurs when $hkl = (0\ 0\ 0)$ (see Example 7.2). Its position is labeled in red in Figure 12.21.

$$F(000) = 4 f_{Si^{+4}}(000) + 8 f_{O^{-2}}(000)$$

Because the atomic scattering factor, $f(000)$, for an ion is equal to the number of electrons in that ion, $F(000)$ is the sum of all the electrons in the unit cell (Section 7.6.4). Table 12.8 shows how the total number of electrons is counted in the unit cell of α-cristobalite to give

$$F(000) = 120 \text{ electrons} \qquad (12.8)$$

12.9.2 Calculate the Contribution of One Ion to a Structure Factor

The calculation in Equation 12.7 can be sampled by calculating only the contribution of one ion, O^{2-}, to the most intense hkl reflection found in the x-ray powder diagram, which is 101.

Table 12.5 gives the fractional coordinates for O^{2-} as

$$x = 0.2392, \quad y = 0.1044, \quad z = 0.1787$$

For the reflection 101

$$F_{O^{-2}}(101) = f_{O^{-2}}(101)e^{2\pi i(0.2392 + 0 \times 0.2392 + 0.1787)} = 8.11\ e^{0.8358\pi i}\text{electrons} \qquad (12.9)$$

Note that this is a complex number. The entire structure factor, including all 12 ions, may be real or complex because the space group is noncentrosymmetric (see Section 7.7.1.2). The importance of the structure factors is that they become coefficients of the Fourier expansion that is used to calculate the electron density (see Section 7.8.2).

DEFINITIONS

Enantiomer
Enantiomorphic or chiral pair of space groups
Enantiomorphic pair of molecules
Polarimeter

EXERCISES

12.1 Suggested crystals from CSD with space group #92, $P4_12_12$, are listed in ascending order of volume in Table 12.9. Construct a paper or a PowerPoint presentation based on the parallel topics in this chapter. The website is http://webcsd.ccdc.cam. ac.uk/teaching_database_demo.php; a shortcut to an individual Reference Code is: http://webcsd.ccdc.cam.ac.uk/display_csd_entry.php?identifier=FURANE10 for a crystal with Reference Code "Furane10." The parallel topics are discussed in Section 8.3.1. Examples are given at https://sites.google.com/a/vt.edu/ foundations_of_crystallography/.

TABLE 12.9   Crystals with Space Group #92, $P4_12_12$

Ref. Code	Chemical Name	Formula	Volume (Å^3)
FURANE10	Furan	C_4H_4O	385.923
CAFORM05	Calcium formate	$2(CHO_2)^-, Ca^{2+}$	434.027
CIMNUH	Cyclopentene-1,2,3-trione	$C_5H_2O_3$	458.692
ALOXAN	Alloxan	$C_4H_2N_2O_4$	488.494
BZONTR	Benzonitrile	C_7H_5N	576.183
KIPPEE	Ethylenediammonium orthofluoroberyllate	$C_2H_{10}N_2^{2+}BeF_4^{2-}$	621.89
ETDAMS	Ethylenediammonium sulfate	$C_2H_{10}N_2^{2+}, O_4S^{2-}$	646.745
BMLTAA	Dibromo-(maleic acid)-thioanhydride	$C_4Br_2O_2S$	691.581
ACAZEK	Catena-[bis(m2-Bromo)-(m2-ethylene-1,2-diamine)-di-silver(i)]	$(C_2H_8Ag_2Br_2N_2)n$	798.786
YAXXAW	tris(Acetamidinium) hexafluoro-aluminum	$3(C_2H_7N_2)^+, AlF_6^{3-}$	1445.23
ASPRIN	Aspirin anhydride	$C_{18}H_{14}O_7$	1656.852

TABLE 12.10   Crystals with Space Group #135, $P4_2/mbc$

Ref. Code	Chemical Name	Formula	Volume (Å³)
HUJLUT	Catena-(bis(m2-Methoxo)-copper(ii))	$(C_2H_6CuO_2)n$	393.706
MECJOT	Catena-((m2-1,4-Dithiacyclohexane-S,S')-(m2-tetrafluoroborato-F,F',F'',F''')-silver(i))	$(C_4H_8AgBF_4S_2)n$	838.537

TABLE 12.11   Crystals with Space Group #141, $I4_1/amd$

Ref. Code	Chemical Name	Formula	Volume (Å³)
GUWWIF	Catena-((m6-acetylene-1,2-dicarboxylato)-calcium)	$(C_4CaO_4)n$	482.556
MUZKAU	Catena-((m6-Acetylenedicarboxylato)-lead)	$(C_4O_4Pb)n$	547.749
KMEIND	Potassium tetramethylindate	$C_4H_{12}In^-$, K^+	797.662
DUFZIO	Guanidinium tetrahydroborate	CH_6N^{3+}, H_4B^-	1100.197
HIBDUR	Catena-(bis(m2-Succinimido-O,O')-lithium iodide)	$(C_8H_{10}LiN_2O_4^+)n$, $n(I^-)$	1178.767

12.2 Suggested crystals from CSD with space group #135, $P4_2/mbc$, are listed in ascending order of volume in Table 12.10. Construct a paper or a PowerPoint presentation based on the parallel topics in this chapter. The website is http://webcsd.ccdc.cam.ac.uk/teaching_database_demo.php; a shortcut to an individual Reference Code is: http://webcsd.ccdc.cam.ac.uk/display_csd_entry.php?identifier=FURANE10 for a crystal with Reference Code "FURANE10." The parallel topics are discussed in Section 8.3.1. Examples are given at https://sites.google.com/a/vt.edu/foundations_of_crystallography/.

12.3 Suggested crystals from CSD with space group #141, $I4_1/amd$, are listed in ascending order of volume in Table 12.11. Construct a paper or a PowerPoint presentation based on the parallel topics in this chapter. The website is http://webcsd.ccdc.cam.ac.uk/teaching_database_demo.php; a shortcut to an individual Reference Code is: http://webcsd.ccdc.cam.ac.uk/display_csd_entry.php?identifier=FURANE10 for a crystal with Reference Code "FURANE10." The parallel topics are discussed in Section 8.3.1. Examples are given at https://sites.google.com/a/vt.edu/foundations_of_crystallography/.

## CHAPTER 13

Trigonal Crystal System

$H_{12}B_{12}^{-2},3K^+,Br^-$

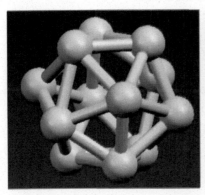

The picture above left shows an ideal icosahedron with the fivefold axis prominently displayed. Boron forms in icosahedrons and related polyhedrons. A fivefold axis is not allowed as a global symmetry for a crystal. In this boron compound, the icosahedron, B_{12}, formed by the boron atoms accommodates itself to the requirement of global symmetry by having its symmetry slightly distorted from the ideal icosahedron. The picture on the right is a statue on the main street in Brisbane, Australia, to honor the truncated B_{60} icosahedron found in β-boron. (Photograph courtesy of Colin Kennard.)

CONTENTS

CHAPTER OBJECTIVES

- Study the crystallographic properties of the boron compound $H_{12}B_{12}^{-2}, 3K^+, Br^-$, space group #166, $R\bar{3}m$, as a representative of the trigonal crystal system.

- Present the parallel crystallographic topics that recur in Chapters 9 through 15. A few examples are given here. A complete list is found in Section 8.3.1.

  - Multiplication table of the associated point group.

  - Diagrams of the unit cell, the asymmetric unit, the populated unit cell, and the special projections.

  - Diagram of the reciprocal cell superimposed on the direct unit cell.

  - Calculation of the powder diffraction pattern.

  - Calculation of the atomic scattering curves.

  - Calculation of the structure factor at 000.

  - Calculation of the contribution of one atom to the structure factor of the largest reflection.

- Present the special crystallographic topics of this chapter, which are listed here. A complete list of the special topics found in Chapters 9 through 15 is in Section 8.3.2.

  - Comparison of rhombohedral and hexagonal axes.

  - Transformation of crystallographic directions.

  - Comparison of space group diagrams for hexagonal and rhombohedral cells.

  - Multiplicity.

  - Boron icosahedron.

  - Space-filling model.

  - Isotypic crystal structures.

13.1 INTRODUCTION

The example chosen for the trigonal crystal system is the boron compound $H_{12}B_{12}^{-2},3K^+,Br^-$. This ionic structure, which is described in the CSD, contains a nonideal icosahedron B_{12} with boron atoms on each of the 12 vertices. Anhydrous alum also forms in the trigonal system (see Table 4.12).

Boron icosahedrons, B_{12}, are found in elemental α-boron (Decker and Kasper, 1959; Donohue, 1974, pp. 48–82; Douglas, 2006, pp. 48–51). β-Boron, which is rhombohedral, forms truncated icosahedrons, B_{60}, (Hoard et al., 1970). α-Boron, β-boron, and $H_{12}B_{12}^{-2},3K^+,Br^-$ all form in the same space group #166, $R\bar{3}m$. Bucky balls or buckminsterfullerene also forms in truncated icosahedrons, C_{60}. The CSD shows over 650 structures with the word "dodecaborate" in the name of the compound and over 1200 structures with the word "fullerene." Thus boron, like carbon, occurs in stable covalently bonded polyhedrons. Table 13.1 contains the crystal data for $H_{12}B_{12}^{-2},3K^+,Br^-$.

TABLE 13.1 $H_{12}B_{12}^{-2},3K^+,Br^-$ Crystal Data

Compound	Tri-potassium dodecahydro-closo-dodecaborate bromide
Formula	$H_{12}B_{12}^{-2},3K^+,Br^-$
Space group	#166, $R\bar{3}m$
Bravais lattice	hR
Crystal system	Trigonal
Ambient temperature	293 K (room temperature)
Diffraction probe	X-rays, diffractometer
Crystal color	Colorless
Reference: Cambridge Structure Database	GAZLEY
Density	1.745 g/cm^3
Rhombohedral primitive cell	$a = b = c = 6.871$ Å; $\alpha = \beta = \gamma = 93.4°$
	Volume (UC) = 322.6 Å^3; Volume (AU) = 26.9 Å^3;
	$Z = 1$; $Z' = 1/12$
Hexagonal triple cell	$a_1 = b_1 = 10.024(11)$ Å, $c_1 = 11.224(21)$ Å; $\alpha_1 = \beta_1 = 90°$; $\gamma_1 = 120°$
	Volume (UC) = 967.8 Å^3; Volume (AU) = 26.9 Å^3;
	$Z = 3$; $Z' = 1/12$

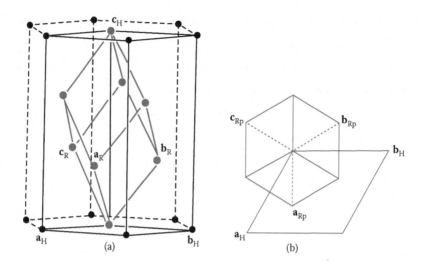

FIGURE 13.1   (a) Hexagonal and rhombohedral reference axes; (b) projection.

13.1.1 Special Topic: Rhombohedral and Hexagonal Axes

There are two ways to index space groups with the Bravais lattice hR. The primitive cell uses rhombohedral axes, and the triple cell (with three lattice points in the unit cell) uses hexagonal axes. With rhombohedral reference axes

$$a = b = c \quad \alpha = \beta = \gamma$$

With hexagonal reference axes

$$a_1 = b_1 \quad \alpha_1 = \beta_1 = 90°; \quad \gamma_1 = 120°$$

Figure 13.1a shows the rhombohedral primitive cell in red with lattice parameters a_R, b_R, and c_R; and the hexagonal triple cell in black with lattice parameters a_H, b_H, and c_H. In Figure 13.1b, the rhombohedral cell is projected down its [111], and the hexagonal cell is projected down its [001]. The dashed lines represent the projections a_{RP}, b_{RP}, and c_{RP} of the basis vectors of the rhombohedral cell. The Starter Program handles either set of axes. Table 13.1 gives lattice parameters, the volumes of both the unit cell and the asymmetric unit, and Z and Z', for both. As expected the volume of the unit cell with the hexagonal axes is exactly three times the volume using rhombohedral axes. On the other hand the volume of the asymmetric unit is the same for both cases. While the Z for hexagonal axes is three times the Z for the rhombohedral axes, the value of Z' is the same, namely 1/12. In general, the number of molecules or formula units in the asymmetric unit as well as the volume of the asymmetric unit are preserved regardless of the indexing of the lattice.

13.1.2 Special Topic: Transforming Crystallographic Directions

In Section 4.10.6, the transformation from rhombohedral to hexagonal coordinates is given as

$$(\mathbf{a}_H \quad \mathbf{b}_H \quad \mathbf{c}_H) = (\mathbf{a}_R \quad \mathbf{b}_R \quad \mathbf{c}_R)P = (\mathbf{a}_R \quad \mathbf{b}_R \quad \mathbf{c}_R)\begin{pmatrix} 1 & 0 & 1 \\ \bar{1} & 1 & 1 \\ 0 & \bar{1} & 1 \end{pmatrix} \qquad (13.1)$$

From Section 1.4, the relationship between the two sets of basis vector is

$$\mathbf{a}_H = \mathbf{a}_R - \mathbf{b}_R \qquad (13.2a)$$

$$\mathbf{b}_H = \mathbf{b}_R - \mathbf{c}_R \qquad (13.2b)$$

$$\mathbf{c}_H = \mathbf{a}_R + \mathbf{b}_R + \mathbf{c}_R \qquad (13.2c)$$

From Section 2.11, a crystallographic direction $[uvw]$ with respect to bases $\mathbf{a}, \mathbf{b}, \mathbf{c}$ can be written as

$$\mathbf{t}(uvw) = u\mathbf{a} + v\mathbf{b} + w\mathbf{c}$$

From Sections 1.6 and 5.3.2, a vector can be written with different sets of basis vectors. For example, using hexagonal (H) axes and rhombohedral (R) axes gives

$$u_H\mathbf{a}_H + v_H\mathbf{b}_H + w_H\mathbf{c}_H = u_R\mathbf{a}_R + v_R\mathbf{b}_R + w_R\mathbf{c}_R \qquad (13.3)$$

Or, in other words

$$[u_H\,v_H\,w_H] \text{ is parallel to } [u_R\,v_R\,w_R]$$

The *International Tables for Crystallography* give the crystallographic directions for the special projections using both hexagonal (H) and rhombohedral (R) axes (Hahn, 2010, pp. 140–143). For example, along either $[001]_H$ or $[111]_R$, which are parallel, the symmetry of the special projection is *p6mm*. Thus, as seen in Equation 13.2c

$$\mathbf{c}_H = \mathbf{a}_R + \mathbf{b}_R + \mathbf{c}_R$$

Along either $[100]_H$ or $[1\bar{1}0]_R$, which are parallel, the symmetry of the special projection is *p2*. Again, as seen in Equation 13.2a

$$\mathbf{a}_H = \mathbf{a}_R - \mathbf{b}_R$$

EXAMPLE 13.1

Use the *P* matrix to convert the crystallographic direction $[110]_H$ into rhombohedral axes.

(continued)

EXAMPLE 13.1 (continued)

Solution

A crystallographic direction [uvw] with respect to bases **a**, **b**, **c** can be written in two different bases (e.g., hexagonal and rhombohedral).

Then rewriting Equation 13.3 gives

$$(\mathbf{a}_H \quad \mathbf{b}_H \quad \mathbf{c}_H)\begin{pmatrix} u_H \\ v_H \\ w_H \end{pmatrix} = (\mathbf{a}_R \quad \mathbf{b}_R \quad \mathbf{c}_R)\begin{pmatrix} u_R \\ v_R \\ w_R \end{pmatrix}$$

Now substitute Equation 13.1 to get

$$(\mathbf{a}_R \quad \mathbf{b}_R \quad \mathbf{c}_R)P\begin{pmatrix} u_H \\ v_H \\ w_H \end{pmatrix} = (\mathbf{a}_R \quad \mathbf{b}_R \quad \mathbf{c}_R)\begin{pmatrix} u_R \\ v_R \\ w_R \end{pmatrix}$$

The equation can be simplified as

$$\begin{pmatrix} u_R \\ v_R \\ w_R \end{pmatrix} = P\begin{pmatrix} u_H \\ v_H \\ w_H \end{pmatrix} \tag{13.4}$$

Thus the crystallographic direction $[110]_H$ may be converted into a rhombohedral basis by using P from Equation 13.1

$$\begin{pmatrix} u_R \\ v_R \\ w_R \end{pmatrix} = P\begin{pmatrix} u_H \\ v_H \\ w_H \end{pmatrix} = \begin{pmatrix} 1 & 0 & 1 \\ \bar{1} & 1 & 1 \\ 0 & \bar{1} & 1 \end{pmatrix}\begin{pmatrix} 1 \\ 1 \\ 0 \end{pmatrix} = \begin{pmatrix} 1 \\ 0 \\ \bar{1} \end{pmatrix}$$

Thus, the crystallographic direction $[110]_H = [10\bar{1}]_R$ *answer.*

See a similar treatment in Sections 2.9.2 and 5.7.3.5 with reciprocal bases.

EXAMPLE 13.2

Along $[2\bar{1}\bar{1}]_R$ the symmetry of the special projection is *p2mm* (Hahn, 2010, p. 143). Calculate the crystallographic direction in hexagonal axes that also has *p2mm* symmetry in projection.

(continued)

EXAMPLE 13.2 (continued)

Solution

To start out with rhombohedral axes and go to hexagonal axes solve Equation 13.4

$$\begin{pmatrix} u_H \\ v_H \\ w_H \end{pmatrix} = P^{-1} \begin{pmatrix} u_R \\ v_R \\ w_R \end{pmatrix}$$

From P in Equation 13.1

$$P^{-1} = \frac{1}{3}\begin{pmatrix} 2 & \bar{1} & \bar{1} \\ 1 & 1 & \bar{2} \\ 1 & 1 & 1 \end{pmatrix}$$

Now calculate the direction parallel to $[2\,\bar{1}\,\bar{1}]_R$

$$\begin{pmatrix} u_H \\ v_H \\ w_H \end{pmatrix} = \frac{1}{3}\begin{pmatrix} 2 & \bar{1} & \bar{1} \\ 1 & 1 & \bar{2} \\ 1 & 1 & 1 \end{pmatrix}\begin{pmatrix} 2 \\ \bar{1} \\ \bar{1} \end{pmatrix} = \begin{pmatrix} 2 \\ 1 \\ 0 \end{pmatrix}$$

Answer

Along $[210]_H$ the symmetry of the special projection is *p2mm* (in agreement with Hahn, 2010, p. 141).

13.2 $H_{12}B_{12}{}^{-2}, 3K^+, Br^-$: POINT GROUP PROPERTIES

Rhombohedral axes are chosen for this crystal study. The choice is used here for indexing the point group.

The associated point group of the boron compound, $H_{12}B_{12}{}^{-2}, 3K^+, Br^-$, is $\bar{3}m$ in the Hermann–Mauguin notation and D_{3d} in the Schönflies notation. This point group has 12 symmetry operations—1, 3^+, 3^-, $\bar{1}$, $\bar{3}^+, \bar{3}^-$, three twofolds perpendicular to $[111]_R$, and three mirror planes containing $[111]_R$. Thus the order, or number of elements, is 12. The trigonal system is characterized by having just one threefold axis, either 3 or $\bar{3}$.

13.2.1 Multiplication Table

The generators from the Appendix 3 are 1, 3^+, 2, and $\bar{1}$. However, the point group can be generated from just two symmetry operations (see Figure 3.49). The first operation has to be either $\bar{3}^+$ or $\bar{3}^-$. The second is one of the three twofolds or one of the three mirrors. For

TABLE 13.2   Point Group $\bar{3}m$ Multiplication Table

	1	3^+	3^-	$2_{\bar{x}0x}$	$2_{x\bar{x}0}$	$2_{0y\bar{y}}$	$\bar{1}$	$\bar{3}^+$	$\bar{3}^-$	m_{xyx}	m_{xxz}	m_{xyy}
1	1	3^+	3^-	$2_{\bar{x}0x}$	$2_{x\bar{x}0}$	$2_{0y\bar{y}}$	$\bar{1}$	$\bar{3}^+$	$\bar{3}^-$	m_{xyx}	m_{xxz}	m_{xyy}
3^+	3^+	3^-	1	$2_{x\bar{x}0}$	$2_{0y\bar{y}}$	$2_{\bar{x}0x}$	$\bar{3}^+$	$\bar{3}^-$	$\bar{1}$	m_{xxz}	m_{xyy}	m_{xyx}
3^-	3^-	1	3^+	$2_{0y\bar{y}}$	$2_{\bar{x}0x}$	$2_{x\bar{x}0}$	$\bar{3}^-$	$\bar{1}$	$\bar{3}^+$	m_{xyy}	m_{xyx}	m_{xxz}
$2_{\bar{x}0x}$	$2_{\bar{x}0x}$	$2_{0y\bar{y}}$	$2_{x\bar{x}0}$	1	3^-	3^+	m_{xyx}	m_{xyy}	m_{xxz}	$\bar{1}$	$\bar{3}^-$	$\bar{3}^+$
$2_{x\bar{x}0}$	$2_{x\bar{x}0}$	$2_{\bar{x}0x}$	$2_{0y\bar{y}}$	3^+	1	3^-	m_{xxz}	m_{xyx}	m_{xyy}	$\bar{3}^+$	$\bar{1}$	$\bar{3}^-$
$2_{0y\bar{y}}$	$2_{0y\bar{y}}$	$2_{x\bar{x}0}$	$2_{\bar{x}0x}$	3^-	3^+	1	m_{xyy}	m_{xxz}	m_{xyx}	$\bar{3}^-$	$\bar{3}^+$	$\bar{1}$
$\bar{1}$	$\bar{1}$	$\bar{3}^+$	$\bar{3}^-$	m_{xyx}	m_{xxz}	m_{xyy}	1	3^+	3^-	$2_{\bar{x}0x}$	$2_{x\bar{x}0}$	$2_{0y\bar{y}}$
$\bar{3}^+$	$\bar{3}^+$	$\bar{3}^-$	$\bar{1}$	m_{xxz}	m_{xyy}	m_{xyx}	3^+	3^-	1	$2_{x\bar{x}0}$	$2_{0y\bar{y}}$	$2_{\bar{x}0x}$
$\bar{3}^-$	$\bar{3}^-$	$\bar{1}$	$\bar{3}^+$	m_{xyy}	m_{xyx}	m_{xxz}	3^-	1	3^+	$2_{0y\bar{y}}$	$2_{\bar{x}0x}$	$2_{x\bar{x}0}$
m_{xyx}	m_{xyx}	m_{xyy}	m_{xxz}	$\bar{1}$	$\bar{3}^-$	$\bar{3}^+$	$2_{\bar{x}0x}$	$2_{0y\bar{y}}$	$2_{x\bar{x}0}$	1	3^-	3^+
m_{xxz}	m_{xxz}	m_{xyx}	m_{xyy}	$\bar{3}^+$	$\bar{1}$	$\bar{3}^-$	$2_{x\bar{x}0}$	$2_{\bar{x}0x}$	$2_{0y\bar{y}}$	3^+	1	3^-
m_{xyy}	m_{xyy}	m_{xxz}	m_{xyx}	$\bar{3}^-$	$\bar{3}^+$	$\bar{1}$	$2_{0y\bar{y}}$	$2_{x\bar{x}0}$	$2_{\bar{x}0x}$	3^-	3^+	1

example, with rhombohedral reference axes, $\bar{3}^+$ and m_{xyx} are a minimal set of generators. Their matrices are

$$\bar{3}^+ = \begin{pmatrix} 0 & 0 & \bar{1} \\ \bar{1} & 0 & 0 \\ 0 & \bar{1} & 0 \end{pmatrix} \quad \text{and} \quad m_{xyx} = \begin{pmatrix} 0 & 0 & 1 \\ 0 & 1 & 0 \\ 1 & 0 & 0 \end{pmatrix}$$

The Starter Program from Chapter 3 calculates the multiplication table in Table 13.2. The rhombohedral reference axes are indicated by the subscripts. Because

$$3^+ \times m_{xyx} \neq m_{xyx} \times 3^+,$$

the matrix of the point group is not symmetric, and the point group is not abelian. The point group has an inversion; thus its symmetry does not permit the piezoelectric effect. The relationships among the symmetry operations are seen in the multiplication table.

13.2.2 Stereographic Projections

For rhombohedral axes, the [111] is chosen normal to the plane of the paper. Figure 13.2a shows the projections of the rhombohedral axes—$\mathbf{a}_{Rp}$, $\mathbf{b}_{Rp}$, $\mathbf{c}_{Rp}$—on the plane normal to [111].

Figure 13.2b is the symbol stereographic projection. The purple triangle with the white dot represents the threefold rotoinversion axis with the cyclic point group $\bar{3}$, which has six symmetry operations, of which three are proper and three are improper. The three mirrors, each with the point group m, are represented by three purple lines. The point group m

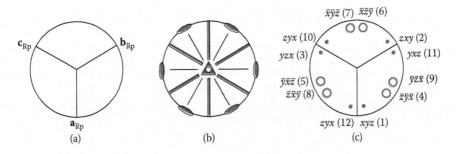

FIGURE 13.2 Point group $\bar{3}m$ stereographic projections: (a) axes; (b) symbol; (c) general position.

is a cyclic group with one proper operation, 1, and one improper operation, m. Finally there are the three twofold axes, each with the point group 2. Each twofold axis is represented by two red ovals connected by a black guide line. Both operations in the point group 2 are proper. The purple symbols indicate that half the operations are proper, or red, and half the operations are improper, or blue.

For rhombohedral axes, the general position stereographic projection for $\bar{3}m$, Figure 13.2c, has a combination of 12 dots and circles corresponding to the 12 symmetry operations. See Section 3.9 for a general description of these projections. The coordinates, xyz, corresponding to the identity operation, are chosen to have xyz positive, as is customary, and are indicated by a dot labeled in the diagram. Each point with positive xyz coordinates is indicated by a dot, and each point with a negative xyz coordinates is indicated by an open circle. The color red is used for the six proper operations: $1, 3^+, 3^-, 2_{\bar{x}0x}, 2_{x\bar{x}0}, 2_{0y\bar{y}}$ A proper operation does not change handedness, and the determinant of its matrix is +1. The color blue is used for the six improper operations: $\bar{1}, \bar{3}^+, \bar{3}^- m_{xyx}, m_{xxz}, m_{xyy}$. An improper operation does change handedness, and the determinant of its matrix is –1. Note that the points, both dots and open circles, are placed inside the circumference to allow any value for z.

Given the symbol diagram, the general position diagram can be created and vice versa.

13.3 $H_{12}B_{12}^{-2},3K^+,Br^-$: SPACE GROUP PROPERTIES

$H_{12}B_{12}^{-2},3K^+,Br^-$ crystallizes in the rhombohedral space group #166, $R\bar{3}m$. The Bravais lattice is hR. With rhombohedral reference axes, the symmetry operations require that

$$a = b = c \quad \alpha = \beta = \gamma$$

13.3.1 Space Group Diagrams

The seven space groups with the Bravais lattice hR have unusual space group diagrams in the *International Tables for Crystallography* that combine hexagonal and rhombohedral cells.

13.3.1.1 Special Topic: Space Group Diagrams for Combined Hexagonal and Rhombohedral Cells

Figure 13.3 gives the symbol diagram for the symmorphic space group #166, $R\bar{3}m$, showing both the hexagonal and rhombohedral cells. The principal axis is a $\bar{3}$ rotoinversion axis. The origin for both cells is chosen at an inversion point on the $\bar{3}$ axis, which is indicated in the figure.

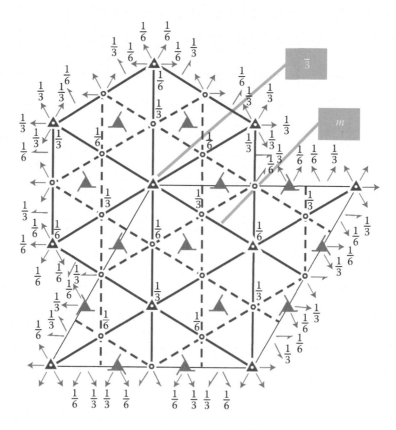

FIGURE 13.3   Symbol diagram for space group #166, $R\bar{3}m$.

An explanation of the Hermann–Mauguin symbol follows. The "R" refers to the center-ing type, for which the cell is a primitive cell with rhombohedral axes (R) or a triple cell with hexagonal axes (H). The "$\bar{3}$" refers to the threefold rotoinversion axes, which are parallel to $[001]_H$ or $[111]_R$. The "m" refers to the mirror planes normal to $[100]_H$, $[010]_H$, and $[110]_H$ or $[1\bar{1}0]_R [01\bar{1}]_R$ and $[10\bar{1}]_R$, respectively. The label, "m", in the figure refers to the mirror perpendicular to $[100]_H$ or $[1\bar{1}0]_R$ For R centering the third place is not used.

Consider the coloring of the symbols. The symbols for the twofold axes and the screw axes—2_1, 3_1, and 3_2—are red because there is no change of handedness. The symbols for the inversion points, $\bar{1}$, the threefold rotoinversion axes, $\bar{3}$, the mirror planes, m, and the glide planes are purple because half the operations change the handedness and half do not.

Figure 13.4a gives the general position diagram for the space group #166, $R\bar{3}m$, show-ing the hexagonal and rhombohedral cells. The lattice point at the origin for both cells is labeled 000. Two more lattice points are labeled. The hexagonal cell has lattice points 000_H, $\frac{1}{3}\frac{2}{3}\frac{2}{3}_H$, and $\frac{2}{3}\frac{1}{3}\frac{1}{3}_H$. For the rhombohedral cell these lattice points are respec-tively labeled 000_R, 110_R, and 100_R. Of course, because the rhombohedral cell is primi-tive, only one lattice point need be identified, which is the 000. All red circles represent proper operations, and all blue circles with commas represent improper operations. The plus signs show that the z-coordinate is positive, and the minus signs show that the z-coordinate is negative.

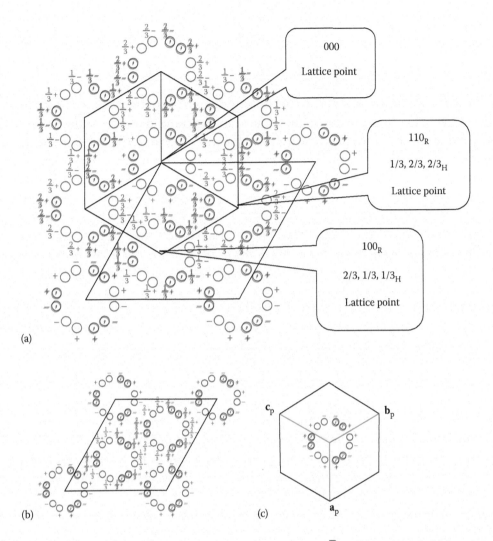

FIGURE 13.4 General position diagram for space group #166, $R\bar{3}m$: (a) combined; (b) hexagonal; (c) rhombohedral.

In Figure 13.4b, the hexagonal unit cell has been extracted from Figure 13.4a. Here the $[001]_H$ is normal to the plane of the paper and points toward the reader. The 36 operations in the hexagonal unit cell are represented by 36 circles. Half are red representing proper operations, and half are blue representing improper operations. Each of the three lattice points in the triple unit cell is surrounded by 12 operations. In the diagram the red circle in the upper left corner with the plus sign represents the coordinates, xyz, of the identity operation.

In Figure 13.4c, the rhombohedral cell is extracted and the projections of the rhombohedral basis vectors are indicated. Here the $[111]_R$ is normal to the plane of the paper and points toward the reader. The rhombohedral cell was chosen for this book because the smaller volume is easier for students, while the Starter Program easily handles either the primitive rhombohedral cell or the triple hexagonal cell.

Given the symbol diagram, the general position diagram can be created and vice versa.

13.3.2 Maximal Subgroups and Minimal Supergroups

There are 230 space groups. Because the space groups contain translation operations, each space group has infinite order. Space groups can be arranged in trees similar to the trees of the point groups. However, there is no unique way of constructing the tree for the space groups. In a tree of Type I, the centering type is kept and the order of the associated point group is changed. (Maximal subgroups and minimal supergroups are discussed in Section 4.8.) For example, a maximal Type I subgroup of the space group #166, $R\bar{3}m$, is the space group #160, $R3m$. Both these space groups have the same centering type, namely R; but for the subgroup the order of the associated point group is decreased. Specifically, the associated point group of the space group #166, $R\bar{3}m$, is $\bar{3}m$ and has order 12 while the associated point group of the space group #160, $R3m$ is $3m$ and has order 6. The space group #166, $R\bar{3}m$, has four maximal subgroups, of which three have associated point groups of order 6 and one has an associated point group of order 4. The space group has five minimal supergroups, all of which have associated point groups of order 48. The *International Tables for Crystallography* give the information necessary to construct Figure 13.5.

13.3.3 Asymmetric Unit

The asymmetric unit is the smallest part of the unit cell that, when operated on by the symmetry operations, produces the whole unit cell. The volume of the asymmetric unit is related to the volume of the unit cell by $V_{AU} = (V_{UC}/nh)$, where n is the number of lattice points in the unit cell and h is the number of operations in the associated point group. (Note that the product nh is equal to the multiplicity of the general position.) Because there are 12 symmetry operations in this primitive rhombohedral cell ($n = 1$, $h = 12$), the volume of the asymmetric unit is 1/12th the volume of the unit cell. Therefore, because the volume of the unit cell is 322.6 Å^3, the volume of the asymmetric unit is 26.9 Å^3.

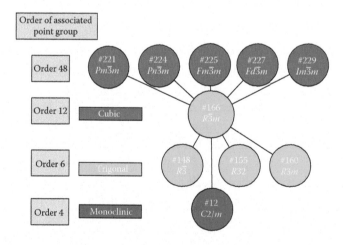

FIGURE 13.5   Maximal subgroups and minimal supergroups for space group #166, $R\bar{3}m$.

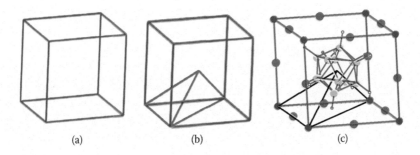

(a) (b) (c)

FIGURE 13.6 $H_{12}B_{12}^{-2}, 3K^+, Br^-$: (a) unit cell; (b) with the asymmetric unit; (c) populated with atoms.

Figure 13.6a gives the unit cell in blue. Figure 13.6b shows the unit cell drawn with the asymmetric unit in red superimposed on it. The asymmetric unit has

$$0 \le x \le 1, \ 0 \le y \le 1, \ 0 \le z \le \tfrac{1}{2}, \ y \le x, \ \text{and} \ z \le \min(y, 1 - x)$$

Vertices 000; 100; 110; and ½ ½ ½

(Hahn, 2010, p. 143). The vertices are given for this complicated asymmetric unit because they are easier to interpret.

Figure 13.6c shows the populated cell with the asymmetric unit. Note how only a small part, exactly 1/12, of the icosahedron (green boron atoms and red bonds) is in the asymmetric unit. The symmetries of the space group generate the rest of the icosahedron. The volume of the triple hexagonal unit cell is three times as large as that of the primitive rhombohedral cell, but the volume of the asymmetric unit is the same for both cells, 26.9 Å³. The shape of the asymmetric unit of the triple hexagonal unit cell (not shown here) is different from that of the primitive rhombohedral cell (see Exercise 13.1).

13.4 $H_{12}B_{12}^{-2}, 3K^+, Br^-$: DIRECT AND RECIPROCAL LATTICES

In addition to the direct lattice, there is the reciprocal lattice, which is the lattice of the diffraction pattern. The diffraction pattern has the intensities of the Bragg reflections superimposed on the reciprocal lattice. Table 13.3 compares the direct and reciprocal lattices, including the lattice parameters, the metric matrix, and the cell volume for both the direct and reciprocal cells. The Starter Program from Chapter 5 produces the reciprocal cell superimposed on the unit cell. The program produces a figure that can be rotated to observe that, for example, **a** is perpendicular to both **b*** and **c***. See Sections 5.2 and 5.3 for all the relevant relationships. Figure 13.7 shows the relationship between the direct unit cell (blue) and the reciprocal unit cell (red) with the volumes normalized to one.

13.5 $H_{12}B_{12}^{-2}, 3K^+, Br^-$: FRACTIONAL COORDINATES AND OTHER DATA FOR THE CRYSTAL STRUCTURE

The crystal structure consists of the placement of the atoms in the unit cell and the bonds between the atoms. The fractional coordinates of the atoms in the asymmetric unit combined with the general position coordinates of the space group and the lattice parameters are sufficient to specify all the atoms in the unit cell.

TABLE 13.3 $H_{12}B_{12}{}^{-2},3K^+,Br^-$: Direct and Reciprocal Lattices with Rhombohedral Axes

Direct lattice	Reciprocal Lattice
$a = b = c = 6.871$ Å	$a^* = b^* = c^* = 0.1461$ Å^{-1}
$\alpha = \beta = \gamma = 93.4°$	$\alpha^* = \beta^* = \gamma^* = 86.39°$
$\mathbf{G} = \begin{pmatrix} 47.21 & -2.80 & -2.80 \\ -2.80 & 47.21 & -2.80 \\ -2.80 & -2.80 & 47.21 \end{pmatrix}$ Å^2	$\mathbf{G^*} = \begin{pmatrix} 0.0213 & 0.0013 & 0.0013 \\ 0.0013 & 0.0213 & 0.0013 \\ 0.0013 & 0.0013 & 0.0213 \end{pmatrix}$ Å^{-2}
$V = 322.60$ Å^3	$V^* = 0.0031$ Å^{-3}

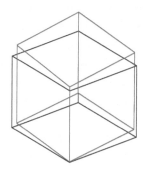

FIGURE 13.7 $H_{12}B_{12}{}^{-2},3K^+,Br^-$: direct and reciprocal cells.

Z equals 1 or, in other words, there is one formula unit of this boron compound in the unit cell. Z' equals 1/12th; that is to say, there is 1/12 of a formula unit of the boron compound in the asymmetric unit. The fractional coordinates of the atoms for the asymmetric unit from the CIF are given in Table 13.4.

13.5.1 Special Topic: Multiplicity

There are 12 boron atoms in the unit cell, and they form an icosahedron. However, there are only two crystallographically independent boron atoms, B1 and B2. This conclusion is reached by considering the multiplicity and the Wyckoff positions given in the *International Tables for Crystallography* (see Section 4.13.6).

Now, in detail, consider Table 13.5. Because $Z = 1$, the multiplicity of each crystallographically independent atom gives the number of atoms in the unit cell related by symmetry. For example, the multiplicity of atom B1 is six; and thus there are six symmetry-related B1 atoms in the unit cell. Note that the fractional coordinates of B1 are 0.437, 0.437, 0.258 corresponding to Wyckoff letter h, which has multiplicity 6 (Hahn, 2010, p. 143). The multiplicity of the general position is 12. The next question is how many B1 atoms are there in the asymmetric unit. In general, the number of atoms of each type, say B1, in the asymmetric unit is the number of these atoms in the unit cell divided by the multiplicity of the general position. For example, the number of B1 atoms in the asymmetric unit is 6/12 = 1/2. That is, one-half of the B1 atom is in the asymmetric unit. The red box outlines the asymmetric unit in Figure 13.8 and shows one-half of the B1 atom is inside the red box. Similar calculations show that one-half each of atoms B2, H1, and H2 are in the asymmetric unit as well as 1/4 of K and 1/12 of Br.

TABLE 13.4 $H_{12}B_{12}^{-2}$,3K$^+$,Br$^-$ Fractional Coordinates, Multiplicity, and Wyckoff Letter

Atom Number	Atom Label	x Fractional Coordinate	y Fractional Coordinate	z Fractional Coordinate	Multiplicity Wyckoff Letter
1	Br1	0	0	0	1a
2	K1	0.5	0	0	3d
3	B1	0.437(1)	0.437(1)	0.258(1)	6h
4	B2	0.665(1)	0.370(1)	0.370(1)	6h
5	H1	0.389	0.389	0.099	6h
6	H2	0.770	0.278	0.278	6h
7	B1A	0.258	0.437	0.437	
8	B1B	0.437	0.258	0.437	
9	B2A	0.370	0.665	0.370	
10	B2H	0.630	0.630	0.335	
11	B1G	0.742	0.563	0.563	
12	B2G	0.630	0.335	0.630	
13	B2F	0.335	0.630	0.630	
14	B2B	0.370	0.370	0.665	
15	H1A	0.099	0.389	0.389	
16	H1B	0.389	0.099	0.389	
17	B1H	0.563	0.742	0.563	
18	H2A	0.278	0.770	0.278	
19	H2H	0.722	0.722	0.230	
20	B1F	0.563	0.563	0.742	
21	H1G	0.901	0.611	0.611	
22	H2G	0.722	0.230	0.722	
23	H2F	0.230	0.722	0.722	
24	H2B	0.278	0.278	0.770	
25	H1H	0.611	0.901	0.611	
26	H1F	0.611	0.611	0.901	

Next, the bonds must be inserted. In this crystal there are crystallographically independent bonds *not* contained within the asymmetric unit. *Crystallographically independent atoms* are atoms not related by symmetries. Examples of crystallographically independent atoms are the atoms in the asymmetric unit. In the case of the boron compound, the six atoms—Br1, K1, B1, B2, H1, and H2—are (at least partially) in the asymmetric unit. When the symmetry operations are applied to these atoms, all the atoms in the unit cell are produced. Next, the bonds need to be produced. The bonds also obey the symmetry

TABLE 13.5 Fraction of Atoms in the Asymmetric Unit

Atom	Multiplicity	Atoms in the Asymmetric Unit
Bromine, Br1	1	1/12
Potassium, K1	3	3/12 = 1/4
B1, boron 1	6	6/12 = 1/2
B2, boron 2	6	6/12 = 1/2
H1, hydrogen 1	6	6/12 = 1/2
H2, hydrogen 2	6	6/12 = 1/2

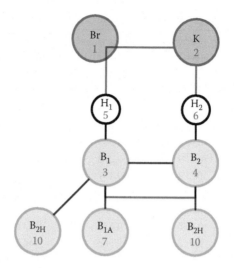

FIGURE 13.8 Schematic showing the atoms and bonds needed to produce the populated cell.

operations. There are two types of bonds, the bonds that are contained entirely within the asymmetric unit and the bonds where one atom is in the asymmetric unit and the other atom is outside the asymmetric unit. The atoms that are outside the asymmetric unit, indicated by a letter added to the label, are related by symmetry to the atoms inside the asymmetric unit. For example, atom B1A is related by symmetry to atom B1. Both B1A and B2H are incorporated in the schematic diagram Figure 13.8. See Section 13.6 for a more detailed explanation of this boron example.

In the Starter Program for populating the unit cell, Figure 13.8 is the key to putting in the intramolecular bonds. Figure 13.6c shows the asymmetric unit in the populated cell.

13.6 SPECIAL TOPIC: BORON ICOSAHEDRON

Figure 13.9 emphasizes the fivefold axis of the ideal icosahedron. This section compares the ideal or Platonic icosahedron with the one in the boron compound. The ideal *icosahedron*

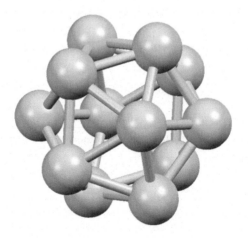

FIGURE 13.9 Fivefold axis of an ideal icosahedron.

is a regular convex polyhedron with 12 vertices, 20 congruent equilateral triangular faces, and 30 edges of equal length. Each vertex is connected to five neighboring vertices. The icosahedron has the noncrystallographic point group $m\overline{3}5$ (Hahn, 2006, p. 801). However, no fivefold symmetry can be part of the global symmetry of a crystal. In the boron compound, the center of the icosahedron is located at 1/2, 1/2, 1/2. The boron atoms are located in the special position with Wyckoff letter h, which has site symmetry m and multiplicity six. (See Section 14.63 for a discussion of site symmetry.) The two sets of boron atoms, B1 and B2, have $2 \times 6 = 12$ atoms that form the vertices of the boron icosahedron.

The 30 edges, or bonds, must also obey the symmetry requirements of the space group. There are four sets of crystallographically distinct bonds that are colored magenta, blue, red, and green in Figure 13.10a. The magenta set, illustrated by B1–B2, has $\overline{3}m$ symmetry, with its inversion point at 1/2, 1/2, 1/2, and has multiplicity 12. The B1–B2 bond length is 1.80 (1) Å. Next, three atoms B1, B1A, and B1B form the top cap and three more atoms B1F, B1G, and B1H form the bottom cap. The caps are equilateral triangles with the bonds having multiplicity six. The six blue bonds are (B1–B1A), (B1A–B1B), (B1B–B1), (B1F–B1G), (B1G–B1H), and (B1H–B1F), each with length 1.79 (1) Å. The next six atoms form a zigzag red belt between the top cap and the bottom cap. These six red bonds have multiplicity six. From the diagram, the cyclic sequence of the atoms is [B2–B2H–B2A–B2F–B2B–B2G–B2]; and every bond has the length 1.84 (1) Å. Finally, there are six green bonds that lie in the three mirror planes. They are (B1–B2H), (B1A–B2F), (B1B–B2G), (B1F–B2B), (B1G–B2), and (B1H–B2A). These six green bonds have multiplicity six and the length 1.850 (9) Å.

The schematic diagram, Figure 13.10b, shows the magenta bond B1–B2 contained entirely within the asymmetric unit, while three bonds, one of each of the three other colors, are attached on one end to an atom in the asymmetric unit and on the other end to an atom outside the asymmetric unit. The diagram also shows four colored boron–boron

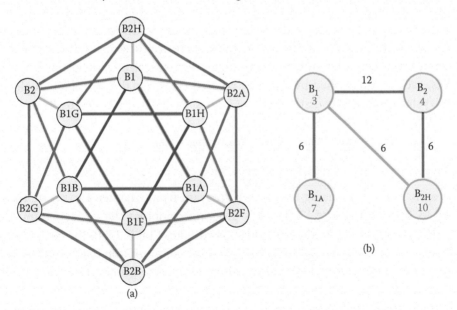

(a)

(b)

FIGURE 13.10   Color coded (a) icosahedron along [111]; (b) boron atoms and bonds.

bonds with their multiplicities written above them. Symmetry-related atoms B1A and B1H, outside the asymmetric unit, have been added so that all the bonds in the icosahedron can be generated by the Starter Program.

The 20 triangular faces of the boron icosahedron obey the symmetry requirements. There are three sets of crystallographically distinct triangles that are color coded according to the color of their sides: blue–blue–blue, magenta–magenta–blue, and red–green–magenta. Two equilateral triangles form the top and bottom caps. Triangle B1–B1A–B1B forms the top cap and triangle B1F–B1G–B1H forms the bottom cap. Both of these triangles have three blue sides. Next, each of six isosceles triangles has two magenta sides and one blue side. These triangles are bisected by mirror planes. The isosceles triangles are illustrated by triangle B1–B1A–B2. Finally, each of 12 scalene triangles has one green side, one magenta side, and one red side. The scalene triangles are illustrated by triangle B2–B2H–B1.

Finally, the 60 angles of the boron icosahedron obey the symmetry requirements. There are six crystallographically distinct sets of angles, which are color coded by the color of their sides: blue–blue, magenta–magenta, magenta–blue, magenta–green, magenta–red, and green–red. The two equilateral triangles have a total of six blue–blue angles, illustrated by angle B1–B1A–B1B, each of which has 60°. Each of the six isosceles triangles has one magenta–magenta angle and two magenta–blue angles. Each magenta–magenta angle has 59.6° and is illustrated by angle B1G–B2H–B1H. Each magenta–blue angle has 60.7° and is illustrated by angle B2H–B1G–B1H. Each of the 12 scalene triangles has one green–red angle, one magenta–green angle, and one magenta–red angle. Each green–red angle has 58.3° and is illustrated by angle B1–B2H–B2. Each magenta–green angle has 60.7° and is illustrated by angle B2–B1–B2H. Each magenta–red angle has 61.0° and is illustrated by angle B1–B2–B2H. Every triangle adds up to 180°, confirming that the angles have been calculated correctly.

In summary, of the 30 bonds in the boron icosahedron, there are four crystallographically distinct sets of bonds. Of the 20 triangles, there are three crystallographically distinct sets of triangles: two equilateral triangles, six isosceles triangles, and 12 scalene triangles, each distinguished by its three colored sides. Finally, of the 60 angles, there are six crystallographically distinct sets of angles, each distinguished by its two colored sides.

Table 13.6 compares the ideal and boron icosahedrons.

13.7 $H_{12}B_{12}^{-2}, 3K^+, Br^-$: CRYSTAL STRUCTURE

Populate the unit cell by using the Starter Program from Chapter 4 to create the entire unit cell as shown in Figure 13.11a. This program combines the lattice parameters—a, b, c, α, β, and γ (Table 13.3), the fractional coordinates (Table 13.4), the general position coordinates of the space group (Table 13.7), and the configuration of bonds (Figure 13.8). The lattice parameters, the fractional coordinates, and the general position coordinates are from the CIF.

Beginning with the general position coordinates the Starter Program automatically includes all the atoms. No special procedure is required for atoms in special positions.

TABLE 13.6 Comparison of Boron and Ideal Icosahedrons

Boron Icosahedron				Ideal Icosahedron
Bonds	**Example, Color**	**Multiplicity**	**Bond Length (Å)**	**Bonds**
Cap atoms (B1 atoms)	B1–B1A, blue	6	1.79 (1)	All bonds equal
Belt atoms (B2 atoms)	B2–B2H, red	6	1.84 (1)	All bonds equal
(B1,B2 atoms in mirror)	B1–B2H, green	6	1.850 (9)	All bonds equal
(B1,B2 atoms not in mirror)	B1–B2, magenta	12	1.80 (1)	All bonds equal
Weighted average		30 bonds total	1.82	All bonds equal
Triangles	**Example, Color of Sides**	**Multiplicity**	**Triangles**	**Triangles**
Triangles on caps	<B1–B1A–B1B Blue–blue–blue	2	Equilateral triangles	Equilateral triangles
Triangles bisected by mirrors	<B1–B1A–B2 Blue–magenta–magenta	6	Isosceles triangles	Equilateral triangles
Triangles with unequal sides	<B2–B2H–B1 Red–green–magenta	12	Scalene triangles	Equilateral triangles
Total number of triangles		20 triangles total		
Bond Angles	**Example, Colors of Adjacent Bonds**	**Multiplicity**	**Bond Angles**	**Bond Angles**
Cap atoms (B1 atoms)	<B1–B1A–B1B Blue–blue	6	Exactly 60°	Exactly 60°
Sum of angles in the triangle			Exactly 180°	Exactly 180°
1. Angles bisected by mirror	<B1G–B2H–B1H Magenta–magenta	6	59.6°	Exactly 60°
2. Two other equal angles in triangle bisected by mirror	<B2H–B1G–B1H Magenta–blue	12	60.2°	Exactly 60°
Sum of angles in the triangle			Exactly 180°	Exactly 180°
	<B1–B2H–B2 Green–red	12	58.3°	Exactly 60°
	<B2–B1–B2H Magenta–green	12	60.7°	Exactly 60°
	<B1–B2–B2H Magenta–red	12	61.0°	Exactly 60°
Sum of angles in the triangle			Exactly 180°	Exactly 180°
Total number of angles			60 angles total	

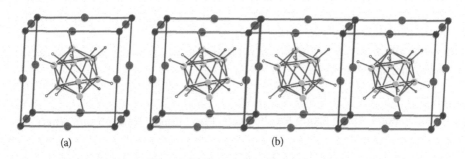

(a)   (b)

FIGURE 13.11 Populated: (a) unit cell; (b) three unit cells.

TABLE 13.7 $H_{12}B_{12}^{-2},3K^+,Br^-$: General Position Coordinates and Symmetry Operations

#	General Position Coordinates	Symmetry Operations
1	x,y,z	1, identity
2	z,x,y	3^- along [111] at x,x,x
3	y,z,x	3^+ along [111] at x,x,x
4	y,x,z	Mirror normal to $[1\bar{1}0]$
5	x,z,y	Mirror normal to $[01\bar{1}]$
6	z,y,x	Mirror normal to $[10\bar{1}]$
7	$\bar{x},\bar{y},\bar{z}$	$\bar{1}$ inversion 000
8	$\bar{z},\bar{x},\bar{y}$	3^+ along [111] at x,x,x, inversion at 000
9	$\bar{y},\bar{z},\bar{x}$	3^- along [111] at x,x,x, inversion at 000
10	$\bar{y},\bar{x},\bar{z}$	2 along $[1\bar{1}0]$ at $x,\bar{x},0$
11	$\bar{x},\bar{z},\bar{y}$	2 along $[01\bar{1}]$ at $0,y,\bar{y}$
12	$\bar{z},\bar{y},\bar{x}$	2 along $[10\bar{1}]$ at $x,0,\bar{x}$

Z equals 1, that is, there is one formula unit in the unit cell. The bromine atoms (blue) are on the vertices of the cell and the potassium atoms (magenta) are on the centers of the edges of the cell. The icosahedron is centered in the middle of the cell with green boron atoms and hydrogen atoms shown in white are outlined in red. The rhombohedral unit cell is a slightly squashed cube with $\alpha = \beta = \gamma = 93.4°$. Figure 13.11b has three unit cells in order to show how the potassium and bromine ions are shared between unit cells.

13.7.1 Symmetry of the Special Projections

The *International Tables for Crystallography* give the symmetry for three special projections for each space group in the standard orientation. These projections are orthogonal, which means that each projection is onto a plane normal to the direction of the projection. The basis vectors of the two-dimensional unit cell are labeled $\mathbf{a}'$ and $\mathbf{b}'$, regardless of which two basis vectors of the three-dimensional unit cell (selected from $\mathbf{a}$, $\mathbf{b}$, $\mathbf{c}$) are projected. A Starter Program from Chapter 4 has a "view" statement that selects the appropriate projections. For rhombohedral axes, the special projections are along [111], $[1\bar{1}0]$, and $[2\bar{1}\bar{1}]$. For $R\bar{3}m$, there is no shift of origin. A careful examination of these projections can demonstrate that the atoms and bond lengths have been correctly implemented.

Quite generally, every two-dimensional projection has the symmetries of one of the 17 two-dimensional space groups.

Along [111]

The two-dimensional space group for the projection along [111] is *p6mm*, where

$$\mathbf{a}' = \tfrac{1}{3}(2\mathbf{a} - \mathbf{b} - \mathbf{c}) \quad \mathbf{b}' = \tfrac{1}{3}(-\mathbf{a} + 2\mathbf{b} - \mathbf{c})$$

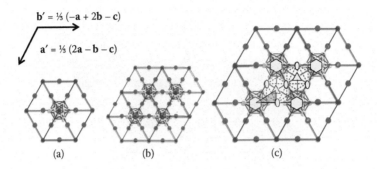

b' = ⅓ (−**a** + 2**b** − **c**)

a' = ⅓ (2**a** − **b** − **c**)

(a) (b) (c)

FIGURE 13.12 $H_{12}B_{12}^{-2}, 3K^+, Br^-$ along [111]: (a) unit cell; (b) expanded unit cell; (c) with graphical symbols and asymmetric unit in light pink.

with the origin at x, x, x. Figure 13.12a is the direct output from the Starter Program. The figure contains six equilateral triangles. Two triangles make up a hexagonal unit cell; thus the direct output *partially* portrays three unit cells (see Figure 4.43). However, the direct output contains enough information to portray the cells completely. Figure 13.12b uses the available information to complete a unit cell. The completion is made by overlapping one partially portrayed unit cell on another. The overlapping is continued to form an array of adjacent figures that exhibit the symmetries of *p6mm*. The projected icosahedron is clearly seen superimposed on the blue bromine atom. (Each bromine atom has an icosahedron associated with it: but, except for the central icosahedron, the icosahedrons are not shown.) The potassium atoms are in magenta. The figure is made so that each lattice point has a projected icosahedron. Only four complete lattice points are shown. To avoid confusion, 10 blue bromine atoms do not have their projected icosahedrons included. The expanded figure also shows that the opposite sides of the unit cell are related by a translation, as always. All the information necessary to create the entire pattern is found in the unit cell. Figure 13.12c shows a diagram with graphical symbols for one unit cell and the asymmetric unit, in light pink. The asymmetric unit is one-twelfth of the unit cell.

13.7.1.1 Special Topic: Space-Filling Model

Figure 13.13 shows a space-filling model projected along [111]. A *space-filling model* is a molecular model in which the atoms are represented by spheres whose radii are proportional to the atomic radii of the atoms. The spheres touch and therefore hide the bonds. The almost spherical $[H_{12}B_{12}]^{2-}$ anions in pink and white and the potassium cations in blue are nearly in a cubic closed-packed structure (see Section 14.1.3). In each layer the boron icosahedron anions are surrounded by six potassium cations. Similarly, each potassium cation is surrounded by two boron icosahedron anions and four potassium cations. Even though the "spheres" are not identical, it is useful to think of this and other similar crystal structures in terms of close packing. The hidden bromine anions fit into the octahedral interstices between the layers of potassium cations and boron icosahedron anions (Tiritiris, 2005).

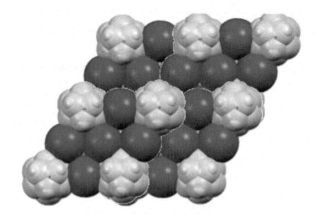

FIGURE 13.13 $H_{12}B_{12}^{-2}, 3K^+, Br^-$ space-filling model along [111].

Along $[1\,\overline{1}\,0]$
The two-dimensional space group along $[1\,\overline{1}\,0]$ is $p2$, where

$$\mathbf{a'} = 1/2(\mathbf{a} + \mathbf{b} - 2\mathbf{c}) \quad \mathbf{b'} = \mathbf{c}$$

with the origin at $x, \overline{x}, 0$. Figure 13.14a is the direct output from the Starter Program. The direct output *partially* portrays two unit cells. However, it contains enough information to portray the cells completely. Figure 13.14b uses the available information to complete a unit cell. The completion is made by overlapping, by integral numbers of $\mathbf{a'}$, one partially portrayed unit cell on another. The overlapping is continued to form an array of adjacent diagrams that exhibit the symmetries of $p2$. Finally, Figure 13.14c shows a diagram with graphical symbols for one unit cell with the asymmetric unit, in pink.

Along $[2\,\overline{1}\,\overline{1}]$
The two-dimensional space group along $[2\,\overline{1}\,\overline{1}]$ is $p2mm$, where

$$\mathbf{a'} = 1/2(\mathbf{b} - \mathbf{c}) \quad \mathbf{b'} = 1/3(\mathbf{a} + \mathbf{b} + \mathbf{c})$$

with the origin at $2x, \overline{x}, \overline{x}$, where x is an arbitrary parameter that could be chosen to be zero. Figure 13.15a is the direct output from the Starter Program.

The direct output contains enough information to portray the unit cell completely. Here Figure 13.15b uses the information to complete a unit cell. The completion is made by overlapping one *partially* portrayed unit cell on another. The overlapping is continued to form an array of adjacent figures that exhibit the symmetries of $p2mm$. Figure 13.15c shows a diagram with graphical symbols for the two-dimensional space group with the asymmetric unit in pale blue.

13.8 $H_{12}B_{12}^{-2}, 3K^+, Br^-$: RECIPROCAL LATTICE AND d-SPACINGS

Planes in the direct lattice correspond to points of the reciprocal lattice. The reciprocal vector

$$\mathbf{H}(hkl) = h\mathbf{a}^* + k\mathbf{b}^* + l\mathbf{c}^*$$

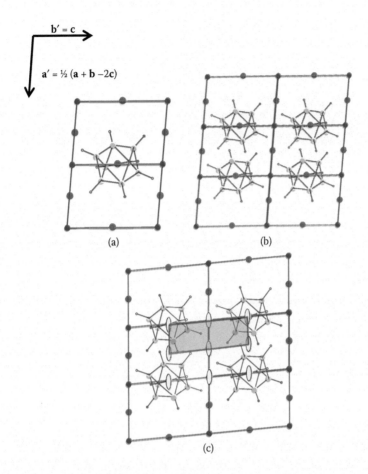

FIGURE 13.14 $H_{12}B_{12}{}^{-2}$,$3K^+$,Br^- along $[1\bar{1}0]$: (a) unit cell; (b) expanded unit cell; (c) with graphical symbols and asymmetric unit in pink.

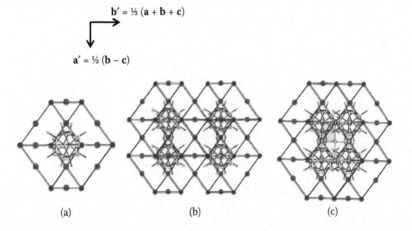

FIGURE 13.15 $H_{12}B_{12}{}^{-2}$,$3K^+$,Br^- along $[2\bar{1}\bar{1}]$: (a) unit cell; (b) expanded unit cell; (c) with graphical symbols and asymmetric unit in pink.

TABLE 13.8 $H_{12}B_{12}{}^{-2}, 3K^+, Br^-$ Intensity versus hkl from CSD

hkl	2θ (degrees)	Intensity
100	12.9	3835
10$\bar{1}$	17.7	4553
110	18.9	2974
11$\bar{1}$	22.0	1191
200	26.0	10,000
20$\bar{1}$	28.4	6504
210	29.9	6600
20$\bar{2}$	35.9	3647
21$\bar{2}$	38.3	6236
300	39.5	2568

where hkl are the indices of the planes of the direct lattice. The d-spacing is the distance from the origin to either of the two closest planes (see Section 5.7.1). In general, the d-spacing can be calculated by the following equation:

$$\frac{1}{d_{hkl}^2} = H^2(hkl) = (h \quad k \quad l)\mathbf{G}^* \begin{pmatrix} h \\ k \\ l \end{pmatrix}$$

The powder diffraction data in Table 13.8 give $hkl = 200$ as the reflection with the maximum intensity. The calculation of the d-spacing for this reflection follows

$$\frac{1}{d_{200}^2} = H^2(200) = (2 \quad 0 \quad 0) \begin{pmatrix} 0.0213 & 0.0013 & 0.0013 \\ 0.0013 & 0.0213 & 0.0013 \\ 0.0013 & 0.0013 & 0.0213 \end{pmatrix} \begin{pmatrix} 2 \\ 0 \\ 0 \end{pmatrix} \quad (13.5)$$

Now the d-spacing for the 200 can be calculated

$$d_{200} = 3.4226 \text{ Å}. \quad (13.6)$$

13.8.1 Powder Diffraction Pattern

The diffraction angle is now calculated as a function of hkl. First, Bragg's law is applied to the d-spacing. Bragg's law is

$$\lambda = 2 d_{hkl} \sin \theta_{hkl}$$

where λ is the wavelength of the incoming x-rays, d_{hkl} is the d-spacing of the hkl plane, and θ_{hkl} is the Bragg angle for the hkl diffraction maximum. The diffraction angle, $2\theta_{hkl}$, is twice the Bragg angle (see Section 6.6.3). Table 13.8 has hkl indices and intensities for the 10 most intense

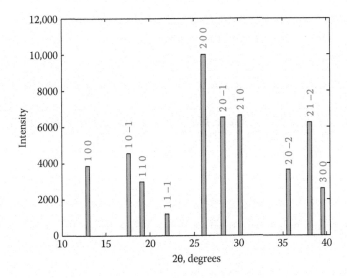

FIGURE 13.16   Powder diffraction pattern for $H_{12}B_{12}^{-2}, 3K^+, Br^-$.

reflections. The intensities are not actual experimental data but were calculated from the CIF by the program *Mercury* by using the CSD. The intensities are normalized so that the highest intensity is 10,000 arbitrary units. This information is the input for the Starter Program in Chapter 6. Figure 13.16 is the resulting powder diffraction pattern for the boron compound.

13.9 $H_{12}B_{12}^{-2}, 3K^+, Br^-$: ATOMIC SCATTERING CURVES

Figure 13.17 shows the atomic scattering curves from the Starter Program in Chapter 7 for the four elements found in $H_{12}B_{12}^{-2}, 3K^+, Br^-$. The atomic scattering curves are calculated from the Cromer–Mann coefficients (see Section 7.3.1). The atomic scattering factor (or form factor) is the ratio of the amplitude of the wave scattered by an atom to the amplitude

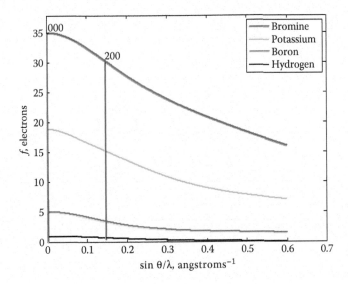

FIGURE 13.17   Atomic scattering curves for $H_{12}B_{12}^{-2}, 3K^+, Br^-$.

of the wave scattered by one electron. Its units are electrons. The atomic scattering curve is a measure of the efficiency of an atom in scattering x-rays.

Bragg's law for (200) is

$$\sin\theta/\lambda = 1/2d = 1/(2 \times 3.42) = 0.1461 \text{ Å}^{-1} \tag{13.7}$$

This reflection is labeled in Figure 13.17 in red. The values from the atomic scattering curves are

$$f_{Br}(200) = 30.29, \; f_K(200) = 15.36, \; f_B(200) = 3.37, \text{ and } f_H(200) = 0.725 \text{ electrons} \tag{13.8}$$

13.10 $H_{12}B_{12}{}^{-2}, 3K^+, Br^-$: STRUCTURE FACTOR

The general formula for the structure factor, from Equation 7.5, is

$$F(hkl) = \sum_{j=1}^{N} f_j \left(\frac{\sin\theta_{hkl}}{\lambda} \right) e^{2\pi i (hx_j + ky_j + lz_j)}$$

where N is the number of atoms in the unit cell. In the case of the boron compound there are 28 atoms in the unit cell (see Table 13.9). The running index j goes from 1 to 28. Table 13.4 includes the fractional coordinates x_j, y_j, z_j for the jth atom in the asymmetric unit.

The formula for the structure factor for the boron compound can be broken down in the following way:

$$F(hkl) = F_H(hkl) + F_B(hkl) + F_K(hkl) + F_{Br}(hkl) \tag{13.9}$$

The first term is a sum over 12 hydrogen atoms, the second term is over 12 boron atoms, the third term is over 3 potassium atoms, and the fourth term has 1 bromine atom

$$F(hkl) = \sum_{j}^{12} F_{Hj}(hkl) + \sum_{j}^{12} F_{Bj}(hkl) + \sum_{j}^{3} F_{Kj}(hkl) + F_{Br}(hkl) \tag{13.10}$$

Now put in the expressions for the individual atomic form factors and the phase factors

$$F(hkl) = \sum_{j}^{12} f_H(hkl) e^{2\pi i (hx_j + ky_j + lz_j)} + \sum_{j}^{12} f_B(hkl) e^{2\pi i (hx_j + ky_j + lz_j)}$$
$$+ \sum_{j}^{3} f_K(hkl) e^{2\pi i (hx_j + ky_j + lz_j)} + f_{Br}(hkl) e^{2\pi i (hx_j + ky_j + lz_j)} \tag{13.11}$$

TABLE 13.9 $H_{12}B_{12}{}^{-2}, 3K^+, Br^-$ Calculation of Number of Electrons in Unit Cell for Primitive Cell

Atom	H	B	K	Br	Total/Molecule
Atomic number	1	5	19	35	
Atoms/formula unit	12	12	3	1	28
Total electrons/formula unit	12	60	57	35	164

13.10.1 Calculate the Structure Factor at 000

The special case of the undeviated beam occurs when $hkl = 000$. See Example 7.2. Its position is labeled in red in Figure 13.17.

$$F(000) = 12F_H(000) + 12F_B(000) + 3F_K(000) + F_{Br}(000)$$

The atomic scattering factor, f, is equal to the number of electrons; therefore $F(000)$ is the sum of all the electrons in the unit cell (Section 7.6.4). Table 13.9 shows how the total number of electrons is counted in the unit cell of the boron compound to give

$$F(000) = 164 \text{ electrons} \tag{13.12}$$

13.10.2 Calculate the Contribution of One Atom to a Structure Factor

The calculation in Equation 13.11 can be sampled by calculating only the contribution of one atom, B1, to the most intense hkl reflection found in the x-ray powder diagram, which is 200. Table 13.4 gives the fractional coordinates for B1 as

$$x = 0.437, \quad y = 0.437, \quad \text{and} \quad z = 0.258.$$

For the reflection 200

$$F_{B1}(200) = f_B(200)e^{2\pi i(2\times0.437+0\times0.437+0\times0.258)} = 3.37\, e^{1.748\pi i} \text{ electrons} \tag{13.13}$$

Note that this is a complex number. The entire structure factor, including all 28 atoms, is real because the space group is centrosymmetric with an inversion point located at the origin (see Section 7.7.2). The importance of the structure factors is that they become coefficients of the Fourier expansion that is used to calculate the electron density (see Section 7.8.2).

TABLE 13.10 Comparison of $H_{12}B_{12}^{-2}, 3K^+, Br^-$ with $H_{12}B_{12}^{-2}, 3Rb^+, Cl^-$

	$H_{12}B_{12}^{-2}, 3K^+, Br^-$	$H_{12}B_{12}^{-2}, 3Rb^+, Cl^-$
Space group	#166, $R\bar{3}m$	#166, $R\bar{3}m$
Z	3	3
a, c	$a = 10.02$ Å, $c = 11.18$ Å	$a = 10.10$ Å, $c = 11.39$ Å
Volume	972.5 Å³	1005.8 Å³
Multiplicity, Wyckoff position, $3b$	Br⁻	Cl⁻
Fractional coordinates	0 0 ½	0 0 ½
Multiplicity, Wyckoff position, $9e$	K⁺	Rb⁺
Fractional coordinates	½ 0 0	½ 0 0
Multiplicity, Wyckoff position, $18h$	B1, B2, H1, H2	B1, B2, H1, H2
Fractional coordinates B1	0.0588, 0.9412, 0.8796	0.0584, 0.9416, 0.8819
Fractional coordinates B2	0.0956, 0.9044, 1.0288	0.0949, 0.9051, 1.0279
Fractional coordinates H1	0.098, 0.902, 0.811	0.096, 0.904, 0.807
Fractional coordinates H2	0.150, 0.870, 1.068	0.156, 0.844, 1.043

Source: Adapted from Tiritiris, I., J. Weidlein, and T. Schleid, 2005. *Zeitschrift Fur Naturforschung* 60, 627–639.

13.11 SPECIAL TOPIC: $H_{12}B_{12}^{-2}, 3K^+, Br^-$ ISOTYPIC CRYSTAL STRUCTURES

In Section 4.13.9, the anhydrous alum isotypic crystal structures are introduced. Here is another series. $H_{12}B_{12}^{-2}, 3K^+, Br^-$ crystals belong to an isotypic series of the form $H_{12}B_{12}^{-2}$, $3X^+, Y^-$, where $3X^+$, Y^- can take on the following substitutions: $(3K^+, Br^-)$, $(3K^+, I^-)$, $(3Rb^+, I^-)$, $(3Cs^+, I^-)$, $(3Rb^+, Cl^-)$, $(3Rb^+, Br^-)$, $(3Cs^+, Br^-)$, and $(3Cs^+, Cl^-)$ (Tiritiris, I., 2005). Table 13.10 compares $H_{12}B_{12}^{-2}, 3K^+, Br^-$ with $H_{12}B_{12}^{-2}, 3Rb^+, Cl^-$ using the hexagonal reference axes for convenience. Note also that the icosahedrons are centered about the origin (see Exercises 4.58 through 4.62).

DEFINITIONS

Crystallographically dependent atoms
Crystallographically independent angles
Crystallographically independent atoms
Crystallographically independent bonds
Icosahedron
Space-filling model

EXERCISES

13.1 Use the Starter Program from Chapter 4, hexagonal lattice parameters from Table 13.1, and the description of the asymmetric unit (Hahn, 2010, p. 140) to create the asymmetric unit superimposed on the unit cell. Compare this figure with Figure 13.6b. This figure can be sketched as well.

13.2 A suggested trigonal crystal from CSD with space group #166, $R\bar{3}m$ is listed in Table 13.11. This crystal has rhombohedral axes. Construct a paper or a PowerPoint presentation based on the parallel topics in this chapter. The website is http://webcsd.ccdc.cam.ac.uk/teaching_database_demo.php; a shortcut to an individual Reference code is: http://webcsd.ccdc.cam.ac.uk/display_csd_entry. php?identifier=MBRMET10 for a crystal with Reference Code "MBRMET10". The parallel topics are discussed in Section 8.3.1. Examples are given at https:// sites.google.com/a/vt.edu/foundations_of_crystallography/.

13.3 Suggested trigonal crystals from CSD with space group #166, $R\bar{3}m$ are listed in ascending order of volume in Table 13.12. These crystals have hexagonal axes. Construct a paper or a PowerPoint presentation based on the parallel topics in this chapter. The website is http://webcsd.ccdc.cam.ac.uk/teaching_database_demo.php; a shortcut to an individual Reference code is: http://

TABLE 13.11 Trigonal Crystal in Space Group #166, $R\bar{3}m$ with Rhombohedral Axes

Refcode	Chemical Name	Formula	Volume (Å³)
SUKFEJ01	(1,4,7,10,13,16-Hexa-azacyclo-octadecane)-chromium(iii) tribromide	$C_{12}H_{30}CrN_6^{3+}$ $3(Br^-)$	461.594

TABLE 13.12   Trigonal Crystals in Space Group #166, $R\bar{3}m$ with Hexagonal Axes

Refcode	Chemical Name	Formula	Volume (Å^3)
DANXUM	Catena-((m6-Carbodiimide)-manganese)	$(CMnN_2)n$	140.128
VEWJAI	Guanidinium hexafluorophosphate	$CH_6N_3^+$, F_6P^-	338.235
OCTRNE	Octa-2,4,6-triyne	C_8H_6	477.587
RESSAJ	1,2,3,4,5,6,7,8-Octamethylpentacyclo(4.2.0.02,5.03,8.04,7)octane	$C_{16}H_{24}$	1068.668
BOBLAG	4,4'-Di-iodobicubyl	$C_{16}H_{12}I_2$	1072.401
YAVZEB01	Diaminomethyleneammonium hexafluoro-tantalum	$CH_6N_3^+$, F_6Ta^-	1100.24
CDSCDS	Cyclodecasulfur carbon disulfide	CS_2, S_{12}	1138.456
TUSBIS	Trimethyloxonium hexafluoro-arsenic	$C_3H_9O^+$, AsF_6^-	1179.873
AJAVUC	1-Phospha-adamantane 1-oxide	$C_9H_{15}OP$	1229.253
ACALDA	Acetaldehyde-ammonia trihydrate	$C_6H_{15}N_3$, $3(H_2O)$	1751.844

webcsd.ccdc.cam.ac.uk/display_csd_entry.php?identifier=MBRMET10 for a crystal with Reference Code "MBRMET10". The parallel topics are discussed in Section 8.3.1. Examples are given at https://sites.google.com/a/vt.edu/foundations_of_crystallography/.

13.4   Suggested primitive trigonal crystals from CSD with space group #164, $P\bar{3}m1$ are listed in ascending order of volume in Table 13.13. Construct a paper or a PowerPoint presentation based on the parallel topics in this chapter. The website is http://webcsd.ccdc.cam.ac.uk/teaching_database_demo.php; a shortcut to an individual Reference code is: http://webcsd.ccdc.cam.ac.uk/display_csd_entry.php?identifier=MBRMET10 for a crystal with Reference Code "MBRMET10". The parallel topics are discussed in Section 8.3.1. Examples are given at https://sites.google.com/a/vt.edu/foundations_of_crystallography/.

TABLE 13.13   Primitive Trigonal Crystals in Space Group #164, $P\bar{3}m1$

Refcode	Chemical Name	Formula	Volume (Å^3)
TAHJAP01	Catena-((μ_8-acetylido)-palladium-di-potassium)	$(C_2K_2Pd)_n$	118.149
EYAQEA	Hexammine-cobalt(iii) chloride bis(methanesulfonate)	$2(CH_3O_3S^-)$, $H_{18}CoN_6^{3+}$, $Cl-$	404.97
WERKIN20	Lithium bis(tetramethylammonium) hexanitro-cobalt(iii)	$2(C_4H_{12}N^+)$, Li^+, $CoN_6O_{12}^{3-}$	476.604
JUCRAA	Catena-(tris(Methylammonium) tri(μ_2-bromo)-hexabromo-di-antimony)	$(CH_6N^+)_{3n}$, $n(Br_9Sb_2^{3-})$	576.374
EBENIJ	Bis(Tetramethylammonium) dodecakis(μ_2-chloro)-hexachloro-hexa-tungsten	$2(C_4H_{12}N^+)$, $Cl_{18}W_6^{2-}$	865.411
IWIRIP	Catena-(bis(μ_3-1,3,5-Benzenetricarboxylato)-octaaqua-di-neodymium(iii) dihydrate)	$(C_{18}H_{22}Nd_2O_{20})n$, $2n(H_2O)$	876.147

## CHAPTER 14

Hexagonal System

Magnesium

The pictures above show hexagonal close-packed (left) and cubic close-packed (right) peaches.

CONTENTS

CHAPTER OBJECTIVES

- Study the crystallographic properties of the element magnesium, space group #194, $P6_3/mmc$, as a representative of the hexagonal crystal system.

- Present the parallel crystallographic topics that recur in Chapters 9 through 15. A few examples are given below. A complete list is found in Section 8.3.1.

  - Multiplication table of the associated point group

  - Diagrams of the unit cell, the asymmetric unit, the populated unit cell, and the special projections

  - Diagram of the reciprocal cell superimposed on the direct unit cell

  - Calculation of the powder diffraction pattern

  - Calculation of the atomic scattering curves

  - Calculation of the structure factor at 000

  - Calculation of the contribution of one atom to the structure factor of the largest reflection

- Present the special crystallographic topics of this chapter, which are listed below. A complete list of the special topics found in Chapters 9 through 15 is in Section 8.3.2.

- Close-packed structures

  - Model of close-packed spheres

  - Hexagonal close-packed structure (hcp)

  - Cubic close-packed structure (ccp)

  - Comparing hcp and ccp structures

- Interstitial spaces in a close-packed structure

- Sixfold axes

- Shifting the origin of the crystal

- Coordination number 12

- Site symmetry

14.1 INTRODUCTION

The example chosen for the hexagonal crystal system is the element magnesium. This structure, found in the American Mineralogists Database, closely approximates the model of hexagonal close-packed, hcp, spheres. The coordination number of crystalline magnesium is 12. The polyhedron with 12 vertices surrounding each magnesium atom is a triangular orthobicupola. In this chapter, that polyhedron is compared with the cuboctahedron in the cubic close-packed structure, which also has 12 vertices. For a classical approach see Megaw (1973, pp. 75–80).

Table 14.1 contains the crystal data for magnesium.

14.1.1 Special Topic: Close-Packed Structures

In addition to magnesium, the crystal structures of many of the elements are approximated by hexagonal close-packed, hcp, structures. These elements include helium, lithium, sodium, beryllium, strontium, titanium, cobalt, zinc, cadmium, and others. Other elements form in cubic close-packed, ccp, structures, among them the noble gases—neon,

TABLE 14.1   Magnesium Data

Compound	Magnesium
Formula	Mg
Space group	#194, $P6_3/mmc$
Bravais lattice	hP
Crystal system	Hexagonal
Packing	Nearly hcp
Ambient temperature for data	25°C (room temperature)
Z, Z′	$Z = 2$, $Z' = 1/12$
Diffraction probe	X-ray powder and rotation photographs
Reference, AMCSD	0012871
Density	1.737 g/cm^3

argon, krypton, and xenon—and metals—lithium, calcium, manganese, iron, copper, gold, silver, aluminum, and lead. Some elements, like lithium, exhibit *polymorphism*. The crystal structure of lithium is hcp at 78 K (Wyckoff, 1963, pp. 7–83), is body-centered cubic at 6.6 GPa, 296 K and ccp at 9.8 GPa, 296 K (Olinger, 1983). For some interesting examples see *The Pursuit of Perfect Packing* (Aste and Weaire, 2008).

14.1.1.1 Special Topic: Model of Close-Packed Spheres

A useful model for studying structures is close-packed identical spheres (Giacovazzo, 2002, p. 510; Basu, 2009, p. 12; Callister, 2010, p. 47; Smart, 2012, p. 2). Try it out with ping-pong balls or marbles. This model is an ideal structure that illuminates real structures and is applicable to a wide range of elements, metals, alloys, inorganic compounds, and organometallics. Often atoms or ions correspond to spheres, with other atoms or ions occupying the voids (or holes) between the spheres. Many structures are referred to as close-packed even though they differ considerably from the ideal close-packed structures.

Figure 14.1a shows one part of a planar layer of spheres in a close-packed structure. Each sphere is touching six other spheres. The two-dimensional layer extends infinitely and is called a *hexagonal close-packed layer*. There are two crystallographically distinct types of triangular spaces between the spheres, red with the triangle pointing upward and blue with the triangle pointing downward. The two-dimensional space group is *p6mm*. Figure 14.1b shows some symmetries of *p6mm*. The sixfold is at the origin; the mirrors are along the edges of the unit cell and along the diagonals; the red threefold is at 2/3, 1/3; and the blue twofold is at 1/3, 2/3. Call this layer A.

Now a second layer is placed on top of layer A, making a three-dimensional structure. The spheres sit in either the red holes or the blue holes, but not both. It does not matter which type of hole is chosen. Figure 14.2 shows a second layer of yellow spheres centered over the red holes. Note again that there are two sets of triangles—the triangles pointing upward, which, as before, have a blue mark in the middle and the triangles pointing downward, which are white. Call this layer B.

14.1.1.2 Special Topic: Hexagonal Close-Packed Structure

Now add a third layer. There are two choices. If the layer is centered over the white triangles, then the new layer is a translation of layer A and is also called A (see Figure 14.3a). The sequence

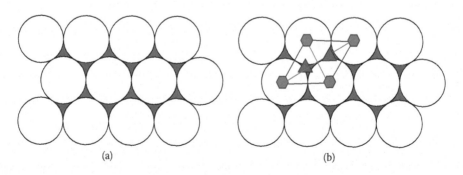

(a) (b)

FIGURE 14.1   (a) Hexagonal close-packed layer; (b) with some symmetries of *p6mm* superimposed.

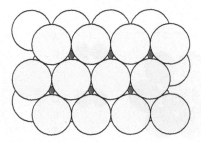

FIGURE 14.2 Second layer centered over the red holes.

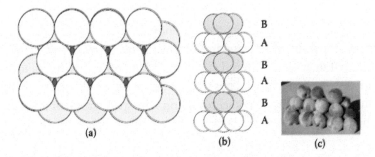

FIGURE 14.3 Layers of *hcp*: (a) along [001]; (b) stacking edge on; (c) peaches.

is ABABAB…. No matter how many layers are added, a wire could be inserted through each blue hole. The white layers alternate with the yellow layers. This packing is called *hexagonal close-packed (hcp)*. Figure 14.3b shows the ABABAB…. stacking of the layers seen edge on.

By continuing through the layers, the blue holes accommodate threefold axes. Also the other threefolds and the mirrors persist. The sixfold becomes a 6_3 screw. The space group of this structure is #194, $P6_3/mmc$. In general, the hexagonal Bravais lattice has two cell parameters a_H and c_H. In hcp,

$$a_H = 2R$$

where R is the radius of each sphere and c_H/a_H must equal 2/3 $\sqrt{6}$, which is approximately 1.633. Another condition for hcp is that $Z = 2$, with the spheres in a special position, Wyckoff letter either c or d. In crystalline magnesium the atoms are good approximations to spheres, and indeed the space group is $P6_3/mmc$ with

$$a_H = 3.209 \text{ Å}, c_H = 5.210 \text{ Å}, c_H/a_H = 1.624, Z = 2, \text{ and } R_{Mg} = a/2 = 1.60 \text{ Å}$$

The magnesium atoms are in the special position with Wyckoff letter c. Magnesium is considered to be an excellent example of a hcp structure, even though the c_H/a_H ratio differs slightly from the ideal.

14.1.1.3 Special Topic: Cubic Close-Packed Structure

Another important type of packing results from a different placement of the third layer. Place the third red layer over blue holes, as in Figure 14.4. The sequence is ABCABC…. Note that neither the blue holes nor the red holes continue through multiple layers. The Bravais

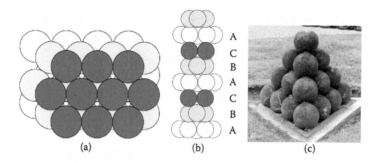

FIGURE 14.4   Layers of ccp (a) along [111], (b) stacking edge on, (c) cannon balls.

lattice is no longer hexagonal; but, surprisingly enough, is now cubic. The white, yellow, and red layers repeat indefinitely. This packing is called *cubic close-packed (ccp)*. Unfortunately ccp is often referred to as face-centered cubic, fcc. Many other structures form in face-centered cubic lattices. For this reason ccp is used in this book. See Megaw (1973, p. 75). Figure 14.4b shows the stacking of the layers seen edge on. The space group of this structure is #225, $Fm\overline{3}m$. In general, the cubic Bravais lattice has one cell parameter a_C. In ccp,

$$a_C = 2\sqrt{2}\,R$$

where R is the radius of the sphere. Another condition for ccp is $Z = 4$ with the spheres in a special position, Wyckoff letter either a or b. In crystalline copper, the atoms are good approximations to spheres. The space group for copper is #225, $Fm\overline{3}m$ with

$$a = 3.60 \text{ Å}, Z = 4, \text{ and } R = \sqrt{2}\,(3.6)/4 = 1.27 \text{ Å}$$

Copper is an example of a ccp structure. The stack of cannon balls in Figure 14.4c is another example as is the stack of peaches in the picture at the beginning of this chapter.

14.1.1.4 Special Topic: Comparison of hcp and ccp Structures

The number of closest neighbors is called the *coordination number, CN*. The coordination number of a sphere is six in a two-dimensional layer with a hcp structure. The three-dimensional configurations, both *hcp* and *ccp*, have the same coordination number, 12.

Crystals form when their free energies are minimized. The free energies of hcp and ccp ideal structures are very close (Zolotoyabko, 2011, p. 131). When the centers of the 12 nearest-neighbor atoms in ccp are connected, they form a cuboctahedron. Figure 14.5

FIGURE 14.5   Comparison of triangular orthobicupola; cuboctahedron.

FIGURE 14.6   Close-packed layer in $H_{12}B_{12}^{-2}, 3K^+, Br$.

shows the cuboctahedron with $[111]_c$ perpendicular to a triangular face. A *cuboctahedron* is an Archimedean polyhedron with 12 identical vertices, 24 identical edges, 8 triangular faces, and 6 square faces. Two triangles and two squares meet at each vertex. While the coordination numbers of both ccp and hcp are identical, namely 12, the number of second and higher nearest neighbors differs, causing one structure or the other to be preferred. In hcp there are two second nearest neighbors at distance c_H and in *ccp* there are 12 second nearest neighbors at distance a_C. For a discussion of the *hcp* polyhydron, the triangular orthobicupola, Figure 14.5, see Section 14.6.2 (see Exercise 14.1).

14.1.1.5 Special Topic: Interstitial Spaces in a Close-Packed Structure

A large number of materials can be described in terms of spheres with close-packed structures, considering that smaller atoms or ions can fit into the spaces between the spheres. The interstitial spaces between the spheres have symmetries that contribute to the overall structure. Both tetrahedral holes, with coordination number four, and octahedral holes with coordination number six are found in hcp and ccp. The boron compound $H_{12}B_{12}^{2-}$, 3 K^+, Br (Chapter 13) is a good example of a generalization of a close-packed structure (see Figure 14.6). The $H_{12}B_{12}^{2-}$ anions, which are distorted icosahedrons shown in pink and white, and the potassium cations, K^+, shown in blue, both approximate spheres of the same size. Together they pack almost in a ccp structure, which seems remarkable. The unit cell has $a = b = c$ and $\alpha = \beta = \gamma = 93.4°$. The unseen bromine anions are hidden in the octahedral holes formed by the potassium cations (Tiritiris, 2005).

14.2   MAGNESIUM: POINT GROUP PROPERTIES

The associated point group for magnesium is 6/*mmm* in the Hermann–Mauguin notation and D_{6h} in the Schönflies notation. This point group has 24 symmetry operations—the identity, 7 twofolds, 2 threefolds, 2 sixfolds, the inversion, 4 mirrors, 3 axial glides, 2 threefold rotoinversions, and 2 sixfold rotoinversions. Thus, the order, or number of elements, is 24. This point group illustrates the rule that whenever there are improper operations there is an equal number of proper operations.

14.2.1 Multiplication Table

The generators shown in Appendix 3 are 1, 3⁺, 2_z, 2_{xx}, and $\bar{1}$. However, the point group can be generated from just three symmetry operations (see Figure 3.58). The first operation has to be any one of $\bar{3}^+$, $\bar{3}^-$, $\bar{6}^+$, or $\bar{6}^-$. The other two depend on the first choice. For example, $\bar{6}^+$, $\bar{1}$, and 2_{xx} are a minimal set of generators. Their matrices are

$$\bar{6}^+ = \begin{pmatrix} \bar{1} & 1 & 0 \\ \bar{1} & 0 & 0 \\ 0 & 0 & \bar{1} \end{pmatrix}, \; \bar{1} = \begin{pmatrix} \bar{1} & 0 & 0 \\ 0 & \bar{1} & 0 \\ 0 & 0 & \bar{1} \end{pmatrix}, \; 2_{xx} = \begin{pmatrix} 0 & 1 & 0 \\ 1 & 0 & 0 \\ 0 & 0 & \bar{1} \end{pmatrix}$$

The Starter Program from Chapter 3 calculates the multiplication table in Table 14.2. Because

$$2_{xx} \cdot 2_x \neq 2_x \cdot 2_{xx}$$

the matrix of the point group is not symmetric, and the point group is not abelian. The point group has an inversion; thus its symmetries do not permit the piezoelectric effect. The relationships among the symmetry operations are seen in the multiplication table.

14.2.2 Stereographic Projections

The general position stereographic projection for 6/mmm, Figure 14.7a, has a combination of 12 dots and 12 circles corresponding to the 24 symmetry operations. See Section 3.9 for a general description of these projections. The coordinates, xyz, corresponding to the identity operation, have a positive z-coordinate indicated by the red dot labeled in the diagram. Each point with a positive z-coordinate is indicated by a dot, and each point with a negative z-coordinate is indicated by an open circle. There are two combinations, a red dot with a blue circle or a blue dot with a red circle. These combinations indicate a mirror in the plane of the figure. The color red is used for the 12 proper operations: 1, 3⁺, 3⁻, 2, 6⁺, 6⁻, 2_{xx}, 2_x, 2_y, $2_{x\bar{x}}$, 2_{x2x}, and 2_{2xx}. The color blue is used for the 12 improper operations: $\bar{1}$, $\bar{3}^+$, $\bar{3}^-$, m_{xy}, $\bar{6}^+$, $\bar{6}^-$, $m_{x\bar{x}z}$, m_{x2xz}, m_{2xxz}, m_{xxz}, m_{xz}, and m_{yz}. A proper operation does not change handedness, and the determinant of the matrix is +1. An improper operation changes handedness and the determinant of the matrix is −1. Note that the points, both dots and open circles, are placed inside the circumference to allow any value for z.

Figure 14.7b is the symbol stereographic projection. The red hexagon at the center indicates the cyclic group 6, which has six proper symmetry operations. There are six twofolds in the xy plane. They are represented by the six pairs of red ovals located on the circumference of the circle. The inversion point at the origin is represented by a purple circle superimposed on the red hexagon. There is a mirror plane in the plane of the paper, and it is indicated by the large purple circle on the circumference of the figure. The additional six mirrors, which are perpendicular to the plane of the paper, are indicated by the six purple lines going under the red hexagon. The point group has a sixfold rotoinversion cyclic subgroup, which is not represented in the symbol diagram. There is, however, sufficient information in the symbol diagram to generate this subgroup.

TABLE 14.2 Point Group 6/mmm Multiplication Table

	1	3^+	3^-	2	6^+	6^-	2_{xx}	2_x	2_y	$2_{\bar{x}\bar{x}}$	2_{x2x}	2_{2xx}	$\bar{1}$	$\bar{3}^+$	$\bar{3}^-$	m_{xy}	$\bar{6}^-$	$\bar{6}^+$	$m_{\bar{x}\bar{x}z}$	m_{x2xz}	m_{2xxz}	m_{xxz}	m_{xz}	m_{yz}
1	1	3^+	3^-	2	6^+	6^-	2_{xx}	2_x	2_y	$2_{\bar{x}\bar{x}}$	2_{x2x}	2_{2xx}	$\bar{1}$	$\bar{3}^+$	$\bar{3}^-$	m_{xy}	$\bar{6}^-$	$\bar{6}^+$	$m_{\bar{x}\bar{x}z}$	m_{x2xz}	m_{2xxz}	m_{xxz}	m_{xz}	m_{yz}
3^+	3^+	3^-	1	6^+	6^-	2	2_x	2_y	2_{xx}	2_{x2x}	2_{2xx}	$2_{\bar{x}\bar{x}}$	$\bar{3}^+$	$\bar{3}^-$	$\bar{1}$	$\bar{6}^-$	$\bar{6}^+$	m_{xy}	m_{x2xz}	m_{2xxz}	$m_{\bar{x}\bar{x}z}$	m_{xz}	m_{yz}	m_{xxz}
3^-	3^-	1	3^+	6^-	2	6^+	2_y	2_{xx}	2_x	2_{2xx}	$2_{\bar{x}\bar{x}}$	2_{x2x}	$\bar{3}^-$	$\bar{1}$	$\bar{3}^+$	$\bar{6}^+$	m_{xy}	$\bar{6}^-$	m_{2xxz}	$m_{\bar{x}\bar{x}z}$	m_{x2xz}	m_{yz}	m_{xxz}	m_{xz}
2	2	6^+	6^-	1	3^+	3^-	$2_{\bar{x}\bar{x}}$	2_{x2x}	2_{2xx}	2_{xx}	2_x	2_y	$\bar{2}$	$\bar{6}^+$	$\bar{6}^-$	$m_{\bar{x}\bar{x}z}$	$\bar{3}^-$	$\bar{3}^+$	m_{xy}	m_{xz}	m_{yz}	m_{2xxz}	m_{x2xz}	m_{xxz}
6^+	6^+	6^-	2	3^+	3^-	1	2_{x2x}	2_{2xx}	$2_{\bar{x}\bar{x}}$	2_x	2_y	2_{xx}	$\bar{6}^+$	$\bar{6}^-$	$\bar{2}$	m_{x2xz}	$\bar{3}^+$	$\bar{3}^-$	m_{xz}	m_{yz}	m_{xy}	m_{xxz}	m_{2xxz}	m_{x2xz}
6^-	6^-	2	6^+	3^-	1	3^+	2_{2xx}	$2_{\bar{x}\bar{x}}$	2_{x2x}	2_y	2_{xx}	2_x	$\bar{6}^-$	$\bar{2}$	$\bar{6}^+$	m_{2xxz}	$\bar{3}^-$	$\bar{3}^+$	m_{yz}	m_{xy}	m_{xz}	m_{x2xz}	m_{xxz}	m_{2xxz}
2_{xx}	2_{xx}	2_x	2_y	$2_{\bar{x}\bar{x}}$	2_{x2x}	2_{2xx}	1	3^+	3^-	2	6^+	6^-	$m_{\bar{x}\bar{x}z}$	m_{xy}	m_{x2xz}	$\bar{1}$	$\bar{3}^+$	$\bar{3}^-$	$\bar{2}$	$\bar{6}^+$	$\bar{6}^-$	2_x	2_y	2_{xx}
2_x	2_x	2_y	2_{xx}	2_{x2x}	2_{2xx}	$2_{\bar{x}\bar{x}}$	3^-	1	3^+	6^-	2	6^+	m_{xy}	m_{x2xz}	m_{2xxz}	$\bar{3}^-$	$\bar{1}$	$\bar{3}^+$	$\bar{6}^-$	$\bar{2}$	$\bar{6}^+$	2_y	2_{xx}	2_x
2_y	2_y	2_{xx}	2_x	2_{2xx}	$2_{\bar{x}\bar{x}}$	2_{x2x}	3^+	3^-	1	6^+	6^-	2	m_{x2xz}	m_{2xxz}	$m_{\bar{x}\bar{x}z}$	$\bar{3}^+$	$\bar{3}^-$	$\bar{1}$	$\bar{6}^+$	$\bar{6}^-$	$\bar{2}$	2_{xx}	2_x	2_y
$2_{\bar{x}\bar{x}}$	$2_{\bar{x}\bar{x}}$	2_{x2x}	2_{2xx}	2_{xx}	2_x	2_y	2	6^+	6^-	1	3^+	3^-	m_{xx}	m_{yz}	m_{xz}	$\bar{2}$	$\bar{6}^+$	$\bar{6}^-$	$\bar{1}$	$\bar{3}^+$	$\bar{3}^-$	2_{x2x}	2_{2xx}	$2_{\bar{x}\bar{x}}$
2_{x2x}	2_{x2x}	2_{2xx}	$2_{\bar{x}\bar{x}}$	2_x	2_y	2_{xx}	6^-	2	6^+	3^-	1	3^+	m_{xz}	m_{xx}	m_{yz}	$\bar{6}^-$	$\bar{2}$	$\bar{6}^+$	$\bar{3}^-$	$\bar{1}$	$\bar{3}^+$	2_{2xx}	$2_{\bar{x}\bar{x}}$	2_{x2x}
2_{2xx}	2_{2xx}	$2_{\bar{x}\bar{x}}$	2_{x2x}	2_y	2_{xx}	2_x	6^+	6^-	2	3^+	3^-	1	m_{yz}	m_{xz}	m_{xx}	$\bar{6}^+$	$\bar{6}^-$	$\bar{2}$	$\bar{3}^+$	$\bar{3}^-$	$\bar{1}$	$2_{\bar{x}\bar{x}}$	2_{x2x}	2_{2xx}
$\bar{1}$	$\bar{1}$	$\bar{3}^+$	$\bar{3}^-$	$\bar{2}$	$\bar{6}^+$	$\bar{6}^-$	$m_{\bar{x}\bar{x}z}$	m_{xy}	m_{x2xz}	m_{xx}	m_{xz}	m_{yz}	1	3^+	3^-	2_{xx}	6^-	6^+	$2_{\bar{x}\bar{x}}$	2_x	2_y	2_{xx}	2_x	2_{xx}
$\bar{3}^+$	$\bar{3}^+$	$\bar{3}^-$	$\bar{1}$	$\bar{6}^+$	$\bar{6}^-$	$\bar{2}$	m_{xy}	m_{x2xz}	m_{2xxz}	m_{yz}	m_{xx}	m_{xz}	3^+	3^-	1	6^-	6^+	2	2_x	2_y	2_{xx}	$2_{x\bar{x}}$	2_{x2x}	$2_{x\bar{x}}$
$\bar{3}^-$	$\bar{3}^-$	$\bar{1}$	$\bar{3}^+$	$\bar{6}^-$	$\bar{2}$	$\bar{6}^+$	m_{x2xz}	m_{2xxz}	$m_{\bar{x}\bar{x}z}$	m_{xz}	m_{yz}	m_{xx}	3^-	1	3^+	6^+	6^-	2	2_y	2_{xx}	2_x	2_{2xx}	$2_{x\bar{x}}$	2_{x2x}
m_{xy}	m_{xy}	m_{x2xz}	m_{2xxz}	$m_{\bar{x}\bar{x}z}$	m_{xz}	m_{yz}	$\bar{1}$	$\bar{3}^+$	$\bar{3}^-$	$\bar{2}$	$\bar{6}^+$	$\bar{6}^-$	2_{xx}	2_x	2_y	1	3^+	3^-	2	6^+	6^-	2_x	2_y	2_{xx}
$\bar{6}^-$	$\bar{6}^-$	$\bar{2}$	$\bar{6}^+$	$\bar{3}^-$	$\bar{1}$	$\bar{3}^+$	m_{yz}	m_{xx}	m_{xz}	m_{2xxz}	$m_{\bar{x}\bar{x}z}$	m_{x2xz}	6^-	2	6^+	3^-	1	3^+	6^-	2	6^+	2_y	2_{xx}	2_x
$\bar{6}^+$	$\bar{6}^+$	$\bar{6}^-$	$\bar{2}$	$\bar{3}^+$	$\bar{3}^-$	$\bar{1}$	m_{xx}	m_{xz}	m_{yz}	m_{x2xz}	m_{2xxz}	$m_{\bar{x}\bar{x}z}$	6^+	6^-	2	3^+	3^-	1	6^+	6^-	2	2_{xx}	2_x	2_y
$m_{\bar{x}\bar{x}z}$	$m_{\bar{x}\bar{x}z}$	m_{x2xz}	m_{2xxz}	m_{xx}	m_{xz}	m_{yz}	$\bar{2}$	$\bar{6}^+$	$\bar{6}^-$	$\bar{1}$	$\bar{3}^+$	$\bar{3}^-$	$2_{\bar{x}\bar{x}}$	2_{x2x}	2_{2xx}	2	6^+	6^-	1	3^+	3^-	m_{x2xz}	m_{2xxz}	$m_{\bar{x}\bar{x}z}$
m_{x2xz}	m_{x2xz}	m_{2xxz}	$m_{\bar{x}\bar{x}z}$	m_{xz}	m_{yz}	m_{xx}	$\bar{6}^-$	$\bar{2}$	$\bar{6}^+$	$\bar{3}^-$	$\bar{1}$	$\bar{3}^+$	2_{x2x}	2_{2xx}	$2_{\bar{x}\bar{x}}$	6^-	2	6^+	3^-	1	3^+	m_{2xxz}	$m_{\bar{x}\bar{x}z}$	m_{x2xz}
m_{2xxz}	m_{2xxz}	$m_{\bar{x}\bar{x}z}$	m_{x2xz}	m_{yz}	m_{xx}	m_{xz}	$\bar{6}^+$	$\bar{6}^-$	$\bar{2}$	$\bar{3}^+$	$\bar{3}^-$	$\bar{1}$	2_{2xx}	$2_{\bar{x}\bar{x}}$	2_{x2x}	6^+	6^-	2	3^+	3^-	1	$m_{\bar{x}\bar{x}z}$	m_{x2xz}	m_{2xxz}
m_{xxz}	m_{xxz}	m_{xz}	m_{yz}	$m_{x\bar{x}z}$	m_{x2xz}	m_{2xxz}	2_x	2_y	2_{xx}	$2_{\bar{x}\bar{x}}$	2_{x2x}	2_{2xx}	m_{xz}	m_{yz}	m_{xxz}	2_x	2_y	2_{xx}	6^+	6^-	2	1	3^+	3^-
m_{xz}	m_{xz}	m_{yz}	m_{xxz}	m_{x2xz}	m_{2xxz}	$m_{\bar{x}\bar{x}z}$	2_y	2_{xx}	2_x	2_{x2x}	2_{2xx}	$2_{\bar{x}\bar{x}}$	m_{yz}	m_{xxz}	m_{xz}	2_y	2_{xx}	2_x	6^-	2	6^+	3^-	1	3^+
m_{yz}	m_{yz}	m_{xxz}	m_{xz}	m_{2xxz}	$m_{\bar{x}\bar{x}z}$	m_{x2xz}	2_{xx}	2_x	2_y	2_{2xx}	$2_{\bar{x}\bar{x}}$	2_{x2x}	m_{xxz}	m_{xz}	m_{yz}	2_{xx}	2_x	2_y	2	6^+	6^-	3^+	3^-	1

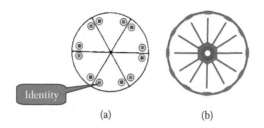

(a)   (b)

FIGURE 14.7   Point group 6/*mmm* stereographic projections: (a) general position; (b) symbol.

The purple symbols indicate that half the operations are proper, or red, and half the operations are improper, or blue.

Given the symbol diagram, the general position diagram can be created and vice versa.

14.3 MAGNESIUM: SPACE GROUP PROPERTIES

Magnesium crystallizes in the hexagonal space group #194, *P6₃/mmc*. The Bravais lattice is *hP*. The symmetry operations require that

$$a = b \quad \text{and} \quad \alpha = \beta = 90° \quad \text{and} \quad \gamma = 120°$$

The sixfold axis characterizes the hexagonal crystal system. There are seven different kinds of sixfold axes, and they are examined in the next section.

14.3.1 Special Topic: Sixfold Axes

The hexagonal point groups are 6, $\bar{6}$, 622, 6*mm*, 6/*m*, $\bar{6}$*m*2, and 6/*mmm*; and all contain a sixfold rotation, 6, or a sixfold rotoinversion, $\bar{6}$ (see Section 3.9). In a space group both 6 and $\bar{6}$ axes, unlike screw axes, have their translations only at unit cell distances. See Figure 14.8, which is to be thought of as three-dimensional drawing in perspective. The operations associated with the sixfold axis, 6, allow no change of handedness; hence all six circles are red. On the other hand, three of the operations associated with the sixfold rotoinversion, $\bar{6}$, have a change of handedness, as indicated by the presence of three red circles and three blue circles. When a translation that is a fraction of the lattice vector along the rotation axis is

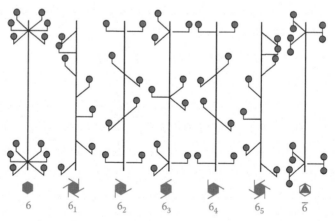

FIGURE 14.8   Sixfold axes.

allowed, there are five hexagonal screw axes—6_1, 6_2, 6_3, 6_4, 6_5. The operations associated with these screw axes allow no change of handedness; hence all six circles are red. Note that $6_6 = 6$.

The symbols for the axes, as used in the symbol space group diagrams, are shown at the bottom of the figure. These symbols are red when all the operations are proper and purple when half the operations are proper and half are improper (see Section 4.11.4.2).

Note that the 6_1 screw axis is a mirror of 6_5 and that the 6_2 screw axis is a mirror of 6_4. Furthermore, the space groups #169, $P6_1$, and #170, $P6_5$, form an enantiomorphic space group pair as do the space groups #171, $P6_2$, and #172, $P6_4$. Note that 6_1 and 6_5 create mirror-image spirals (see Section 12.1.2).

14.3.2 Space Group Diagrams

Figure 14.9a shows the symbol diagram for the nonsymmorphic space group #194, $P6_3/mmc$. The full form of the Hermann–Mauguin symbol is $P6_3/m\ 2/m\ 2/c$ (see Section 11.3.1). An explanation of the full Hermann–Mauguin symbol follows. The capital letter refers to the centering type. The next three parts of the Hermann–Mauguin symbol describe the symmetries in their relationship to the crystallographic directions in the space group. In the hexagonal system, the first place describes a sixfold axis parallel to [001], the second place describes symmetries along [100], [010], and [110] and the third place describes symmetries perpendicular to [100], [010], and [110] as well as the principal sixfold axis.

For space group #194, $P6_3/m\ 2/m\ 2/c$, the "P" means that the cell is primitive. The "6_3" refers to the 6_3 screw axis parallel to [001], which is shown at the origin. The "$/m$" refers to the mirror whose normal is parallel to this 6_3 screw axis. The mirror, labeled in blue, is represented in the diagram by the two purple solid lines in the upper left-hand corner at $z = 1/4$. The "$2/m$" refers to the twofold axes along [100], [010], and [110], with mirrors having normals parallel to these twofolds. The twofold along [100] and its mirror are labeled in green.

The "$2/c$" refers to the twofold axes perpendicular to [100], [010], and [110] as well as to [001], with c-glide planes having normals parallel to these twofolds. The twofold perpendicular to [100] at $z = 1/4$ and its purple dotted c-glide plane are labeled in orange. Note also there is an inversion point at the origin that is represented by a purple circle superimposed upon the red symbol for the axis 6_3. The short symbol for this space group is #194, $P6_3/mmc$.

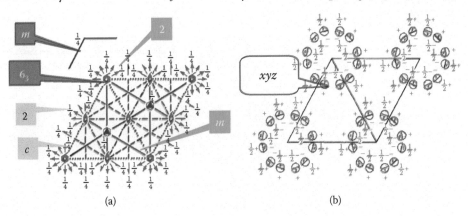

FIGURE 14.9 Space group #194, $P6_3/mmc$ (a) symbol diagram; (b) general position diagram.

Figure 14.9b gives the general position diagram for space group #194, $P6_3/mmc$. Each circle is split into a red half and a blue half with a comma. In addition, each circle has a "+" sign, a "½+" sign, a "−" sign, or "½−" sign nearby. These correspond to the four options for the z-coordinate of the general position point—z, $z + 1/2$, $\bar{z}$, $\bar{z}+ 1/2$. The split-circle symbol means that there is a mirror plane perpendicular to the z-axis. Furthermore, the mirror plane is located at $z = ¼$, as indicated in the upper left-hand corner of the symbol diagram. In the diagram the red half-circle in the upper left corner with the plus sign represents the coordinates, xyz, of the identity operation.

Given the symbol diagram, the general position diagram can be created and vice versa.

14.3.3 Maximal Subgroups

There are 230 space groups. Because the space groups contain translation operations, each space group has infinite order. Space groups can be arranged in trees similar to the trees of the point groups. However, there is no unique way of constructing the tree for the space groups. In a tree of type I, the centering type is kept and the order of the associated point group is changed. (Maximal subgroups and minimal supergroups are discussed in Section 4.8.) For example, a maximal type I subgroup of the space group #194, $P6_3/mmc$, is the space group #176, $P6_3/m$. Both these space groups have the same centering type, namely P; but for the subgroup the order of the associated point group are decreased. Specifically, the associated point group of the space group #194, $P6_3/mmc$, is $6/mmm$ and has order 24, while the associated point group of the space group #176, $P6_3/m$, is $6/m$ and has order 12.

The space group #194, $P6_3/mmc$, has no supergroups, but has eight maximal subgroups, of which seven have associated point groups of order 12 and one has an associated point group of order eight. The *International Tables for Crystallography* give the information necessary to construct Figure 14.10.

14.3.4 Asymmetric Unit

The asymmetric unit is the smallest part of the unit cell that, when operated on by the symmetry operations, produces the whole unit cell. The volume of the asymmetric unit

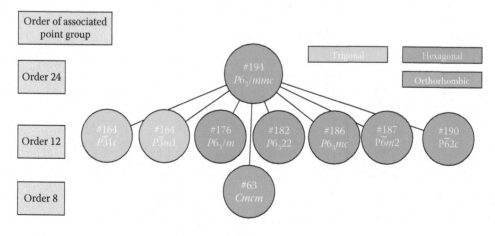

FIGURE 14.10 Maximal subgroups for space group #194, $P6_3/mmc$.

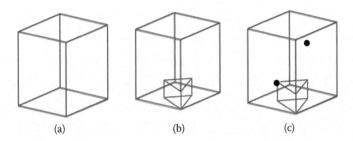

FIGURE 14.11   (a) Unit cell; (b) with asymmetric unit; (c) populated with atoms.

is related to the volume of the unit cell by $V_{AU} = V_{UC}/nh$, where n is the number of lattice points in the unit cell and h is the number of operations in the associated point group. (Note that the product nh is equal to the multiplicity of the general position.) Because there are 24 operations and the unit cell is primitive ($n = 1$, $h = 24$), the volume of the asymmetric unit is 1/24th the volume of the unit cell. Therefore, because the volume of the unit cell is 47.47 Å³, the volume of the asymmetric unit is 1.94 Å³.

Figure 14.11a shows the unit cell drawn in blue. Figure 14.11b shows the unit cell drawn with the asymmetric unit in red superimposed on it. Figure 14.11c shows the populated cell with the asymmetric unit. Because $Z = 2$, there is 1/12 of an atom in the asymmetric unit. The symmetries of the space group generate the other 11/12 of this magnesium atom as well as the complete second atom.

The asymmetric unit has

$$0 \leq x \leq 2/3, 0 \leq y \leq 2/3, 0 \leq z \leq 1/4, x \leq 2y, y \leq \text{minimum } (1 - x, 2x)$$

Vertices: 000,   ⅔ ⅓ 0,   ⅓ ⅔ 0

00¼,   ⅔ ⅓ ¼,   ⅓ ⅔ ¼

(Hahn, 2010, p. 144). The vertices are given for this complicated asymmetric unit because they are easier to interpret.

14.4 MAGNESIUM: DIRECT AND RECIPROCAL LATTICES

In addition to the direct lattice, there is the reciprocal lattice, which is the lattice of the diffraction pattern. The diffraction pattern has the intensities of the Bragg reflections superimposed on the reciprocal lattice. Table 14.3 compares the direct and reciprocal lattices, including the lattice parameters, the metric matrix, and the cell volume for both the direct and reciprocal cells.

The Starter Program from Chapter 5 produces the reciprocal cell superimposed on the unit cell. The program produces a figure that can be rotated to observe that, for example, **a** is perpendicular to both **b*** and **c***. See Sections 5.2 and 5.3 for all the relevant relationships. Figure 14.12 shows the relationship between the direct unit cell (blue) and the reciprocal unit cell (red) with the volumes normalized to one. Note that **c** and **c*** point in the same direction.

TABLE 14.3   Magnesium: Comparison of Direct and Reciprocal Lattices

Direct Lattice	Reciprocal Lattice
$a = b = 3.2093$ Å	$a^* = b^* = 0.3598$ Å^{-1}
$c = 5.2103$ Å	$c^* = 0.1919$ Å^{-1}
$\alpha = \beta = 90°$	$\alpha^* = \beta^* = 90°$
$\gamma = 120°$	$\gamma^* = 60°$

$$\mathbf{G} = \begin{pmatrix} 10.30 & -5.15 & 0 \\ -5.15 & 10.30 & 0 \\ 0 & 0 & 27.15 \end{pmatrix} \text{Å}^2 \qquad \mathbf{G}^* = \begin{pmatrix} 0.130 & 0.0647 & 0 \\ 0.0647 & 0.130 & 0 \\ 0 & 0 & 0.0368 \end{pmatrix} \text{Å}^{-2}$$

$V = 46.474$ Å^3 $V^* = 0.0215$ Å^{-3}

FIGURE 14.12   Magnesium direct cell (blue) and reciprocal cell (red) with volumes normalized to one.

14.5 MAGNESIUM: FRACTIONAL COORDINATES AND OTHER DATA FOR THE CRYSTAL STRUCTURE

The crystal structure consists of the placement of the atoms in the unit cell and their nearest neighbors. See Section 14.6.2 for a discussion of nearest neighbors. The fractional coordinates of the atom in the asymmetric unit combined with the general position coordinates of the space group and the lattice parameters are sufficient to specify both atoms in the unit cell.

Z equals two; that is, there are two magnesium atoms in the unit cell. Z', the number of formula units in the asymmetric unit, equals 1/12th; that is to say, there is 1/12th magnesium atom in the asymmetric unit (see Figure 14.13). The fractional coordinates of this atom from the CIF are given in Table 14.4. The atom has Wyckoff letter c and a multiplicity of 2.

14.6 MAGNESIUM: CRYSTAL STRUCTURE

Populate the unit cell by using the Starter Program from Chapter 4 to create the entire unit cell as shown in Figure 14.14a. This program combines the lattice parameters—a, b, c, α, β, and γ (Table 14.3), the fractional coordinates (Table 14.4), and the general position

FIGURE 14.13   Schematic showing the atom needed to produce populated cell.

TABLE 14.4   Magnesium Fractional Coordinates, Multiplicity, and Wyckoff Letter for the Atom in Asymmetric Unit

Atom	x Fractional Coordinate	y Fractional Coordinate	z Fractional Coordinate	Multiplicity Wyckoff Letter
Mg	1/3	2/3	1/4	2c

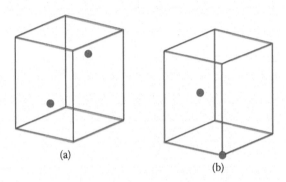

(a)

(b)

FIGURE 14.14   Populated unit cell with atoms (a) in Wyckoff c; (b) origin shifted.

coordinates of the space group (Table 14.5). The lattice parameters, the fractional coordinates, and the general position coordinates are from the CIF.

Beginning with the general position coordinates the Starter Program automatically includes both atoms. No special procedure is required for atoms in special positions.

14.6.1 Special Topic: Shifting the Origin of the Crystal

The hcp crystal structure is easier to describe if a magnesium atom is placed at the origin. This point can be appreciated by remembering the picture at the beginning of this chapter showing peaches arranged in the hcp structure. For peaches the contact forces together with gravity are sufficient to maintain the structure.

However, in the *International Tables for Crystallography* the inversion point is placed at the origin, exactly between two magnesium atoms. The Starter Program supplied with the general position coordinates from the *International Tables for Crystallography* produces Figure 14.14a.

The Starter Program is next adapted to place an atom at the origin. A similar shift is described in Section 2.3. The two magnesium atoms, in Wyckoff c, have coordinates from the *International Tables for Crystallography* (1/3, 2/3, 1/4) and (2/3, 1/3, 3/4). If both atoms are moved by adding (−1/3, −2/3, −1/4) to each, then their new coordinates are (0, 0, 0) and (1/3, −1/3, 1/2). The first atom is now at the origin, but the second atom is outside the unit cell. To bring it back into the unit cell, add (0, 1, 0) to the second atom giving (1/3, 2/3, 1/2). (Remember that any atom can be shifted along the direction of any basis vector by an integral number of unit cells.) These two atoms are plotted in red in Figure 14.14b. Of course, all the symmetries have been shifted as well. There are two ways of looking at the result. Either move the atoms or move the unit cell.

TABLE 14.5   General Position Coordinates and Symmetry Operations

#	General Position Coordinates	Symmetry Operations
1	x, y, z	Identity
2	$\bar{y}, x+\bar{y}, z$	3^+ along [0,0,1] at 0, 0, z
3	$\bar{x}+y, \bar{x}, z$	3^- along [0,0,1] at 0, 0, z
4	$\bar{x}, \bar{y}, 1/2+z$	2_1 along [0,0,1] at 0, 0, z
5	$y, \bar{x}+y, 1/2+z$	6_3^- along [0,0,1] at 0, 0, z
6	$x-y, x, 1/2+z$	6_3^+ along [0,0,1] at 0, 0, z
7	$\bar{y}, \bar{x}, z$	m normal to [1,1,0]
8	$\bar{x}+y, y, z$	m normal to [1, 0, 0]
9	$x, x-y, z$	m normal to [0,1,0]
10	$y, x, 1/2+z$	c-glide normal to $[1, \bar{1}, 0]$
11	$x-y, \bar{y}, 1/2+z$	c-glide normal to [1, 2, 0]
12	$\bar{x}, \bar{x}+y, 1/2+z$	c-glide normal to [2, 1, 0]
13	$\bar{x}, \bar{y}, \bar{z}$	Inversion at 000
14	$y, \bar{x}+y, \bar{z}$	$\bar{3}^+$ along [0, 0, 1] at 0, 0, z with inversion at 0, 0, 0
15	$x-y, x, \bar{z}$	$\bar{3}^-$ along [0, 0, 1] at 0, 0, z with inversion at 0, 0, 0
16	$x, y, 1/2-z$	m normal to [0, 0, 1]
17	$\bar{y}, x-y, 1/2-z$	$\bar{6}^-$ along [0, 0, 1] at 0, 0, z + 1/4 with inversion at 0, 0, 1/4
18	$\bar{x}+y, \bar{x}, 1/2-z$	$\bar{6}^+$ along [0, 0, 1] at 0, 0, z + 1/4 with inversion at 0, 0, 1/4
19	$y, x, \bar{z}$	2 along [1, 1, 0] at x, x, 0
20	$x-y, \bar{y}, \bar{z}$	2 along [1, 0, 0] at x, 0, 0
21	$\bar{x}, \bar{x}+y, \bar{z}$	2 along [0, 1, 0] at 0, y, 0
22	$\bar{y}, \bar{x}, 1/2-z$	2 along $[1, \bar{1}, 0]$ at x, $\bar{x}$, 1/4
23	$\bar{x}+y, y, 1/2-z$	2 along [1, 2, 0] at x, 2x, 1/4
24	$x, x-y, 1/2-z$	2 along [2, 1, 0] at 2x, x, 1/4

14.6.2 Special Topic: Coordination Number 12

The coordination number for hcp is 12. Because, with the origin shifted, there is a magnesium atom at the origin, the translations generate atoms at each of the eight vertices of the unit cell. Six of these atoms are indicated in Figure 14.15a. There is another atom at 1/3, 2/3, 1/2, which has three nearest neighbors outlined in green above and another three below. Take this second atom as the central atom. The distance from the central atom to each of the six atoms indicated by the two green triangles is

$$\sqrt{\left(\frac{2}{3} \times \frac{\sqrt{3}}{2}a\right)^2 + \left(\frac{1}{2}c\right)^2}$$

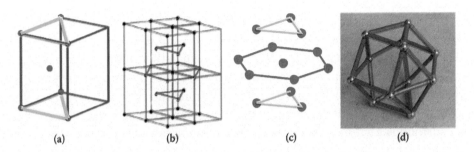

FIGURE 14.15 (a) Six nearest neighbors of populated unit cell; (b) twelve nearest neighbors; (c) without unit cell; (d) triangular orthobicupola.

There are six more nearest-neighbor atoms at $z = 1/2$, forming a hexagon about the central atom as shown in Figure 14.15b. The centers of these six atoms are connected with a red line, and the unit cell has been eliminated in Figure 14.15c. The distance between the central atom and each of the six atoms in the hexagon is a_H. If all 12 atoms were equally distant from the central atom, then c/a would be $\sqrt{(8/3)}$, which is the requirement for the ideal hcp.

Now connect the atoms in the triangles to the atoms in the hexagon to form a triangular orthobicupola shown in Figure 14.15d. This polyhedron of 12 atoms surrounds each atom in hcp. A *triangular orthobicupola* or an *anticuboctahedron* is a Johnson solid with twelve vertices, 8 triangular faces and 6 square faces and 24 edges. Each vertex has two triangles and two squares. Its point group is $\bar{6}m2$. Compare this polyhedron with the boron isocahedron (Chapter 12) and the cuboctahedron (see Exercise 14.2). See Exercise 14.3.

14.6.3 Special Topic: Site Symmetry

The *site symmetry* is the symmetry of the point group of the crystal from the view of a special or general position in a space group. For example, in Figure 14.16a, the magnesium atom at 1/3, 2/3, 1/4, in special position Wyckoff *c*, views the world of the crystal as having point group symmetry $\bar{6}m2$ with itself at the center. The symbol "$\bar{6}$" refers to the sixfold

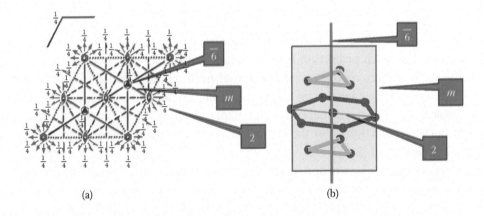

FIGURE 14.16 (a) Symbol space diagram with the magnesium atom at 1/3, 2/3, 1/4; (b) vertices of the triangular orthobicupola with same symmetry as the magnesium atom.

rotoinversion axis operating at 1/3, 2/3, 1/4. The symbol "*m*" refers to the mirrors normal to [100], [010], and [110] and containing the point 1/3, 2/3, 1/4. The mirror perpendicular to [100] and containing the point 1/3, 2/3, 1/4 is shown in this figure. The symbol "2" refers to the twofolds perpendicular to [100], [010], and [110] and containing the point 1/3, 2/3, 1/4. The twofold perpendicular to [100] and containing the point 1/3, 2/3, 1/4 is shown in this figure also. Figure 14.16b shows that the triangular orthobicupola, whose vertices are shown, surrounds each magnesium atom and has point group symmetry $\bar{6}m2$ with the symmetries seen by the magnesium atom. The $\bar{6}$ and the twofold are both contained in the mirror indicated in the diagram.

Also look at Figure 14.10 which shows the maximal subgroups for space group #194, $P6_3/mmc$. Note that the associated point groups of the symorphic maximal subgroups are $\bar{6}m2$ and $\bar{3}m1$. Then note that the site symmetry for special positions with Wyckoff letters *b*, *c*, and *d* is $\bar{6}m2$ and that the site symmetry for special position with Wyckoff letter *a* is $\bar{3}m$. Now look at the point group tree in Figure 3.54, and see that the subgroups of $\bar{6}m2$ are $mm2$, $3m$, 32, $\bar{6}$, m, 2, 3, and 1. The subgroups of $\bar{3}m$ are $2/m$, $3m$, $\bar{3}$, 32, $\bar{1}$, m, 2, 3, and 1. All of the site symmetries for this space group are found among these two lists. Note especially that the general position has site symmetry 1. In fact, the general position can be defined as the position with site symmetry 1. Every point in the crystal has at least site symmetry 1.

In general, if the space group is symmorphic, then position *a* has the site symmetry of the associated point group of that space group. There may be more than one special position with that symmetry. All the rest of the special positions have site symmetries that are subgroups of the associated point group. The general position has site symmetry 1.

On the other hand, if the group is nonsymmorphic, then consider the site symmetry of position *a*. This has the point group of the associated point group of the largest symmorphic subgroup. There may be more than one point group involved such as in #194, $P6_3/mmc$. All the rest of the special positions have site symmetries that are subgroups of these associated point groups. Again, the general position has site symmetry 1. The concept of site symmetry is applied in solid state physics (Evarestov, 1997).

14.6.4 Symmetry of the Special Projections

The *International Tables for Crystallography* give the symmetry for three special projections for each space group in the standard orientation. These projections are orthogonal, which means that each projection is onto a plane normal to the direction of the projection. The basis vectors of the two-dimensional unit cell are labeled **a′** and **b′**, regardless of which two basis vectors of the three-dimensional unit cell (selected from **a**, **b**, **c**) are projected. A Starter Program from Chapter 4 has a "view" statement that selects the appropriate projections. For the hexagonal crystal system the special projections are along [001], [100], and [210].

Quite generally, every two-dimensional projection has the symmetries of one of the 17 two-dimensional space groups.

In this case, to be consistent with the descriptions of the special projections in the *International Tables for Crystallography*, the atoms are placed at 1/3, 2/3, 1/4 and 2/3, 1/3, 3/4; that is, with the inversion point at the origin. This placement is shown in Figure 14.13a.

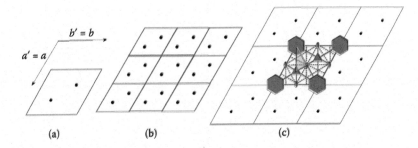

FIGURE 14.17 Magnesium along [001] (a) unit cell; (b) expanded unit cell; (c) with symmetries and asymmetric unit in pale blue.

Along [001]

The two-dimensional space group for the projection along [001] is *p6mm*, where

$$\mathbf{a}' = \mathbf{a} \quad \mathbf{b}' = \mathbf{b}$$

with the origin at 0, 0, z. Figure 14.17a is taken directly from the Starter Program. Here, Z equals two, and the figure shows both atoms. The enhanced unit cell in Figure 14.17b shows the original unit cell surrounded by eight unit cells in order to make clear the sixfold symmetry at the corners. The expanded figure also shows that the opposite sides of the unit cell are identical translations of one and other. Hence, as always when done correctly, all the information necessary to create the entire pattern is found in the unit cell.

Figure 14.17c shows a diagram with the graphical symbols for one unit cell and the asymmetric unit, in light blue. The asymmetric unit is 1/12th of the unit cell. The simplicity of the projection masks the elaborate symmetries involved!

Along [100]

The two-dimensional space group along [100] is *p2gm*, where

$$\mathbf{a}' = 1/2 \, (\mathbf{a} + 2\mathbf{b}) \quad \mathbf{b}' = \mathbf{c}$$

with the origin at x, 0, 0. Figure 14.18a is taken directly from the Starter Program. Here, Z equals two and the figure shows both atoms. The enhanced unit cell in Figure 14.18b shows the original unit cell surrounded by eight unit cells in order to make clear the twofold

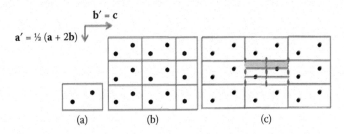

FIGURE 14.18 Magnesium along [100]: (a) unit cell; (b) expanded unit cell; (c) with symmetries and asymmetric unit in pink.

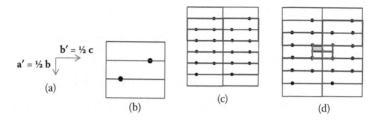

FIGURE 14.19 Magnesium along [210] (a) from Starter Program; (b) enhanced diagram; (c) with symmetries and asymmetric unit in pink. See Exercise 14.1.

symmetries. Finally, Figure 14.18c shows a diagram with the graphical symbols for this two-dimensional space group and the asymmetric unit, in pink (see Exercise 14.4).

Along [210]

Finally, the two-dimensional space group along [210] is *p2mm*, where

$$\mathbf{a}' = 1/2 \, \mathbf{b} \quad \mathbf{b}' = 1/2 \, \mathbf{c}$$

with the origin at x, $1/2 \, x$, 0. Figure 14.19a is the direct output from the Starter Program. The direct output partially portrays three unit cells. However, it contains enough information to portray the cells completely. Figure 14.19b uses the available information to complete a unit cell. The completion is made by overlapping, by integral numbers of $\mathbf{a}'$, one partially portrayed unit cell on another. The overlapping is continued to form an array of adjacent figures that exhibit the symmetries of *p2mm*. Finally, Figure 14.19c shows a diagram with the graphical symbols for the two-dimensional space group and the asymmetric unit, in pink (see Exercise 14.5).

14.7 MAGNESIUM: RECIPROCAL LATTICE AND *d*-SPACINGS

Planes in the direct lattice correspond to points of the reciprocal lattice. The reciprocal vector

$$\mathbf{H}(hkl) = h \, \mathbf{a}^* + k \, \mathbf{b}^* + l \, \mathbf{c}^*$$

where *hkl* are the indices of the planes of the direct lattice. The *d*-spacing is the distance from the origin to either of the two closest planes. In general, the *d*-spacing can be calculated by the following equation (See Section 5.7.1):

$$\frac{1}{d_{hkl}^2} = H^2(hkl) = (h \quad k \quad l) \, \mathbf{G}^* \begin{pmatrix} h \\ k \\ l \end{pmatrix}$$

The powder diffraction data in Table 14.6 gives *hkl* = 101 as the reflection with the maximum intensity. The calculation for the *d*-spacing for this reflection follows:

TABLE 14.6   Magnesium *hkl* and Intensity

h	k	l	Intensity
1	0	0	24.17
0	0	2	26.65
1	0	1	100.00
1	0	2	14.63
1	1	0	15.74
1	0	3	16.81
2	0	0	2.21
1	1	2	16.44
2	0	1	11.46
0	0	4	2.24
2	0	2	2.61
1	0	4	2.23

$$\frac{1}{d_{101}^2} = H^2(101) = \begin{pmatrix} 1 & 0 & 1 \end{pmatrix} \begin{pmatrix} 0.130 & 0.0647 & 0 \\ 0.0647 & 0.130 & 0 \\ 0 & 0 & 0.0368 \end{pmatrix} \begin{pmatrix} 1 \\ 0 \\ 1 \end{pmatrix} \tag{14.1}$$

Now the *d*-spacing for the 101 can be calculated as

$$d_{101} = 2.45 \text{ Å} \tag{14.2}$$

14.7.1 Powder Diffraction Pattern

The diffraction angle is now calculated as a function of *hkl*. First, Bragg's law is applied to the *d*-spacing. Bragg's law is

$$\lambda = 2d_{hkl} \sin \theta_{hkl}$$

where λ is the wavelength of the incoming x-rays, d_{hkl} is the *d*-spacing of the *hkl* plane, and θ_{hkl} is the Bragg angle for the *hkl* diffraction maximum. The diffraction angle, $2\theta_{hkl}$, is twice the Bragg angle (see Section 6.6.3). Table 14.6 has *hkl* indices and intensities for the 10 most intense reflections from the American Mineralogical Crystal Structure Database. The intensities are normalized so that the highest intensity is 100 arbitrary units. This information is the input for the Starter Program in Chapter 6. Figure 14.20 is the resulting powder diffraction pattern for magnesium.

14.8 MAGNESIUM: ATOMIC SCATTERING CURVE

Figure 14.21 shows the atomic scattering curve from the Starter Program in Chapter 7 for the element magnesium. The atomic scattering curve is calculated from the Cromer–Mann coefficients. The atomic scattering factor (or form factor) is the ratio of the amplitude of

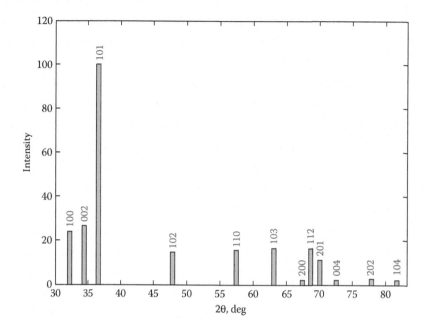

FIGURE 14.20   Magnesium powder diffraction pattern.

the wave scattered by an atom to the amplitude of the wave scattered by one electron. Its units are electrons. The atomic scattering curve is a measure of the efficiency of an atom in scattering x-rays.

Bragg's law for (101) is

$$\sin \theta/\lambda = 1/2\ d = 1/(2 \times 2.45) = 0.204\ \text{Å}^{-1} \tag{14.3}$$

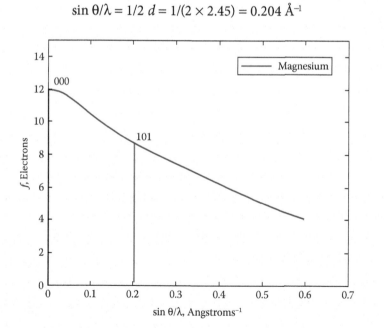

FIGURE 14.21   Magnesium atomic scattering curve.

This reflection is labeled in Figure 14.21 in red. The value from the atomic scattering curve is

$$f_{Mg}(101) = 8.67 \text{ electrons} \qquad (14.4)$$

14.9 MAGNESIUM: STRUCTURE FACTOR

The general formula for the structure factor, from Equation 7.5, is

$$F(hkl) = \sum_{j=1}^{N} f_j \left(\frac{\sin \theta_{hkl}}{\lambda} \right) e^{2\pi i(hx_j + ky_j + lz_j)}$$

where N is the number of atoms in the unit cell. In the case of magnesium there are two atoms in the unit cell (see Table 14.7). The running index j goes from 1 to 2. Table 14.4 includes the fractional coordinates x_1, y_1, z_1 for the atom in the asymmetric unit.

The formula for the structure factor for magnesium is

$$F(hkl) = F_{Mg}(hkl) \qquad (14.5)$$

This term is summed over both atoms

$$F(hkl) = \sum_{j=1}^{2} F_{Mgj}(hkl) \qquad (14.6)$$

Now put in the expressions for the atomic form factor and the phase factors

$$F(hkl) = \sum_{j=1}^{2} f_{Mg}(hkl) e^{2\pi i(hx_j + ky_j + lz_j)} \qquad (14.7)$$

14.9.1 Calculate the Structure Factor at 000

The special case of the undeviated beam occurs when $hkl = 000$. Its position is labeled in red in Figure 14.21 (see Example 7.2).

$$F(000) = 2F_{Mg}(000)$$

TABLE 14.7   Magnesium: Calculation of Number of Electrons in Unit Cell

Atom	Mg
Atomic number	12
Atom/ (formula unit)	1
Total electrons/formula unit	12
$Z = 2$	
Atoms/unit cell	2
Electrons/unit cell	24

The atomic scattering factor, f, is equal to the number of electrons; therefore $F(000)$ is the sum of all the electrons in the unit cell (Section 7.6.4). Table 14.7 shows how the total number of electrons is counted in the unit cell of magnesium to give

$$F(000) = 24 \text{ electrons} \tag{14.8}$$

14.9.2 Calculate the Contribution of One Atom to a Structure Factor

The calculation in Equation 14.7 can be sampled by calculating only the contribution of one atom, Mg1, to the most intense hkl reflection found in the x-ray powder diagram, which is 101. Table 14.4 gives the fractional coordinates for Mg1 as

$$x = 1/3, \quad y = 2/3, \quad z = 1/4$$

For the reflection 101

$$F_{Mg1}(101) = f_{Mg}(101)e^{2\pi i(0.333+0\times.667+0.25)} = 8.67 \, e^{1.1667\pi i} \text{electrons} \tag{14.9}$$

Note that this is a complex number. The entire structure factor, including both atoms, is real because the space group is centrosymmetric with an inversion point located at the origin (see Section 7.7.2). The importance of the structure factors is that they become coefficients of the Fourier expansion used to calculate the electron density (see Section 7.8.2 and Exercises 14.6 and 14.7).

DEFINITIONS

Coordination number
Cubic close-packed structure (ccp)
Cuboctahedron
Hexagonal close-packed layer
Hexagonal close-packed structure (hcp)
Polymorphism
Site symmetry
Triangular orthobicupola or anticuboctahdron

EXERCISES

14.1   Calculate the distance for nearest neighbors in hcp and ccp.

14.2   Compare the number of edges and faces of the cuboctahedron, the triangular orthobicupola or anticuboctahdron and the boron icosahedron from Chapter 12. All have 12 vertices. Look on the Internet and see what other information you can find.

14.3   Calculate the distance for the second nearest neighbors in ccp and hcp.

14.4   Calculate the scalar values for $\mathbf{a}'$ and $\mathbf{b}'$ for magnesium along [100].

TABLE 14.8   Crystals with Space Group #194, $P6_3/mmc$

Ref. Code	Chemical Name	Formula	Volume(Å^3)
NIDQEX	Catena-[(μ_6-Carbodiimido)-cobalt(ii)]	$(CCoN_2)_n$	83.945
METPOT	Methyl potassium	CH_3^-, K^+	131.28
EVUJEL	Dichloro-trimethyl-tantalum	$C_3H_9Cl_2Ta$	385.347
QEKBAK	Catena-((μ_3-oxo)-tris(μ_3-1,2,4-triazole)-trichloro-tetra-copper(ii))	$(C_6H_6Cl_3Cu_4N_9O)_n$	697.494
IJUDIA	Catena-((μ_2-Hydrogen)-1,4-diazabicyclo(2.2.2) octane tris(μ_2-methoxo)-hexacarbonyl-di-rhenium)	$(C_6H_{13}N^{2+})_n$, $n(C_9H_9O_9Re^{2-})$	1047.21
LOFRII02	Tris(Guanidinium) tris(μ_2-iodo)-hexaiodo-di-antimony(iii)	$3(CH_6N_3^+)$, $I_9Sb_2^{3-}$	1608.92
PEKGUH	Tris(Tetramethylammonium) tris(μ_2-chloro)-hexachloro-di-bismuth(iii)	$3(C_4H_{12}N^+)$, $Bi_2Cl_9^{3-}$	1628.62
MNBISA01	Tris(Tetramethylammonium) tris(μ_2-bromo)-hexabromo-di-bismut	$3(C_4H_{12}N^+)$, $Bi_2Br_9^{3-}$	1762.03

14.5   Calculate the scalar values for $\mathbf{a'}$ and $\mathbf{b'}$ for magnesium along [210] (see Section 14.6.4).

14.6   Calculate the total structure factor for the Mg 101 reflection when the magnesium atoms have coordinates (1/3, 2/3, 1/4) and (2/3, 1/3, 3/4).

14.7   Calculate the total structure factor for the Mg 101 reflection when the magnesium atoms have coordinates (0, 0, 0) and (1/3, −1/3, 1/2).

14.8   Suggested crystals from CSD with space group #194, $P6_3/mmc$ are listed in ascending order of volume in Table 14.8. Construct a paper or a PowerPoint presentation based on the parallel topics in this chapter. The website is http://webcsd.ccdc.cam.ac.uk/teaching_database_demo.php; a shortcut to an individual Reference code is: http://webcsd.ccdc.cam.ac.uk/display_csd_entry.php?identifier=MBRMET10 for a crystal with Reference Code "MBRMET10". The parallel topics are discussed in Section 8.3.1. Examples are given at https://sites.google.com/a/vt.edu/foundations_of_crystallography/.

## CHAPTER 15

Cubic System

Acetylene

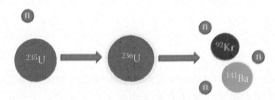

Neutron diffraction, which was used in this study of crystalline acetylene, complements x-ray analysis. However, the neutron experiments are very expensive and are conducted at national or international reactor facilities. The neutrons are produced by a nuclear chain reaction. A uranium 235 nucleus absorbs a neutron and forms an unstable isotope, uranium 236. In a fission reaction this nucleus splits into a barium 141 nucleus, a krypton 92 nucleus, and three neutrons. Hence with every event there is a net gain of two neutrons. Neutrons exhibit wave–particle duality. Neutrons can be diffracted by crystals and produce a neutron diffraction pattern.

CONTENTS

CHAPTER OBJECTIVES

- Study the crystallographic properties of solid acetylene, space group #205, $Pa\overline{3}$ as a representative of the cubic crystal system.

- Present the parallel crystallographic topics that recur in Chapters 9 through 15. A few examples are given below. A complete list is found in Section 8.3.1.

  - Multiplication table of the associated point group

  - Diagrams of the unit cell, the asymmetric unit, the populated unit cell, and the special projections

  - Diagram of the reciprocal cell superimposed on the direct unit cell

  - Calculation of the powder diffraction pattern

  - Calculation of the atomic scattering curves

  - Calculation of the structure factor at 000

  - Calculation of the contribution of one atom to the structure factor of the largest reflection

- Present the special crystallographic topics of this chapter, which are listed below. A complete list of the special topics found in Chapters 9 through 15 is in Section 8.3.2.

  - Comparison of x-ray, neutron, and electron diffraction

  - Neutron diffraction

  - Neutron structure factors

  - Symbols for threefold symmetry axes parallel to body diagonals

TABLE 15.1   Acetylene Data

Compound	Acetylene
Formula	C_2H_2
Space group	#205, $Pa\bar{3}$
Bravais lattice	cP
Crystal system	Cubic
Diffraction ambient temperature	141 K
Z, Z'	$Z = 4$, $Z' = 4/24 = 1/6$
Diffraction probe	Neutron beam at 1.0404 Å from Be crystal
Reference, CSD	ACETYL03
Reference	McMullan, R.K., A. Kvick, and P. Popelier. 1992. Structures of cubic and orthorhombic phases of acetylene by single-crystal neutron diffraction. *Acta Crystallographia*, 1348, 726–731.
Density	0.76 g/cm³

15.1 INTRODUCTION

The example chosen for the cubic crystal system is crystalline acetylene. At room temperature acetylene is a colorless hazardous gas used in blow torches and as a starting material in many chemical processes. Its triple point is 192.4 K (−80.8°C) at 1.27 atm. In this study, the gas sample was condensed with liquid nitrogen in a vacuum line. Platinum pins served both as crystal nucleation sites and as an experimental technique for controlling the temperature of the crystal. The crystal was kept at a constant temperature of 144 K (McMullan, 1992).

Table 15.1 contains the crystal data for acetylene.

15.1.1 Special Topic: Comparison of X-Ray, Neutron, and Electron Diffraction

X-ray diffraction is the most common technique for studying crystal structures. There are several other useful techniques. See Table 15.2 for a comparison of x-ray, neutron, and electron diffraction. For a review, see deGraeh and McHenry (2007, Chapter 13); and for a more advanced review see Giacovazzo (2002, 214–219, 313–318) (see Exercises 15.1 and 15.2).

The acetylene crystal in this chapter was analyzed with neutron diffraction. Bragg's law and the structure factors are applicable in neutron diffraction. The Cramer–Mann atomic

TABLE 15.2   Comparison of Diffraction Techniques

Diffraction	X-Ray (X-Ray Tube)	Neutron	Electron	High-Energy X-Ray, Synchrotron
Location	Universities, industry	National or international reactor facilities	Selected universities, industries	National or international reactor facilities
Cost	Least expensive	Very expensive	More expensive	Very expensive
Type	Electromagnetic radiation	Neutrons	Electrons	Electromagnetic radiation
Charge	None	None	Charged	None
Magnetic dipole moment	No	Yes	Yes	No
Wavelength	≈1 Å	≈1 Å	≈0.02 Å	≈1 Å

scattering curves for x-rays are replaced by neutron scattering lengths, *b*. The following section is an introduction to neutron diffraction.

15.1.2 Special Topic: Neutron Diffraction

A *neutron* is a subatomic particle with zero electrical charge and a mass approximately equal to that of a proton. The neutron is found in the nucleus of the atom. The *atomic mass* of an atom is the total mass of protons, neutrons, and electrons. The *atomic number*, *Z*, is the number of protons in the nucleus. The *mass number*, *A*, is the number of protons plus the number of neutrons. There are *A–Z* neutrons in the nucleus.

Neutrons can be produced by nuclear fission reactions with, for example, uranium or plutonium. After the neutrons are produced they have to be slowed down, as can be done with a moderator like water, heavy water, or graphite. A *monochromatic beam* is a beam of a single wavelength. To create a monochromatic beam, the neutron beam is scattered by a single crystal. The resulting Bragg reflection becomes the new beam. For acetylene the neutron beam was from the 002 plane of a beryllium crystal. That beam bathed the crystal of acetylene, and the resulting pattern was analyzed. This neutron beam had a wavelength of 1.0404 Å (R. K. McMullan, 1992).

The *neutron scattering length*, *b*, is a measure of the scattering of a neutron beam by a nucleus. It relates to the apparent size of a nucleus as seen by a neutron on a collision course with the nucleus and is measured in *femtometers* (fm). One femtometer equals 10^{-15} m. By definition, *b* is positive if there is a phase change of 180° between the incident and scattered waves. The quantity *b* can be positive or negative and can have an imaginary component. The symbol 17O refers to the oxygen isotope, *A* = 17, *Z* = 8, for which there are 9 or *A–Z* neutrons. Boron has a scattering length of 5.30 − 0.213*i*. Although the atomic scattering factor, *f*, for x-rays increases with increasing atomic number, *Z*, the neutron scattering length, *b*, is not so easily correlated to the atomic number, *Z*. Isotopes of the same element can have different values for *b*. From Table 15.3, the values of *b* for hydrogen, deuterium, and tritium are −3.7406, 6.671, and 4.792, respectively. The table gives the atomic number, *Z*; the isotope; and the neutron scattering length, *b*, for the first eight elements. A complete list of neutron scattering data is given by the National Institute for Standards and Technology (NIST) at the website http://www.ncnr.nist.gov/resources/n-lengths/list.html.

Figure 15.1 compares the scattering factors for x-ray and neutron diffraction. While x-rays scatter off the electron cloud of an atom, neutrons interact with the nucleus. The nucleus is so small that it acts like a point source. In other words, the scattering is isotropic (or uniform) unlike the scattering of x-rays, which is dependent on the scattering angle, 2θ. Also the x-ray curves vary monotonically as the scattering atoms go from *Z* equals 1 for hydrogen to *Z* equals 19 for potassium. On the other hand, for neutrons the values of *b* hop around as *Z* increases and differ for different isotopes (see also Table 15.2).

15.1.3 Special Topic: Neutron Structure Factors

The equations for the structure factors for x-ray and neutron diffraction are formally similar. Bragg's law applies equally to both types of coherent scattering. For x-rays, the structure factor, *F*, for a given reflection *hkl* and number of atoms *N* of the unit cell, is

TABLE 15.3   Atomic Number, Isotope, and Neutron Scattering Length

Atomic Number, Z	Isotope	b, Femtometers (fm)
1	1H, Hydrogen	−3.7406
1	2H, Deuterium	6.671
1	3H, Tritium	4.792
2	He*	3.26(3)
2	3He	5.74−1.483i
2	4He	3.26
3	Li	−1.90
4	Be	7.79
5	B	5.30−0.213i
6	C*	6.6460
6	12C	6.6511
6	13C	6.19
7	N*	9.36
7	14N	9.37
7	15N	6.44
8	O*	5.803
8	16O	5.803
8	17O	5.78
8	18O	5.84

* Natural Abundance.

$$F(hkl) = \sum_{j=1}^{N} f_j \left(\frac{\sin \theta_{hkl}}{\lambda} \right) e^{2\pi i (hx_j + ky_j + lz_j)}$$

where f_j is the atomic scattering factor for atom j.

For neutrons, the structure factor, F, for a given reflection hkl and number of atoms N in the unit cell, is

$$F(hkl) = \sum_{j=1}^{N} b_j e^{2\pi i (hx_j + ky_j + lz_j)}$$

where b_j is the neutron scattering length for atom j.

EXAMPLE 15.1

Calculate the structure factors for both x-ray and neutron diffraction for a face-centered copper crystal for reflections 000, 111, 020, and 220. Create a graph of the structure factors vs. 2θ, the diffraction angle. Assume that the wavelengths of the x-ray and the neutron beams are equal, namely 1.54 Å. The neutron scattering length, b, for copper is 7.718 fm.

Solution

In Section 7.6.3, the structure factors are derived for face-centered copper crystals. The x-ray structure factors are

(*continued*)

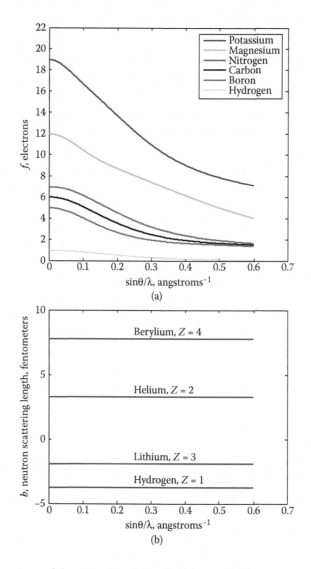

FIGURE 15.1 Comparison of the scattering factor for (a) x-rays, (b) neutrons.

EXAMPLE 15.1 (continued)

$$F(hkl) = 4 f_{Cu} \text{ for } hkl \text{ unmixed}$$

$$F(hkl) = 0 \quad \text{for } hkl \text{ mixed.}$$

The neutron structure factors are

$$F(hkl) = 4 b_{Cu} \text{ for } hkl \text{ unmixed}$$

$$F(hkl) = 0 \quad \text{for } hkl \text{ mixed.}$$

See Table 15.4, Figure 15.2, and Exercises 15.1 through 15.6.

TABLE 15.4 Structure Factors for a Copper Crystal for Both X-Ray and Neutron Diffraction

		X-Ray Diffraction	Neutron Diffraction
hkl	2θ	$F(hkl)$ Electrons	$F(hkl)$ Femtometers
000	0	116.0	30.872
111	43.35	88.31	30.872
020	50.49	82.89	30.872
220	74.20	67.13	30.872

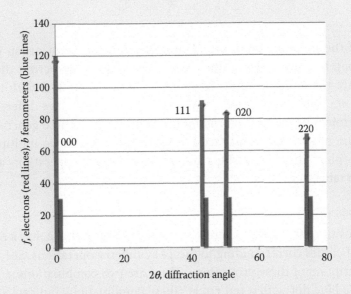

FIGURE 15.2 Structure factors versus 2θ, the diffraction angle, for (a) x-rays, f, red, (b) neutrons, b, blue.

15.2 ACETYLENE: POINT GROUP PROPERTIES

The associated point group of crystalline acetylene is $m\overline{3}$ in the Hermann–Mauguin notation and T_h in the Schönflies notation. This point group has 24 symmetry operations—$1, 2_z, 2_y, 2_x, 3^+_{xxx}, 3^+_{\overline{x}x\overline{x}}, 3^+_{x\overline{x}\overline{x}}, 3^+_{\overline{x}\overline{x}x}, 3^-_{xxx}, 3^-_{\overline{x}x\overline{x}}, 3^-_{x\overline{x}\overline{x}}, 3^-_{\overline{x}\overline{x}x}, \overline{1}, m_{xy}, m_{xz}, m_{yz}, \overline{3}^+_{xxx}, \overline{3}^+_{\overline{x}x\overline{x}}, \overline{3}^+_{x\overline{x}\overline{x}}, \overline{3}^+_{\overline{x}\overline{x}x}, \overline{3}^-_{xxx}; \overline{3}^-_{\overline{x}x\overline{x}}, \overline{3}^-_{x\overline{x}\overline{x}}$, and $\overline{3}^-_{\overline{x}\overline{x}x}$. Thus the order, or number of elements, is 24. This point group illustrates the rule that whenever there are improper operations there is an equal number of proper operations.

15.2.1 Multiplication Table

The generators shown in Appendix 3 are $1, 2_z, 2_y, 3^+_{xxx}$, and $\overline{1}$. However, the point group can be generated from just two symmetry operations (see Figure 3.58). The first operation has to be any one of the eight threefold rotoinversion operations. For example, if $\overline{3}^+_{xxx}$ is chosen, its group has elements $\{1, 3^+_{xxx}, 3^-_{xxx}, \overline{1}, \overline{3}^+_{xxx},$ and $\overline{3}^-_{xxx}\}$. The second operation can be any operation included in $m\overline{3}$ but not in the group generated by $\overline{3}^+_{xxx}$. For example, $\overline{3}^+_{xxx}$ and 2_z are a minimal set of generators. Their matrices are

$$3^+_{xxx} = \begin{pmatrix} 0 & 0 & \bar{1} \\ \bar{1} & 0 & 0 \\ 0 & \bar{1} & 0 \end{pmatrix}, \; 2_z = \begin{pmatrix} \bar{1} & 0 & 0 \\ 0 & \bar{1} & 0 \\ 0 & 0 & 1 \end{pmatrix}$$

The Starter Program from Chapter 3 calculates the multiplication table in Table 15.5. Because

$$3^+_{xxx} \cdot 3^+_{\bar{x}x\bar{x}} \neq 3^+_{\bar{x}x\bar{x}} \cdot 3^+_{xxx}$$

the matrix of the point group is not symmetric, and the point group is not abelian. The point group has an inversion; thus its symmetries do not permit the piezoelectric effect. Compare the multiplication table for $6/mmm$ in Table 14.2 with the multiplication table for $m\bar{3}$. Both point groups have order 24, but they have different numbers of minimal generators. Point group $6/mmm$ has minimum of three generators while $m\bar{3}$ has a minimum of two generators (see Figure 3.58). Thus the two multiplication tables are not isomorphic. The relationships among the symmetry operations are seen in the multiplication table.

15.2.2 Stereographic Projections

The general position stereographic projection for $m\bar{3}$ Figure 15.3a, has a combination of 12 dots and 12 circles corresponding to the 24 symmetry operations. See Section 3.9 for a general description of these projections. There are two combinations, a red dot with a blue circle or a blue dot with a red circle. These combinations indicate a mirror in the plane of the figure. The dot, either blue or red, indicates a positive z-axis; and the circle, either blue or red, indicates a negative z-axis. The color red is used for the 12 proper operations—1, 2_z, 2_y, 2_x, 3^+_{xxx}, $3^+_{\bar{x}x\bar{x}}$, $3^+_{x\bar{x}\bar{x}}$, $3^+_{\bar{x}\bar{x}x}$, 3^-_{xxx}, $3^-_{\bar{x}x\bar{x}}$, $3^-_{x\bar{x}\bar{x}}$, and $3^-_{\bar{x}\bar{x}x}$. The color blue is used for the 12 improper operations—$\bar{1}$, m_{xy}, m_{xz}, m_{yz}, $\bar{3}^+_{xxx}$, $\bar{3}^+_{\bar{x}x\bar{x}}$, $\bar{3}^+_{x\bar{x}\bar{x}}$, $\bar{3}^+_{\bar{x}\bar{x}x}$, $\bar{3}^-_{xxx}$, $\bar{3}^-_{\bar{x}x\bar{x}}$, $\bar{3}^-_{x\bar{x}\bar{x}}$, and $\bar{3}^-_{\bar{x}\bar{x}x}$. A proper operation does not change handedness, and the determinant of the matrix is +1. An improper operation changes handedness and the determinant of the matrix is −1. Note that the points, both dots and open circles, are placed inside the circumference to allow any value for z.

Figure 15.3b is the symbol stereographic projection. The three twofolds are red with two proper operations each. There are two twofolds in the xy-plane. They are represented by the two pairs of red ovals located on the circumference of the circle. The third twofold is along the z-axis and is represented by a single red oval at the center. The inversion is represented by the purple circle superimposed on this red oval. There are four threefold rotoinversion axes; and they each contain six operations, three proper and three improper, including the inversion. The $\bar{3}_{xxx}$ rotoinversion is pointed out in the diagram. There are mirror planes normal to the x-axis, the y-axis, and the z-axis. The mirror plane normal to the z-axis, m_{xy}, is pointed out in the diagram. The additional two mirrors, normal to the x-axis and the y-axis, are indicated by perpendicular purple lines. The black lines are guide

TABLE 15.5 Point Group $m\bar{3}$ Multiplication Table

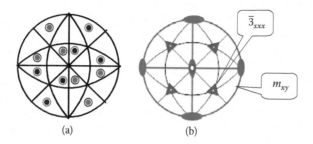

FIGURE 15.3 Point group $m\overline{3}$ stereographic projections. (a) general position, (b) symbol.

lines. The purple symbols indicate that half the operations are proper, or red, and half the operations are improper, or blue.

Given the symbol diagram, the general position diagram can be created and *vice versa*.

15.3 ACETYLENE SPACE GROUP PROPERTIES

Acetylene crystallizes in the cubic space group #205, $Pa\overline{3}$. The Bravais lattice is cP. The symmetry operations require that

$$a = b = c \text{ and } \alpha = \beta = \gamma = 90°$$

15.3.1 Special Topic: Symbols for Threefold Symmetry Axes Parallel to Body Diagonals

A characteristic of cubic space groups is the presence of threefold symmetry axes each of which is parallel to a body diagonal of the cube. Column 2 of Table 15.6 shows the graphic symbols for the threefold symmetry axes along the body diagonals—3, 3_1, 3_2, and $\overline{3}$. Compare these symbols with those in column 3 showing the corresponding threefold axes when they are parallel to the z-axis. Note that the symbol for the inversion point is included in the symbol for $\overline{3}$.

15.3.2 Symbol Diagram

Figure 15.4 shows the symbol diagram for the nonsymmorphic space group #205, $Pa\overline{3}$. An explanation of the Hermann–Mauguin symbol follows. The capital letter refers to the centering type. The next two parts of the Hermann–Mauguin symbol describe the symmetries in their relationship to the crystallographic directions in the space group. In the cubic system, the first place describes symmetries related to [100], [010], and [001], and the second place refers to 3 or $\overline{3}$ along the body diagonals.

For space group #205, $Pa\overline{3}$, the "P" means that the cell is primitive. The "a" refers to the a-glide planes normal to [001], [100], and [010] (see Table 4.8). The a-glide plane normal to [001] is pointed out in the diagram. The "$\overline{3}$" refers to the threefold rotoinversion axes that are parallel to body diagonals. The threefold rotoinversion axis, $\overline{3}_{xxx}$, in purple, along [111] with an inversion point at 000, is pointed out.

Consider the coloring of the symbols. The symbols for the screw axes—2_1, 3_1, and 3_2—are red because there is no change of handedness. The symbols for the threefold rotoinversion

TABLE 15.6   Threefold Symmetry Axes of the Cube

| | Graphical Symbol | |
Symmetry Axis	Parallel to Body Diagonals	Along [001]
Threefold rotation, 3		
Threefold screw, 3_1		
Threefold screw, 3_2		
Threefold rotoinversion, $\overline{3}$		

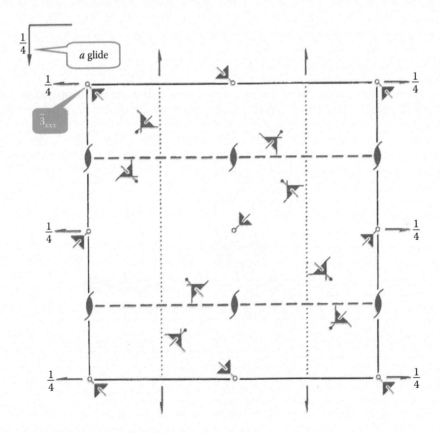

FIGURE 15.4   Space group #205, $Pa\overline{3}$ symbol diagram.

axes, $\bar{3}$, and the glide planes are purple because half the operations change the handedness and half do not.

The *International Tables for Crystallography* do not give general position diagrams for the cubic systems. Instead they give stereo diagrams with the points of the general position diagram given as vertices of transparent polyhedrons.

15.3.3 Maximal Subgroups

There are 230 space groups. Because the space groups contain translation operations, each space group has infinite order. Space groups can be arranged in trees similar to the trees of the point groups. However, there is no unique way of constructing the tree for the space groups. In a tree of type I, the centering type is kept and the order of the associated point group is changed. (Maximal subgroups and minimal supergroups are discussed in Section 4.8.) For example, a maximal type I subgroup of the space group #205, *Pa*$\bar{3}$, is the space group #61, *Pbca*. Both these space groups have the same centering type, namely *P*; but for the subgroup the order of the associated point group is decreased. Specifically, the associated point group of the space group #205, *Pa*$\bar{3}$, is *m*$\bar{3}$ and has order 24, while the associated point group of the space group #61, *Pbca*, is *mmm* and has order 8.

The space group #205, *Pa*$\bar{3}$, has no type 1 supergroups, but has three maximal type 1 subgroups of which the associated point groups have orders 12, 8, and 6. The *International Tables for Crystallography* give the information necessary to construct Figure 15.5.

15.3.4 Asymmetric Unit

The asymmetric unit is the smallest part of the unit cell that, when operated on by the symmetry operations, produces the whole unit cell. The volume of the asymmetric unit is related to the volume of the unit cell by $V_{AU} = V_{UC}/(nh)$, where *n* is the number of lattice

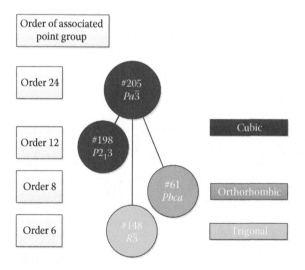

FIGURE 15.5 Maximal subgroups for space group #205, *Pa*$\bar{3}$.

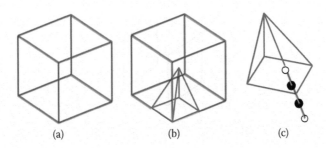

(a) (b) (c)

FIGURE 15.6 (a) Unit cell, (b) unit cell with asymmetric unit, (c) asymmetric unit together with one molecule.

points in the unit cell and h is the number of operations in the associated point group. (Note that the product nh is equal to the multiplicity of the general position.) Because there are 24 operations and the unit cell is primitive ($n = 1$, $h = 24$), the volume of the asymmetric unit is one twenty-fourth the volume of the unit cell. Therefore, because the volume of the unit cell is 227.54 $Å^3$, the volume of the asymmetric unit is 9.48 $Å^3$.

Figure 15.6a shows the unit cell drawn in blue. Figure 15.6b shows the unit cell drawn with the asymmetric unit in red superimposed on it. Because $Z = 4$ and the volume of the asymmetric unit is one twenty-fourth of the volume of the unit cell, there is 1/6 of an acetylene molecule in the asymmetric unit. Figure 15.6c shows a different view of the asymmetric unit together with one molecule of acetylene. A single carbon atom and a single hydrogen atom each lie on an edge. Exactly 1/3 of each those two atoms is in the asymmetric unit. The symmetries of the space group generate the other 2/3 of the two atoms on the edge of the asymmetric unit as well as all the other atoms in the unit cell.

The asymmetric unit has

$$0 \leq x \leq 1/2, \quad 0 \leq y \leq 1/2, \quad 0 \leq z \leq 1/2, \quad z \leq \text{mínimum } (x, y)$$

Vertices: 000,   1/2, 0, 0,   1/2, 1/2, 0   0, 1/2, 0   1/2, 1/2, 1/2

(Hahn, 2010, p. 148). The vertices are given for this complicated asymmetric unit because they are easier to interpret.

15.4 ACETYLENE: DIRECT AND RECIPROCAL LATTICES

In addition to the direct lattice, there is the reciprocal lattice, which is the lattice of the diffraction pattern. The diffraction pattern has the intensities of the Bragg reflections superimposed on the reciprocal lattice.

Figure 15.7 shows the reciprocal unit cell (red) superimposed on the direct unit cell (blue). Note that when the volumes are normalized to one, the direct unit cell and reciprocal unit cell are superimposed. Only the red reciprocal unit cell shows because it completely covers up the blue direct unit cell. The Starter Program from Chapter 5 plots the blue direct unit cell first and then plots the red reciprocal unit cell on top of it.

FIGURE 15.7   Reciprocal unit cell (red) and direct unit cell (blue and invisible) with the volumes normalized to 1.

15.5 ACETYLENE: FRACTIONAL COORDINATES AND OTHER DATA FOR THE CRYSTAL STRUCTURE

The crystal structure consists of the placement of the atoms in the unit cell and the bonds between the atoms. The fractional coordinates of the atoms in the asymmetric unit combined with the general position coordinates of the space group and the lattice parameters are sufficient to specify all the atoms in the unit cell.

Z equals four or, in other words, there are four acetylene molecules in the unit cell. Z' equals 1/6; that is to say, there is 1/6 acetylene molecule in the asymmetric unit. The fractional coordinates of the atoms in the molecule are given in Table 15.8.

Table 15.9 gives the symmetry operations that are needed to populate the unit cell. There are four atoms in the molecule, of which parts of two are in the asymmetric unit. Both these crystallographically independent atoms have multiplicity 8 and have Wyckoff letter c, that is, they are in the same special position. The multiplicity 8 means there are 8 carbon atoms and 8 hydrogen atoms corresponding to four C_2H_2 molecules.

Now the bonds must be inserted. Figure 15.8 shows how the atoms in the asymmetric unit are connected using the labeling from Table 15.8. In addition to the two crystallographically independent atoms—C1 and H1—two more atoms—C1L and H1L—are given to facilitate drawing the bonds. Atoms C1L and H1L are related to C1 and H1 by the inversion point between the atoms C1 and C1L. For emphasis, the red box outlines the asymmetric unit. In the Starter Program from Chapter 4 for populating the unit cell, this diagram is the key to putting in the intramolecular bonds.

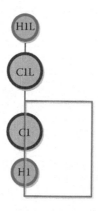

FIGURE 15.8   Schematic showing the partial atoms needed to produce the populated cell.

15.6 ACETYLENE: CRYSTAL STRUCTURE

Populate the unit cell by using the Starter Program from Chapter 4 to create the entire unit cell as shown in Figure 15.9a. This program combines the lattice parameters—a, b, c, α, β, and γ—(Table 15.7), the fractional coordinates (Table 15.8), the general position coordinates of the space group (Table 15.9), and the schematic of the bonds (Figure 15.8). The lattice parameters, the fractional coordinates, and the general position coordinates are from the CIF.

Beginning with the general position coordinates the Starter Program automatically includes all the atoms. No special procedure is required for atoms in special positions.

Figure 15.9b shows two populated cells.

15.6.1 Symmetry of the Special Projections

The *International Tables for Crystallography* give the symmetry for three special projections for each space group in the standard orientation. These projections are orthogonal, which means that each projection is onto a plane normal to the direction of the projection. The basis

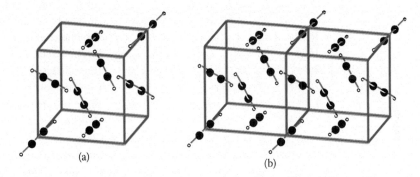

(a)   (b)

FIGURE 15.9   (a) Populated unit cell from the Starter Program, (b) two populated unit cells.

TABLE 15.7   Acetylene: Comparison of Direct and Reciprocal Lattices

Direct Lattice	Reciprocal Lattice
$a = b = c = 6.105\text{Å}$	$a^* = b^* = c^* = 0.1638 \text{ Å}^{-1}$
$\alpha = \beta = \gamma = 90°$	$\alpha^* = \beta^* = \gamma^* = 90°$
$G = \begin{pmatrix} 37.27 & 0 & 0 \\ 0 & 37.27 & 0 \\ 0 & 0 & 37.27 \end{pmatrix} \text{Å}^2$	$G^* = \begin{pmatrix} 0.0268 & 0 & 0 \\ 0 & 0.0268 & 0 \\ 0 & 0 & 0.0268 \end{pmatrix} \text{Å}^{-2}$
$V = 227.54 \text{ Å}^3$	$V^* = 0.0044 \text{ Å}^{-3}$

TABLE 15.8   Acetylene Fractional Coordinates, Multiplicity, and Wyckoff Letter

Atom	x Fractional Coordinate	y Fractional Coordinate	z Fractional Coordinate	Multiplicity and Wyckoff Letter
C1	0.0556(3)	0.0556(3)	0.0556(3)	8c
H1	0.1525(7)	0.1525(7)	0.1525(7)	8c
C1L	−0.0556(3)	−0.0556(3)	−0.0556(3)	
H1L	−0.1525(7)	−0.1525(7)	−0.1525(7)	

TABLE 15.9   General Position Coordinates and Symmetry Operations from CIF

Number CIF	General Position Coordinates			Symmetry Operations
1	x	y	z	Identity
2	z	x	y	Threefold rotation
3	y	z	x	Threefold rotation
4	$1/2 + x$	y	$1/2 - z$	Glide
5	$1/2 + z$	x	$1/2 - y$	Threefold rotoinversion
6	$1/2 + y$	z	$1/2 - x$	Threefold rotoinversion
7	$1/2 - x$	$1/2 + y$	z	Glide
8	$1/2 - z$	$1/2 + x$	y	Threefold rotoinversion
9	$1/2 - y$	$1/2 + z$	x	Threefold rotoinversion
10	$-x$	$1/2 + y$	$1/2 - z$	Twofold screw
11	$-z$	$1/2 + x$	$1/2 - y$	Threefold rotation
12	$-y$	$1/2 + z$	$1/2 - x$	Threefold screw
13	$-x$	$-y$	$-z$	$\bar{1}$, inversion operation
14	$-z$	$-x$	$-y$	Threefold rotoinversion
15	$-y$	$-z$	$-x$	Threefold rotoinversion
16	$1/2 - x$	$-y$	$1/2 + z$	Twofold screw
17	$1/2 - z$	$-x$	$1/2 + y$	Threefold rotation
18	$1/2 - y$	$-z$	$1/2 + x$	Threefold screw
19	$1/2 + x$	$1/2 - y$	$-z$	Twofold screw
20	$1/2 + z$	$1/2 - x$	$-y$	Threefold rotation
21	$1/2 + y$	$1/2 - z$	$-x$	Threefold screw
22	x	$1/2 - y$	$1/2 + z$	Glide
23	z	$1/2 - x$	$1/2 + y$	Threefold rotoinversion
24	y	$1/2 - z$	$1/2 + x$	Threefold rotoinversion

vectors of the two-dimensional unit cell are labeled $\mathbf{a}'$ and $\mathbf{b}'$, regardless of which two basis vectors of the three-dimensional unit cell (selected from $\mathbf{a}$, $\mathbf{b}$, $\mathbf{c}$) are projected. The Starter Program from Chapter 4 has a "view" statement that selects the appropriate projections. For the cubic crystal system the special projections are along [001], [111], and [110]. Here the [010] and [100] are not useful because identical information is found along the [100] projection. In this space group there is no shift of origin. A careful examination of these projections can demonstrate that the atoms and bond lengths have been correctly implemented.

Quite generally, every two-dimensional projection has the symmetries of one of the 17 two-dimensional space groups.

Along [001]

The two-dimensional space group along [001] is *p2gm*, where

$$\mathbf{a}' = 1/2\,\mathbf{a} \quad \mathbf{b}' = \mathbf{b}$$

with the origin at $0,0,z$. Figure 15.10a is the direct output from the Starter Program. Seven molecules are shown. However, inside the blue outline of the unit cell there are eight carbon atoms and eight hydrogen atoms. The enhanced diagram in Figure 15.10b is obtained by translating a copy of the diagram in Figure 15.10a to form adjacent diagrams that exhibit

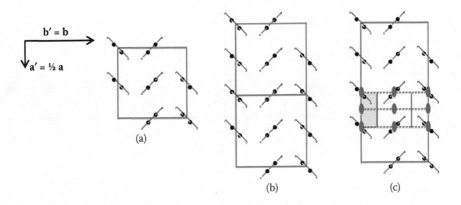

FIGURE 15.10 Acetylene along [001] (a) unit cell, (b) enhanced unit cell, (c) with symmetries and asymmetric unit.

the symmetries of *p2gm*. Note that **a′** equals one-half **a**. The expanded figure also shows that the opposite sides of the unit cell are related by a translation, as always. All the information necessary to create the entire pattern is found in the unit cell. Figure 15.10c shows a diagram with graphical symbols for one unit cell with the asymmetric unit in pink.

Along [111]

The two-dimensional space group along [111] is *p6*, where

$$\mathbf{a}' = 1/3(2\mathbf{a} - \mathbf{b} - \mathbf{c}) \qquad \mathbf{b}' = 1/3(-\mathbf{a} + 2\mathbf{b} - \mathbf{c})$$

with the origin at *x,x,x*. Figure 15.11a is the direct output from the Starter Program. The figure contains six equilateral triangles. Two triangles make up a hexagonal unit cell; thus, the direct output *partially* portrays three unit cells (see Figure 4.43). However, the direct output contains enough information to portray the cells completely. Figure 15.11b uses the available information to complete a unit cell. The completion is made by overlapping one partially portrayed unit cell on another. The overlapping is continued to form an array of adjacent figures that exhibit the symmetries of *p6*. Only four complete lattice points are shown. The expanded

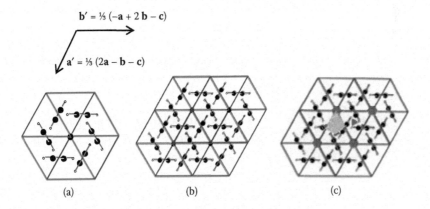

FIGURE 15.11 Acetylene along [111] (a) unit cell, (b) enhanced unit cell, (c) with symmetries and asymmetric unit in pink.

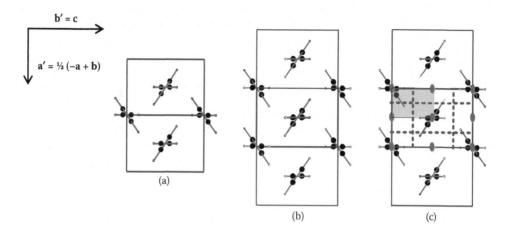

b′ = c

a′ = ½ (−a + b)

(a)

(b)   (c)

FIGURE 15.12   Acetylene along [110] (a) from Starter Program, (b) enhanced cells, (c) with symmetries and asymmetric unit.

figure also shows that the opposite sides of the unit cell are related by a translation, as always. All the information necessary to create the entire pattern is found in the unit cell. Finally, Figure 15.11c shows a diagram with graphical symbols for one unit cell with the asymmetric unit in pink. The asymmetric unit is one-sixth of the unit cell.

   Along [110]

   Finally, the two-dimensional space group along [110] is *p2gg*, where

$$\mathbf{a}' = 1/2\,(-\mathbf{a} + \mathbf{b}) \quad \mathbf{b}' = \mathbf{c}$$

with the origin at $x,x,0$. Figure 15.12a is the direct output from the Starter Program. The direct output *partially* portrays two unit cells. However, it contains enough information to portray the cells completely. Figure 15.12b uses the available information to complete a unit cell. The completion is made by overlapping, by integral numbers of $\mathbf{a}'$, one partially portrayed unit cell on another. The overlapping forms an array of adjacent diagrams that exhibit the symmetries of *p2gg*. Figure 15.12c shows a diagram with graphical symbols for one unit cell with the asymmetric unit in pink.

15.7 ACETYLENE: RECIPROCAL LATTICE AND *d*-SPACINGS

Planes in the direct lattice correspond to points of the reciprocal lattice. The reciprocal vector

$$\mathbf{H}(hkl) = h\mathbf{a}^* + k\mathbf{b}^* + l\mathbf{c}^*$$

where *hkl* are the indices of the planes of the direct lattice. The *d*-spacing is the distance from the origin to either of the two closest planes (see Section 5.7.1). In general, the *d*-spacing can be calculated by the following equation:

$$\frac{1}{d_{hkl}^2} = H^2(hkl) = (h \quad k \quad l)\mathbf{G}^* \begin{pmatrix} h \\ k \\ l \end{pmatrix}$$

TABLE 15.10   Acetylene *hkl* and Intensity

hkl	Intensity
111	10,000
200	2969
210	861
211	723
220	956
300	101
311	397

The powder diffraction data in Table 15.10 give *hkl* = 111 as the reflection with the maximum intensity. The calculation of the *d*-spacing for this reflection follows

$$\frac{1}{d_{111}^2} = H^2(111) = (1 \quad 1 \quad 1) \begin{pmatrix} 0.0268 & 0 & 0 \\ 0 & 0.0268 & 0 \\ 0 & 0 & 0.0268 \end{pmatrix} \begin{pmatrix} 1 \\ 1 \\ 1 \end{pmatrix} \tag{15.1}$$

Now, the *d*-spacing for the 111 can be calculated

$$d_{111} = 3.52 \text{ Å}. \tag{15.2}$$

15.7.1 Powder Diffraction Pattern

The diffraction angle is now calculated as a function of *hkl*. First, Bragg's law is applied to the *d*-spacing. Bragg's law is

$$\lambda = 2 d_{hkl} \sin \theta_{hkl}$$

where λ is the wavelength of the incoming x-rays, d_{hkl} is the *d*-spacing of the *hkl* plane, and θ_{hkl} is the Bragg angle for the *hkl* diffraction maximum. The diffraction angle, $2\theta_{hkl}$, is twice the Bragg angle (see Section 6.6.3).

This acetylene crystal structure was done with neutron diffraction at a wavelength of 1.0404 Å from a beryllium crystal. However, in order to compare acetylene with the other crystals in this book, simulated x-ray diffraction data are presented.

Table 15.10 has *hkl* indices and intensities for the seven most intense reflections using x-ray diffraction with a wavelength of 1.54 Å. The intensities are not actual experimental data but were calculated from the CIF by the program *Mercury* by using the CSD. The intensities are normalized so that the highest intensity is 10,000 arbitrary units. This information is the input for the Starter Program in Chapter 6. Figure 15.13 is the resulting powder diffraction pattern for acetylene.

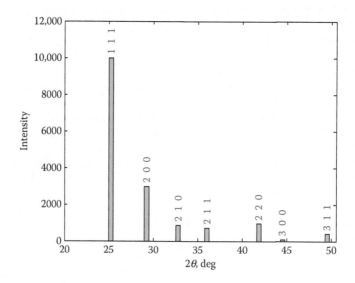

FIGURE 15.13   Acetylene powder diffraction pattern.

15.8 ACETYLENE: ATOMIC SCATTERING CURVES

Figure 15.14 shows the atomic scattering curves from the Starter Program in Chapter 7 for the two elements found in acetylene. The atomic scattering curves are calculated from the Cromer–Mann coefficients (see Section 7.3.1). The atomic scattering factor (or form factor) is the ratio of the amplitude of the wave scattered by an atom to the amplitude of the wave scattered by one electron. Its units are electrons. The atomic scattering curve is a measure of the efficiency of an atom in scattering x-rays.

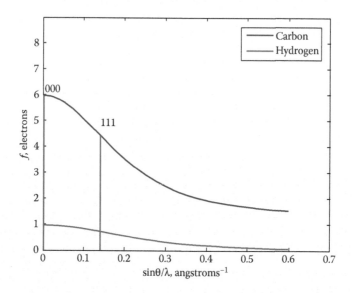

FIGURE 15.14   Atomic scattering curves for acetylene.

Bragg's law for (111) is

$$\sin\theta/\lambda = 1/(2d) = 1/(2 \times 3.52) = 0.141 \text{ Å}^{-1}. \tag{15.3}$$

This reflection is labeled in Figure 15.14 in red. The values from the atomic scattering curves are

$$f_C(111) = 4.413, \qquad f_H(111) = 0.737 \text{ electrons} \tag{15.4}$$

15.9 ACETYLENE: STRUCTURE FACTOR

The general formula for the structure factor, from Equation 7.5, is

$$F(hkl) = \sum_{j=1}^{N} f_j \left(\frac{\sin\theta_{hkl}}{\lambda} \right) e^{2\pi i (hx_j + ky_j + lz_j)}$$

where N is the number of atoms in the unit cell. In the case of acetylene there are 16 atoms in the unit cell (see Table 15.11). The running index j goes from 1 to 16. Table 15.8 includes the fractional coordinates x_j, y_j, z_j for the jth atom in the asymmetric unit.

The formula for the structure factor for acetylene can be broken down in the following way:

$$F(hkl) = F_H(hkl) + F_C(hkl) \tag{15.5}$$

The first term is a sum over 8 hydrogen atoms and the second term is a sum over 8 carbon atoms

$$F(hkl) = \sum_{j}^{8H} F_{Hj}(hkl) + \sum_{j}^{8C} F_{Cj}(hkl) \tag{15.6}$$

Now put in the expressions for the individual atomic form factors and the phase factors

$$F(hkl) = \sum_{j}^{8} f_{Hj}(hkl) e^{2\pi i (hx_j + ky_j + lz_j)} + \sum_{j=1}^{8} f_{Cj}(hkl) e^{2\pi i (hx_j + ky_j + lz_j)} \tag{15.7}$$

TABLE 15.11   Acetylene Calculation of Number of Electrons in Unit Cell

Atom	C	H	Total/Molecule
Atomic number	6	1	
Atom/molecule	2	2	4
Total electrons/molecule	12	2	14
$Z = 4$			Total/unit cell
Atoms/unit cell	8	8	16
Electrons/unit cell	48	8	56

15.9.1 Calculate the Structure Factor at 000

The special case of the undeviated beam occurs when $hkl = 000$. Its position is labeled in red in Figure 15.14 (see Example 7.2).

$$F(000) = 8f_H(000) + 8f_C(000)$$

The atomic scattering factor, f, is equal to the number of electrons; therefore, $F(000)$ is the sum of all the electrons in the unit cell (Section 7.6.4). Table 15.11 shows how the total number of electrons is counted in the unit cell of acetylene to give

$$F(000) = 56 \text{ electrons} \tag{15.8}$$

15.9.2 Calculate the Contribution of One Atom to a Structure Factor

The calculation in Equation 15.7 can be sampled by calculating only the contribution of one atom, C1, to the most intense hkl reflection found in the x-ray powder diagram, which is 111. Table 15.8 gives the fractional coordinates for C1 as

$$x = .0556, \quad y = .0556, \quad z = .0556$$

For the reflection 111

$$F_{C1}(111) = f_C(111)e^{2\pi i(0.0556+0.0556+0.0556)} = 4.4413\, e^{0.3336\pi i} \text{ electrons} \tag{15.9}$$

Note that this is a complex number. The entire structure factor, including all 16 atoms, is real because the space group is centrosymmetric with an inversion point located at the origin (see Section 7.7.2). The importance of the structure factors is that they become coefficients of the Fourier expansion that is used to calculate the electron density (see Section 7.8.2 and Exercise 15.4).

DEFINITIONS

Atomic mass
Atomic number
Femtometer
Mass number
Monochromatic beam
Neutron
Neutron scattering length

EXERCISES

15.1 Look up electron scattering and compare it to x-ray and neutron scattering.

15.2. Compare x-rays from synchrotrons with x-rays from x-ray tubes.

15.3 Compare x-ray and neutron structure factors for NaCl.

TABLE 15.12 Crystal Suggestions for Space Group #205, $Pa\overline{3}$

Refcode	Chemical Name	Formula	Volume (Å³)
DTFMHG02	bis(Trifluoromethyl)mercury	C_2F_6Hg	536.773
ZZZWOU01	Hexa-aminobenzene	$C_6H_{12}N_6$	776.658
DBENCR03	bis(η^6-Benzene)-chromium(0)	$C_{12}H_{12}Cr$	871.805
VOMBEF	Di-lithium closo-dodecaborate	$H_{12}B_{12}^{2-}, 2(Li^+)$	878.428
LUVJIV	bis(Methyl isocyanide)-gold(i) hexafluorophosphate	$C_4H_6AuN_2^+, F_6P^-$	943.99
HCCYHB	β-(e,e,e,e,e,e)-1,2,3,4,5,6-Hexachlorocyclohexane	$C_6H_6Cl_6$	1024.19
BUKNAX	bis(Acetonitrile)-iodonium hexafluoroarsenate	$C_4H_6IN_2^+, AsF_6^-$	1032.97
CONGRS	Congressane	$C_{14}H_{20}$	1033.06
PANCUC	bis(Acetonitrile)-gold(i) hexafluoroantimonate(v)	$C_4H_6AuN_2^+, F_6Sb^-$	1076.89
JACPEI	tris(1,1-Di-t-butyl-2-dimethylphosphino-ethoxy-O,P)-yttrium	$C_{36}H_{78}O_3P_3Y$	1080.2
HEVRUV	Trinitromethane	CHN_3O_6	1111.29
TIVWOL	Tetramethylsilane	$C_4H_{12}Si$	1236.34
BZCBNL01	Benzenehexacarbonitrile	$C_{12}N_6$	1253.08
FOJBUB02	Tetracarbonyl-nickel	C_4NiO_4	1270.94
VADRAU	Tetramethyl-lead(iv)	$C_4H_{12}Pb$	1394.6
BATLEN	Sodium (m2-hydrido)-bis(trimethyl-aluminum)	$C_6H_{19} \mid Al_2^-, Na^+$	1419.66
CADNOK	bis(Trimethylammonium) hexachloro-tin(iv)	$2(C_3H_{10}N^+), Cl_6Sn_2^-$	1820.32

15.4 Calculate $F_{Cl}(111)$ using neutron diffraction with a wavelength of 1.0404 Å.

15.5 Calculate the 2θ diffraction angle for acetylene reflections 000, 111, 020, and 220 using neutron diffraction with a wavelength of 1.0404 Å.

15.6 Compare x-ray and neutron structure factors for NaCl.

15.7 Suggested cubic crystals from CSD with space group #205, $Pa\overline{3}$ are listed in ascending order of volume in Table 15.12. The website is http://webcsd. ccdc.cam.ac.uk/teaching_database_demo.php; a shortcut to an individual Reference code is: http://webcsd.ccdc.cam.ac.uk/display_csd_entry. php?identifier=MBRMET10 for a crystal with Reference Code "MBRMET10". Construct a paper or a PowerPoint presentation based on the parallel topics in this chapter. The parallel topics are discussed in Section 8.3.1. Examples are given at https://sites.google.com/a/vt.edu/foundations_of_crystallography/.

15.8 Suggested cubic crystals from CSD with space group #199, $I2_13$, are listed in ascending order of volume in Table 15.13. The website is http://webcsd.ccdc.cam. ac.uk/teaching_database_demo.php; a shortcut to an individual Reference code

TABLE 15.13 Crystal Suggestions for Space Group #199, $I2_13$

Refcode	Chemical Name	Formula	Volume (Å³)
ESOTAI	Carbonyl-trifluoromethyl-gold(i)	C_2AuF_3O	453.759
BUZFUX	tris(Hydroxymethyl)aminomethane hydroiodide	$C_4H_{12}NO_3^+, I^-$	1571.82

is:http://webcsd.ccdc.cam.ac.uk/display_csd_entry.php?identifier=MBRMET10 for a crystal with Reference Code "MBRMET10". Construct a paper or a PowerPoint presentation based on the parallel topics in this chapter. The parallel topics are discussed in Section 8.3.1. Examples are given at https://sites.google.com/a/vt.edu/foundations_of_crystallography/.

References

Angel, R. J., D. R. Allen, R. Miletich, and L. W. Finger. 1977. The use of quartz as an internal pressure standard in high-pressure crystallography. *Journal of Applied Crystallography* 30, 461–466.

Aroyo, M. I., A. Kirov, C. Capillas, J. M. Perez-Mato, and H. Wondratschek. 2006. Bilbao crystallographic server. II. Representations of crystallographic point groups and space groups. *Acta Crystagraphia,* A 62, 115–128.

Aste, T. and D. Weaire. 2008. *The Pursuit of Perfect Packing,* 2nd edition. Taylor & Francis, Boca Raton, FL.

Authier, A., Editor. 2013. *International Tables for Crystallography,* Vol. D. *Physical Properties of Crystals.* John Wiley & Sons, Chichester, UK.

Basu, P. K. and H. Dhasmana. 2009. *Solid State Engineering Physics.* Taylor & Francis, Boca Raton, FL.

Battle, G. M. and F. H. Allen. 2012. Learning about intermolecular interactions from the Cambridge Structural Database. *Journal of Chemical Education* 89, 38–44.

Battle, G. M., G. M. Ferrence, and F. H. Allen. 2010. Applications of the Cambridge structural database in chemical education. *Journal of Applied Crystallography* 43, 1208–1223.

Beardon, A. F. 2005. *Algebra and Geometry* (Chapter 3). Cambridge University Press, Cambridge.

Bernal, I., W. Hamilton, and J. S. Ricci. 1972. *Symmetry: A Stereoscopic Guide for Scientists.* W. H. Freeman and Company, San Francisco.

Bloss, F. D. 1971. *Crystallography and Crystal Chemistry: An Introduction.* Holt, Rinehart and Winston, Inc., New York.

Boisen, M. B. and G. V. Gibbs. 1985. *Mathematical Crystallography.* Mineral Society of America, Virginia Tech.

Bradley, C. J. and A. P. Cracknell, 2011. *The Mathematical Theory of Symmetry in Solids,* Clarendon Press, Oxford.

Brown, I. D. and McMahon, B. 2002. CIF: The computer language of crystallography. *Acta Crystallographia* B58, 317–324.

Bunn, C. W. 1939. The crystal structure of long-chain normal paraffin hydrocarbons. *Transactions of the Faraday Society* 35, 482–491.

Burns, G. and A. M. Glazer. 2013. *Space Groups for Solid State Scientists,* 3rd edition. Academic Press, Waltham, Massachusetts.

Callister, W. D. Jr and D. G. Rethwisch. 2010. *Materials Science and Engineering: An Introduction,* 8th edition. J. Wiley and Sons, Hoboken, NJ.

Chapman, S. J. 2002. *Matlab Programming for Engineers,* 2nd edition. Brooks/Cole, California.

Chung, D. D. L., P. W. DeHaven, H. Arnold, and D. Ghosh, 1993. *X-ray Diffraction at Elevated Temperatures; A Method for In Situ Process Analysis.* VCH Publishers, NYC.

Clegg, W., A. J. Blake, R. O. Gould, and P. Main 2001. *Crystal Structure Analysis: Principles and Practice.* W. Clegg, Editor. Oxford University Press, Oxford.

Collins, G. P. 2004. Next search for plastic electronics, conductive plastic. *Scientific American* 291, 74–81.

Cromer, D. T., M. I. Kay, and A. C. Larson. 1967. Refinement of the alum structures. *Acta Crystallographia* 22, 182.

Cromer, D. T. and J. B. Mann. 1967. Compton scattering factors for spherically symmetric free atoms. *Journal of Chemical Physics* 47, 1892–1983.

Cullity, B. D. and S. R. Stock. 2001. *Elements of X-ray Diffraction*, 3rd edition. Prentice Hall, Upper Saddle River, NJ.

David, W. I .F., W. T. A. Harrison, J. M. F. Gunn, O. Moze, A. K. Soper, P. Day, J. D. Jorgensen et al. 1987. *Nature* 327, 310–312.

Davis, T. A. and K. Sigmon. 2005. *Matlab® Primer*, 7th edition. Chapman & Hall/CRC, Boca Raton, FL.

Decker, B. F. and J. Kasper. 1959. The crystal structure of a simple rhombohedral form of boron. *Acta Crystallographia* 12, 503–506; database code_amcsd 0009230.

Deer, W. A., R. A. Howie, and J. Zussman, 1992. *An Introduction to the Rock Forming Minerals*, Longman, London.

deGraef, M. 1998. A Novel Way to Represent the 32 Crystallographic Point Groups. *Journal of Materials Education* 20, 31–42.

deGraef, M. and M. E. McHenry. 2007. *Structure of Materials, an Introduction to Crystallography, Diffraction, and Symmetry*. Cambridge University Press, Cambridge.

deWolff, P. M. et al. 1989. Definition of symmetry elements in space groups and point groups. *Acta Crystagraphia A* 45, 494–499.

deWolff, P. M. et al. 1992. Symbols for symmetry elements and symmetry operations. *Acta Crystagraphia A* 48, 727.

Donohue, J. 1974. *The Structures of the Elements*. John Wiley and Sons, Hoboken, NJ.

Douglas, B. E. and S.-M. Ho, 2006. *Structure and Chemistry of Crystalline Structures*. Springer, New York City.

Downs, R. T. and M. Hall-Wallace, 2003. The American Mineralogist Crystal Structure Database. *American Mineralogist* 88, 247–250.

Downs, R. T. and Palmer, D. C. 1994. The pressure behavior of α cristobalite. *American Mineralogist* 79, 9–14.

Evarestov. R. A. and V. P. Smirnov. 1997. *Site Symmetry in Crystals, Theory and Applications*, Second Enlarged Edition. Springer, New York City.

Ewald, P. P. 1962. *Fifty Years of X-ray Diffraction*, N. V. A. Oosthoek's Uitgeversamaatschappij, Utrecht, The Netherlands.

Fay, T. H. 1989. The butterfly curve. *American Mathematics Monthly* 96(5), 442–443.

Flack H. et al. 2000. Symmetry elements in space groups and point groups. Addenda to two IUCr Reports on the Nomenclature of Symmetry. *Acta Crystagraphia A* 56, 96–98.

Giacovazzo, C. 2002. Crystallographic computing. In: *Fundamentals of Crystallography*, 2nd Edition, Chapter 2. C. Giacovazzo, Editor, Oxford University Press, Oxford, pp. 67–151.

Giester, G. 1994. Crystal structure of anhydrous alum $RbFe^{3+}(SeO_4)_2$. *Monatshefte für Chemie/Chemical Monthly* 125(11), 1223–1228.

Glusker, J. P. and K. N. Trueblood. 2010. *Crystal Structure Analysis*, 3rd edition. Oxford University Press, Oxford.

Goff, A. C., A. O. Aning, and S. L. Kampe. 2004. A model to predict the damping characteristics of piezoelectric-reinforced metal matrix composites. *TMS Letters*, TMS (The Minerals, Metals, and Materials Society) 1.3, 59–60.

Golubitsky, M. and I. Stewart. 2002. *The Symmetry Perspective*. Verlag, Birkhäuser.

Hahn, T. ed. 2006. *International Tables for Crystallography, Volume A: Space-Group Symmetry*. John Wiley & Sons, Chichester, UK.

Hahn, T. ed. 2010. *International Tables for Crystallography: Brief Teaching Edition*. John Wiley & Sons, Chichester, UK.

Hammond, C. 2009. *The Basics of Crystallography and Diffraction*, 3rd edition. Oxford University Press, Oxford.

Hammond, R., K. Pencheva, K. J. Roberts, P. Mougin, and D. Wilkinson. 2005. An examination of the thermal expansion of urea using high-resolution variable-temperature x-ray powder diffraction. *Journal of Applied Crystallography* 38 (Part 6), 1038–1039.

Hazen, R. M. 1982. *Comparative Crystal Chemistry, Temperature, Pressure, Composition and the Variation of Crystal Structure*. Wiley-Interscience Publication, John Wiley & Sons, Chichester.

Higham, D. J. and N. J. Higham, M. 2000. *MATLAB Guide*. Siam, Society for Industrial and Applied Mathematics, Philadelphia.

Hill IV, V. E and E. Victor. 2000. *Groups and Characters*. Chapman & Hall CRC, Boca Raton, FL.

Hoard, J. L., D. B. Sullenger, C. H. L. Kennard, and R. E. Hughes, 1970. The structure analysis of β-rhombohedral boron. *Journal of Solid State Chemistry* 1, 268–277.

Honess, A. P. 1927. *The Nature, Origin and Interpretation of the Etch Figures on Crystals*, John Wiley & Sons, London.

Ibers, J. A. and W. C. Hamilton. 1963. Xenon tetrafluoride: Crystal structure. *Science* 139 (3550), 106–107.

International Tables for Crystallography. 2002. *Vol. A: Space Group Symmetry*, 5th Edition. In: T. Hahn (ed.), *International Union of Crystallography*, Kluwer Academic Publishers, Dordrecht, the Netherlands.

International Tables for Crystallography. 2003. *Vol. D: Physical Properties of Crystals*, 1st edition. In: A. Authier (ed.), *International Union of Crystallography*. Kluwer Academic Publishers, Dordrecht, the Netherlands.

International Tables for Crystallography. 2005. *Brief Teaching Edition of Vol. A: Space Group Symmetry*, 5th edition. In: T. Hahn (ed.), *International Union of Crystallography*. Springer, Dordrecht, the Netherlands.

Jablan, S. V. 1995. Symmetry and ornament electronic reprint book *Theory of Symmetry and Ornament* originally published on paper by the Mathematical Institute of the Serbian Academy of Science and Arts, Belgrade, Yugoslavia, 1995, ISBN 86-80593-17-6.

Jablan, S. V. 2002. *Symmetry, Ornament and Modularity*. World Scientific, NJ.

Jaradat, D. M. M., S. Mebs, L. Checinska, and P. Luger. 2007. Experimental charge density of sucrose at 20 K: Bond topological, atomic, and intermolecular quantitative properties. *Carbohydrate Research* 342, 1480–1489.

Joyce, D. E., Clark University, http://www.clarku.edu/~djoyce/wallpaper/

Julian, M. M. 1980. Crystals and x-rays, a demonstration. *Journal of Chemical Education*

Julian, M. M. 1980. Kathleen Lonsdale and the planarity of the benzene ring. *Journal of Chemical Education* 58(4), 365.

Julian, M. M. 1997. The early days of the x-ray revolution. *Chemical Heritage* 14(2), 38–39.

Julian, M. M. 1999. Comparison of stretching force constants in symmetry coordinates between T_d and C_{3v} point groups. *Journal of Chemical Education* 76, 679–683.

Julian, M. M. 2005. Dorothy Wrinch and a search for the structure of proteins. In: R. E. Dickerson (ed.), *Present at the Flood: How Structural Molecular Biology Came About*. Sinauer Associates, Inc. Publishers, Sunderland, Massachusetts.

Julian, M. M. 2011. Crystallography. In: C. Cooper (ed.), *Organic Chemist's Desk Reference*, 2nd edition, Chapter 14, pp. 225–232. CRC, Boca Raton, FL.

Julian, M. M. and F. D. Bloss. 1987. Matrix solution of optical indicatrix parameters from central cross sections through the index ellipsoid. *American Mineralogist* 72, 612–616.

Julian, M. M. and G. V. Gibbs. 1985. Bonding in the silicon nitrides. *Journal of Physical Chemistry* 89, 5476–5480.

Julian, M. M. and G. V. Gibbs. 1988. Modeling the configuration about the nitrogen atom in methyl and silyl substituted amines. *Journal of Physical Chemistry* 92, 1444–1451.

Julian, M. M., W. E. Horner, and S. A. Alexander, 1986. Qualitative determination of cellulose in the cell walls of *Verticicladiella Procera*. *Mycologia* 78, 300–303.

Julian, M. M. and M. V. Orna. 1985. Synthetic medieval blue pigments. *Studies in Conservation* 30, 155–160.

Klein, C. 1989. *Minerals and Rocks: Exercises in Crystallography, Mineralogy, and Hand Specimen Petrology.* John Wiley & Sons, Hoboken, NJ.

Ladd, M. and R. Palmer. 2003. *Structure Determination by X-ray Crystallography*, 4th edition. Plenum Publishers, New York City.

Lang, A. R. 1965. The orientation of the Miller–Bravais axes of α-quartz. *Acta Crystallographia* 19, 290–291.

Liboff, R. L. 2004. *Primer for Point and Space Groups.* Springer, New York.

Lin, Y.-Y., Gundlach, D. J. Nelson, S. F., and Jackson, T. N. 1997. Stacked pentacene layer organic thin-film transistors with improved characteristics. *Electron Device Letters, IEEE* 18(12), 606–608.

Lipson, H. 1935. The relation between the alum structures. *Proceedings of the Royal Society of London, Series A* 151, 347–356.

Lonsdale, K. 1929a. The structure of the benzene ring in $C_6(CH_3)_6$. *Proceedings of the Royal Society of London A* 123, 494–515.

Lonsdale, K. 1929b. X-ray evidence on the structure of the benzene nucleus. *Transactions of the Faraday Society* 25, 352–366.

Lonsdale, K. 1948. *Crystals and X-rays.* Bell and Sons, London.

Ma, Z.-Q. and X.-Y. Gu. 2004. *Problems and Solutions in Group Theory for Physicists.* World Scientific, New Jersey.

MacGillavry, C. H. 1976. *Fantasy and Symmetry: Periodic Drawings of M. C. Escher.* Harry N. Abrams, Inc., New York.

Mak, T. and C. W. Gong-Du Zhou. 1992. *Crystallography in Modern Chemistry: A Resource Book of Crystal Structures.* John Wiley & Sons, Hoboken, NJ.

Manoli, J.-M., P. Herpin, and G. Pannetier. 1970. No. 21. Structure Crystalline du sulfate double d'aluminium et de potassium. *Bulletin de la Société Chimique de France*, 98.

McKie, D. and C. McKie. 1986. *Essentials of Crystallography.* Blackwell Scientific Publications, Oxford.

McMullan, R. K., A. Kvick, and P. Popelier. 1992. Structures of cubic and orthorhombic phases of acetylene by single-crystal neutron diffraction. *Acta Crystallographia* 1348, 726–731.

McPherson, A. 2003. *Macromolecular Crystallography.* Wiley & Sons, Hoboken NJ.

Megaw, H. 1973. *Crystal Structures: A Working Approach.* W. B. Saunders Company, Philadelphia.

National Institute of Standards and Technology, last updated October 1, 2012, About Cryogenics, http://www.nist.gov/mml/properties/cryogenics/aboutcryogenics.cfm

Nitske, W. R. 1971. *Wilhelm Conrad Röntgen; Discoverer of X-rays.* University of Arizona Press, Tucson.

Nye, J. F. 2010. *Physical Properties of Crystals.* Oxford University Press, Oxford, UK.

Oehzelt, M., R. Resel, and A. Nakayama. 2002. High pressure structural properties of anthracene up to 10 GPa. *Physical Review* B66, 174104.

Olinger, B. and J. W. Shaner. 1983. Lithium, compression and high-pressure structure. *Science* 219, 1071–1072.

Olver, P. J. and C. Shakiban. 2006. *Applied Linear Algebra.* Prentice-Hall, NJ.

Prince, E. 2004. *Mathematical Techniques in Crystallography and Materials Science.* Springer-Verlag.

Rasmussen, S. E. and R. G. Hazell. 1979. Structure and order parameter of an A15-type niobium germanium single crystal. *Acta Crystallographia* B25, 1677–1679.

Recktenwald, G. W. 2000. *Numerical Methods with MATLAB, Implementations and Applications.* Prentice-Hall, NJ.

Robertson, J. M. 1953. *Organic Crystals and Molecules: Theory of X-ray Structure Analysis with Applications to Organic Chemistry.* Cornell University Press, Ithaca, New York.

Rupp, B. 2010. *Biomolecular Crystallography, Principles, Practice, and Application to Structural Biology.* Garland Science, Taylor & Francis Group, New York.

Savytskii, D. I., A. O. Matkovskii, I. M. Solskii, F. Wallrafen, A. Suchocki, L. O. Vasylechko. 2000. *Research Technology* 35(2), 197–205.

Senechal, M. 1990. *Crystalline Symmetries: An Informal Mathematical Introduction.* Adam Hilger, Bristol.

Shakiban, C., No date. *Geometric Symmetry,* Chapter 5. http://courseweb.stthomas.edu/c9shakiban/ SymMath.pdf

Smart, L. E. and E. A. Moore. 2012. *Solid State Chemistry, an Introduction,* 4th edition. CRC Press, Boca Raton, FL.

Sorrell, C. 1973. *A Guide to Field Identification Minerals of the World.* Golden Press, Racine Wisconsin.

Stride, J. A. 2005. Determination of low-temperature structure of hexamethylbenzene. *Acta Crystallographica,* B61, 200–206.

Sperling, L. H. 2001. *Introduction to Physical Polymer Science,* 3rd edition. Wiley-Interscience, NY.

Stróz, K. 2003. 'Plane groups from basic to advanced crystallographic concepts'. *Zeitschrift für Kristallographie* 218, 642–649.

Suryanarayana, C. and M. G. Norton. 1998. *X-ray Diffraction, a Practical Approach.* Plenum Press, New York.

Sutor, D. J. 1958. The structures of the pyrimidines and purines. VII. The crystal structure of caffeine. *Acta Crystallographia* 11, 453.

Szebenyi, D. M. E., Liu, X., Kriksunov, I., Stover, P. J., and Thiel, D. 2000. Structure of a murine cytoplasmic serine hydroxymethyltransferase quinonoid ternary complex: Evidence for asymmetric obligate dimers. *Biochemistry* 39, 13313–13323.

Tadokoro, H. 1990. *Structure of Crystalline Polymers.* Robert E. Krieger Publishing Company, Malabar, Florida.

Takahashi, Y. 1998. Neutron structure analysis of polyethylene-D4. *Macromolecules* 31, 3868–3871.

Tiekink, E. R. T. and J. J. Vittal. 2006. *Frontiers in Crystal Engineering.* Wiley & Sons, Hoboken, NJ.

Tilley, R. 2006. *Crystals and Crystal Structures.* John Wiley & Sons, Ltd, Chichester, England.

Tiritiris, I., J. Weidlein, and Schleid. 2005. Dodecahydro-closo-dodecaborate halides of the heavy alkali metals with the formula $M_3X[B_{12}H_{12}]$ (M = K-Cs, NH4; X = Cl and Br). *Zeitschrift Fur Naturforschung* 60, 627–639.

Van Valkenburg, Jr., A. and B. F. Buie, 1945. Octahedral cristobalite with quartz paramorphs from Ellora Caves, Hyderabad State, India. *American Mineralogist* 30:526–535.

Vegard, L. and A. Maurstad. 1929. Die Kristallstruktur der wasserfreien Alaune R'R"$(SO_4)_2$. *Z. Kristallogr. Dtsch* 69, 519–532.

Welberry, T. R. 2004. *Diffuse X-ray Scattering and Models of Disorder.* Oxford University Press, Oxford.

Weng, N. S. and Hu Sheng-Zhi. 2003. On chiral space groups and chiral molecules. *Chinese Journal of Structural Chemistry* 22, 37–38.

Wyckoff, R. W. G., 1968. *Crystal Structures,* 2nd edition. Interscience, New York City.

Zachariasen, W. H. 1967. *Theory of X-ray Diffraction in Crystals.* Dover Publications, Inc., New York.

Zoltai, T. and J. H. Stout. 1984. *Mineralogy, Concepts and Principles.* Burgess Publishing Company, Minneapolis, Minnesota.

Zolotoyabko, E. 2011. *Basic Concepts of Crystallography, An Outcome from Crystal Symmetry,* John Wiley & Sons, Weinheim, Germany.

Appendix 1: Definitions

Abelian group: See **commutative group**.

Amplitude: Displacement from the average or mean value of a wave. (C-6)

Ångström, Å: A unit of length used in x-ray crystallography and spectroscopy, $1 \text{ Å} = 10^{-10} \text{ m} = 0.1$ nanometers $= 10^{-8}$ cm. (C-1)

Anisotropic properties: Properties that depend on the crystallographic direction. (C-1)

Anode: Positively charged electrode. (C-6)

Associative law: Given elements a, b, and c of a group, then (a b) c = a (b c). (C-3)

Asymmetric unit: Smallest part of the unit cell which when operated on by the symmetry operations produces the whole unit cell. (C-4)

Atomic mass: Total mass of protons, neutrons, and electrons in a single atom. (C-15)

Atomic number, Z: is the number of protons in the nucleus. (C-15)

Atomic scattering factor f (also known as the atomic form factor): Amplitude of the wave scattered by the atom measured relative to the amplitude of the wave scattered by one electron. It is measured in electrons. The atomic scattering factor is a measure of the efficiency of the atom in scattering x-rays. The Cromer–Mann coefficients fit an analytical form of the atomic scattering factor. (C-7)

Basis vectors: Linearly independent vectors **a**, **b**, and **c** that generate the direct lattice. See **lattice parameters** and **lattice constants**. (C-1).

Body-centered unit cell, I: Unit cell with two lattice points whose fractional coordinates are at 0,0,0; and 1/2, 1/2, 1/2. (C-4)

Bragg's angle: Angle θ in Bragg's equation $n\lambda = 2\,d \sin\theta$, where λ is the wavelength of the x-rays, n is an integer, d is the distance between planes in a crystal. (C-6)

Bragg's law: $n\lambda = 2\,d_{hkl} \sin\theta_{hkl}$, where n is an integer which is the order of the diffracted beam, λ is the wavelength of the incoming beam, d_{hkl} is the d-spacing and θ_{hkl} is the Bragg angle defined in Figure 5.9 for the (hkl) planes. (C-5), (C-6)

Bravais lattice: Classification of 5 two-dimensional lattices or 14 three-dimensional lattices based on primitive and nonprimitive unit cells. See Tables 4.1 and 4.3. Named after Auguste Bravais who first used them. (C-4)

Bremsstrahlung radiation: X-rays produced when electrons with sufficiently high kinetic energy are decelerated by collision with the atoms of the cathode. (Bremsstrahlung means braking radiation.) Another name is *white* or *continuous radiation* because many different wavelengths are produced. (C-6)

Bulk modulus, K: The reciprocal of the compressibility. (C-2)

Cathode: Negatively charged electrode. (C-6)

Centrosymmetric crystals: Crystals which contain an inversion point. (C-3)

C-faced-centered unit cell: Unit cell with two lattice points, one at the origin and another at position 1/2, 1/2, 0. (C-4)

Characteristic x-ray spectrum: Discrete x-ray lines superimposed on the bremsstrahlung. Characteristic radiation depends on the atomic number, Z, of the anode. (C-6)

Chiral carbon atom: A carbon atom, in a molecule, that has four different atoms or groups of atoms attached to it. (C-9)

Chiral molecule: A molecule whose handedness changes under an improper crystallographic operation to an image that cannot be superimposed on the original. (See **Enantiomers**) (C-9)

Chiral point group: See **proper point group**. This is a term used by biochemists. (C-10)

Chiral space group: See **proper space group**. This is a term used by biochemists. (C-10)

CIF see crystallographic information file.

Clathrate: A structure in which one molecule traps and contains a second molecule. (C-3)

Closure: Property of an operation between any two elements a and b in group G, such that $ab = c$ where c is also an element in G. (C-3)

Commutative group: Group that has a symmetric multiplication table. (Point group 3 is an example.) Also known as an **Abelian group**. (C-3)

Constructive interference: Phenomenon that occurs when superimposed waves are in phase. A new wave with the largest possible amplitude is produced. (C-6)

Coordinate matrix: Matrix $X = \begin{pmatrix} x \\ y \\ z \end{pmatrix}$ where x, y, and z are the fractional coordinates of an atom. (C-2)

Coordinate vector r: $\mathbf{r} = x\mathbf{a} + y\mathbf{b} + z\mathbf{c} = (\mathbf{a}\ \mathbf{b}\ \mathbf{c}) \begin{pmatrix} x \\ y \\ z \end{pmatrix}$

where x, y, z are the fractional coordinates and **a**, **b**, **c** are the basis vectors. (C-2)

Coordination number, CN: Number of closest neighbors to a given atom in a crystal. (C-14)

Cryogenics: The study of materials at temperatures below 123 K. (C11)

Crystal: A solid composed of atoms arranged in a periodic array (C-1); a material that produces a diffraction pattern.

Crystal structure: Structure consisting of the placement of atoms in the unit cell and the bonds between the atoms. Sometimes more than one cell is used to make the chemical sense of the crystal structure obvious. (C-4)

Crystal system: A classification of point groups as oblique, rectangular, square, or hexagonal in two dimensions or as triclinic, monoclinic, orthorhombic, trigonal, tetragonal, hexagonal, or cubic in three dimensions. (C-3)

Crystallographic direction is a vector between two lattice points where the direction is indicated by [u v w] where u, v, and w do not contain a common integer. The integers *u*, *v*, and *w* are called the *indices* of the crystallographic direction and specify an infinite set of parallel vectors. (C-2)

Crystallographic Information File (CIF): An electronic archiving file for crystallographic data. (C-9)

Crystallographic plane: Set of parallel *lattice planes* labeled by three integers, *hkl*, called the **Miller indices**. (C-2)

Crystallographic point groups: Point groups consistent with a lattice. The rotations in these point groups are limited to the crystallographic rotations. (C-3)

Crystallographic rotations: Rotations consistent with a lattice. They are limited to five cases, one-, two-, three-, four-, and sixfold rotation. (C-3)

Crystallographically dependent atoms: Atoms related by symmetries. (C-13)

Crystallographically independent angles: Angles not related by symmetries. (C-13)

Crystallographically independent atoms: Atoms not related by symmetries. (C-13)

Crystallographically independent bonds: Bonds not related by symmetries. (C-13)

Cubic close-packed structure (ccp): Packing of identical spheres in a layer sequence ABCABCABC..., resulting in cubic symmetry. The space group of the ensemble is #225, $Fm\bar{3}m$ with the spheres in a special position, Wyckoff letter either *a* or *b*. Furthermore, $Z = 4$, and $R = \sqrt{2}a/4$. (C-14)

Cubic *F* lattice: Face-centered Bravais lattice in three dimensions that has point group $m\bar{3}m$ symmetry. The lattice points are at 0, 0, 0; 0, 1/2, 1/2; 1/2, 0, 1/2; 1/2, 1/2, 0. (C-4)

Cubic *I* lattice: Body-centered Bravais lattice in three dimensions that has point group $m\bar{3}m$ symmetry. The lattice points are at 0, 0, 0; 1/2, 1/2, 1/2. (C-4)

Cubic *P* lattice: Primitive Bravais lattice in three dimensions that has point group $m\bar{3}m$ symmetry. The lattice point is at 0, 0, 0. (C-4)

Cuboctahedron: An Archimedean polyhedron with 12 identical vertices, 24 identical edges, eight triangular faces, and six square faces. Two triangles and two squares meet at each vertex. (C-14)

Cyclic group: Group with a single generator. (C-3)

Density, δ: $\delta = (\text{mass/volume}) = (Z(\Sigma A))/(V \cdot N_A)$, where Z is the number of molecules (or the number of formula units) in the unit cell, A is the atomic mass of each of the atoms or ions in the molecule or formula unit, V is the volume of the unit cell, and N_A is Avogadro's number. (C-2)

Destructive interference: Phenomenon that occurs when amplitude of the final wave for a set of superimposed waves is zero. (C-6)

Diffraction angle: The angle between the undeviated beam and the diffracted beam, 2θ, where θ is the Bragg angle. (C-6)

Direct lattice: Lattice of the pattern associated with the atoms of the crystal. See **reciprocal lattice**. (C-5)

Discrepancy index: See *R*-value. (C-7)

Electromagnetic waves: Waves that transmit energy through space. They have an electric component, *E*, and a perpendicular magnetic component, *H* and travel at the speed of light, *c* in vacuum. (C-6)

Electron volt: eV, is the energy acquired by an electron if its potential is increased by 1 V. 1 eV = 1.602×10^{-19} C (charge on the electron) $\times$ 1 V = 1.602×10^{-19} J. (C-6)

Enantiomer: A member of an enantiomorphic pair. (C-12)

Enantiomers: Two molecules consisting of a chiral molecule and its image under a change of handedness. (See **chiral**.) (C-9)

Enantiomorphic or chiral pair of space groups: A pair of space groups that are mirror images of each other. An example is the pair #144, $P3_1$, and #145, $P3_2$. (C-12)

Enantiomorphic pair of molecules: A pair of molecules that are mirror images of each other. (C-12)

Face-centered cell, F: Unit cell with four lattice points at 0, 0, 0; 0, 1/2, 1/2; 1/2, 0, 1/2; 1/2, 1/2, 0. (C-4)

Family of planes: Planes related by symmetry and indicated by curly brackets {}. For example, in point group $2/m$ with the b-axis unique, the family $\{hkl\}$ has four equivalent planes (hkl), $(\bar{h}k\bar{l})$, $(\bar{h}\bar{k}l)$, and $(h\bar{k}l)$. See **forms**. (C-3)

Femtometer: Unit of length equal to 10^{-15}m. (C-15)

Forms: Families that make up the external shape or morphology of the crystal. (C-3)

Fourfold rotation: An operation in which a 90° rotation is a symmetry operation. (C-3)

Fourier series: Representation of a periodic function expressed as a sum of a series of sine or cosine terms. It is useful for the calculation of electron density. (C-7)

Fractional coordinates: The atomic coordinates of the atoms within the cell written as fractions of the basis vectors. (C-2)

Friedel's law: All x-ray diffraction spectra have an inversion point. (C-7)

Full form of the Hermann–Mauguin symbol: The *Hermann–Mauguin symbol* that lists the symmetry axes and the symmetry planes of each symmetry direction. The information is given for the x-, y-, and the z-axis in that order. For example, in space group #62, $P2_1/n2_1/m2_1/a$ is the full form. Compare with the **short form of the Hermann–Mauguin symbol**. (C11)

General position: Position of a space group with site symmetry 1. There is always exactly one general position and it has the greatest multiplicity in the space group. Compare **special position**. (C-4)

Generators: A set of elements of a group such that every group element is obtained as an ordered product of these generators. (C-3)

Glide (three-dimensional): Operation that is a product of a mirror and a translation of some distance that is a fraction of the unit cell in the plane of the mirror. There are axial glides, double glides, diagonal glides, and diamond glides. (C-4)

Glide (two-dimensional): Operation that is a product of a mirror and a translation of one-half a unit cell parallel to the mirror. (C-4)

Graphical symbol: Symbol that shows both where the geometric object is and what the related group of operations is. See Table 3.15. (C-3)

Group: Set of elements with an operation between the elements that has the following four properties: closure, identity, inverse, and the associative law. (C-3)

Hexagonal close-packed layer: Layer of close-packed spheres. The two-dimensional space group of the ensemble is $p6mm$. (C-14)

Hexagonal close-packed structure (hcp): Packing of identical spheres in a layer sequence ABABAB…, resulting in hexagonal symmetry. The space group of the ensemble is #194, $P6_3/mmc$ with the spheres in a special position, Wyckoff letter either c or d. Furthermore, $c_H/a_H = (2/3)\sqrt{6} \approx 1.6323$, $Z = 2$, and $R = a_H/2$. (C-14)

Hexagonal P lattice (three dimensions): Primitive Bravais lattice in three dimensions that has point group $6/mmm$ symmetry. The lattice point is at 0, 0, 0. (C-4)

Hexagonal P lattice (two dimensions): Primitive Bravais lattice that has point group $6mm$ symmetry. (C-4)

Hexagonal R lattice (rhombohedral): Bravais lattice in three dimensions that has point group $\bar{3}m$ symmetry. The lattice points are at 0, 0, 0; 1/3, 2/3, 2/3; 2/3, 1/3, 1/3 with hexagonal reference axes. Another choice is a primitive rhombohedral cell with the lattice constants $a' = b' = c'$ and $\alpha' = \beta' = \gamma'$. This cell has rhombohedral reference axes. (C-4)

Icosahedron: A Platonic solid, that is a regular convex polyhedron with 12 vertices, 20 congruent equilateral triangular faces, and 30 edges of equal length. (C-13)

Identity E: Element, E, of a group such that, if a and b are elements in the group, then E a = a E = a. (C-3)

Identity operation, 1: A symmetry operation that in two dimensions takes a point with coordinates x, y into itself, or x, y, and in three dimensions takes x, y, z into x, y, z. (C-3)

Improper operation: Operation whose matrix has a determinant of –1; the operation causes the handedness to change. (C-3)

Interatomic bond length: $(r_{12})^2 = X_{12}\,\mathbf{G}\,X_{12}^t$ where $\mathbf{G}$ is the metric matrix, and $X_{12} = (x_2 - x_1\ \ y_2 - y_1\ z_2 - z_1)$, where $(x_1\ y_1\ z_1)$ and $(x_2\ y_2\ z_2)$ are the fractional coordinates of atoms 1 and 2, respectively. (C-2)

Interfacial crystal angle: The angle between the two normals to the crystal planes or crystal faces. (C-5)

Invariant properties: Properties that are the same before and after a transformation. (C-2)

Inverse, a^{-1}: For every element a in a group, there is an inverse, a^{-1}, such that a a^{-1} = a^{-1} a = E. (C-3)

Inversion operation: Three-dimensional operation that takes x, y, z into $\bar{x}, \bar{y}, \bar{z}$. (C-3)

Isomorphic crystals: See **isotypic crystals**. (C-4)

Isomorphic groups: Groups with the same multiplication table. (C-3)

Isotropic properties: Properties that do not depend on crystallographic direction. (C-1)

Isotypic crystals (also called isomorphic crystals): Crystals of different chemical composition with the same structure. An example of isotypic crystals are K^+Al^{3+} $(SO_4)_2$ and Rb^+Fe^{3+} $(SeO_4)_2$. (C-4)

k-subgroups (*klassengleiche*): Subgroups that keep the associated point group, but reduce the translations. These are known in the *International Tables for Crystallography* as type II. Contrast *t*-**subgroups**. (C-4)

k-supergroups (*klassengleiche*): Supergroups that keep the associated point group, but increase translations. These are known in the *International Tables for Crystallography* as type II. Contrast *t*-**supergroups**. (C-4)

Lattice: (1) An array of points in a crystal with identical neighborhoods; (2) an array defined by vector $\mathbf{t} = u\mathbf{a} + v\mathbf{b} + w\mathbf{c}$ where u, v, w are integers and $\mathbf{a}$, $\mathbf{b}$, $\mathbf{c}$ are basis vectors. (C-1)

Lattice constants: See **lattice parameters**. (C-1)

Lattice of a pattern: An array of points in space associated with a repeating pattern such that each point has an identical neighborhood. (C-1)

Lattice parameters: The scalar values a, b, c, α, β, and γ. Also called **lattice constants.** (C-1)

Lattice plane: Plane defined by three noncollinear lattice points. (C-2)

Lattice vector: Vector that connects any two lattice points, provided there is no intermediate lattice point between the two end points. (C-4)

Laue index, *hkl*: Method of labeling diffraction maxima where the order is incorporated in the symbol. Parentheses are not used. For example, the fifth order of a plane with Miller indices (111) has Laue indices (555). See ***n*, order of diffracted beam.** (C-5)

Linear compressibility or **coefficient of pressure expansion along a particular crystallographic axis at constant temperature:** $\beta_l = -\dfrac{1}{l}\left(\dfrac{\partial l}{\partial P}\right)_T$ where l is the lattice parameter, T the temperature, and P the pressure. (C-1)

Linear thermal expansion coefficient: $\alpha = \dfrac{1}{l}\left(\dfrac{\partial l}{\partial T}\right)_P$ where l is the lattice parameter, T the temperature, and P the pressure. (C-1)

Mass number, *A*: The number of protons plus the number of neutrons. This number is always a whole number. (C-15)

Maximal subgroup: Subgroup of a given group which is not contained in another subgroup. It forms the next layer down in the tree. The maximal subgroup of a finite group always has a lower order than the reference group. (C-4)

Mean compressibility, $\bar{\beta}$, between any two pressures P_1 and P_2 is

$$\bar{\beta} = \frac{-2}{V_1 + V_2}\left[\frac{V_2 - V_1}{P_2 - P_1}\right]_T \quad \text{(C-2)}$$

Mer: The repeat unit in a polymer. (C11)

Metric matrix, G: $G = \begin{pmatrix} \mathbf{a}\cdot\mathbf{a} & \mathbf{a}\cdot\mathbf{b} & \mathbf{a}\cdot\mathbf{c} \\ \mathbf{a}\cdot\mathbf{b} & \mathbf{b}\cdot\mathbf{b} & \mathbf{b}\cdot\mathbf{c} \\ \mathbf{a}\cdot\mathbf{c} & \mathbf{b}\cdot\mathbf{c} & \mathbf{c}\cdot\mathbf{c} \end{pmatrix}$ where $\mathbf{a}$, $\mathbf{b}$, $\mathbf{c}$ are the basis vectors of the unit cell. **G** is also known as the metric tensor. (C-2)

Miller indices, *hkl*: Three relatively prime integers, *hkl*, that are reciprocals of the fractional intercepts that the crystallographic plane makes with the crystallographic axes. The *crystallographic plane (hkl)* is described by its Miller indices. (C-2)

Minimal supergroup: Supergroup that completely contains the given group and is in a next generation. It forms the "next layer of branches" in the tree. The minimal supergroup of a finite group has a higher order than the reference group. (C-4)

Mirror, *m*: An operation where reflection is a symmetry operation. (C-3)

Monochromatic beam: A beam of a single wavelength. (C-15)

Monoclinic C lattice: *C*-centered Bravais lattice in three dimensions that has point group $2/m$ symmetry. The lattice points are at 0, 0, 0; 1/2, 1/2, 0. (C-4)

Monoclinic P lattice: Primitive Bravais lattice in three dimensions that has point group $2/m$ symmetry. The lattice point is at 0, 0, 0. (C-4)

Morphology: External shape of the crystal. (C-3)

Multiple unit cell: Unit cell with more than one lattice point. (C-1)

Multiplication: Operation between the elements in a group. (C-3)

Multiplication table: A table for a finite group giving all possible combinations relating two elements by an operation. (C-3)

Multiplicity: The number of symmetry-equivalent points in unit cell. (C-4)

n, order of the diffracted beam: Integer in the Bragg's law, $n\lambda = 2\,d_{hkl} \sin \theta_{hkl}$. $n = 1$ is the first-order diffracted beam satisfying $\lambda = 2\,d_{hkl} \sin \theta_{hkl}$. (C-5)

Neumann's principle: The point group of a crystal is a *subgroup* of the symmetry group of any of its physical properties. (C-3)

Neutron: A subatomic particle with zero electrical charge and a mass approximately equal to that of a proton. (C-15)

Neutron scattering length, b: A measure of the scattering of a neutron beam by the nucleus. It is related to the apparent size of the nucleus as seen by the neutron on a collision course with the nucleus and is measured in femtometers (fm). The neutron scattering length plays a role for neutron scattering analogous to the role played by the *atomic scattering factor* for x-ray scattering. (C-15)

n-fold rotation: A $360°/n$ rotation, where n is an integer, that is a symmetry operation. (C-3)

Niggli reduced cell: A cell that uniquely describes the lattice. The cell is primitive. In Chapter 5 Milledge chose the negatively reduced Niggli cell meaning that the angles α, β, and γ are all greater than 90°. Other conditions are that a, b, c are listed in increasing order. (There is also a positively reduced Niggli cell where the angles α, β, and γ are all less than 90°.) (C-5)

Nonabelian group: See **noncommutative group**.

Noncommutative group: A group that does not have a symmetric multiplication table. Point group $3m$ is an example. Same as **nonabelian group**. (C-3)

Noncrystallographic point group: Point group that is not a crystallographic point group. An example is point group 5. (C-3)

Nonsymmorphic space group: Space group that does not have a point in the unit cell with the point symmetry of its point group. See **symmorphic space group**. (C-4)

Oblique P lattice: Primitive Bravais lattice in two dimensions that has point group 2 symmetry. (C-4)

Order of a group: The number of elements in the group. (C-3)

Orthorhombic C lattice: *C*-centered Bravais lattice in three dimensions that has point group mmm symmetry. The lattice points are at 0, 0, 0 and 1/2, 1/2, 0. (C-4)

Orthorhombic F lattice: Face-centered Bravais lattice in three dimensions that has point group mmm symmetry. The lattice points are at 0, 0, 0; 0,1/2, 1/2; 1/2, 0, 1/2; and 1/2, 1/2, 0. (C-4)

Orthorhombic *I* lattice: Body-centered Bravais lattice in three dimensions that has point group *mmm* symmetry. The lattice points are at 0, 0, 0 and 1/2, 1/2, 1/2. (C-4)

Orthorhombic *P* lattice: Primitive Bravais lattice in three dimensions that has point group *mmm* symmetry. The lattice point is at 0, 0, 0. (C-4)

Partial interference: Phenomenon that occurs when the amplitudes of the superimposed waves are less than the maximum possible value and not zero. (C-6)

Photon: Quantum of electromagnetic radiation. Its energy, $E = h\,v$ where h is Planck's constant and v is the frequency of the wave. (C-6)

Piezoelectric material: A material that produces electricity when placed under mechanical stress. (C-3)

Plane polarized waves: Transverse waves that vibrate in a single direction. (C-6)

Point group: A group whose symmetry operations leave at least one point unmoved. (C-3)

Point stereographic projection: What happens to a point under the symmetry operations of the point group. (C-3)

Polarimeter: An instrument that measures the angle of rotation of polarized light after it has passed through an optically active material. (C-10)

Polymorphism: Presence of several different structures of the same material under different conditions such as temperature and pressure. (C-14)

Position (singular): A set of symmetry equivalent points. (C-4)

Powder Diffraction File, PDF: Library of x-ray patterns run by the Joint Committee on Powder Diffraction Standards (JCPDS) (C-5).

Primitive unit cell, *P*: Unit cell with one lattice point whose fractional coordinate is 0, 0, 0. (C-4)

Principle of superposition: Principle that states when two or more waves combine, the new wave is the algebraic sum of the original waves. (C-6)

Proper operation: An operation whose matrix has a determinant of +1; the operation retains the original handedness. (C-3)

Proper point group: A point group that has only proper operations. Point group 222 is an example of a proper point group. Biochemists use the term *chiral point groups* for what is called, in this book, the *proper point groups*. Another term is *pure rotational point groups*. (C-10)

Proper space group: A space group that has only proper operations. Space group $P2_1$ is an example of a proper space group. Biochemists use the term *chiral space group* for what is called, in this book, a *proper space group*. (C-10)

Racemic mixture: Mixture containing equal amounts of two enantiomers. (C-9)

Reciprocal lattice basis: Given the basis vectors **a, b, c** in direct space, then the reciprocal basis vectors **a*, b*, c*** are defined by the equation **(a b c)**^{t} **(a* b* c*)** = *I*, where *I* is the identity matrix. (C-5)

Rectangular *C* lattice: Centered Bravais lattice in two dimensions that has point group 2*mm* symmetry. (C-4)

Rectangular *P* lattice: Primitive Bravais lattice in two dimensions that has point group 2*mm* symmetry. (C-4)

Rotoinversion: Three-dimensional symmetry operation composed of a rotation followed by the inversion operation. (C-3)

Rotoreflection: Three-dimensional symmetry operation composed of a rotation followed by a mirror. (C-3)

R-value or **discrepancy index:** A measure of the correctness of the structure defined as

$$R = \frac{\Sigma \left| \left(|F_o| - |F_c| \right) \right|}{\Sigma |F_o|}.$$

where $|F_c|$ is the amplitude of the structure factor calculated from the model and $|F_o|$ is calculated from the diffraction pattern. (C-7)

Scattering by an electron: Is the process by which an electron, exposed to an x-ray beam, emits x-rays of the same frequency. (C-7)

Screw operation: Operation that is a combination of a rotation and a translation of some distance that is a fraction of the unit cell along the rotation axis and is found only in three dimensions. There are 11 screw operations—2_1, 3_1, 3_2, 4_1, 4_2, 4_3, 6_1, 6_2, 6_3, 6_4, and 6_5. (C-4)

Set: A collection of elements. (C-3)

Short form of the Hermann–Mauguin symbol: The Hermann–Mauguin symbol that compresses the symbols for the symmetry axes as much as possible. The information is given for the *x*-, *y*-, and the *z*-axis in that order. For example in space group #62, *Pnma* is the short form. Compare with the *full form of the Hermann–Mauguin symbol.* (C11)

Site symmetry: Set of all operations in a space group that leave a point fixed. (C-4)

Sixfold rotation: An operation where a 60° rotation is a symmetry operation. (C-3)

Space-filling model: Molecular model in which the atoms are represented by spheres whose radii are proportional to the atomic radii of the atoms. (C-13)

Space group: Symmetry group of a three-(two-) dimensional (crystal) pattern. Each group has an infinite set of translations. (C-4)

Special position: A position of a space group with multiplicity less than that of the general position. This position is associated with a site symmetry other than the identity. Compare with *general position.* (C-4)

Square *P* lattice: Primitive Bravais lattice in two dimensions that has point group *4mm* symmetry. (C-4)

Structure factor *F(hkl)*: $\sum_{i=1}^{N} f_i e^{2\pi i H \cdot R_i}$, where *N* is the number of atoms in the unit cell. The structure factor represents the scattering of x-rays at reciprocal lattice point *hkl* from all the atoms in the unit cell. (C-7)

Subgroup: A group wholly contained in a larger group. Point group 2 is a subgroup of point group *2mm*. (C-3)

Supergroup: A group that wholly contains a smaller group. Point group *2mm* is a supergroup of point group 2. (C-3)

Symbol stereographic projection: A graphic symbol for the point group. (C-3)

Symmetry operation for an object: An operation in which the object looks the same before and after the operation. (C-3)

Symmorphic space group: Space group that has at least one point in the unit cell with the symmetry of the associated point group. This space group is generated by combining a point group with the appropriate Bravais lattice. (C-4)

Synchrotron: Electron accelerator that uses a magnetic field to guide electrons in a circular orbit. Emitted x-rays used in protein studies. (C-6)

Tetragonal I lattice: Body-centered Bravais lattice in three dimensions that has point group $4/mmm$ symmetry. The lattice points are at 0, 0, 0 and 1/2, 1/2, 1/2. (C-4)

Tetragonal P lattice: Primitive Bravais lattice in three dimensions that has point group $4/mmm$ symmetry. The lattice point is at 0, 0, 0. (C-4)

Threefold rotation: An operation where a 120° rotation is a symmetry operation. (C-3)

Trace of a matrix: The sum of the diagonal elements. (C-3)

Translation: Operation in which every point of an object is displaced so that its position after the displacement minus the original position equals $t(u, v, w) = u\mathbf{a} + v\mathbf{b} + w\mathbf{c}$. The translations $t(1,0,0)$, $t(0,1,0)$, and $t(0,0,1)$ generate an infinite three-dimensional lattice. (C-4)

Triclinic P lattice: Primitive Bravais lattice in three dimensions that has point group $\bar{1}$ symmetry. The lattice point is at 0,0,0. (C-4)

A **triangular orthobicupola** or an **anticuboctahedron:** A Johnson solid with twelve vertices, 8 triangular faces, 6 square faces, and 24 edges. Each vertex has two triangles and two squares. Its point group is $\bar{6}m2$. (C-14)

t-**subgroups** (*translationgleiche*): Subgroups that keep the translations, but reduce the order of the associated point group. These are known in the *International Tables for Crystallography* as type I. Contrast k-**subgroup**. (C-4)

t-**supergroups** (*translationgleiche*): Supergroups that keep translations, but increase the order of the associated point group. These are known in the *International Tables for Crystallography* as type I. Contrast k-**supergroups**. C-4

Twofold rotation: An operation where a 180° rotation is a symmetry operation. (C-3)

Unit cell: Two-dimensional parallelogram defined by basis vectors $\mathbf{a}$ and $\mathbf{b}$; three-dimensional parallelepiped defined by basis vectors $\mathbf{a}$, $\mathbf{b}$, and $\mathbf{c}$. Unit cell fills space under translation. (C-1)

Unpolarized waves: Transverse waves that vibrate in any direction in the plane perpendicular to the direction of motion. (C-6)

Volumetric compressibility:

$$\beta_V = -\frac{1}{V}\left(\frac{\partial V}{\partial P}\right)_T = -\left(\frac{\partial \ln V}{\partial P}\right)_T,$$ where V is the volume of the unit cell, T the temperature, and P the pressure. (C-2)

Volumetric thermal expansion coefficient:

$$\alpha_V = \frac{1}{V}\left(\frac{\partial V}{\partial T}\right)_P = \left(\frac{\partial \ln V}{\partial T}\right)_P,$$ where V is the volume of the unit cell, T the temperature, and P the pressure. (C-2)

Wavelength: Distance between any two successive points with identical neighborhoods in a periodic function. The slope, as well as the amplitude, of the selected points must be the same for the two points to have identical neighborhoods. (C-6)

Wyckoff letter: A symbol used to identify each position in the unit cell. (C-4)

Zwitterion: A neutral molecule with a positive and a negative electrical charge at different places within the molecule. Leucine is an example. (C-9)

Appendix 2: The Ten Two-Dimensional Point Groups

These point groups are in order of their associated symmorphic space group—1, 2, *m*, 2*mm*, 4, 4*mm*, 3 (3*m*1, 31*m*), 6, and 6*mm*. Note that 3*m*1 and 31*m* represent the same point group. This is indicated by a blue box around them.

System	Oblique		
Point group	1		
Generator	(1)		
	(1)		
Symmetry operation	1		
Coordinates	x, y		

System	Oblique		
Point group	2		
Generators	(1), (2)		
	(1)		(2)
Symmetry operations	1		2
Coordinates	x, y		$-x, -y$

System	Rectangular		
Point group	*m*		
Generators	(1), (2)		
	(1)		(2)
Symmetry operations	1		m_y
Coordinates	x, y		$-x, y$

System	Rectangular		
Point group	$2mm$		
Generators	(1), (2), (3)		
	(1)	(2)	(3)	(4)
Symmetry operations	1	2	m_y	m_x
Coordinates	x, y	$-x, -y$	$-x, y$	$x, -y$

System	Square		
Point group	4		
Generators	(1), (2), (3)		
	(1)	(2)	(3)	(4)
Symmetry operations	1	2	4^+	4^-
Coordinates	x, y	$-x, -y$	$-y, x$	$y, -x$

System	Square		
Point group	$4mm$		
Generators	(1), (2), (3), (5)		
	(1)	(2)	(3)	(4)
Symmetry operations	1	2_z	4^+	4^-
Coordinates	x, y	$-x, -y$	$-y, x$	$y, -x$
	(5)	(6)	(7)	(8)
Symmetry operations	m_y	m_x	m_{xx}	$m_{x\bar{x}}$
Coordinates	$-x, y$	$x, -y$	y, x	$-y, -x$

System	Hexagonal	
Point group	3	
Generators	(1), (2)	
	(1)	(2)	(3)
Symmetry operations	1	3^+	3^-
Coordinates	x, y	$-y, x - y$	$-x + y, -x$

System	Hexagonal					
Point group	$3m1$					
Generators	(1), (2), (4)					
	(1)	(2)	(3)	(4)	(5)	(6)
Symmetry operations	1	3^+	3^-	m_{x-x}	m_{x2x}	m_{2xx}
Coordinates	x, y	$-y, x-y$	$-x+y, -x$	$-y, -x$	$-x, +y, y$	$x, x-y$

System	Hexagonal					
Point group	$31m$					
Generators	(1), (2), (4)					
	(1)	(2)	(3)	(4)	(5)	(6)
Symmetry operations	1	3^+	3^-	m_{xx}	m_x	m_y
Coordinates	x, y	$-y, x-y$	$-x+y, -x$	y, x	$x-y, -y$	$-x, -x+y$

System	Hexagonal					
Point group	6					
Generators	(1), (2), (4)					
	(1)	(2)	(3)	(4)	(5)	(6)
Symmetry operations	1	3^+	3^-	2	6^-	6^+
Coordinates	x, y	$-y, x-y$	$-x+y, -x$	$-x, -y$	$y, -x+y$	$x-y, x$

System	Hexagonal					
Point group	$6mm$					
Generators	(1), (2), (4), (7), (13)					
	(1)	(2)	(3)	(4)	(5)	(6)
Symmetry operations	1	3^+	3^-	2	6^-	6^+
Coordinates	x, y	$-y, x-y$	$-x+y, -x$	$-x, -y$	$y, -x+y$	$x-y, x$
	(7)	(8)	(9)	(10)	(11)	(12)
Symmetry operations	$m_{x\bar{x}}$	m_{x2x}	m_{2xx}	m_{xx}	m_x	m_y
Coordinates	$-y, -x$	$-x+y, y$	$x, x-y$	y, x	$x-y, -y$	$-x, -x+y$

Appendix 3: The Thirty-Two Three-Dimensional Point Groups

These point groups are in order of their associated symmorphic space group. Note that for some point groups, there is more than one representation, for example, 312, 321, and 32. This is indicated by a blue box around them.

Point Group	Triclinic 1 C_1	
Generator	(1)	
	(1)	
Symmetry operation	1	
Coordinates	x, y, z	

Point Group	Triclinic $\bar{1}$ C_i	
Generators	(1), (2)	
	(1)	(2)
Symmetry operations	1	$\bar{1}$
Coordinates	x, y, z	$-x, -y, -z$

	Monoclinic 2 C_2 **Unique Axis b**	
Point Group		
Generators	(1), (2)	
	(1)	(2)
Symmetry operations	1	2_y
Coordinates	x, y, z	$-x, y, -z$

	Monoclinic 2 C_2 **Unique Axis c**	
Point Group		
Generators	(1), (2)	
	(1)	(2)
Symmetry operations	1	2_z
Coordinates	x, y, z	$-x, -y, z$

	Monoclinic m C_s **Unique Axis b**	
Point Group		
Generators	(1), (2)	
	(1)	(2)
Symmetry operations	1	m_{xz}
Coordinates	x, y, z	$x, -y, z$

	Monoclinic m C_s **Unique Axis c**	
Point Group		
Generators	(1), (2)	
	(1)	(2)
Symmetry operations	1	m_{xy}
Coordinates	x, y, z	$x, y, -z$

Point Group	Monoclinic 2/m C_{2h} Unique Axis b			
Generators	(1), (2), (3)			
	(1)	(2)	(3)	(4)
Symmetry operations	1	2_y	$\bar{1}$	m_{xz}
Coordinates	x, y, z	$-x, y, -z$	$-x, -y, -z$	$x, -y, z$

Point Group	Monoclinic 2/m C_{2h} Unique Axis c			
Generators	(1), (2), (3)			
	(1)	(2)	(3)	(4)
Symmetry operations	1	2_z	$\bar{1}$	m_{xy}
Coordinates	x, y, z	$-x, -y, z$	$-x, -y, -z$	$x, y, -z$

Point Group	Orthorhombic 222 D_2			
Generators	(1), (2), (3)			
	(1)	(2)	(3)	(4)
Symmetry operations	1	2_z	2_y	2_x
Coordinates	x, y, z	$-x, -y, z$	$-x, y, -z$	$x, -y, -z$

Point Group	Orthorhombic $mm2$ C_{2v}			
Generators	(1), (2), (3)			
	(1)	(2)	(3)	(4)
Symmetry operations	1	2_z	m_{xz}	m_{yz}
Coordinates	x, y, z	$-x, -y, z$	$x, -y, z$	$-x, y, z$

Point Group	**Orthorhombic** mmm D_{2h}		
Generators	(1), (2), (3), (5)		
	(1)	(2)	(3)	(4)
Symmetry operations	1	2_z	2_y	2_x
Coordinates	x, y, z	$-x, -y, z$	$-x, y, -z$	$x, -y, -z$
	(5)	(6)	(7)	(8)
Symmetry operations	$\bar{1}$	m_{xy}	m_{xz}	m_{yz}
Coordinates	$-x, -y, -z$	$x, y, -z$	$x, -y, z$	$-x, y, z$

Point Group	**Tetragonal 4** C_4		
Generators	(1), (2), (3)		
	(1)	(2)	(3)	(4)
Symmetry operations	1	2_z	4^+	4^-
Coordinates	x, y, z	$-x, -y, z$	$-y, x, z$	$y, -x, z$

Point Group	**Tetragonal** $\bar{4}$ S_4		
Generators	(1), (2), (3)		
	(1)	(2)	(3)	(4)
Symmetry operations	1	2_z	$\bar{4}^+$	$\bar{4}^-$
Coordinates	x, y, z	$-x, -y, z$	$y, -x, -z$	$-y, x, -z$

Point Group		Tetragonal 4/m C_{4h}		
Generators		(1), (2), (3), (5)		
	(1)	(2)	(3)	(4)
Symmetry operations	1	2_z	4^+	4^-
Coordinates	x, y, z	$-x, -y, z$	$-y, x, z$	$y, -x, z$
	(5)	(6)	(7)	(8)
Symmetry operations	$\bar{1}$	m_{xy}	$\bar{4}^+$	$\bar{4}^-$
Coordinates	$-x, -y, -z$	$x, y, -z$	$y, -x, -z$	$-y, x, -z$

Point Group		Tetragonal 422 D_4		
Generators		(1), (2), (3), (5)		
	(1)	(2)	(3)	(4)
Symmetry operations	1	2_z	4^+	4^-
Coordinates	x, y, z	$-x, -y, z$	$-y, x, z$	$y, -x, z$
	(5)	(6)	(7)	(8)
Symmetry operations	2_y	2_x	2_{xx}	$2_{x\bar{x}}$
Coordinates	$-x, y, -z$	$x, -y, -z$	$y, x, -z$	$-y, -x -z$

Point Group		Tetragonal 4mm C_{4v}		
Generators		(1), (2), (3), (5)		
	(1)	(2)	(3)	(4)
Symmetry operations	1	2_z	4^+	4^-
Coordinates	x, y, z	$-x, -y, z$	$-y, x, z$	$y, -x, z$
	(5)	(6)	(7)	(8)
Symmetry operations	m_{xz}	m_{yz}	$m_{x\bar{x}z}$	m_{xxz}
Coordinates	$x, -y, z$	$-x, y, z$	$-y, -x, z$	y, x, z

Point Group		**Tetragonal $\bar{4}2m$ D_{2d}**		
Generators		(1), (2), (3), (5)		
	(1)	(2)	(3)	(4)
Symmetry operations	1	2_z	$\bar{4}^+$	$\bar{4}^-$
Coordinates	x, y, z	$-x, -y, z$	$y, -x, -z$	$-y, x, -z$
	(5)	(6)	(7)	(8)
Symmetry operations	2_y	2_x	$m_{x\bar{x}z}$	m_{xxz}
Coordinates	$-x, y, -z$	$x, -y, -z$	$-y, -x, z$	y, x, z

Point Group		**Tetragonal $\bar{4}m2$ D_{2d}**		
Generators		(1), (2), (3), (5)		
	(1)	(2)	(3)	(4)
Symmetry operations	1	2_z	$\bar{4}^+$	$\bar{4}^-$
Coordinates	x, y, z	$-x, -y, z$	$y, -x, -z$	$-y, x, -z$
	(5)	(6)	(7)	(8)
Symmetry operations	m_{xz}	m_{yz}	2_{xx}	$2_{x\bar{x}}$
Coordinates	$x, -y, z$	$-x, y, z$	$y, x, -z$	$-y, -x, -z$

Point Group		**Tetragonal $4/mmm$ D_{4h}**		
Generators		(1), (2), (3), (5), (9)		
	(1)	(2)	(3)	(4)
Symmetry operations	1	2_z	4^+	4^-
Coordinates	x, y, z	$-x, -y, z$	$-y, x, z$	$y, -x, z$
	(5)	(6)	(7)	(8)
Symmetry operations	2_y	2_x	2_{xx}	$2_{x\bar{x}}$
Coordinates	$-x, y, -z$	$x, -y, -z$	$y, x, -z$	$-y, -x -z$
	(9)	(10)	(11)	(12)
Symmetry operations	$\bar{1}$	m_{xy}	$\bar{4}^+$	$\bar{4}^-$
Coordinates	$-x, -y, -z$	$x, y, -z$	$y, -x, -z$	$-y, x, -z$
	(13)	(14)	(15)	(16)
Symmetry operations	m_{xz}	m_{yz}	$m_{x\bar{x}z}$	m_{xxz}
Coordinates	$x, -y, z$	$-x, y, z$	$-y, -x, z$	y, x, z

Point Group	Trigonal 3 C_3 Hexagonal Axes		
Generators	(1), (2)		
	(1)	(2)	(3)
Symmetry operations	1	3_z^+	3_z^-
Coordinates	x, y, z	$-y, x-y, z$	$-x+y, -x, z$

Point Goup	Trigonal 3 C_3 Rhombohedral Axes		
Generators	(1), (2)		
	(1)	(2)	(3)
Symmetry operations	1	3_{xxx}^+	3_{xxx}^-
Coordinates	x, y, z	z, x, y	y, z, x

Point Group	Trigonal $\bar{3}$ C_{3i} Hexagonal Axes					
Generators	(1), (2), (4)					
	(1)	(2)	(3)	(4)	(5)	(6)
Symmetry operations	1	3_z^+	3_z^-	$\bar{1}$	$\bar{3}_z^+$	$\bar{3}_z^-$
Coordinates	x, y, z	$-y, x-y, z$	$-x+y, -x, z$	$-x, -y, -z$	$y, -x+y, -z$	$x-y, x, -z$

Point Group	Trigonal $\bar{3}$ C_{3i} Rhombohedral Axes					
Generators	(1), (2), (4)					
	(1)	(2)	(3)	(4)	(5)	(6)
Symmetry operations	1	3_{xxx}^+	3_{xxx}^-	$\bar{1}$	$\bar{3}_{xxx}^+$	$\bar{3}_{xxx}^-$
Coordinates	x, y, z	z, x, y	y, z, x	$-x, -y, -z$	$-z, -x, -y$	$-y, -z, -x$

Point Group	**Trigonal 312 D_3** Hexagonal Axes					
Generators	(1), (2), (4)					
	(1)	(2)	(3)	(4)	(5)	(6)
Symmetry operations	1	3_z^+	3_z^-	$2_{x\bar{x}}$	2_{x2x}	2_{2xx}
Coordinates	x, y, z	$-y, x-y, z$	$-x+y, -x, z$	$-y, -x, -z$	$-x+y, y, -z$	$x, x-y, -z$

Point Group	**Trigonal 321 D_3** Hexagonal Axes					
Generators	(1), (2), (4)					
	(1)	(2)	(3)	(4)	(5)	(6)
Symmetry operations	1	3_z^+	3_z^-	2_{xx}	2_x	2_y
Coordinates	x, y, z	$-y, x-y, z$	$-x+y, -x, z$	$y, x, -z$	$x-y, -y, -z$	$-x, -x+y, -z$

Point Group	**Trigonal 32 D_3** Rhombohedral Axes					
Generators	(1), (2), (4)					
	(1)	(2)	(3)	(4)	(5)	(6)
Symmetry operations	1	3_{xxx}^+	3_{xxx}^-	2_{-x0x}	2_{x-x0}	2_{0y-y}
Coordinates	x, y, z	z, x, y	y, z, x	$-z, -y, -x$	$-y, -x, -z$	$-x, -z, -y$

Point Group	**Trigonal 3*m*1 C_{3v}** **Hexagonal Axes**					
Generators	(1), (2), (4)					
	(1)	(2)	(3)	(4)	(5)	(6)
Symmetry operations	1	3_z^+	3_z^-	$m_{x\bar{x}}$	m_{x2x}	m_{2xx}
Coordinates	x, y, z	$-y, x-y, z$	$-x+y, -x, z$	$-y, -x, z$	$-x+y, y, z$	$x, x-y, z$

Point Group	**Trigonal 31*m* C_{3v}** **Hexagonal Axes**					
Generators	(1), (2), (4)					
	(1)	(2)	(3)	(4)	(5)	(6)
Symmetry operations	1	3_z^+	3_z^-	m_{xxz}	m_{xz}	m_{yz}
Coordinates	x, y, z	$-y, x-y, z$	$-x+y, -x, z$	y, x, z	$x-y, -y, z$	$-x, -x+y, z$

Point Group	**Trigonal 3*m* C_{3v}** **Rhombohedral Axes**					
Generators	(1), (2), (4)					
	(1)	(2)	(3)	(4)	(5)	(6)
Symmetry operations	1	3_{xxx}^+	3_{xxx}^-	m_{xyx}	m_{xxz}	m_{xyy}
Coordinates	x, y, z	z, x, y	y, z, x	z, y, x	y, x, z	x, z, y

Point Group	**Trigonal $\bar{3}1m\,D_{3d}$** **Hexagonal Axes**					
Generators	(1), (2), (4), (7)					
	(1)	(2)	(3)	(4)	(5)	(6)
Symmetry operations	1	3_z^+	3_z^-	2_{x-x}	2_{x2x}	2_{2xx}
Coordinates	x, y, z	$-y, x-y, z$	$-x+y, -x, z$	$-y, -x, -z$	$-x+y, y, -z$	$x, x-y, -z$
	(7)	(8)	(9)	(10)	(11)	(12)
Symmetry operations	$\bar{1}$	$\bar{3}_z^+$	$\bar{3}_z^-$	m_{xxz}	m_{xz}	m_{yz}
Coordinates	$-x, -y, -z$	$y, -x+y, -z$	$x-y, x, -z$	y, x, z	$x-y, -y, z$	$-x, -x+y, z$

Point Group	**Trigonal $\bar{3}m1\,D_{3d}$** **Hexagonal Axes**					
Generators	(1), (2), (4), (7)					
	(1)	(2)	(3)	(4)	(5)	(6)
Symmetry operations	1	3_z^+	3_z^-	2_{xx}	2_x	2_y
Coordinates	x, y, z	$-y, x-y, z$	$-x+y, -x, z$	$y, x, -z$	$x-y, -y, -z$	$-x, -x+y, -z$
	(7)	(8)	(9)	(10)	(11)	(12)
Symmetry operations	$\bar{1}$	$\bar{3}_z^+$	$\bar{3}_z^-$	m_{x-xz}	m_{x2xz}	m_{2xxz}
Coordinates	$-x, -y, -z$	$y, -x+y, -z$	$x-y, x, -z$	$-y, -x, z$	$-x+y, y, z$	$x, x-y, z$

Point Group	**Trigonal $\bar{3}m\,D_{3d}$** **Rhombohedral Axes**					
Generators	(1), (2), (4), (7)					
	(1)	(2)	(3)	(4)	(5)	(6)
Symmetry operations	1	3_{xxx}^+	3_{xxx}^-	$2_{\bar{x}x}$	$2_{x\bar{x}}$	$2_{y\bar{y}}$
Coordinates	x, y, z	z, x, y	y, z, x	$-z, -y, -x$	$-y, -x, -z$	$-x, -z, -y$
	(7)	(8)	(9)	(10)	(11)	(12)
Symmetry operations	$\bar{1}$	$\bar{3}_{xxx}^+$	$\bar{3}_{xxx}^-$	m_{xyx}	m_{xxz}	m_{xyy}
Coordinates	$-x, -y, -z$	$-z, -x, -y$	$-y, -z, -x$	z, y, x	y, x, z	x, z, y

Point Group	**Hexagonal 6 C_6**					
Generators	(1), (2), (4)					
	(1)	(2)	(3)	(4)	(5)	(6)
Symmetry operations	1	3^+	3^-	2	6^-	6^+
Coordinates	x, y, z	$-y, x-y, z$	$-x+y, -x, z$	$-x, -y, z$	$y, -x+y, z$	$x-y, x, z$

Point Group	**Hexagonal $\bar{6}$ C_{3h}**					
Generators	(1), (2), (4)					
	(1)	(2)	(3)	(4)	(5)	(6)
Symmetry operations	1	3^+	3^-	m	$\bar{6}^-$	$\bar{6}^+$
Coordinates	x, y, z	$-y, x-y, z$	$-x+y, -x, z$	$x, y, -z$	$-y, x-y, -z$	$-x+y, -x, -z$

Point Group	**Hexagonal 6/m C_{6h}**					
Generators	(1), (2), (4), (7)					
	(1)	(2)	(3)	(4)	(5)	(6)
Symmetry operations	1	3^+	3^-	2	6^-	6^+
Coordinates	x, y, z	$-y, x-y, z$	$-x+y, -x, z$	$-x, -y, z$	$y, -x+y, z$	$x-y, x, z$
	(7)	(8)	(9)	(10)	(11)	(12)
Symmetry operations	$\bar{1}$	$\bar{3}^+$	$\bar{3}^-$	m_{xy}	$\bar{6}^-$	$\bar{6}^+$
Coordinates	$-x, -y, -z$	$y, -x+y, -z$	$x-y, x, -z$	$x, y, -z$	$-y, x-y, -z$	$-x+y, -x, -z$

Point Group	**Hexagonal 622 D_6**					
Generators	(1), (2), (4), (7)					
	(1)	(2)	(3)	(4)	(5)	(6)
Symmetry operations	1	3^+	3^-	2_z	6^-	6^+
Coordinates	x, y, z	$-y, x-y, z$	$-x+y, -x, z$	$-x, -y, z$	$y, -x+y, z$	$x-y, x, z$
	(7)	(8)	(9)	(10)	(11)	(12)
Symmetry operations	2_{xx}	2_x	2_y	$2_{x\bar{x}}$	2_{x2x}	2_{2xx}
Coordinates	$y, x, -z$	$x-y, -y, -z$	$-x, -x+y, -z$	$-y, -x, -z$	$-x+y, y, -z$	$x, x-y, -z$

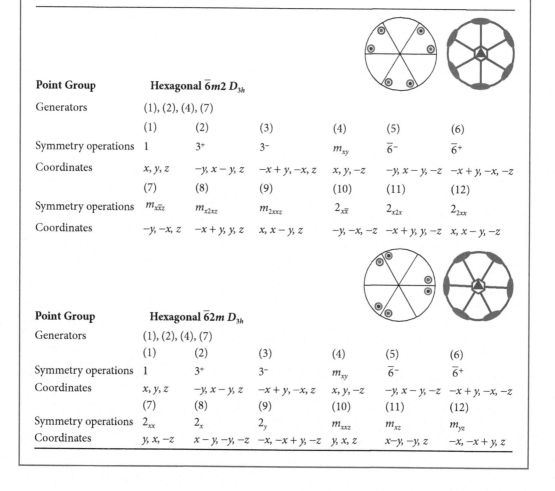

Point Group **Hexagonal 6mm C_{6v}**

Generators (1), (2), (4), (7)

	(1)	(2)	(3)	(4)	(5)	(6)
Symmetry operations	1	3^+	3^-	2	6^-	6^+
Coordinates	x, y, z	$-y, x-y, z$	$-x+y, -x, z$	$-x, -y, z$	$y, -x+y, z$	$x-y, x, z$
	(7)	(8)	(9)	(10)	(11)	(12)
Symmetry operations	$m_{x\bar{x}z}$	m_{x2xz}	m_{2xxz}	m_{xxz}	m_{xz}	m_{yz}
Coordinates	$-y, -x, z$	$-x+y, y, z$	$x, x-y, z$	y, x, z	$x-y, -y, z$	$-x, -x+y, z$

Point Group **Hexagonal $\bar{6}m2\ D_{3h}$**

Generators (1), (2), (4), (7)

	(1)	(2)	(3)	(4)	(5)	(6)
Symmetry operations	1	3^+	3^-	m_{xy}	$\bar{6}^-$	$\bar{6}^+$
Coordinates	x, y, z	$-y, x-y, z$	$-x+y, -x, z$	$x, y, -z$	$-y, x-y, -z$	$-x+y, -x, -z$
	(7)	(8)	(9)	(10)	(11)	(12)
Symmetry operations	$m_{x\bar{x}z}$	m_{x2xz}	m_{2xxz}	$2_{x\bar{x}}$	2_{x2x}	2_{2xx}
Coordinates	$-y, -x, z$	$-x+y, y, z$	$x, x-y, z$	$-y, -x, -z$	$-x+y, y, -z$	$x, x-y, -z$

Point Group **Hexagonal $\bar{6}2m\ D_{3h}$**

Generators (1), (2), (4), (7)

	(1)	(2)	(3)	(4)	(5)	(6)
Symmetry operations	1	3^+	3^-	m_{xy}	$\bar{6}^-$	$\bar{6}^+$
Coordinates	x, y, z	$-y, x-y, z$	$-x+y, -x, z$	$x, y, -z$	$-y, x-y, -z$	$-x+y, -x, -z$
	(7)	(8)	(9)	(10)	(11)	(12)
Symmetry operations	2_{xx}	2_x	2_y	m_{xxz}	m_{xz}	m_{yz}
Coordinates	$y, x, -z$	$x-y, -y, -z$	$-x, -x+y, -z$	y, x, z	$x-y, -y, z$	$-x, -x+y, z$

Point Group	Hexagonal 6/*mmm* D$_{6h}$					
Generators	(1), (2), (4), (7), (13)					
	(1)	(2)	(3)	(4)	(5)	(6)
Symmetry operations	1	3$^+$	3$^-$	2	6$^-$	6$^+$
Coordinates	x, y, z	$-y, x-y, z$	$-x+y, -x, z$	$-x, -y, z$	$y, -x+y, z$	$x-y, x, z$
	(7)	(8)	(9)	(10)	(11)	(12)
Symmetry operations	2$_{xx}$	2$_x$	2$_y$	2$_{x\text{-}x}$	2$_{x2x}$	2$_{2xx}$
Coordinates	$y, x, -z$	$x-y, -y, -z$	$-x, -x+y, -z$	$-y, -x, -z$	$-x+y, y, -z$	$x, x-y, -z$
	(13)	(14)	(15)	(16)	(17)	(18)
Symmetry operations	$\bar{1}$	$\bar{3}^+$	$\bar{3}^-$	m_{xy}	$\bar{6}^-$	$\bar{6}^+$
Coordinates	$-x, -y, -z$	$y, -x+y, -z$	$x-y, x, -z$	$x, y, -z$	$-y, x-y, -z$	$-x+y, -x, -z$
	(19)	(20)	(21)	(22)	(23)	(24)
Symmetry operations	$m_{x\bar{x}z}$	m_{x2xz}	m_{2xxz}	m_{xxz}	m_{xz}	m_{yz}
Coordinates	$-y, -x, z$	$-x+y, y, z$	$x, x-y, z$	y, x, z	$x-y, -y, z$	$-x, -x+y, z$

Point Group	Cubic 23 *T*			
Generators	(1), (2), (3), (5)			
	(1)	(2)	(3)	(4)
Symmetry operations	1	2$_z$	2$_y$	2$_x$
Coordinates	x, y, z	$-x, -y, z$	$-x, y, -z$	$x, -y, -z$
	(5)	(6)	(7)	(8)
Symmetry operations	3$^+_{xxx}$	3$^+_{\bar{x}\bar{x}x}$	3$^+_{x\overline{xx}}$	3$^+_{\bar{x}x\bar{x}}$
Coordinates	z, x, y	$z, -x, -y$	$-z, -x, y$	$-z, x, -y$
	(9)	(10)	(11)	(12)
Symmetry operations	3$^-_{xxx}$	3$^-_{x\overline{xx}}$	3$^-_{\bar{x}x\bar{x}}$	3$^-_{\bar{x}\bar{x}x}$
Coordinates	y, z, x	$-y, z, -x$	$y, -z, -x$	$-y, -z, x$

Point Group	Cubic $m\bar{3}$ *T$_h$*		
Generators	(1), (2), (3), (5), (13)		
	(1)	(2)	(3)	(4)
Symmetry operations	1	2$_z$	2$_y$	2$_x$

Coordinates	x, y, z	$-x, -y, z$	$-x, y, -z$	$x, -y, -z$
	(5)	(6)	(7)	(8)
Symmetry operations	3^+_{xxx}	$3^\pm_{\bar{x}x\bar{x}}$	$3^+_{x\bar{x}\bar{x}}$	$3^\pm_{\bar{x}\bar{x}x}$
Coordinates	z, x, y	$z, -x, -y$	$-z, -x, y$	$-z, x, -y$
	(9)	(10)	(11)	(12)
Symmetry operations	3^-_{xxx}	$3^-_{x\bar{x}\bar{x}}$	$3^-_{\bar{x}\bar{x}x}$	$3^-_{\bar{x}x\bar{x}}$
Coordinates	y, z, x	$-y, z, -x$	$y, -z, -x$	$-y, -z, x$
	(13)	(14)	(15)	(16)
Symmetry operations	$\bar{1}$	m_{xy}	m_{xz}	m_{yz}
Coordinates	$-x, -y, -z$	$x, y, -z$	$x, -y, z$	$-x, y, z$
	(17)	(18)	(19)	(20)
Symmetry operations	$\bar{3}^+_{xxx}$	$\bar{3}^+_{\bar{x}x\bar{x}}$	$\bar{3}^+_{x\bar{x}\bar{x}}$	$\bar{3}^+_{\bar{x}\bar{x}x}$
Coordinates	$-z, -x, -y$	$-z, x, y$	$z, x, -y$	$z, -x, y$
	(21)	(22)	(23)	(24)
Symmetry operations	$\bar{3}^-_{xxx}$	$\bar{3}^-_{x\bar{x}\bar{x}}$	$\bar{3}^-_{\bar{x}x\bar{x}}$	$\bar{3}^-_{\bar{x}\bar{x}x}$
Coordinates	$-y, -z, -x$	$y, -z, x$	$-y, z, x$	$y, z, -x$

Point Group	**Cubic 432 O**			
Generators	(1), (2), (3), (5), (13)			
	(1)	(2)	(3)	(4)
Symmetry operations	1	2_z	2_y	2_x
Coordinates	x, y, z	$-x, -y, z$	$-x, y, -z$	$x, -y, -z$
	(5)	(6)	(7)	(8)
Symmetry operations	3^+_{xxx}	$3^\pm_{\bar{x}x\bar{x}}$	$3^+_{x\bar{x}\bar{x}}$	$3^\pm_{\bar{x}\bar{x}x}$
Coordinates	z, x, y	$z, -x, -y$	$-z, -x, y$	$-z, x, -y$
	(9)	(10)	(11)	(12)
Symmetry operations	3^-_{xxx}	$3^-_{x\bar{x}\bar{x}}$	$3^-_{\bar{x}\bar{x}x}$	$3^-_{\bar{x}x\bar{x}}$
Coordinates	y, z, x	$-y, z, -x$	$y, -z, -x$	$-y, -z, x$
	(13)	(14)	(15)	(16)
Symmetry operations	2_{xx0}	$2_{x\bar{x}0}$	4^-_z	4^+_z
Coordinates	$y, x, -z$	$-y, -x, -z$	$y, -x, z$	$-y, x, z$
	(17)	(18)	(19)	(20)
Symmetry operations	4^-_x	2_{yy}	$2_{y\bar{y}}$	4^+_x
Coordinates	$x, z, -y$	$-x, z, y$	$-x, -z, -y$	$x, -z, y$
	(21)	(22)	(23)	(24)
Symmetry operations	4^-_y	2_{x0x}	4^+_y	$2_{\bar{x}0x}$
Coordinates	$z, y, -x$	$z, -y, x$	$-z, y, x$	$-z, -y, -x$

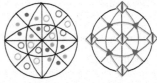

Point Group	**Cubic $\overline{4}3m$ T_d**		
Generators	(1), (2), (3), (5), (13)		
	(1)	(2)	(3)	(4)
Symmetry operations	1	2_z	2_y	2_x
Coordinates	x, y, z	$-x, -y, z$	$-x, y, -z$	$x, -y, -z$
	(5)	(6)	(7)	(8)
Symmetry operations	3^+_{xxx}	$3^\pm_{\overline{x}x\overline{x}}$	$3^+_{x\overline{x}\overline{x}}$	$3^+_{\overline{x}\overline{x}x}$
Coordinates	z, x, y	$z, -x, -y$	$-z, -x, y$	$-z, x, -y$
	(9)	(10)	(11)	(12)
Symmetry operations	3^-_{xxx}	$3^-_{x\overline{x}\overline{x}}$	$3^-_{\overline{x}\overline{x}x}$	$3^-_{\overline{x}x\overline{x}}$
Coordinates	y, z, x	$-y, z, -x$	$y, -z, -x$	$-y, -z, x$
	(13)	(14)	(15)	(16)
Symmetry operations	m_{xxz}	$m_{x\overline{x}z}$	$\overline{4}^+_z$	$\overline{4}^-_z$
Coordinates	y, x, z	$-y, -x, z$	$y, -x, -z$	$-y, x, z$
	(17)	(18)	(19)	(20)
Symmetry operations	m_{xyy}	$\overline{4}^+_x$	$\overline{4}^-_x$	$m_{xy\overline{y}}$
Coordinates	x, z, y	$-x, z, -y$	$-x, -z, y$	$x, -z, -y$
	(21)	(22)	(23)	(24)
Symmetry operations	m_{xyx}	$\overline{4}^-_y$	$m_{\overline{x}yx}$	$\overline{4}^+_y$
Coordinates	z, y, x	$z, -y, -x$	$-z, y, -x$	$-z, -y, x$

Point Group	**Cubic $m\overline{3}m$ O_h**		
Generators	(1), (2), (3), (5), (13), (25)		
	(1)	(2)	(3)	(4)
Symmetry operations	1	2_z	2_y	2_x
Coordinates	x, y, z	$-x, -y, z$	$-x, y, -z$	$x, -y, -z$
	(5)	(6)	(7)	(8)
Symmetry operations	3^+_{xxx}	$3^+_{\overline{x}x\overline{x}}$	$3^+_{x\overline{x}\overline{x}}$	$3^+_{\overline{x}\overline{x}x}$
Coordinates	z, x, y	$z, -x, -y$	$-z, -x, y$	$-z, x, -y$
	(9)	(10)	(11)	(12)
Symmetry operations	3^-_{xxx}	$3^-_{x\overline{x}\overline{x}}$	$3^-_{\overline{x}\overline{x}x}$	$3^-_{\overline{x}x\overline{x}}$
Coordinates	y, z, x	$-y, z, -x$	$y, -z, -x$	$-y, -z, x$
	(13)	(14)	(15)	(16)
Symmetry operations	2_{xx0}	$2_{x\overline{x}0}$	4^-_z	4^+_z

Coordinates	$y, x, -z$ (17)	$-y, -x, -z$ (18)	$y, -x, z$ (19)	$-y, x, z$ (20)
Symmetry operations	4_x^-	2_{yy}	$2_{y\bar{y}}$	4_x^+
Coordinates	$x, z, -y$ (21)	$-x, z, y$ (22)	$-x, -z, -y$ (23)	$x, -z, y$ (24)
Symmetry operations	4_y^-	2_{x0x}	4_y^+	$2_{\bar{x}0x}$
Coordinates	$z, y, -x$ (25)	$z, -y, x$ (26)	$-z, y, x$ (27)	$-z, -y, -x$ (28)
Symmetry operations	$\bar{1}$	m_{xy}	m_{xz}	m_{yz}
Coordinates	$-x, -y, -z$ (29)	$x, y, -z$ (30)	$x, -y, z$ (31)	$-x, y, z$ (32)
Symmetry operations	$\bar{3}_{xxx}^+$	$\bar{3}_{x\bar{x}\bar{x}}^+$	$\bar{3}_{\bar{x}x\bar{x}}^+$	$\bar{3}_{\bar{x}\bar{x}x}^+$
Coordinates	$-z, -x, -y$ (33)	$-z, x, y$ (34)	$z, x, -y$ (35)	$z, -x, y$ (36)
Symmetry operations	$\bar{3}_{xxx}^-$	$\bar{3}_{x\bar{x}x}^-$	$\bar{3}_{\bar{x}x\bar{x}}^-$	$\bar{3}_{xx\bar{x}}^-$
Coordinates	$-y, -z, -x$ (37)	$y, -z, x$ (38)	$-y, z, x$ (39)	$y, z, -x$ (40)
Symmetry operations	$m_{x\bar{x}z}$	m_{xxz}	$\bar{4}_z^-$	$\bar{4}_z^+$
Coordinates	$-y, -x, z$ (41)	y, x, z (42)	$-y, x, -z$ (43)	$y, -x, -z$ (44)
Symmetry operations	$\bar{4}_x^-$	$m_{xy\bar{y}}$	m_{xyy}	$\bar{4}_x^+$
Coordinates	$-x, -z, y$ (45)	$x, -z, -y$ (46)	x, z, y (47)	$-x, z, -y$ (48)
Symmetry operations	$\bar{4}_y^+$	$m_{\bar{x}yx}$	$\bar{4}_y^-$	m_{xyx}
Coordinates	$-z, -y, x$	$-z, y, -x$	$z, -y, -x$	z, y, x

Index of Molecular and/or Crystal Structures by Figure Number or Chapter

Many of the molecular structures are shown in projection to emphasize various symmetries. The Cambridge Structural Database identifier is in parenthesis; proteins are from the Protein Data Base and other figures by the author.

Molecular and Crystal Index

Subject Index